Concept of One Health and Zoonoses

Ripan Biswas • Ajay Pratap Singh
Maninder Singh • Indranil Samanta
Editors

Concept of One Health and Zoonoses

Breaking Silos for the Promotion of Healthier World

Springer

Editors
Ripan Biswas
Department of Veterinary Public Health and Epidemiology, Faculty of Veterinary and Animal Sciences
West Bengal University of Animal and Fishery Sciences
Kolkata, West Bengal, India

Maninder Singh
Department of Veterinary Public Health and Epidemiology,
College of Veterinary Science
Guru Angad Dev Veterinary and Animal Sciences University
Bathinda, Punjab, India

Ajay Pratap Singh
Department of Veterinary Microbiology, College of Veterinary Science and Animal Husbandry
Uttar Pradesh Pandit Deen Dayal Upadhyaya Pashu Chikitsa Vigyan Vishwavidyalaya Evam Go-Anusandhan Sansthan (DUVASU)
Mathura, Uttar Pradesh, India

Indranil Samanta
Department of Veterinary Microbiology, Faculty of Veterinary and Animal Sciences
West Bengal University of Animal and Fishery Sciences
Kolkata, West Bengal, India

ISBN 978-981-95-6626-6 ISBN 978-981-95-6627-3 (eBook)
https://doi.org/10.1007/978-981-95-6627-3

This Springer imprint is published by the registered company Springer Nature Singapore Pte Ltd.
The registered company address is: 152 Beach Road, #21-01/04 Gateway East, Singapore 189721, Singapore

To the unsung heroes on the frontlines of One Health—the scientists, veterinarians, doctors, ecologists, public health workers, and all those who tirelessly bridge the gaps between human, animal, and environmental health.
Your courage, devotion, and unwavering commitment to safeguard the interconnected well-being of our world inspire us all. In the face of emerging zoonotic threats and global challenges, you stand as beacons of hope, collaboration, and resilience.
This book is dedicated to you, for your relentless efforts to protect life in all its forms and for reminding us that the health of one is indeed the health of all.
With gratitude and respect,
The Authors

Preface

The global landscape has become increasingly intertwined and interdependent. As human populations expand and occupy more land, natural habitats contract, and climate conditions undergo shifts, the distinctions between the realms of humans, animals, and the environment become less defined. This interconnectedness has illuminated the imperative for a comprehensive health framework that recognizes the complex interdependencies among human, animal, and environmental domains. This is the essence of **One Health**, a paradigm that has gained global recognition as a critical framework for addressing complex health challenges, particularly **Zoonoses**—diseases that are naturally transmitted from vertebrate animals to humans or vice versa.

The One Health concept is not a novel idea, but its significance has become increasingly evident. The emergence and the persistent threats posed by zoonotic diseases such as avian influenza, Nipah virus infection, and tuberculosis have repeatedly underscored their capacity to disrupt societies, economies, and ecological systems. These are quintessential One Health challenges, as they emerge and proliferate at the intersection of human activities, animal health, and environmental transformations. Addressing them requires breaking down silos, fostering collaboration across disciplines, and embracing a unified approach to health.

This book, *Concept of One Health and Zoonoses*, is designed to provide a comprehensive exploration of these interconnected themes. It is structured into 8 parts, each delving into a critical aspect of the One Health framework and its application to zoonotic diseases. From the historical roots of One Health to the latest scientific advancements, from case studies of zoonotic outbreaks to strategies for prevention and control, this book aims to equip readers with a deep understanding of the challenges and opportunities at the human-animal-environment interface.

The journey through these pages will take the readers from the microscopic world of pathogens to the global stage of public health policy. Readers will be able to explore how deforestation, urbanization, and climate change drive zoonotic disease emergence, and how collaborative efforts across human medicine, veterinary science, and environmental health can mitigate these risks. Readers will also encounter real-world examples of One Health in action, showcasing the power of interdisciplinary approaches to solve some of the most pressing health issues of our time.

This book is aimed at a diverse readership, including students, researchers, policymakers, healthcare professionals, veterinarians, ecologists, and others, with an interest in the future of global health. Whether one is new to the principles of One Health or an experienced practitioner, this work aims to encourage readers to adopt a fresh perspective on health, recognize the interconnections between humans, animals, and the environment, and acknowledge the shared responsibility in protecting the health of our planet.

Amidst the confluence of unparalleled global complexities, the imperative for a comprehensive One Health framework has become increasingly vital. By elucidating and addressing the fundamental drivers of zoonotic diseases, we can not only mitigate the risk of future pandemics but also foster the development of a healthier and more sustainable global ecosystem for all.

We encourage the readers to embark on this exploration—to investigate, acquire knowledge, and participate in advancing the principles of One Health.

Kolkata, West Bengal, India — Ripan Biswas
Mathura, Uttar Pradesh, India — Ajay Pratap Singh
Bathinda, Punjab, India — Maninder Singh
Kolkata, West Bengal, India — Indranil Samanta

Acknowledgments

We would like to extend our heartfelt gratitude to all the individuals who contributed to the creation of this book, *Concept of One Health and Zoonoses*. This work would not have been possible without their expertise, dedication, and support.

First and foremost, we thank the contributors and researchers who shared their invaluable knowledge and insights, enriching the content with their diverse perspectives and experiences. Your contributions have been instrumental in shaping this comprehensive exploration of One Health and zoonotic diseases.

We are deeply grateful to the reviewers and editors who meticulously reviewed the manuscript, providing constructive feedback and ensuring the accuracy and clarity of the information presented. Your commitment to excellence has greatly enhanced the quality of this book.

We also acknowledge the efforts of the publishing team, whose professionalism and dedication have brought this book to life. Your guidance and support throughout the process have been invaluable.

Finally, we extend our appreciation to our families, friends, and colleagues for their unwavering encouragement and patience during the creation of this book. Your support has been a constant source of motivation.

This book is a collective effort, and we are profoundly grateful to everyone who played a part in its realization. Together, we hope to advance the understanding of One Health and zoonoses, contributing to a healthier and more sustainable future for all.

Thank you.

Ripan Biswas
Ajay Pratap Singh
Maninder Singh
Indranil Samanta

Competing Interests The authors have no competing interests to declare that are relevant to the content of this manuscript.

Contents

Contributors

Ambika Arun Department of Veterinary Microbiology, College of Veterinary Science & Animal Husbandry, Pandit Deen Dayal Upadhyaya Pashu Chikitsa Vigyan Vishwavidyalaya Evam Go-Anusandhan Sansthan, Mathura, Uttar Pradesh, India

Tuhina Banerjee Department of Microbiology, Institute of Medical Sciences, Banaras Hindu University, Varanasi, Uttar Pradesh, India

Ripan Biswas Department of Veterinary Public Health and Epidemiology, Faculty of Veterinary & Animal Sciences, West Bengal University of Animal & Fishery Sciences, Kolkata, India

Suman Biswas Department of Avian Sciences, West Bengal University of Animal and Fishery Sciences, Mohanpur, West Bengal, India

Sandhya Chaurasiya Department of Microbiology, Institute of Medical Sciences, Banaras Hindu University, Varanasi, Uttar Pradesh, India

Subhasree Das Department of Veterinary Biochemistry, West Bengal University of Animal and Fishery Sciences, Kolkata, West Bengal, India

Molla Zakirul Haque Department of Veterinary Medicine, West Bengal University of Animal and Fishery Sciences, Kolkata, West Bengal, India

Department of Microbiology, Institute of Medical Sciences, Banaras Hindu University, Varanasi, Uttar Pradesh, India

Priscilla Kerketta Department of Veterinary Microbiology, Institute of Veterinary Sciences and Animal Husbandry (IVSAH), Siksha O Anusandhan (SOA), Bhubaneswar, Odisha, India

Pesingi Pavan Kumar Faculty of Veterinary and Animal Sciences, Department of Veterinary Public Health, Rajiv Gandhi South Campus, Banaras Hindu University, Mirzpur, Uttar Pradesh, India

Krishnendu Kundu Department of Veterinary Parasitology, FVAS, Institute of Agricultural Sciences, Banaras Hindu University, South Campus, Mirzapur, Uttar Pradesh, India

Ayan Mukherjee Department of Animal Biotechnology, West Bengal University of Animal and Fishery Sciences, Kolkata, West Bengal, India

Anil Patyal Department of Veterinary Public Health and Epidemiology, College of Veterinary and Animal Sciences, Sardar Vallabhbhai Patel University of Agriculture and Technology, Meerut, Uttar Pradesh, India

Shyama N. Prabhu Department of Veterinary Pathology, College of Veterinary Science & Animal Husbandry, U.P. Pandit Deen Dayal Upadhyaya Pashu Chikitsa Vigyan Vishwavidyalaya Evam Go-Anusandhan Sansthan, Mathura, Uttar Pradesh, India

Ajay Pratap Singh Department of Veterinary Microbiology, College of Veterinary Science & Animal Husbandry, U.P. Pandit Deen Dayal Upadhyaya Pashu Chikitsa Vigyan Vishwavidyalaya Evam Go-Anusandhan Sansthan, Mathura, Uttar Pradesh, India

Bhavani Puvvala Department of Veterinary Microbiology, Institute of Veterinary Sciences and Animal Husbandry (IVSAH), Siksha O Anusandhan (SOA), Bhubaneswar, Odisha, India

Shriya Rawat Department of Veterinary Public Health and Epidemiology, College of Veterinary and Animal Sciences, Sardar Vallabhbhai Patel University of Agriculture and Technology, Meerut, Uttar Pradesh, India

Indranil Samanta Department of Veterinary Microbiology, West Bengal University of Animal and Fishery Sciences, Kolkata, West Bengal, India

Sukhen Samanta Department of Microbiology, University of Kalyani, Nadia, West Bengal, India

Souti Prasad Sarkhel Department of Veterinary Parasitology, FVAS, Institute of Agricultural Sciences, Banaras Hindu University, South Campus, Mirzapur, Uttar Pradesh, India

Sunanda Sarkhel Department of Optometry and Vision Sciences, Amity University, Gurgaon, Haryana, India

Aradhana Singh Ashworth Laboratories, Institute of Immunology and Infection Research, School of Biological Sciences, University of Edinburgh, Edinburgh, UK

Maninder Singh Department of Veterinary Public Health and Epidemiology, College of Veterinary Science, Guru Angad Dev Veterinary and Animal Sciences University, Bathinda, Punjab, India

Prachi Singh Department of Veterinary Microbiology, College of Veterinary Science & Animal Husbandry, U.P. Pandit Deen Dayal Upadhyaya Pashu Chikitsa Vigyan Vishwavidyalaya Evam Go-Anusandhan Sansthan, Mathura, Uttar Pradesh, India

Shiv Varan Singh Department of Veterinary Microbiology, College of Veterinary and Animal Sciences, Bihar Animal Sciences University, Kishanganj, Bihar, India

Vikrant Sudan Department of Veterinary Parasitology, College of Veterinary Science, Rampura Phul, Guru Angad Dev Veterinary and Animal Sciences University, Punjab, India

Piyush Tomar Department of Veterinary Public Health and Epidemiology, College of Veterinary and Animal Sciences, Sardar Vallabhbhai Patel University of Agriculture and Technology, Meerut, Uttar Pradesh, India

Prasanna Vadhana A Department of Veterinary Microbiology, College of Veterinary and Animal Sciences, Rani Lakshmi Bai Central Agricultural University, Jhansi, Uttar Pradesh, India

Jay Prakash Yadav Department of Veterinary Public Health and Epidemiology, College of Veterinary Science, Guru Angad Dev Veterinary and Animal Sciences University, Bathinda, Punjab, India

Part I

Zoonoses at the Crossroads of One Health

One Health Concept and Its Evolution

1

Anil Patyal, Shriya Rawat, Piyush Tomar, and Ripan Biswas

Abstract

In the twenty-first century, the One Health framework, which recognizes the mutual reliance of human, animal, and environmental health, has become a vital method for resolving intricate health problems. This chapter explores the historical development of One Health, from its roots in the nineteenth-century recognition of zoonotic diseases to its modern-day emphasis on transdisciplinary collaboration and systems thinking. The chapter also considers the key milestones in its development, including the contributions of early pioneers like Rudolf Virchow and Calvin Schwabe, the increasing recognition of newly arising infectious illnesses and the spread of antimicrobial resistance, along with the formal establishment of One Health programs by global entities. Furthermore, the expanding scope of One Health, encompassing not only zoonotic diseases but also food security, environmental sustainability, and social determinants of health have been discussed.

Keywords

One Health · Zoonoses · Evolution

A. Patyal (✉) · S. Rawat · P. Tomar
Department of Veterinary Public Health and Epidemiology, College of Veterinary and Animal Sciences, Sardar Vallabhbhai Patel University of Agriculture and Technology, Meerut, Uttar Pradesh, India

R. Biswas
Department of Veterinary Public Health and Epidemiology, Faculty of Veterinary & Animal Sciences, West Bengal University of Animal & Fishery Sciences, Kolkata, India

R. Biswas et al. (eds.), *Concept of One Health and Zoonoses*,
https://doi.org/10.1007/978-981-95-6627-3_1

With the beginning of twenty-first century, the whole world has encountered many emerging life-threatening diseases such as SARS, Classical H1N1 flu, highly pathogenic avian influenza (HPAI H5N1), MERS, COVID-19, etc. that had the potential to cause pandemics, created several international crises. Due to rapid urbanization, industrialization, globalization, population growth and environmental degradation, diseases/ health threats, whether communicable or non-communicable, become more complex and multifaceted. So, solutions against them cannot be found by one sector alone, and more holistic approach involving different sectors and disciplines is mandatory to tackle such health threats. The "One Health Concept" emerged as a result of the global recognition by various international and national organizations, governments, and scientists that interdisciplinary collaboration was essential for the prevention and control of these complex health threats.

"One Health" Definitions

Defining the scope of the One Health Concept is a complex task, yet at its core, One Health seeks to enhance health through a cooperative approach that spans various disciplines and sectors, covering all animal species. Different organizations around the world define "One Health Concept" in their own way, which reflects the mission of the respective organizations. Some of the definitions are given below:

> One Health is the collaborative effort of multiple health science professions, together with their related disciplines and institutions—working locally, nationally, and globally—to attain optimal health for people, domestic animals, wildlife, plants, and our environment (One Health Commission).

Food and Agriculture Organization (FAO) defined 'One Health is a collaborative, international, cross sectoral, multidisciplinary mechanism to address threats and reduce risks of detrimental infectious diseases at the animal-human-ecosystem interface.'

World Organization for Animal Health (WOAH) or OIE recommended One Health approach as 'a collaborative and all-encompassing way to address, when relevant, animal and public health globally. This collaboration should not be confined solely to the international domain, but rather must be transposed as a novel and integral paradigm at the national level.'

One Health Global Network defined One Health is aimed at to 'improve health and wellbeing through the prevention of risks and the mitigation of effects of crises that originate at the interface between humans, animals and their various environments.'

One Health Initiative considered One Health is 'a worldwide strategy for expanding interdisciplinary collaborations and communications in all aspects of health care for humans, animals, and the environment.'

Center for Disease Control and Prevention (CDC) defined that One Health is a collaborative, multisectoral, and transdisciplinary approach—working at the local, national, regional, and global levels—with the goal of achieving best health outcomes that identify the interconnection between people, animals, plants, and their common environment.

The overarching "One Health Concept" acknowledges the interconnectedness of human, animal, and environmental health, implying that a danger to one element can rapidly jeopardize all. Consequently, the complex health challenges impacting humans, animals, vegetation, and the environment necessitate enhanced coordination, communication, and cooperative endeavors across all sectors. In recent years, the One Health approach has been utilized to explore the complex interconnections between human, animal, plant, and environmental health and well-being. Fundamentally, this concept advocates for interdisciplinary and intersectoral collaboration to comprehensively comprehend and mitigate complex health risks at the human–animal–environment interface. The One Health approach has gained prominence in national and international frameworks, facilitating the management of zoonotic diseases, food safety concerns, and antimicrobial resistance on a global scale.

Evolution of One Health Concept

The observation of parallels between diseases affecting animals and humans, a practice that existed independently until the twentieth century, dates back to the 1800s when scientists from various international locales made note of these shared characteristics. The One Health concept does not have a singular origination in human thought. This concept represents a basic and widespread element of terrestrial life, a phenomenon that has been consistently recognized and investigated throughout human history.

For centuries, human health and welfare have been inextricably connected to animals and their shared environment. The One Health paradigm, emphasizing the interdependence of humans, animals, and environmental elements, such as land and water, is deeply rooted in the cultural and spiritual practices of numerous ancient and contemporary indigenous communities. Reflecting its multifaceted nature, the principles of One Health are discernible in various historical perspectives within the Western intellectual tradition.

Era of "Comparative Medicine"

Prior to the eighteenth century, the understanding of human and animal health, as well as their relationship to the environment, was governed by the medical theory of humoralism. This theory suggested that all bodies consisted of four humors, which were affected by elements like diet, climate, and physical activity; illness occurred when these humors were out of balance. The core tenets of the One Health paradigm are evident in Hippocrates' 'On Airs, Waters and Places,' where he established a link between environmental conditions and human health. He determined that environmental fluctuations contributed to the outbreak and decline of epidemics, earning him the title of the "father of modern medicine."

During the seventeenth and eighteenth centuries, the shift away from traditional Greek thought brought humans and animals into closer association. Rene Descartes' (1596–1650), experimental philosophy, which viewed animals as 'automata' (self-operating machines) led to a greater reliance on animal dissection in the fields of medical research and education. For instance, Swiss physiologist Albrecht von Haller (1708–1777) employed live animals to study the functions of the human nervous system. Carolus Linnaeus's (1707–1778) novel system of animal classification grouped humans, apes, monkeys, and bats within the primate order. This classification, which placed humans in the genus "Homo" alongside orangutans, contested the concept of a distinct "human-animal divide."

Giovanni Maria Lancisi (1654–1720), an Italian physician and early proponent of epidemiology, medicine, and veterinary science, emphasized the crucial role of the environment in the propagation of illnesses affecting both humans and animals. He advocated for animal depopulation and quarantine to manage Rinderpest in cattle.

The surgeon John Hunter (1728–1793) advanced comparative anatomy, establishing it as a key medical practice. Hunter's influence ensured that animals remained central to medical research in the years following his work. Edward Jenner, a disciple of John Hunter, discovered that introducing cowpox could prevent smallpox in humans.

Vicq d'Azyr (1749–1794), a physician, made significant contributions by advancing comparative anatomy and developing a genuinely comparative medical perspective. This approach was exemplified by his focus on Rinderpest, a prevalent disease affecting cattle throughout eighteenth-century Europe. Vicq d'Azyr sought to manage Rinderpest through quarantine protocols, mirroring strategies previously employed to control bubonic plague in humans. He also explored medical meteorology and topology to examine the relationship between climate, geography, and the incidence of both human and animal epidemics. He asserted that the principles governing human ailments are universally applicable to animal diseases, recognizing no fundamental distinction between the two. The essential principles of medicine are universally applicable and can be easily adapted for various contexts and species.

During the 1830s, the medical field's interest in animal health intensified, with studies suggesting causal links between canine rabies, equine glanders, animal anthrax, and corresponding human illnesses. Veterinarians significantly contributed to all historical investigations that utilized various animals for research, diagnosis, and the creation of vaccines and sera. The discovery of germs led to a redefinition of the existing causal relationships between human and animal diseases. This resulted in the creation of a new classification of diseases, termed "zoonoses," to encompass such illnesses. Moreover, Rudolf Virchow (1821–1902), a German physician and pathologist credited with coining the term 'zoonosis,' highlighted the interconnectedness of animal and human health while studying Trichinella spiralis in swine, stating that 'Between animal and human medicines there are no dividing lines—nor should there be.' Although the subject matter varies, the knowledge gained forms the foundation of all medical practices. Despite variations in subject matter, the insights acquired establish the foundation of all medical practices. The emergence of these two scientific advancements led to the establishment of a new discipline known as veterinary public health (VPH). William Osler, a former student of Rudolf Virchow and often cited as the "father of veterinary pathology," investigated animal diseases, highlighting the value of comparative medicine for the medical field.

Beginning in the 1920s, researchers globally posited that comparative medicine held the potential to address a broader spectrum of illnesses and yield fundamental insights applicable across all species. In 1958, medical and veterinary experts gathered in Washington for a meeting sponsored by the World Health Organization and the Pan-American Sanitary Bureau, where they suggested initiating a new program in the field of comparative medicine. The objective was to increase the diversity of animals and animal diseases involved in basic medical research. This method aims to close the professional and practical gaps between veterinary and human medicine. These methods for integrating veterinary and human medicine were based on the concept of "One medicine."

"Comparative Medicine" to "One Medicine"

The concept of "One Medicine" is now linked to Calvin Schwabe (1927–2006), a significant contributor to contemporary epidemiology and an advocate for comparative medicine. He popularized the term in the third edition of his book, *Veterinary Medicine and Human Health*. However, this term "One Medicine" was used many times in literature earlier to demonstrate the nature and value of comparative medicine. In the mid-twentieth century, this term was chiefly connected with a group of authors affiliated with the University of Pennsylvania's veterinary school. Although currently attributed to Calvin Schwabe, the concept of "One Medicine" predates his usage and was utilized to highlight the principles and value of comparative medicine. In the mid-twentieth century, this terminology was prevalent, especially among scholars from the University of Pennsylvania's veterinary program. Therefore, it is plausible that Calvin Schwabe appropriated the term "One Medicine" from the intellectual trends in comparative medicine during the mid-twentieth century.

Additionally, in the twentieth century, various individuals and organizations focused on specific diseases, fostering a closer relationship between human and animal health. A significant organization in this context was the Rockefeller Foundation, which incorporated the study of animal pathology into its medical, scientific, and public health initiatives. Theobald Smith, along with Richard E. Shope, utilized a comparative ecological perspective to examine the incidence of Texas fever and suggested the pig influenza virus's involvement in human influenza. As medical professionals, they both acknowledged animal pathology as fundamental to all medical practices. Several institutions in the United States, including the Mayo Clinic at the University of Minnesota, University of Pennsylvania, and the Hooper Foundation for Medical Research at the University of California, facilitated collaborations between medical and veterinary professionals in research and postgraduate studies. In France and Germany, institutions such as the Pasteur and Koch Institutes continued to embrace a comparative medicine approach, mirroring the practices of European medical research centers.

In the twentieth century, the connections between health and the environment were marked by uncertainty and lack of clarity. The development of vaccines and antibiotics enabled the targeted management of infectious agents, diverting focus from the environmental factors influencing disease emergence and propagation. James Harlan Steele (1913–2013), an American veterinarian hailed as the "father of veterinary public health," gained recognition for his pioneering work in emphasizing the ecological relationship between animal and human health. Steele initiated early efforts to prevent the spread of zoonotic diseases. In 1947, he established the Veterinary Public Health division at the Centers for Disease Control and Prevention in the United States, and also contributed to the establishment of postgraduate public health education as a novel veterinary specialty. Steele's work encompassed the study and management of various zoonotic diseases, including brucellosis, bovine tuberculosis, rabies, and Venezuelan equine encephalitis. Likewise, in 1948, the World Health Organization created a Veterinary Public Health unit under the leadership of Martin Kaplan within its Division of Communicable Diseases to advance human health through disease control. It also fostered close partnerships with the Food and Agriculture Organization of the United Nations and the Office International des Epizooties. Collaborative programs focusing on zoonoses, meat hygiene, and veterinary education were established in the 1950s through a series of joint meetings between the World Health Organization and the Food and Agriculture Organization, particularly in developing nations.

"One Medicine" to "One Health"

During the 2000s, the established concepts of One Medicine and VPH converged, leading to a more refined vision of the "One Medicine, One Health" concept. This concept encompassed the integration or alignment of veterinary and human medical disciplines in research and clinical practice, including collaborative research programs, vaccination strategies, and drug development initiatives. Concurrently, another group of researchers and entities championed the "One World, One Health" concept. In contrast to the OMOH concept's focus on veterinary and medical fields, the OWOH concept broadened its scope to include a wider array of fields and sectors within the life and environmental sciences, while still focusing specifically on emerging infectious diseases (EIDs). The "One World" concept can be traced back to discussions about international relations and the establishment of the United Nations Educational, Scientific and Cultural Organization (UNESCO) in the mid-twentieth century. The "One World" concept gained significant traction within the health sector during the 1990s, prompted by the recognition of the global scale and potential wildlife origins of the human immunodeficiency virus/acquired immune deficiency syndrome pandemic.

In 2004, the Wildlife Conservation Society in New York City hosted a symposium titled “Building Interdisciplinary Bridges to Health in a Globalized World,” which convened experts from public health, conservation, and infectious disease fields to address the theme of OWOH. The assembly or symposium culminated in the creation of 12 recommendations, collectively termed the “Manhattan Principles.” These principles called upon global leaders, civil society, the international health sector, and scientific organizations to adopt a unified strategy for preventing and managing epidemic and epizootic diseases, all while safeguarding ecosystem integrity. Over time, these principles evolved into what is now recognized as the OWOH concept.

In December 2007, an international gathering of representatives from over 100 nations and specialists from various global organizations convened in New Delhi, India, for the International Ministerial Conference on Avian and Pandemic Influenza. At the conference, the “One Health” concept was endorsed for pandemic preparedness. Furthermore, governments were urged to broaden the concept by fostering connections between medical and veterinary fields to enhance pandemic readiness and human security.

In 2008, following the guidance from the 2007 International Ministerial Conference on Avian and Pandemic Influenza, the WHO, OIE, and FAO, with backing from the United Nations Children’s Fund, World Bank, and United Nations System Influenza Coordination (UNSIC), established a groundbreaking tripartite agreement to collaborate more closely on issues at the intersection of animal, human, and ecosystem health. A document titled “Contributing to One World, One Health: A Strategic Framework for Reducing Risks of Infectious Diseases at the Animal-Human-Ecosystems Interface” was created. This initiative drew upon lessons from the early 2000s’ countering to the highly pathogenic H5N1 avian influenza. The tripartite agreement was formally unveiled in October 2008 during the International Ministerial Conference which took place in Sharm el-Sheikh, Egypt, on Avian and Pandemic Influenza, convened experts from approximately 120 countries and various international and national organizations. In 2008, the “One Health” concept was adopted as a political and practical strategy to combat avian influenza and other infectious diseases.

In 2010, a United Nations and World Bank report titled “Fifth Global Progress Report on Animal and Pandemic Influenza” underscored the importance of a One Health approach for continued pandemic preparedness. Furthermore, the report advised that nations and regional organizations should enhance their One Health capabilities to address a wide array of both novel and existing disease threats, rather than concentrating solely on controlling avian influenza through emergency measures.

In 2011, the inaugural International One Health Congress convened in Melbourne, Australia, drawing over 600 delegates from diverse fields and 60 nations to explore the benefits of collaborative efforts in promoting the One Health approach. Experts additionally advocated for the inclusion of disciplines like economics, food security, and safety, complementing the established understanding of the interdependence among human, animal, and environmental health. That same year, the

Tripartite held a high-level technical meeting in Mexico City. The meeting focused on diverse health risks across geographic regions, prioritizing three key One Health concerns: rabies, influenza, and antimicrobial resistance.

The inaugural One Health Summit, also known as the "Davos One Health Action Plan," convened in Davos, Switzerland, in 2012. This summit introduced the One Health concept, advocating for multi-sectoral and multi-stakeholder collaborations to address diverse health threats and enhance public health.

The second International One Health Congress, themed "A world united against infectious disease: cross–sectoral solutions," took place in Bangkok, Thailand, in 2013. Attended by over 1000 individuals from approximately 70 nations, this gathering represented the largest One Health conference at that time. The conference fostered interdisciplinary cooperation and communication to facilitate sound policy formulation concerning human, animal, and environmental well-being.

Latest About One Health

Till date, total seven One Health Congress have been organized in different parts of the world and next eighth one will be organized in Cape Town, South Africa in 2024. Earlier One health concept was "Tripartite (including three organizations WHO, WOAH, FAO)" but now it becomes "Quadripartite (including four organizations WHO, WOAH, FAO and UNEP)."

Recognizing the global imperative to mitigate future infectious disease pandemics and foster sustainable health through the One Health framework, the Quadripartite has formulated the One Health Joint Plan of Action. The Quadripartite reaffirms its dedication to advancing the One Health approach through the One Health Joint Plan of Action. The One Health Joint Plan of Action primarily emphasizes enhancing capabilities to manage intricate, multifaceted health risks by establishing more robust health systems at regional, national, and global scales.

History and Scope of Veterinary Public Health

2

Ripan Biswas and Indranil Samanta

Abstract

Veterinary public health, a crucial intersection of human, animal, and environmental health, has a rich history that is intensely intertwined with the progress of public health. From the initial acknowledgements of zoonotic ailments to the creation of official veterinary provisions, this domain has been consistently adapted to tackle novel issues and adjust to shifting realities. In this chapter, the historical trajectory of veterinary public health, tracing its origins from the control of animal diseases to its modern-day focus on One Health initiative, has been discussed. This chapter also examines significant progress in several areas, such as the integration of veterinary expertise into public health agencies, the implementation of food safety protocols, and the shared recognition of the interconnectedness of human and animal health. The expanding scope of veterinary public health encompasses emerging infectious diseases, antimicrobial resistance, food security, and environmental sustainability, highlighting its vigorous role in safeguarding global health in the twenty-first century.

Keywords

Veterinary Public Health · Zoonoses · One Health

R. Biswas (✉)
Department of Veterinary Public Health and Epidemiology, West Bengal University of Animal and Fishery Sciences, Kolkata, West Bengal, India

I. Samanta
Department of Veterinary Microbiology, West Bengal University of Animal and Fishery Sciences, Kolkata, West Bengal, India

R. Biswas et al. (eds.), *Concept of One Health and Zoonoses*,
https://doi.org/10.1007/978-981-95-6627-3_2

Genesis of Veterinary Public Health

Dr. James Harlan Steele, who is considered the founder of Veterinary Public Health, was the first veterinarian to earn a master's degree in public health from Harvard University. This was the result of his interest in zoonotic pathogens. Dr. Steele's fascination with zoonoses developed during his internship in the Michigan Health Department. While pursuing his undergraduate education at Michigan State University, Dr. Steele became intrigued by the epidemiological factors contributing to the transmission of Brucella after noticing the brucellosis epidemic impacting veterinary students and staff in the bacteriology department at Michigan State College. During World War II, Dr. Steele served as a sanitary officer in the United States Public Health Service in Puerto Rico and the Virgin Islands, where he oversaw milk and food sanitation efforts as well as the control of brucellosis, Venezuelan equine encephalitis, rabies, and bovine tuberculosis. A pivotal moment in Dr. Steele's career was a discussion with Joseph Mountin, MD, then the U.S. Assistant Surgeon General Mountin's query, "What role will veterinarians play in public health now that the war has concluded?" spurred Dr. Steele's exploration of veterinary public health. Dr. Steele's report introduced the term "Veterinary Public Health," detailing the risks of zoonotic diseases and the advantages of using veterinarians to study and combat the pathogens. Following a positive evaluation of the veterinary public health program, Dr. Steele recommended the implementation of a Veterinary Medical Officer role within the United States Public Health Service, a proposal that was subsequently approved and implemented in 1947.

The establishment of Veterinary Public Health was recognized as crucial by European nations, which led to the formation of a Veterinary Public Health Unit within the WHO in 1949. Dr. Steele worked with the WHO in Geneva, where he supported Dr. Martin Kaplan, the organization's inaugural director of veterinary public health. In 1950, Dr. Steele represented the Centers for Disease Control and Prevention at the inaugural WHO Expert Committee on Zoonoses. After 1950, several expert groups concentrated on various problems involving zoonoses and their associated topics.

In December 1950, the WHO/FAO expert group on zoonoses collaborated to create recommendations for zoonoses, offering guidance to the FAO and WHO. The expert group defined VPH as "Veterinary Public Health comprises all the community efforts influencing and influenced by the Veterinary-Medical arts and sciences applied to the prevention of diseases, protection of life, and promotion of the well-being and efficiency of mankind."

In 1956, WHO/FAO expert group on zoonoses and redefined the definition as "Veterinary Public Health is the field of activities which protects and advances the human well-being by utilizing the combined knowledge and resources of all those concerned with human and animal health and their interrelationship."

In 1975, again WHO/FAO expert group modified the definition as "Veterinary Public Health is a component of public health activities devoted to the application of professional veterinary skills, knowledge and resources to the protection and improvement of human health."

Again in 1999, the latest definition given by the WHO/FAO expert group in "VPH is the sum of all contributions to the physical, mental and social well-being of humans through understanding and application of veterinary science." This definition encompasses a focus on risks associated with both the production and consumption of animal-derived food.

Scope of Veterinary Public Health

Veterinary Public Health safeguards human health by mitigating exposure to risks originating from animals, animal-derived products, and their surroundings. The scope and structure of Veterinary Public Health differ across nations.

Veterinary Public Health clearly encompasses a wide range of disciplines, bringing together veterinarians, human health specialists, scientists, and paraprofessionals who are involved in the treatment, control, or prevention of diseases originating from animals. The discipline of veterinary medicine, together with public health, holds the capacity to exert a global influence and make significant contributions worldwide. Veterinarians are witnessing a rise in both requirements and career prospects within numerous sectors, including governmental organizations dedicated to the health of humans and animals. Veterinary science covers all facets of veterinary practices, which include animal health and production. It stands as a vital discipline that fulfills key roles in public health, thereby exerting a direct influence on human health through a variety of mechanisms.

Zoonotic Disease Control and Prevention

Prevention and Control: A fundamental principle of veterinary public health is the prevention and control of zoonotic diseases. Veterinarians play a crucial role in detecting, monitoring, and mitigating the dissemination of infectious agents transmissible from animals to humans, thereby protecting public health. The importance of this endeavor has become even more pronounced in recent decades, as a confluence of factors, such as population growth, urbanization, and environmental deterioration, has resulted in the appearance and reappearance of various zoonotic illnesses that present substantial hazards to human populations.

Recognizing the indispensable role of veterinarians in combating zoonotic diseases, the United States Centers for Disease Control and Prevention forged a robust partnership with the veterinary community, drawing upon their specialized knowledge and capabilities to strengthen national disease surveillance and response systems.

Surveillance and Monitoring: The importance of veterinary professionals in the surveillance and monitoring of emerging and re-emerging zoonotic diseases is paramount. As frontline witnesses in animal populations, veterinarians are uniquely positioned to recognize early warning signs of novel or evolving pathogens, such as unusual patterns of illness or mortality, and promptly communicate this critical information to public health authorities. Moreover, integrated human–animal surveillance systems are being developed, in which data from both human and animal populations are systematically collected and analyzed to provide valuable insights into the epidemiological dynamics of zoonotic diseases. This approach has been advocated by different countries and can significantly enhance our understanding of disease transmission pathways, providing more effective prevention and control strategies.

One Health Approach

One represents a holistic framework acknowledging the interdependence of human, animal, and environmental well-being, advocating interdisciplinary cooperation to address multifaceted Health issues. In recent years, this strategy has garnered increased recognition as a method for addressing novel and resurgent zoonotic diseases, which constitute up to 75% of human infectious illnesses.

One Health issue is the collaborative work of multiple health science fields. Collaboration among these fields at the local, national, and global scales is essential for optimizing health outcomes for all living beings and their shared environments. This requires cross-sectoral cooperation and communication as well as the integration of various disciplines, including infectious diseases, chronic diseases, toxicology, ecology, agriculture, and social sciences. Numerous national and international entities have advocated the One Health approach, recognizing its potential to enhance public health strategies and tackle intricate health issues. The successful implementation of this approach can be seen in Qatar, where the government has established a coordinated effort in the realm of human and animal health authorities to prevent and control the Middle East Respiratory Syndrome Coronavirus outbreak. Likewise, in the United States, the formation of the One Health Federal Interagency COVID-19 Coordination Group has played a crucial role in guaranteeing an effective and harmonized, whole-of-government reaction to the COVID-19 pandemic. However, the adoption and implementation of the One Health approach is not without challenges, as it requires overcoming institutional, disciplinary, and cultural barriers to facilitate effective communication, collaboration, and cooperation among diverse stakeholders.

The 3Cs of One Health-Communication, Collaboration, and Cooperation are the keywords of the One Health Concept. These 3Cs are essential for mitigating the multifaceted threats posed by zoonotic diseases, environmental degradation, climate change, and other complex public health challenges that transcend the sectoral boundaries. However, researchers advise that public health stakeholders may not always fully recognize the vital contributions of the field of animal health in addressing these pressing issues. Therefore, bridging this gap and fostering greater awareness and collaboration among humans, animals, and the environment will be indispensable for realizing the full potential of the One Health approach, as veterinarians possess unique insights and capabilities that can greatly enhance the public health community's ability to tackle emerging and enduring global health threats.

Antimicrobial Resistance (AMR)

Surveillance Monitoring the use of antibiotics in animals and tracking the emergence of antibiotic-resistant bacteria. Another critical area of veterinary public health is the mitigation of antimicrobial resistance, a growing global threat to human and animal health. Veterinarians, working collaboratively with physicians and other stakeholders, have implemented judicious antimicrobial use practices in animal, agricultural, and companion animal settings, representing a vital effort to control the development and spread of resistant pathogens.

Education and Policy Promoting the responsible use of antimicrobials in veterinary practices and contributing to policy development to address AMR at a global level. To address the multifaceted challenge of antimicrobial resistance, veterinary education has evolved to incorporate comprehensive training on antimicrobial stewardship, infection prevention and control, and the holistic One Health approach.

Food Safety and Security

- *Food Inspection*: Veterinarians are crucial in maintaining the safety and quality of the food chain, from production to consumption. Veterinary Public Health Specialists are responsible for ensuring the safety and quality of animal-derived products such as meat, milk, and eggs through inspection and regulatory measures. This work is essential for protecting consumer health and maintaining integrity and trust in the food system, which are critical for food security and economic stability.
- *Risk Assessment*: Recognizing and justifying risks associated with the production, processing, and distribution of animal-derived food products. Comprehensive disease surveillance systems, biosecurity measures, and robust emergency response plans are needed to swiftly detect, contain, and mitigate foodborne illness outbreaks. It is also important to enforce evidence-based food safety standards and policies that safeguard consumer health and welfare while also supporting sustainable production and responsible consumption of animals.

Animal Health and Welfare

- *Disease Management*: Veterinary Public Health specialists diagnose animal diseases that have public health consequences, confirming the health of livestock and companion animals.
- *Animal Welfare*: Veterinary public health prioritizes animal welfare and recognizes the inextricable link between animal, human, and environmental health. Veterinarians are responsible for safeguarding animal welfare across diverse contexts, including agriculture, research, and companionship by providing suitable environments, sustenance, and medical attention. This is why they uphold

the ethical obligations of the profession and contribute to the overall health and resilience of the ecosystems in which they function.

Environmental Health

- *Waste Management*: Proper disposal of animal waste to prevent environmental contamination and human exposure to harmful pathogens. Waste, whether generated from domestic, industrial, or agricultural sources, poses significant risks to the health and stability of interconnected ecosystems, necessitating a comprehensive approach that recognizes the intricate interconnections between humans, animals, and ecological welfare.
- *Ecosystem Health*: Addressing the health of ecosystems, including wildlife populations and their habitats, to prevent disease transmission between animals and humans. The delicate balance and intertwined nature of the natural environment, human health, and animal welfare have become increasingly recognized as crucial considerations in addressing the complex challenges facing our global ecosystem.

The anticipated effects of climate change include an increase in the risk posed by vector-borne diseases and other illnesses affecting both humans and animals. Diseases, such as Rift Valley fever and schistosomiasis, are linked to these climatic changes. The full impact of climate change may not become apparent for years or decades.

Emergency Preparedness and Response

- *Disaster Response*: Coordinating responses to crises, such as natural disasters and disease outbreaks, impacts both human and animal populations.
- *Biosecurity*: Employing biosecurity measures to prevent the spread of infectious diseases during calamities.

Veterinary Public Health will continue to develop in response to the dynamic global landscape marked by population expansion, escalating urbanization, disparities in wealth and technology between developed and developing countries, and changes in land use, environmental factors, and climate patterns. These advancements will present novel obstacles for Veterinary Public Health in the larger framework of the human health infrastructure. Furthermore, substantial transformations will occur within the direct domain of Veterinary Public Health, focusing on pivotal areas such as agricultural practices, the food supply network, international commerce, human and animal mobility, and the One Health concept.

The future of veterinary public health hinges on the profession's capacity to adapt, encourage teamwork across disciplines, and proactively address new difficulties.

Zoonoses

3

Ripan Biswas

Abstract

Zoonoses, the diseases transmitted between animals and humans, pose a significant global health concern. This chapter travels the assorted realm of zoonotic diseases, probing into their multifaceted classifications based on pathogen type, transmission modes, and ecological characteristics. The chapter strains on the major categories of zoonotic pathogens, including viruses, bacteria, parasites, and fungi, and provides examples of prominent diseases within each group. Furthermore, the chapter digs into the various transmission pathways, including direct contact, vector-borne transmission, foodborne transmission, and environmental exposure, highlighting the complexities inherent in zoonotic spillover events.

Keywords

Zoonoses · Classification of Zoonoses

Overview of Zoonoses

The term 'zoonoses' comes from the Greek words zōon, meaning 'animal', and nosos, meaning 'disease'. The singular term is 'zoonosis', while the plural is 'zoonoses'. Zoonoses are diseases that can be transmitted from animals to humans, and they represent a significant public health concern globally. Rudolf Virchow coined the term 'zoonoses' in 1855 during his study on Trichinella, to show the link between animal and human diseases. WHO defines "*A zoonosis is any disease or*

R. Biswas (✉)
Department of Veterinary Public Health and Epidemiology, West Bengal University of Animal and Fishery Sciences, Kolkata, West Bengal, India

R. Biswas et al. (eds.), *Concept of One Health and Zoonoses*,
https://doi.org/10.1007/978-981-95-6627-3_3

infection that is naturally transmissible from vertebrate animals to humans." Louise H. Taylor from Centre for Tropical Veterinary Medicine, University of Edinburgh, analyzed by a comprehensive literature review that more than 60% of human infectious diseases are caused by pathogens shared with wild or domestic animals. He also documented that there are 1415 species of infectious organisms known to be pathogenic to humans. This includes 217 viruses and prions, 538 bacteria and rickettsia, 307 fungi, 66 protozoa, and 287 helminths. Of these, 868 are zoonotic. Additionally, 175 pathogenic species are associated with 'emerging' diseases. Over 30 new human pathogens have been detected in the last three decades, 75% of which have originated in animals. Zoonoses account for 60% of known infectious diseases and up to 75% of emerging infectious diseases, with around 2.5 billion infection incidents and 2.7 million deaths annually. The list of zoonotic diseases is extensive and continually expanding, with the COVID-19 being the most recent emergence.

Classifying zoonoses is crucial for effective prevention, control, and management of these diseases, as it helps identify their underlying causes, transmission patterns, and appropriate mitigation strategies.

Classification of Zoonoses

Zoonotic diseases can be classified based on various factors, including the mode of transmission, the type of infectious agent and the animal reservoir. The World Health Organization Expert Committee on Zoonoses recommends the following categories:

1. *Based on the causative agent-.*
   (a) Bacterial: Tuberculosis, Brucellosis, Salmonellosis, Anthrax, Leptospirosis.
   (b) Viral: Japanese encephalitis, Rabies, Ebola, Hantavirus, Influenza.
   (c) Mycotic: Histoplasmosis, dermatophytosis, Cryptococcosis, Aspergillosis.
   (d) Rickettsial: Scrub typhus, Q fever, Ehrlichiosis, Anaplasmosis.
   (e) Parasitic: Echinococcosis, Toxoplasmosis, Taeniasis, Dracunculiasis, Leishmaniasis.
   (f) Prion: Creutzfeldt-Jakob disease, Bovine spongiform encephalopathy.
   (g) Chlamydial: Psittacosis.

2. Based on transmission cycle-.
   A. Direct Zoonoses: These diseases are transmitted directly from an infected vertebrate host to a susceptible vertebrate host through direct contact, fomites, or a mechanical vector. The transmission process does not involve any developmental changes or propagation of the pathogenic organism. Examples include rabies, which is transmitted through the bite or scratch of an infected animal; anthrax, which can be contracted by handling or inhaling spores from infected animal products; and tuberculosis, which can be passed from livestock such as cattle to humans through close contact. These diseases are

transmitted directly from an infected animal host to a susceptible human host through direct contact, contaminated surfaces, or a mechanical vector, without requiring any developmental changes or propagation of the pathogenic organism.

B. Cyclozoonoses: Zoonotic diseases that involve two or more vertebrate hosts to complete their transmission cycle can be further categorized as follows.
   (i) Obligatory cyclozoonoses: Humans are mandatory or essential hosts to complete the life cycle of the agent. This is why they are considered compulsory hosts. Examples include *Taenia solium* and *Taenia saginata*.
   (ii) Non-obligatory cyclozoonoses: In non-obligatory cyclozoonoses, humans become infected accidentally in the transmission cycle and represent a dead-end or cul-de-sac host. Examples include hydatidosis, a parasitic infection caused by the larval stage of the tapeworm *Echinococcus granulosus*. In this type of cyclozoonosis, humans are accidental or dead-end hosts, as the parasite's life cycle typically involves canids, such as dogs, as the definitive host and herbivores, such as sheep or cattle, as the intermediate host.

C. Metazoonoses: In metazoonoses, the causative agent must multiply or develop within an invertebrate host before transmission to a vertebrate host is possible. Transmission requires a specific incubation period to be completed before it can occur. Examples include Yellow fever, Plague. Metazoonoses can be further classified based on the number of vertebrate or invertebrate hosts involved in the transmission cycle.
   (i) Metazoonoses type I: One vertebrate + one invertebrate host are involved. Examples include Plague, Yellow fever.
   (ii) Metazoonoses type II: One vertebrate + two invertebrates host are involved. Examples include Paragonimiasis, also known as lung fluke infection, is a type of metazoonoses that involves two invertebrate hosts in its transmission cycle. The causative agents are parasitic lung flukes of the genus *Paragonimus*, which require a freshwater snail as the first intermediate host and a freshwater crab or crayfish as the second intermediate host before they can infect a vertebrate definitive host, such as humans or other mammals.
   (iii) Metazoonoses type III: Two vertebrates + one invertebrate host are involved. Examples like Clonorchiasis, also known as Chinese liver fluke infection, is a type of parasitic infection caused by the liver fluke *Clonorchis sinensis*. This disease requires three hosts to complete its life cycle: an aquatic snail as the first intermediate host, a freshwater fish as the second intermediate host, and a vertebrate like humans or other mammals as the definitive host.
   (iv) Metazoonoses type IV: In this category, transovarian transmission of the pathogen is occurred, passing from a female invertebrate host to its offspring. An example is tick-borne encephalitis.

D. Saprozoonoses: These infections require a non-animal or non-animate environment, such as food plants, soil, or other organic matter, to complete the life cycle of the causative agent. Histoplasmosis, caused by the fungus *Histoplasma capsulatum*, is an example of a saprozoonotic disease. Larva migrans is another example of a saprozoonotic infection caused by parasites like Toxocara which can be transmitted to humans through ingesting or coming into contact with contaminated soil or other environmental sources that harbor the infective stages of these parasites. This type can also be subcategorized into three types.
   (i) Saproanthrapozoonoses: Zoonotic diseases in this category are transmitted from animals to humans through non-living or inanimate objects. Examples include ancylostomiasis and cutaneous larva migrans.
   (ii) Saproamphixenoses: These infections are shared by humans and animals in nature, but are transmitted through non-living sources. Examples include fungal diseases.
   (iii) Saprometaanthrapozoonoses: These diseases require a vertebrate host, an invertebrate host, and a non-living object to complete the transmission cycle of the causative agent. An example is fascioliasis.

3. On the basis of reservoir host-.
   (i) Anthrapozoonoses: Humans can acquire these infections naturally in certain infrequent circumstances, such as through occupational exposure from animals. Examples of anthrapozoonoses include rabies, which can be transmitted from infected wild or domestic animals, and leptospirosis, which can be contracted from exposure to the urine of infected animals.
   (ii) Zooanthroponoses: These are diseases of animals that are normally transmitted from humans to animals naturally. These zoonotic diseases are typically spread from humans to animals through direct contact, contaminated surfaces, or via a mechanical vector, without requiring any developmental changes or propagation of the pathogenic organism. An example of a zooanthroponoses is amoebiasis, which is caused by the parasite *Entamoeba histolytica* and can be transmitted from infected humans to animals.
   (iii) Amphixenoses: These diseases are omnipresent in nature, which allows for bidirectional transmission between animal and human hosts. An example of an amphixenotic disease is staphylococcosis, which can be transmitted between animals and humans through direct contact or contaminated surfaces.

Some Other Vital Class of Zoonoses

- Nosocomial Zoonoses: Healthcare settings can provide opportunities for the transmission of zoonotic pathogens between animals and humans, or vice versa, leading to the acquisition of nosocomial zoonotic infections. Example:Methicillin-resistant *Staphylococcus aureus* spread from livestock to humans in healthcare settings.

- Xenozoonoses: It refers to the infectious diseases transmitted from animals to humans through the transplantation of animal tissues or organs into the human body. Example: Porcine endogenous retrovirus transmission during xenotransplantation.
- Synanthropic zoonoses: These are infectious diseases that are shared between wild animal species that have adapted to thrive in close proximity to human populations and the humans themselves. Hantavirus is an example of a synanthropic zoonosis, which can be transmitted to humans through contact with rodent urine, droppings, or saliva in areas where rodents have adapted to live in close proximity to human habitats.
- Exoanthropic zoonoses: These are diseases caused by pathogens that exist primarily in wildlife reservoirs but can be transmitted to humans when people have contact with the wildlife, either directly or indirectly through an intermediate host or vector. Lyme disease is an example of an exoanthropic zoonosis, where the bacteria causing it exist in wild rodent populations and are passed to humans by ticks.
- Euzoonoses: Obligatory cyclozoonoses are altenatively known as euzoonoses where man is mandatory to complete the life cycle of the causative pathogen. Euzoonoses, such as *Taenia solium* and *Taenia saginata* infections, require the presence of humans to host the adult tapeworm stage in order to complete their life cycles.
- Orthozoonoses: It is also known as direct zoonoses, involving the direct transmission of diseases from an infected vertebrate to another susceptible vertebrate host. Examples of orthozoonoses include rabies, which is primarily transmitted through contact with infected saliva of a rabid animal, as well as anthrax, which can be contracted through direct contact with infected animals or contaminated animal products.
- Phanerozonoses: Zoonoses where clinical signs are observed in both the animal hosts and human hosts are designated as phanerozoonoses. These zoonoses are readily recognizable and identifiable due to the manifestation of observable symptoms in both animal and human populations. Example: rabies and brucellosis.
- Isosymptomatic zoonoses: Zoonoses in which the disease manifestation is similar in both animal and human hosts are called isosymptomatic zoonoses. Rabies and tuberculosis are examples of isosymptomatic zoonoses where the clinical symptoms are very similar in infected animals and humans.
- Anisosymptomatic zoonoses: Some zoonoses may present with varying clinical presentations in the animal host versus the human host, leading to different symptoms or disease manifestations. Example Brucellosis, where the clinical presentation in animals may differ from that observed in humans.
- Cryptozoonoses: In some instances, animals may have subclinical infections while humans develop clinical illness, or vice versa. In ornithosis pneumonia, humans develop clinical illness, but pigeons and turkeys often have a latent infection without overt symptoms.

- Latent zoonoses: Latent zoonotic infections in animals may lack overt clinical signs, yet still have the potential to cause disease in humans. For example, Toxoplasma gondii can persist as a latent infection in animals, such as cats, yet still be transmitted to and cause disease in pregnant women.
- Endemic zoonoses: That are prevalent in various locations and impact sizable populations of both people and animals. Example: Leptospirosis, which is endemic in many tropical and subtropical regions.
- Epidemic zoonoses: Which occur sporadically in terms of time and location. Example: Ebola virus disease, which causes periodic outbreaks in central Africa.
- Emerging Zoonoses: Diseases that are caused by either a previously unknown agent or a known organism appearing in places or in species where it was not previously observed is considered an emerging zoonotic disease. Nipah virus is the classical example, which was first detected in 1998–1999 in Malaysia when pig farmers contracted the virus from fruit bats, leading to a severe outbreak with high morbidity and mortality rates.
- Reemerging Zoonoses: Zoonotic diseases that were once widespread but declined and have since returned or become more common are considered re-emerging zoonoses. Example tuberculosis. Tuberculosis was once a major public health issue, but declined for a time before resurging in many regions. This resurgence is attributed to factors like drug-resistant strains, HIV/AIDS, and more interactions between humans and animals.

Reference

Chomel B. B. (2009). Zoonoses. *Encyclopedia of Microbiology,* 820–829. https://doi.org/10.1016/B978-012373944-5.00213-3

Emerging, Re-emerging, and Occupational Zoonoses

4

Prasanna Vadhana A, Pesingi Pavan Kumar, Priscilla Kerketta, and Bhavani Puvvala

Abstract

Emerging, re-emerging, and occupational zoonoses epitomize significant threats to global public health. This chapter compiles the dynamic landscape of these zoonotic diseases, exploring the factors driving their emergence, resurgence, and impact on specific occupational groups. We discuss the ecological, social, and economic drivers contributing to the emergence of novel zoonotic pathogens, such as deforestation, wildlife trade, and climate change. Moreover, we analyse the reasons behind the resurgence of previously controlled or eradicated diseases, including antimicrobial resistance and breakdowns in public health infrastructure. The chapter also focuses on occupational zoonoses, highlighting the risks faced by individuals in professions like agriculture, veterinary medicine, and wildlife management. Finally, we discuss strategies for the prevention, surveillance, and control of these zoonotic diseases, emphasizing the importance of interdisciplinary collaboration and a One Health approach to mitigate their impact on human health.

Keywords

Emerging Zoonoses · Re-emerging Zoonoses · Occupational Zoonoses

---

Prasanna Vadhana A (✉)
Department of Veterinary Microbiology, College of Veterinary and Animal Sciences, Rani Lakshmi Bai Central Agricultural University, Jhansi, Uttar Pradesh, India

P. P. Kumar
Faculty of Veterinary and Animal Sciences, Department of Veterinary Public Health, Rajiv Gandhi South Campus, Banaras Hindu University, Mirzpur, Uttar Pradesh, India

P. Kerketta · B. Puvvala
Department of Veterinary Microbiology, Institute of Veterinary Sciences and Animal Husbandry (IVSAH), Siksha O Anusandhan (SOA), Bhubaneswar, Odisha, India

R. Biswas et al. (eds.), *Concept of One Health and Zoonoses*,
https://doi.org/10.1007/978-981-95-6627-3_4

Introduction

A zoonosis is any disease or infection naturally transmissible between vertebrate animals and humans. Zoonotic diseases existed for centuries, playing a significant role in global health. Animals including domestic and wildlife play a major role in the emergence and re-emergence of infectious zoonotic disease. According to the World Organisation for Animal Health (WOAH), about 60% of pathogens that cause human disease originate from animals, 75% of emerging infectious human diseases have an animal origin, and 80% of pathogens of bioterrorism concern originate in animals. Spillover of the pathogen between species, especially from animal species to humans, is responsible for the emergence of new diseases. The consequences of zoonotic spillover can be devastating, leading to localized outbreaks, epidemics, or even global pandemics. Many of the most significant emerging infectious diseases of the twenty-first century—such as SARS, MERS, Ebola, and COVID-19—are the result of zoonotic spillover events. These events have shown how interconnected human, animal, and environmental health is and highlight the importance of the One Health approach to prevent future zoonotic diseases from spreading. In recent decades, the frequency of zoonotic outbreaks has increased, driven by urbanization, ecological disruption, climate change, international travel, and greater interaction between wildlife and human populations. The COVID-19 pandemic serves as a powerful reminder of the catastrophic impact zoonotic diseases can have on global health and economies.

Emerging Diseases/Infections: Definition and Classification of Zoonoses

- Emerging infectious diseases are infections that have recently appeared in a population or whose incidence or geographic range is rapidly increasing. These diseases are caused by pathogens that were previously unknown or have newly evolved.
- Re-emerging infectious diseases are diseases that were previously under control but have reappeared in populations, often with increased incidence or geographical spread. This can occur due to factors like changes in public health measures, resistance to treatments, or shifts in the pathogen's characteristics.

Drivers of Emergence and Re-emergence

Several interconnected factors contribute to the emergence and re-emergence of zoonoses (Judson & Rabinowitz, 2021).

Environmental and Ecological Changes

Deforestation, agricultural expansion, and urbanization disrupt natural ecosystems, bringing humans into closer contact with wildlife reservoirs of zoonotic pathogens. As habitats shrink, viruses adapt to new hosts, raising the risk of spillover events. For example, KFD, Ebola disease, Hanta viral disease, etc. have emerged as result of deforestation. On the other hand, climate change further exacerbates these issues by altering the distribution of disease vectors such as mosquitoes, increasing the prevalence of diseases like malaria, dengue, Zika, etc.

Globalization and Human Mobility

The rapid movement of people and goods across countries accelerates the spread of zoonotic pathogens. For instance, the SARS outbreak in 2002–2003 spread from China to multiple countries within weeks through air travel networks. Similarly, the global trade of used tires has contributed to spreading *Aedes* mosquitoes, vectors for dengue, and Zika viruses. The movement of people and goods on a global scale has increased the risk of zoonotic diseases spreading across borders. International trade in livestock and wildlife products has facilitated the transmission of diseases from one country to another, as seen in the Nipah virus outbreak, which spread from Malaysia to Singapore through pig trade.

Microbial Evolution

Pathogens evolve constantly to adapt to changing environments. RNA viruses, such as influenza, have particularly high mutation rates, allowing them to evade immune responses through mechanisms like *antigenic drift* and *antigenic shift*. These mutations can result in new strains capable of infecting previously resistant populations.

Wildlife Trade and Wet Markets

The trade of wildlife and the operation of wet markets increase the likelihood of cross-species transmission. COVID-19, for example, is believed to have originated from a market in Wuhan, China, where live animals were sold in close proximity to humans. Similarly, the Nipah virus outbreak in Malaysia emerged from interactions between fruit bats, pigs, and humans. The trade of wildlife in live animal markets, as seen with the initial outbreak of SARS and potentially COVID-19, presents a significant risk for zoonotic spillover. These markets bring different animal species into close proximity with humans, increasing the potential for interspecies pathogen transmission.

Bushmeat Trade and Hunting In many parts of the world, particularly in regions with food insecurity, humans rely on hunting wildlife for food (bushmeat). This practice brings humans into direct contact with animals that may harbor pathogens, such as Ebola virus or HIV, which are linked to the consumption of bushmeat from non-human primates and bats.

Agricultural Intensification The intensification of agriculture (especially in tropical regions) has contributed significantly to the emergence of zoonotic diseases. The use of antibiotics and growth promoters in livestock, along with increased contact between humans and domestic animals, has created conditions that facilitate the spread of zoonotic pathogens.

Human-Animal Interface The likelihood of zoonotic spillover increases when humans invade natural habitats, bringing them closer to wildlife. The destruction of forests and expansion of agricultural lands into previously uninhabited areas disturbs natural ecosystems and increases interactions between humans, livestock, and wild animals.

Notable Examples of Emerging and Re-emerging Zoonoses

Viral Zoonoses

- *COVID-19*:
  The origin of SARS-CoV-2 has been closely examined, with strong evidence suggesting it is zoonotic, having originated in bats and potentially transmitted to humans via an intermediate host. Horseshoe bats (Rhinolophus species) are considered primary reservoirs for the virus, and genomic studies have identified close relatives of SARS-CoV-2 in these bats. For example, the RaTG13 coronavirus found in horseshoe bats shares 96% genetic similarity with SARS-CoV-2. The virus likely crossed from bats to humans, either directly or through an intermediary animal, a path similar to that of SARS-CoV, another coronavirus with zoonotic origins. This zoonotic spillover may have occurred in wildlife markets, such as the Huanan Seafood Wholesale Market in Wuhan, where a range of wild animals are sold in close proximity to humans, facilitating cross-species transmission. SARS-CoV-2 utilizes its spike (S) protein to bind with high affinity to the ACE2 receptor on human cells, allowing it to enter cells efficiently and replicate. The ACE2 receptor, found in various tissues including the lungs, heart, kidneys, and intestines, plays a key role in viral transmission and symptom presentation. This widespread expression explains the multi-organ impact of COVID-19 and its wide range of clinical manifestation. Since its initial outbreak, SARS-CoV-2 has undergone genetic mutations, resulting in various strains and variants with unique characteristics. These mutations often affect the spike protein, enhancing the virus's ability to bind to the ACE2 receptor, increasing

transmissibility, or helping it evade immune detection. Some of the most notable variants include:

1. *Alpha (B.1.1.7)*: Identified in the United Kingdom, this variant demonstrated a significant increase in transmissibility and contributed to a substantial wave of infections in early 2021. The Alpha variant has mutations that increase its binding to ACE2, allowing it to spread more effectively among people.
2. *Beta (B.1.351)*: Originating in South Africa, the Beta variant contains mutations in the spike protein's receptor-binding domain, enabling it to partially evade the immune response from natural infection and some vaccines. This immune escape ability heightened public health concerns, particularly for reinfection and vaccine efficacy.
3. *Gamma (P.1)*: First detected in Brazil, the Gamma variant has mutations associated with increased transmissibility and immune escape, contributing to reinfections even among those previously infected with earlier strains of SARS-CoV-2.
4. *Delta (B.1.617.2)*: This variant, first identified in India, became dominant worldwide due to its exceptionally high transmissibility and ability to cause severe disease. Delta's mutations enhance its replication and binding capacity, leading to faster and broader transmission.
5. *Omicron (B.1.1.529)*: Emerging in late 2021, Omicron introduced numerous mutations in the spike protein, many within the receptor-binding domain. These mutations allow Omicron to partially evade both natural and vaccine-induced immunity, although it is generally associated with less severe symptoms than previous variants. Subvariants of Omicron continued to evolve, increasing transmission rates and posing challenges to pandemic control measures.

SARS-CoV-2 spreads primarily through respiratory droplets and aerosols, making close contact and poorly ventilated spaces high-risk environments. Studies show that the virus can also survive on surfaces for hours to days, although surface transmission is less significant than respiratory transmission. The high mutation rate of SARS-CoV-2, particularly within variants like Delta and Omicron, has increased the virus's basic reproductive number (R0) from early estimates of 2.5 to as high as 6–8 for Omicron, complicating containment and control efforts. The gold standard for diagnosing active infection is reverse real-time PCR (rRT-PCR), which detects viral RNA with high sensitivity and specificity (Fiegler-Rudol et al., 2024).

Ebola Virus Disease

Ebola Virus Disease (EVD), caused by members of the *Ebolavirus* genus, exemplifies one of the deadliest zoonotic diseases known to date. First identified in 1976 in simultaneous outbreaks in the Democratic Republic of the Congo (DRC) and Sudan, Ebola has since caused periodic and often devastating outbreaks, especially in West and Central Africa. With a mortality rate ranging from 25% to 90%, Ebola is considered a formidable public health challenge that has highlighted global vulnerabilities in infectious disease control and emergency preparedness. The Ebola virus belongs to the family *Filoviridae*, which includes other deadly pathogens like the Marburg virus. There are five known Ebola virus species: Zaire, Sudan, Bundibugyo, Taï Forest, and Reston viruses. Among these, the Zaire Ebola virus (EBOV) is the most virulent, with a case fatality rate (CFR) of approximately 66.6%, while the Sudan virus has a CFR of about 48.5%. Ebola is thought to be a zoonotic virus, with fruit bats acting as its primary natural reservoir. The virus's transmission to humans likely occurs through direct contact with bats or through handling and consumption of bushmeat, particularly from non-human primates infected by the virus. Bushmeat consumption is common in West and Central Africa, where communities depend on the forest for subsistence. This zoonotic spillover, facilitated by human interaction with wildlife habitats, is an ongoing risk factor that underscores the need for enhanced surveillance at the human–animal interface. Ebola spreads through direct human-to-human contact. Transmission can occur through contact with bodily fluids (such as blood, saliva, urine, sweat, breast milk, and semen) of infected individuals or contaminated surfaces and objects. This mode of transmission necessitates stringent hygiene and protective measures, especially for healthcare workers who handle patients in the later, more infectious stages of illness. A key factor in human-to-human transmission is cultural practices, particularly traditional burial rites that involve close contact with the deceased. Traditional burials in affected regions often involve washing or touching the body, which carries a high risk of exposure to the virus. Health organizations have worked to promote safe burial practices to mitigate this risk, along with stringent infection control protocols that include the use of personal protective equipment (PPE).

- The epidemiology of Ebola is influenced by several socio-cultural and ecological factors. Population movement, deforestation, and increased human–wildlife contact all contribute to the risk of zoonotic spillover. Additionally, societal practices such as hunting, bushmeat consumption, and traditional funeral rites create pathways for the virus to jump from animal hosts to humans and subsequently spread within communities (Jones et al., 2008).

Nipah Virus

Nipah virus (NiV) is an emerging zoonotic pathogen responsible for severe respiratory and neurological disease in humans, with a high case fatality rate and substantial public health implications. First identified during an outbreak in Malaysia in 1998, NiV has since caused multiple outbreaks across South and Southeast Asia, with primary cases occurring in Malaysia, Bangladesh, and India. Nipah virus was first discovered in the village of Sungai Nipah, Malaysia, during an outbreak that significantly impacted pig farmers. Fruit bats of the *Pteropus* genus, known carriers of Henipaviruses, were identified as the virus's natural reservoir. The initial spillover from bats to pigs occurred due to the close interactions between these animals in intensive pig farming environments, where pigs became intermediate hosts. Infected pigs then transmitted the virus to humans through direct contact. NiV outbreaks are concentrated primarily in South and Southeast Asia. After its initial emergence in Malaysia, outbreaks have been documented in Bangladesh, India, and the Philippines, with high mortality rates, particularly in Bangladesh and India. In these regions, NiV has shown a case fatality rate of over 70%, influenced by factors like healthcare access and rapid human-to-human transmission in close-contact settings. The primary natural reservoir of NiV is the fruit bat (*Pteropus* spp.), which carries the virus asymptomatically. Human infection typically occurs through direct or indirect exposure to bats or intermediate hosts, such as pigs. Studies also highlight the adaptability of the virus to human hosts, enabling limited human-to-human transmission, particularly in healthcare settings. Nipah virus spreads through various routes depending on the host interaction and regional factors:

Animal-to-Human Transmission: Occurs through close contact with infected animals, typically pigs, or consumption of food contaminated by bat saliva or urine, such as date palm sap.

Human-to-Human Transmission: Primarily observed in healthcare settings and among close personal contacts, particularly in Bangladesh and India. Human-to-human transmission significantly increases the risk of rapid outbreak spread and high mortality rates in densely populated regions. The transmission patterns vary by country: Malaysia: Infected pigs acted as amplifying hosts, facilitating virus transmission to humans, primarily affecting pig farm workers. In Bangladesh and India, direct bat-to-human transmission occurs through the consumption of contaminated date palm sap, while human-to-human transmission is common in healthcare environments (Fiegler-Rudol et at., 2024).

Global Outbreaks

Malaysia and Singapore (1998–1999): The first recorded outbreak in Malaysia and Singapore marked the discovery of NiV. This outbreak affected pig farmers and abattoir workers, leading to widespread morbidity and mortality. The virus was spread primarily through close contact with infected pigs. Rigorous biosecurity measures, including the mass culling of pigs, were implemented to contain the virus. These actions effectively controlled the outbreak but highlighted the need for ongoing surveillance in regions with bat and livestock interactions.

Bangladesh (2001–Present): Since 2001, annual NiV outbreaks have been reported in Bangladesh, characterized by exceptionally high case fatality rates. In contrast to Malaysia, NiV transmission in Bangladesh typically occurs directly from bats to humans via the consumption of date palm sap contaminated with bat saliva and excreta. Human-to-human transmission is also common in this region, further complicating containment efforts. The recurrent nature of outbreaks in Bangladesh emphasizes the critical role of public health interventions and education to prevent consumption of contaminated sap.

India: India has experienced sporadic yet highly lethal outbreaks of NiV, beginning with a 2001 outbreak in West Bengal. Other significant outbreaks occurred in 2007 and 2018–2019 in Kerala. These outbreaks were associated with high case fatality rates and highlighted the importance of rapid containment and public health response. In Kerala, a coordinated response from state health departments and national health agencies helped limit the spread of the virus, underscoring the importance of preparedness for zoonotic diseases.

The Philippines (2014): A smaller outbreak in the Philippines linked NiV transmission to contact with infected horses, suggesting that NiV can infect various species and that vigilance is needed beyond traditional outbreak areas. This incident underscored the importance of monitoring diverse animal populations for zoonotic pathogens (Nipah2).

Hendra Virus Hendra virus is a zoonotic virus that emerged in Australia in 1994. Fruit bats, particularly species of the genus *Pteropus*, are the natural reservoir for this virus. The virus spills over to horses, and from horses, it can infect humans. Human infections have occurred mainly through close contact with infected horses, underscoring the risk posed by animal farming practices in regions where wildlife is prevalent.

Middle East Respiratory Syndrome (MERS) MERS, caused by the MERS-CoV coronavirus, emerged in 2012 and is believed to have spilled over from bats to camels before reaching humans. The virus has since caused sporadic outbreaks in the Middle East, with human-to-human transmission being facilitated in healthcare settings. MERS has a high case fatality rate, but it has not achieved sustained human-to-human transmission on a global scale like SARS or COVID-19.

Severe Acute Respiratory Syndrome (SARS)

SARS was first identified in Guangdong Province, China, in November 2002, and emerged as a mysterious respiratory illness. The disease spread globally in 2003, affecting over 8000 individuals across five continents with a fatality rate of about 10%. SARS is caused by the SARS-associated coronavirus (SARS-CoV), a member of the Coronaviridae family. This novel virus, identified in 2003, is genetically distinct from previously known coronaviruses. SARS-CoV is an enveloped RNA virus associated with zoonotic origins, likely transmitted from wild animals such as civet cats and bats. SARS spreads primarily through respiratory droplets during close contact with infected individuals. Aerosol-generating procedures in healthcare settings increase the risk of transmission. There have also been rare instances of fecal-oral and fomite transmission. SARS symptoms include fever, chills, malaise, headache, dry cough, and difficulty breathing. Severe cases can progress to pneumonia and acute respiratory distress syndrome (ARDS). Gastrointestinal symptoms, including diarrhoea, are also reported in some patients. The SARS epidemic was controlled through rigorous infection control measures, including Quarantine and isolation of suspected cases, travel advisories and airport screenings, use of personal protective equipment (PPE) in healthcare settings, and education on hygiene and respiratory precautions (Fiegler-Rudol et at., 2024).

Zika Virus Infection

Zika virus infection is an arthropod-borne viral disease caused by the Zika virus of the genus flavivirus, Flaviviridae. It has two major phylogenetic lineages: African and Asian, which exhibit distinct genetic and geographical characteristics. The African lineage is predominantly found in sub-Saharan Africa. Historically, this lineage has been associated with a sylvatic cycle involving non-human primates and *Aedes* mosquitoes, with humans serving as incidental hosts. The Asian lineage diverged from the African lineage over time and spread to Southeast Asia and the Pacific islands responsible for major human outbreaks. This lineage is more efficient in human-to-human transmission due to urban mosquito vectors.

The virus was first isolated in 1947 from a sentinel rhesus monkey in the Zika Forest of Uganda during yellow fever surveillance. It was subsequently found in humans, sporadic cases reported across Africa and Southeast Asia for decades. The first major outbreak occurred in 2007 in Yap Island, Micronesia, followed by significant epidemics in French Polynesia in 2013–2014 and Brazil in 2015. The latter brought Zika to global attention due to its link with congenital microcephaly and other neurological disorders. Zika virus transitioned from localized endemic patterns to widespread pandemics. The disease was limited to Africa and Southeast Asia, with silent transmission and minor outbreaks. In Yap Island, the disease infected 73% of the population and French Polynesia noticed the emergence of neurological complications like Guillain-Barré syndrome. Brazil (2015) marked explosive spread through the Americas, involving millions. By 2017, Zika transmission was reported in 84 countries, with significant outbreaks in the Americas, Southeast Asia, and the Pacific. The disease is transmitted by *Aedes aegypti* and *Aedes albopictus* mosquitoes. Vertical transmission, sexual transmission, and rare cases via blood transfusion have also been documented. Urbanization, globalization, and ecological changes facilitated mosquito population growth and viral spread. Zika virus is maintained in a sylvatic cycle involving primates and mosquitoes in Africa. Outside Africa, humans serve as the primary hosts. Most Zika virus infections are mild or asymptomatic. When symptoms manifest, they typically occur 3–12 days post-infection and last for 2–7 days. In acute conditions, fever, rash, conjunctivitis, arthralgia, myalgia, and headache may be noticed. Zika virus is association with severe congenital defects collectively termed Congenital Zika Syndrome (CZS). These defects arise from maternal Zika virus infection during pregnancy, especially in the first trimester. It increases risks of miscarriage, stillbirth, and preterm birth. Microcephaly in the fetus is the hallmark of CZS and refers to abnormal smallness of the head due to impaired brain development due to its neurotropic nature. The other abnormalities in fetus include intracranial calcifications, cerebral atrophy, ventriculomegaly (enlargement of the brain's ventricles), retinal scarring, optic nerve hypoplasia, macular abnormalities, which can lead to blindness, limb contractures, joint abnormalities, dysphagia (difficulty swallowing), and growth restriction in utero. Guillain-Barré Syndrome is a neurological complication of post-infectious autoimmune condition causing muscle weakness and paralysis. Several outbreaks of Guillain-Barré Syndrome were reported during Zika epidemics, with strong temporal associations. Effective prevention and control of Zika virus (ZIKV) rely on a combination of personal protection, vector control, and public health strategies. Avoid traveling to Zika-endemic areas, especially for pregnant women or those planning pregnancy. Adhere to health advisories and guidelines from organizations like the CDC and WHO when visiting areas with active transmission. No licensed vaccine is available for public use.

West Nile Fever

West Nile fever is caused by West Nile virus a single-stranded RNA virus belonging to the Flavivirus genus of the Flaviviridae family and the sharing lineage with the Japanese encephalitis antigenic complex. WNV was first isolated in the West Nile district of Uganda in 1937. Initially associated with sporadic cases in Africa and the Middle East, and now become endemic in several regions, including Europe, Asia, and Oceania. Global travel and trade, climate change, human encroachment, urbanization, and agricultural practices are the major contributing factors to increased vector range and created habitats conducive to mosquito breeding. The disease is maintained in a mosquito–bird–mosquito cycle, primarily involving *Culex* species mosquitoes and passerine birds as amplifying hosts. Birds, particularly passerines such as house sparrows, crows, and American robins act as primary reservoirs. Humans and horses are incidental dead-end hosts due to insufficient viremia for onward transmission. The disease also can be spread through blood transfusions, organ transplants, transplacental transmission, and rare cases of breastfeeding and needle-stick injuries. In humans, the disease is asymptomatic and may cause flu-like symptoms, headache, Skin rash, primarily on the chest, back, and abdomen, myalgia, swollen lymph nodes, loss of appetite, and vomiting. In severe cases, the disease progresses to meningitis, encephalitis, or acute flaccid paralysis in <1%, with a fatality rate of ~10%. Birds especially corvids are highly susceptible to the infection and show neurological symptoms (incoordination and seizures) and sudden death, often within 24 h of symptom onset. Horses show primary neurological signs including ataxia (lack of coordination), weakness, especially in the hind limbs, and difficulty standing or recumbency. Mortality rates may reach up to 30–50% in unvaccinated animals. No licensed vaccine is available for human use. In the absence of a vaccine, prevention relies on vector control and personal protective measures (e.g., use of repellents and protective clothing). Several vaccines (killed virus vaccines and recombinant Canarypox vector vaccines) are commercially available for horses in regions where West Nile fever is endemic.

Mpox

Mpox, formerly known as monkeypox, is a zoonotic disease first identified in 1958 during outbreaks among laboratory monkeys in Denmark. The first human case was reported in 1970 in the Democratic Republic of Congo (DRC). Initially, mpox was endemic to Central and West Africa, with sporadic cases linked to contact with wildlife and small outbreaks in remote regions. The eradication of smallpox and subsequent cessation of vaccination contributed to increased susceptibility to mpox in unvaccinated populations. The causative agent of mpox is the monkeypox virus (MPXV), a double-stranded DNA virus in the genus Orthopoxvirus within the family Poxviridae of two clades. Clade I (formerly Congo Basin clade) is associated with more severe disease and higher mortality rates and Clade II (formerly West African clade) is a less severe with lower mortality. Clade II is further subdivided into IIa and IIb, with IIb responsible for the global outbreaks in 2022. The disease is originally confined to Central and West Africa, and now mpox has become a global concern. The 2022 outbreak reported over 87,000 cases across 111 countries, predominantly among men who have sex with men. Zoonotic transmission is mainly through close contact with infected animals, such as rodents or monkeys, and their bodily fluids or lesions. Human-to-human transmission occurs in direct contact with infected individuals, especially lesions, respiratory secretions, or contaminated materials. The 2022 outbreak highlighted potential transmission during sexual contact. Animals, particularly rodents and primates, are reservoirs of the monkeypox virus. Gambian pouched rats, rope squirrels, and other small mammals are implicated in zoonotic spillovers. In human fever, lymphadenopathy, and rash are classic symptoms. Rash progresses through macular, papular, vesicular, and pustular stages before crusting over. Anogenital lesions alongside fever and lymphadenopathy were predominant in 2022, particularly when transmitted through a sexual route. Infected animals may show pox-like lesions, respiratory distress, and lethargy. The disease can be prevented by isolating infected individuals, educating people on avoiding contact with wildlife and suspected cases, using personal protective equipment (PPE) for healthcare workers, antiviral treatments with tecovirimat, and vaccination in endemic areas. The vaccinia-based smallpox vaccines, including ACAM2000 and JYNNEOS, have been repurposed for mpox.

Bacterial Zoonoses

Plague Plague is an important re-emerging zoonotic disease caused by *Yersinia pestis*, it is a Gram-negative, rod-shaped bacterium. Despite advances in medical science, its periodic re-emergence, often after decades of quiescence, continues for reasons that remain enigmatic. Adding to the challenge is the alarming occurrence of antibiotic-resistant *Yersinia pestis* isolates, which complicates treatment and control efforts. The causative agent is an evolutionary clone of *Yersinia pseudotuberculosis*, acquiring plasmids that enhance its flea-borne transmission. It is the most devastating zoonotic disease caused three major pandemics in the past: The Justinian Plague (541–542 AD) originating from Ethiopia or Central Asia, which contributed to the decline of the Roman Empire, The Black Death (1347–1351), which killed over 30% of Europe's population and The Third Pandemic, starting in the Yunnan province of China in the 1850s, spread globally and caused millions of deaths. The primary mode of transmission is through the bite of infected fleas, predominantly *Xenopsylla cheopis*. These fleas become vectors after feeding on infected rodents and subsequently transmit the bacteria to humans during subsequent feedings.

Rodents, such as rats (*Rattus rattus* and *Rattus norvegicus*), prairie dogs, marmots, and other small mammals, act as natural reservoirs for the bacterium. The cycle of infection between rodents and fleas sustains the plague in endemic areas. Human-to-human transmission can spread via respiratory droplets from people infected with pneumonic plague. The disease is also transmitted by scratches from infected animals, handling contaminated materials, and consumption of infected animals like camels or contaminated food. The disease appears in two forms based on the mode of transmission. The Pneumonic form is the most severe and contagious form of plague primarily spread through respiratory droplets from an infected individual, making it the only form of plague that can spread directly from person to person. Symptoms include high fever, chills, headache, cough, chest pain, hemoptysis (coughing up blood), and difficulty breathing. If untreated, it rapidly progresses to respiratory failure and septic shock, often resulting in death within 24–48 h. The bubonic form is the most common form of plague which occurs through the bite of an infected flea, which introduces the bacterium into the bloodstream. The symptoms are sudden onset of fever, chills, headache, muscle aches, and the characteristic swollen and painful lymph nodes (buboes), typically found in the groin, armpit, or neck. If untreated, it can lead to septicaemia and spread to other forms, including pneumonic plague.

Control and Prevention Strategies

One Health Approach The *One Health* approach recognizes the interconnectedness of human, animal, and environmental health. A coordinated response involving public health authorities, veterinary services, and environmental agencies is essential for effective zoonotic disease management.

Surveillance and Early Detection Strengthening surveillance systems is vital for detecting outbreaks early and preventing their spread. Organizations such as the WHO and the World Organisation for Animal Health (WOAH) play critical roles in coordinating international surveillance and response efforts.

Vaccination and Public Health Interventions Vaccines are crucial tools in controlling zoonotic diseases. For instance, rabies vaccination programs have significantly reduced human and animal cases in many regions. Similarly, the rapid development and deployment of COVID-19 vaccines have mitigated the pandemic's impact.

Regulation of Wildlife Trade Stricter regulations are needed to reduce the risks associated with wildlife trade and wet markets. Public awareness campaigns can also educate communities about the dangers of close contact with wild animals and the consumption of bushmeat.

Prevention and Mitigation Strategies

To mitigate the risks of zoonotic spillover, several strategies need to be implemented:

One Health Approach The "One Health" concept emphasizes the interconnectedness of human, animal, and environmental health. This approach advocates for coordinated efforts to monitor and address zoonotic diseases at the human–animal–environment interface. Collaboration between veterinarians, ecologists, and public health officials is essential to reduce the risk of zoonotic spillover.

Wildlife Conservation and Habitat Protection Protecting wildlife habitats and minimizing human encroachment into these areas can reduce the likelihood of zoonotic spillover. Habitat destruction forces animals into closer contact with humans, increasing the chances of disease transmission.

Improved Agricultural Practices Sustainable farming practices and the separation of livestock from wildlife can help prevent zoonotic spillovers. This includes regulating the use of antibiotics and reducing the intensity of animal farming in regions where wildlife is prevalent.

Surveillance and Early Detection Enhanced surveillance of wildlife populations, particularly species known to harbour zoonotic pathogens (like bats and rodents), is critical for early detection of potential spillover events. Genetic sequencing and pathogen monitoring in wildlife, livestock, and humans can help identify emerging threats before they lead to widespread outbreaks.

Public Health Education and Risk Communication Educating communities about the risks of zoonotic diseases, particularly in regions where bushmeat consumption and wildlife trade are common, can help reduce behaviours that contribute to spillover events. Public health campaigns can also promote safer agricultural and hunting practices.

Occupational Zoonoses

Zoonoses are diseases that are naturally transmissible between vertebrate animals and humans. These diseases pose significant health risks, especially to individuals whose work involves close interaction with animals or animal products. These infections disproportionately affect veterinarians, livestock handlers, wildlife workers, farmers, slaughterhouse employees, and others working in fields where they are exposed to animals, animal tissues, or waste (Hidajat & de Vocht, 2020).

The first recognized zoonoses with occupational associations were those with shorter incubation periods, primarily affecting the skin, such as dermatophytosis, glanders, and cutaneous anthrax. In contrast, other occupational diseases like tuberculosis, cystic echinococcosis, and deep mycosis went largely unidentified for some time due to their longer incubation periods and their tendency to primarily affect internal organs rather than external lesions. The diseases that first led to the concept of occupational zoonosis are rabies, anthrax, and glanders because of their nature of acuteness and severity. However, the importance of occupational zoonoses was in the limelight when the World Health Organization took important initiatives and expert meetings in consultation with the Food and Agriculture Organization (FAO) and Office International des Epizooties (presently The World Organisation for Animal Health). For instance, in 1975, a meeting of veterinary public health experts, organized by the WHO and Food and Agriculture Organization (FAO) recognized zoonoses and traumas caused by different species of animals as occupational risks and stressed the need for specific knowledge to ensure they were prevented and controlled. In 1977, the WHO in Rome discussed occupational diseases, and recommendations were made on the 'protection of workers at special risk to zoonoses'. In 1982, WHO meeting on bacterial and viral zoonoses, Calvin W. Schwabe classified the occupational groups and populations at high risk of zoonotic infection and the zoonotic infections that represented a high risk to different occupational groups and populations. In the year 1999, WHO expert group in Teramo, Italy discussed 'the future of Veterinary Public Health in the 21st Century' and the issues of occupational diseases in relation to work with animals and their products are included in the final document.

Occupational zoonoses are transmitted through various pathways including direct contact with animals, tissues, or waste, especially during veterinary procedures, slaughtering, or research, inhalation of aerosols generated during animal handling, such as during parturition, cleaning of barns, or laboratory work, ingestion of contaminated animal products or accidental ingestion during work with animals or in the field, vector-borne transmission through flies, ticks, flea and mosquitoes, and fomites through tools, vehicles, etc. Table 4.1 shows a list of some of the occupational zoonotic diseases.

Common Occupational Zoonotic Diseases

Anthrax

Synonyms Wool sorter's disease (pneumonic anthrax or pulmonary anthrax), Hide-porter's disease (cutaneous anthrax), Rag-picker's disease, Malignant edema, Malignant carbuncle, Malignant pustule, Charbon, Milz brand disease, Splenic fever, Siberian ulcer.

Anthrax ('Anthrakis' is the Greek word, that means 'coal', referring to the blackened eschar skin lesion) is a bacterial infection caused by *Bacillus anthracis. Bacillus anthracis* is a Gram-positive, aerobic, rod-shaped bacterium that is encapsulated, spore-forming, and non-motile. The bacteria appear in nature in two forms, a vegetative form (capsular form) within the host and a spore form in the environment. The disease primarily affects herbivorous animals, swine, dog and horse. It is a major occupational disease as humans can contract anthrax through skin contact,

Table 4.1 List of Occupational Zoonotic Diseases

S. No.	Zoonosis	Biological agent	Animal(s) involved	Risk occupations
1.	Anthrax	*Bacillus anthracis*	Cattle, goats, sheep, pigs, environment	Farming, meat processing, military
2.	Bovine tuberculosis	*Mycobacterium bovis*	Cattle	Farming, meat, & dairy processing
3.	Brucellosis	*Brucella abortus, B. melitensis, B. suis*	Cattle, goats, sheep, pigs	Farming, meat processing
4.	Campylobacteriosis	*Campylobacter jejuni*	Poultry, farm animals	Poultry farming, meat processing
5.	Erysipeloid	*Erysipelothrix rhusiopathiae*	Pigs, fish, birds	Farming, meat processing, fishing
6.	Leptospirosis	*Leptospira* species	Rodents, cattle	Watersports instructors, waste water processing, cattle farmers
7.	Listeriosis	*Listeria monocytogenes*	Cattle, sheep, pigs, goats	Veterinarians and farm workers
8.	Lyme disease	*Borrelia burgdorferi*	Birds, mammals	Hill farming, outdoor leisure, forestry workers, military,
9.	Salmonellosis	*Salmonella* species	Wild & domestic animals	Poultry farmers, technicians, veterinary medicine professionals, farm workers, snake farm employees, pet industry staff, restaurant employees, construction workers, workers exposed to raw meat, military, healthcare workers, nursing home staff and childcare staff
10.	Streptococcal sepsis	*Streptococcus suis, S. zooepidemicus*	Pigs, cattle	Farming
11.	Tularemia	*Francisella tularensis*	Rodents, rabbits, hares	Outdoor leisure, wastewater processing
12.	Verotoxigenic E. coli	VTEC (including *E. coli* $O_{157:}$ H_7)	Mammals	Farming, vegetable picking, childcare staff, hospital staff (nurses), nursing home staff, school-based employees, military personnel
13.	Avian influenza	Influenza A viruses (Orthomyxoviridae)	Birds	Poultry farming and processing

(continued)

Table 4.1 (continued)

S. No.	Zoonosis	Biological agent	Animal(s) involved	Risk occupations
14.	Eastern equine encephalitis	Eastern equine encephalomyelitis virus, alphavirus genus (Togaviridae)	Wild birds (e.g., passerines)	Agriculture workers, veterinary technicians.
15.	Crimean-Congo haemorrhagic fever (CCHF)	CCHF virus (Bunyaviridae)	Wild & domestic animals	Farming, meat processing, outdoor leisure
16.	Hantavirus disease	Hantaviruses	Rodents	Farming, wastewater processing, Forest workers, military personnel
17.	Louping ill	Louping ill virus (Flaviviridae)	Sheep, grouse	Hill farming
18.	Orf	Orf virus (Poxviridae)	Sheep, goats	Sheep & goat farming
19.	Rabies	Rabies virus, European bat lyssavirus 2 (Rhabdoviridae)	Dogs, bats, cats, foxes	Dog handing, bat handling, veterinarians, military personnel, Forest workers
20.	Swine influenza	Influenza A & C viruses (Orthomyxoviridae)	Pigs	Pig farming
21.	Tickborne encephalitis (TBE)	TBE virus (Flaviviridae)	Small mammals, livestock, birds	Hill farming, forestry, outdoor leisure
22.	West Nile virus	West Nile virus (Flaviviridae)	Birds	Outdoor leisure, stabling, animal control officers, veterinary officers
23.	Q fever	*Coxiella burnetti*	Sheep, cattle	Farming, slaughterhouse workers, military personnel, cosmetics industry workers,
24.	Psittacosis	*Chlamydophila psittaci*	Birds	Poultry farming and processing, bird-keeping
25.	Ovine Chlamydiosis	*Chlamydophila abortus*	Sheep, goats	Sheep & goat farming
26.	Toxoplasmosis	*Toxoplasmosis gondii*	Mammals	Farming, gardening, vegetable picking
27.	Giardiasis	Giardia species	Farm & domestic animals	Outdoor leisure, wastewater processing, childcare staff
28.	Alveolar echinococcosis	*Echinococcus multilocularis*	Foxes	Hunting, outdoor leisure

(continued)

Table 4.1 (continued)

S. No.	Zoonosis	Biological agent	Animal(s) involved	Risk occupations
29.	Hydatid disease	*Echinococcus granulosus*	Dogs	Sheep farming, dog handling
30.	Ringworm	Dermatophytes	Cattle, horses, cats	Farming, working with animals

inhalation, or ingestion of spores from infected animals, animal hides, wool, or tissues or their products. People engaged in various occupations such as farming, meat processing, military, wool processors, veterinarians, laboratory workers, and livestock handlers are at risk of acquiring the disease. The industrial anthrax is associated with occupations involving the processing of animal by-products for manufacturing goods such as tanning, woollen, animal hair, carpet, bone processing, and other such industries, where the potential for aerosolization of substantial numbers of spores increases the risk of exposure to infectious doses. The disease appears in different forms in humans via cutaneous form, pneumonic, and gastrointestinal depending upon the portal of entry. The cutaneous form is the most common, occurring when spores enter the skin through cuts or abrasions following contact with contaminated animal products. It forms a raised vesiculated edge, inflamed, and with a black base to the ulcer and forms an Escher which is called a Charbon, Siberian ulcer. The pneumonic form of anthrax is a highly fatal disease caused by the inhalation of anthrax spores, commonly associated with workers involved in wool-processing activities. Within 2–3 days of the incubation period, the infected person develops a high fever, cough, myalgia, malaise, chest pain, acute respiratory distress, dyspnea, cyanosis, hemorrhagic mediastinitis, rapid progression shock, and death. Gastrointestinal anthrax develops due to ingestion of the spores via contaminated meat or accidental intake where it forms intestinal eschar similar to cutaneous anthrax lesion. The infected person may develop high fever, acute gastroenteritis, vomiting, bloody diarrhea, and progression to generalized toxemia. Anthrax prevention in workplaces involves several critical measures, including wearing gloves and appropriate protective clothing, with full personal protective equipment (PPE) used as necessary; avoiding touching the face or eyes during work; strictly adhering to standard operating procedures; and educating workers on risks and prevention strategies. The use of air-blowing machinery for cleaning equipment should be avoided to prevent spore dispersal. Additionally, vaccinating high-risk personnel is essential. Clinical specimens and cultures of *Bacillus anthracis* must be handled following appropriate biosafety protocols, with strict containment procedures in place to minimize exposure risks.

Brucellosis

Synonyms In animals: Bang's disease, Contagious abortion, Epizootic abortion; In humans: Melitococcosis, Malta fever, Mediterranean fever, Undulant fever.

Brucellosis is one of the most widely recognized occupational zoonoses and is caused by the bacteria of the genus *Brucella. Brucella* spp. are Gram-negative, facultative intracellular bacteria belonging to the family *Brucellaceae.* Among the various Brucella species, *Brucella melitensis* primarily infects sheep and goats and is highly pathogenic to humans; *Brucella abortus* mainly affects cattle, causing bovine brucellosis and can also infect humans; *Brucella suis* infects pigs and occasionally cattle and humans; *Brucella canis* infects dogs and occasionally humans; *Brucella ovis* infects sheep but is not known to infect humans; and *Brucella neotomae* is found in desert woodrats and is not pathogenic to humans. The highest occupational risk is faced by abattoir employees (through skin abrasions, conjunctival inoculations, and aerosol), farm workers, and veterinary workers (during rectal examinations without gloves, post-mortem examination, skin aberration, etc.) who directly handle infected animals and their tissues, while veterinarians may also contact the infection through accidental inoculation with live attenuated vaccine strains intended for animal vaccination. Transmission of the disease to humans occurs through direct contact with infected animals or tissues (e.g., excreta, placentas, foetuses, vaginal secretions, or aborted materials) or ingestion or consuming unpasteurized dairy products or undercooked meat or inhalation of aerosolized bacteria, especially in laboratory or abattoir settings. Vertical transmission can occur from infected mothers to offspring during birth or via milk.

Acute brucellosis manifests with non-specific symptoms, including fever, malaise, chills, profuse sweating, weight loss, insomnia, nausea, headache, anorexia, arthralgia, muscle and body pain, sexual impotence, and orchitis in males. These vague symptoms make early diagnosis difficult and can mimic other febrile illnesses. In chronic brucellosis, multiple organ systems may be affected, resulting in conditions such as osteomyelitis, orchitis, and endocarditis. Symptoms in chronic cases are typically undulant, with periods of normal temperature alternating with acute episodes. These symptoms may persist for years, either continuously or intermittently, and are often associated with significant morbidity, necessitating prolonged treatment.

The key control and preventive measures in occupational risk groups are

1. *Personal Protective Equipment (PPE)*: Proper use of gloves, masks, goggles, aprons, and boots is essential during high-risk activities such as handling animal fluids, placenta, or fetuses. Educational programs should emphasize the consistent use of PPE.
2. *Vaccination and Handling of Live Vaccines*: Workers administering live attenuated *Brucella* vaccines such as *Brucella abortus* S-19 must follow strict biosafety protocols to prevent accidental self-inoculation. In high-risk areas, vaccination of personnel with available vaccines is essential.
3. *Biosafety in Laboratories*: Laboratory workers should handle *Brucella* cultures in Biosafety Level 3 facilities to minimize exposure. Proper training on the handling of cultures should be given to the working personnel. Should follow good laboratory procedures and standard operating procedures in the labs.
4. *Hygienic Practices*: Avoiding consumption of raw milk or unpasteurized dairy products. Cleaning and disinfecting equipment and workspaces.

5. *Education and Awareness*: Training occupational groups about brucellosis transmission, symptoms, and prevention strategies is crucial. Highlighting risks such as sniffing culture plates or unsafe disposal of biological waste.
6. *Surveillance and Reporting*: Regular health screenings for individuals in high-risk occupations. Immediate reporting and investigation of suspected brucellosis cases to contain outbreaks.

Bovine Tuberculosis

Bovine tuberculosis is a chronic bacterial disease of animals caused by members of the *Mycobacterium tuberculosis* complex, primarily *M. bovis*, followed by *M. caprae* and, *M. tuberculosis*. It is a significant infectious disease in cattle, which serves as the primary reservoir and the main source of infection for humans. The disease also affects other domesticated animals and numerous wildlife species including buffaloes, sheep, goats, equines, camels, pigs, bison, deer, wild boars, antelopes, primates, etc. The disease has also been detected in exotic and predatory species such as lions, tigers, leopards, lynx, rhinoceroses, kudus, elands, oryxes, and addaxes, among others. This broad host range highlights the significant challenge of controlling bovine tuberculosis, given its ability to infect a diverse array of species across different ecosystems (Woolhouse & Gowtage-Sequeria, 2005). Risk occupations include farmers, veterinarians, and dairy workers. Transmission occurs in inhalation of aerosols from infected animals or consumption of unpasteurized dairy products. Symptoms in humans pulmonary issues resembling human tuberculosis, including chronic cough and fever.

Leptospirosis

Leptospirosis is caused by spirochetes of the genus *Leptospira* and affects both animals and humans. It is primarily spread through the urine of infected animals, contaminating water, or soil.

- *Occupational Groups at Risk*: Farmers, sewage workers, veterinarians, and slaughterhouse workers.
- *Symptoms in Humans*: Fever, headache, muscle aches, and, in severe cases, kidney damage, liver failure, or meningitis.

Q Fever

Q fever is caused by *Coxiella burnetii*, a bacterium found in cattle, sheep, and goats. The bacteria can survive in the environment for long periods and infect humans via inhalation.

- *High-Risk Professions*: Veterinary professionals, livestock farmers, and abattoir workers.
- *Symptoms in Humans*: Flu-like symptoms, including fever, headache, and muscle pain. Chronic infections may result in endocarditis.

Prevention and Control of Occupational Zoonoses

Prevention and control strategies for occupational zoonoses involve a multi-faceted approach encompassing policy, workplace hygiene, protective measures, and education.

Personal Protective Equipment (PPE): The use of PPE such as gloves, face shields, and gowns is critical in preventing direct contact with zoonotic pathogens.

Vaccination: Vaccines are available for certain occupational zoonoses, such as rabies and anthrax. Ensuring high-risk workers are vaccinated can greatly reduce the risk of infection.

Hygiene Practices: Regular hand washing, disinfection of equipment, and proper waste disposal are essential in minimizing transmission.

Biosafety Measures: Workplace safety protocols, such as the use of containment levels in laboratories and proper ventilation, are crucial in preventing airborne transmission.

Training and Education: Regular training on zoonotic disease risks and safety measures is essential for workers in high-risk professions. Awareness of the routes of transmission and proper response to exposures can significantly reduce the risk of occupational zoonoses.

Environmental Controls: Proper management of animal waste, regular cleaning of animal enclosures, and minimizing exposure to potential vectors like ticks or rodents can further control zoonotic risks in occupational settings (Morse et al., 2012).

Surveillance and Reporting: A robust system for reporting zoonotic outbreaks in occupational settings is crucial. Continuous surveillance helps identify emerging threats, while prompt reporting facilitates rapid intervention (Morse et al., 2012).

Collaboration Across Disciplines: Effective control of occupational zoonoses requires collaboration between multiple sectors, including veterinary services, public health authorities, and environmental agencies. Multidisciplinary teams can develop comprehensive risk assessments and prevention strategies.

Research and Development: Investing in research to better understand zoonotic transmission mechanisms, pathogen evolution, and resistance is crucial for developing more effective diagnostic tools, vaccines, and treatment protocols for occupational zoonoses.

References

Fiegler-Rudol, J., Lau, D. N. M. K., & Kasperczyk, D. H. N. M. J. (2024). Public health threat of novel zoonotic diseases: Literature review. Zagrożenie związane z nowymi chorobami odzwierzęcymi: Przegląd literatury. *Przeglad epidemiologiczny, 78*(1), 69–80. https://doi.org/10.32394/pe/188161

Hidajat, M., & de Vocht, F. (2020). Occupational zoonoses potential in Southeast Asia. *Occupational Medicine (Oxford, England), 70*(5), 323–326. https://doi.org/10.1093/occmed/kqaa082

Jones, K. E., Patel, N. G., Levy, M. A., Storeygard, A., Balk, D., Gittleman, J. L., & Daszak, P. (2008). Global trends in emerging infectious diseases. *Nature, 451*(7181), 990–993. https://doi.org/10.1038/nature06536

Judson, S. D., & Rabinowitz, P. M. (2021). Zoonoses and global epidemics. *Current Opinion in Infectious Diseases, 34*(5), 385–392. https://doi.org/10.1097/QCO.0000000000000749

Morse, S. S., Mazet, J. A., Woolhouse, M., Parrish, C. R., Carroll, D., Karesh, W. B., Zambrana-Torrelio, C., Lipkin, W. I., & Daszak, P. (2012). Prediction and prevention of the next pandemic zoonosis. *Lancet (London, England), 380*(9857), 1956–1965. https://doi.org/10.1016/S0140-6736(12)61684-5

Woolhouse, M. E., & Gowtage-Sequeria, S. (2005). Host range and emerging and reemerging pathogens. *Emerging Infectious Diseases, 11*(12), 1842–1847. https://doi.org/10.3201/eid1112.050997

References

[illegible]

Application of Bioinformatics Resources in Zoonotic Disease Research

5

Sukhen Samanta, Ayan Mukherjee, Subhasree Das, and Molla Zakirul Haque

Abstract

Emerging zoonotic diseases hold the potential to cause a major global public health threat; hence, there is an urgent need for their rapid and accurate detection, monitoring, and understanding of the pathogens at the molecular level. This chapter hence highlights the pivotal role of computational biology and bioinformatics resources in synergy with Next Generation Sequencing (NGS) technologies in combating such zoonotic diseases. This includes but is not limited to whole-genome sequencing, whole-exome sequencing, and metagenome sequencing; the technology provides deep insights into the ecology of pathogen evolution and transmission patterns as well as into genetic diversity. The technologies involved are indeed changing the complete scenario of disease surveillance, allowing, for instance, tracking in real-time pathogens and their markers of resistance to antibiotics. As part of the book, it will also go through some of the bioinformatics tools involved in the analysis of large-scale sequencing data understanding sequence alignment, variant calling, and functional annotation. For instance, NGS applications in zoonotic spill-over prediction, personalized therapeutic development, and

S. Samanta
Department of Microbiology, University of Kalyani, Nadia, West Bengal, India

A. Mukherjee (✉)
Department of Animal Biotechnology, West Bengal University of Animal and Fishery Sciences, Kshudiram Bose Sarani, Kolkata, West Bengal, India
e-mail: dramukherjee@wbuafscl.ac.in

S. Das
Department of Veterinary Biochemistry, West Bengal University of Animal and Fishery Sciences, Kshudiram Bose Sarani, Kolkata, West Bengal, India

M. Z. Haque
Department of Veterinary Medicine, West Bengal University of Animal and Fishery Sciences, Kshudiram Bose Sarani, Kolkata, West Bengal, India

R. Biswas et al. (eds.), *Concept of One Health and Zoonoses*,
https://doi.org/10.1007/978-981-95-6627-3_5

response to public health measures will be enumerated. However, issues such as variability in data, high requirements for expertise in bioinformatics, and integration into public health systems continue. These can be mitigated by the standardization of data protocols, an increase in the bioinformatics training curriculum, and an improvement in interdisciplinary research collaboration. The global community will be best prepared to respond to emerging zoonotic disease threats by exploiting computational biology resources.

Keywords

Zoonotic diseases · Next-generation sequencing · Bioinformatics · Pathogen surveillance · Antimicrobial resistance

Introduction

The incidences of zoonotic diseases have increased dramatically in recent years as a result of expanding agriculture, deforestation and urbanization, rising globalization, ecosystem disruption, and changing climate. Emerging zoonotic illnesses that can cause severe sickness and high death rates include the coronavirus disease 2019 (COVID-19), severe acute respiratory syndrome (SARS), Middle East respiratory syndrome (MERS), and Ebola virus disease (Di Bari et al., 2023). Moreover, they result in financial losses by interfering with supply chains, travel, and trade. As the risk of zoonotic infections keeps rising, the need for early detection has grown. Early detection and containment of outbreaks can stop the disease progress and lessen its damage (Sim & Cho, 2023). Furthermore, diseases that are spread from animals to people can be stopped from spreading widely among populations by early discovery and prompt use of control measures (Jung Kjær et al., 2023). Thus, it is essential to set up reliable surveillance and response mechanisms that allow for the early identification and prompt handling of outbreaks of zoonotic diseases.

Next-generation sequencing (NGS) is a molecular technology that has revolutionized the surveillance of zoonotic diseases. NGS makes it possible to quickly sequence complete genome of pathogens, giving researchers a thorough insight of the genetic background of the disease (Chiara et al., 2021). Since last two decades numerous next-generation sequencing (NGS) techniques have been developed and put forth to answer any genomics-related question or DNA-related clinical activity. NGS offers a revolutionary method of sequencing that combines a number of techniques based on assembling the genome, template preparation, figuring out the base order, alignment and assembly of the sequences. The combination of bioinformatics methodologies with these high-throughput sequencing technologies has made it possible to identify, analyze, and detect zoonotic infections with great power. Compared to traditional method of DNA sequencing, it has an array of advantages including more precise pathogen identification and characterization, screening for the presence of resistance mutations or genes, genome rearrangement, and determination of virulence and pathogenicity factors (Maljkovic Berry et al., 2020).

Application of this combinatorial approach in zoonotic disease research can illuminate the transmission dynamics, contact tracing, source, evolutionary history, geographical origin of the pathogen. Pathogen genomics is emerging as useful tool to improve surveillance programs, enabling both the timely identification of zoonotic outbreaks and epidemics and accurate diagnosis at the individual level (Pronyk et al., 2023). Recently, zoonotic threats like SARS-CoV-2 and monkeypox viruses have emerged, highlighting the need for NGS capacity to be implemented globally to enhance health systems (Suminda et al., 2022). However, applying NGS and bioinformatics methods for investigating zoonotic disease involves several challenges, including selecting the appropriate sequencing platform, wet lab sequencing technique, bioinformatics algorithms, personnel with the appropriate training and experience, and computational infrastructure to support the analysis of the massive volumes of data generated by NGS and bioinformatics.

In this chapter, we summarize different NGS platforms, bioinformatics support, and use of NGS to identify different zoonotic pathogens. We also focus on various bioinformatics resources in combating zoonotic diseases. We also recommend the potential modern technologies that can be judiciously amalgamated with NGS to make the investigation of zoonosis more effective.

An Overview of NGS Technologies and Platform

Since its advent in 2005, NGS technology has evolved dramatically with development of three generations of sequencing technologies. Several platforms varying in costs, chemistries, capacities, and applications have emerged with the simultaneous progress of computational tools and techniques.

Different NGS Platforms

Several NGS platforms have been developed, each with distinct features and applications. The following are the most commonly used platforms in zoonotic disease research:

Illumina

Illumina's cutting-edge sequencing (NGS) stage overwhelms the market because of its unrivalled precision, adaptability, and flexibility. The stage utilizes sequencing-by-synthesis (SBS) innovation, where fluorescently labelled nucleotides are integrated into the developing DNA strand, and the not entirely set in stone by recognizing the produced light (Bentley et al., 2008). Illumina's different scope of instruments, including the MiSeq, NextSeq, and NovaSeq, take care of different applications, for example, entire genome sequencing (WGS), metagenomics, and transcriptomics (Quail et al., 2012). In the field of zoonotic infection research, Illumina innovation has demonstrated importance for following the development of microorganisms like SARS-CoV-2, empowering scientists to screen viral changes continuously (Hadfield et al., 2018).

Ion Torrent

Ion Torrent, developed by Thermo Fisher Scientific, uses a semiconductor sequencing innovation, which discriminates hydrogen particles delivered during DNA polymerization (Rothberg et al., 2011) This step typically provides faster sequencing at a lower cost, making it suitable for short genome projects focused on sequencing (Liu et al., 2012). The Ion Torrent PGM and S5 systems are normally utilized for microorganism identification and antimicrobial obstruction profiling in zoonotic illness studies (Greninger et al., 2017). In any case, Ion Torrent is by and large less exact than Illumina, especially in homopolymeric regions (Loman et al., 2012).

Pacific Biosciences (PacBio)

Pacific Biosciences (PacBio) Single Molecule Real-Time (SMRT) sequencing advancement provides long-read sequencing, which is useful in establishing accurate genomic regions including repeat sequences and structural varieties (Eid et al., 2009). Long peruses of PacBio are significantly valuable for repeat genome get-togethers and research of heterogenic viral communities (Ardui et al., 2018). The PacBioSequel II stage has been used in the study of zoonotic disease to characterize the genomes of such organisms such as Ebola and bird flu virus, giving insights into their genetic polymorphisms and evolution (Gire et al., 2014).

Oxford Nanopore

Oxford Nanopore Technologies (ONT) offers a one of a kind NGS stage that gives constant, versatile sequencing with long peruses. ONT's MinION and PromethION gadgets use nanopore sequencing, where single DNA strands go through a nanopore, and the subsequent changes in ionic current are utilized to decide the succession (Jain et al., 2016). ONT is especially valuable in field-based exploration and episode circumstances because of its convenience and fast sequencing abilities. In zoonotic illness research, ONT has been utilized in the fast identification of arising microorganisms, for example, during the Zika infection and Ebola infection flare-ups (Faria et al., 2017).

Application of NGS to Understand the Zoonotic Pathogen

Next-Generation Sequencing (NGS) has changed the scene of microorganism research, giving exceptional insights of knowledge into zoonotic infections. Zoonotic microorganisms, which originate in animals and spread to people, present huge dangers to worldwide well-being. The fast identification, interpretation, and observing of these microbes are fundamental for successful avoidance and control methodologies. NGS stages offer numerous ways to deal with comprehend zoonotic microorganisms, including Whole Genome Sequencing (WGS), Whole Exome Sequencing (WES), and Metagenome Sequencing. Every one of these strategies gives extraordinary benefits in concentrating on microorganism development, variety, and communication which has been working with a more profound comprehension of zoonotic sicknesses.

Whole Genome Sequencing (WGS)

One of the mightiest tools of NGS in studying zoonotic pathogens is whole genome sequencing. Sequencing the whole genome of a pathogen ensures garnering an ultimate understanding of a pathogen its genetics and its ability to evolve and spread (Goodwin et al., 2016). WGS has been very commonly used for various researches involved with zoonotic pathogens in the following ways:

Genomic Surveillance and Tracking of Pathogen Evolution

WGS permits researchers to follow mutations, recombinations, and genomic variations in zoonotic microbes, which can have huge significance for their contagiousness and destructiveness. For example, during the Coronavirus pandemic, WGS has been utilized to screen the development of SARS-CoV-2, the zoonotic coronavirus liable for the illness. By sequencing the genomes of viral isolates from various areas of the planet, researchers had the option to recognize mutations, for example, those found in the Delta and Omicron variations, which added to changes in contagiousness and immune evasion (Hadfield et al., 2018). Likewise, WGS has been utilized to concentrate on the hereditary variety of the Ebola infection during flare-ups, assisting specialists with understanding the starting points and transmission courses of the infection (Dudas et al., 2018).

Identifying Zoonotic Spillovers and Cross-Species Transmission

WGS has greatly aided in discovering zoonotic spillover events, whereby microbes directly jump from animal hosts to humans. It enables scientists to trace the origin of zoonotic diseases by comparing the genome of a microbe in the animal reservoir with that infecting the human. For example, WGS was applied in tracing the source of the coronavirus that causes Middle East Respiratory Syndrome (MERS); this was identified to be originating from dromedary camels and being transferred into humans (Corman et al., 2016). A new understanding of zoonotic spillovers opens a door for identifying reservoirs in animals and prevention before the emergence of outbreaks.

Antimicrobial Resistance (AMR) Profiling

Antimicrobial resistance (AMR) genes are commonly carried by zoonotic bacteria, which presents challenges for treatment. WGS facilitates the identification of AMR genes in bacterial zoonotic pathogens, including methicillin-resistant *Staphylococcus aureus* (MRSA), *Campylobacter*, and *Salmonella*, with widespread recognition (Larsen et al., 2012). Through genome sequencing, researchers can identify distinct mutations and resistance mechanisms that play a role in the development of drug resistance in these bacteria. For instance, WGS was utilized to portray AMR genes in zoonotic *Escherichia coli* strains disconnected from animals and people, assisting with explaining the transmission dynamics of resistant strains among animals and humans (Zankari et al., 2012). The information produced by WGS likewise helps in the advancement of powerful anti-infection stewardship programs and designated treatments.

Whole Exome Sequencing (WES)

Whole Exome Sequencing (WES) places a strong emphasis on sequencing a genome's exons—the sections that code for proteins. WES is an economical and effective method for finding mutations in the coding sections of zoonotic diseases, especially when functional variants are of interest. However, it does not offer as much detailed information as WGS (Ng et al., 2009). WES is particularly useful in the following applications:

Recognizing Functional Variants in Viral and Bacterial Pathogens

WES allows the identification of mutations that at once affect protein function, along with those concerned with pathogen virulence or immune evasion. For zoonotic viral pathogens just as the Zika virus and influenza, WES has been used to pinpoint mutations in key viral proteins, which include floor glycoproteins that make contributions to viral attachment, access, and replication (Faria et al., 2017). By specializing in coding regions, WES facilitates researchers to understand how those mutations modify the fitness and pathogenicity of the virus.

For bacterial zoonotic pathogens, WES has been used to pick out mutations that impact toxin manufacturing, secretion structures, and different virulence elements. In instances that include cholera and anthrax outbreaks, WES has provided perception into the genetic basis of virulence in *Vibrio cholerae* and *Bacillus anthracis*, respectively, taking into account greater targeted public fitness responses (Chin et al., 2011).

Exploring Host–Pathogen Interactions

WES has also been contracted to conduct studies on host pathogen interactions particularly in cases of zoonotic diseases, where pathogen ability to infect and replicate in the host is important. Due to the exomes of each pathogen as well as the host being sequence, the researchers are able to identify mutations which affect the susceptibility of the host to contamination. For instance, WES has been used to look at host genetic elements that influence susceptibility to zoonotic infections like tuberculosis and HIV, revealing key mutations in immune device genes that regulate host responses to these pathogens.

Studying Genetic Determinants of Drug Resistance

Like WGS, WES may be used to pick out drug-resistance mutations in zoonotic pathogens. WES offers a more targeted technique by specializing in coding areas that might be more likely to harbor mutations affecting protein targets of antimicrobial pills. In zoonotic pathogens which include *Mycobacterium tuberculosis*, WES has been used to discover resistance mutations in genes encoding drug targets like RNA polymerase that is implicated in rifampicin resistance (Witney et al., 2016). This software is, in particular, vital for growing new therapeutic techniques for treating zoonotic sicknesses.

Metagenome Sequencing

Metagenome sequencing, also simply known as metagenomics, is the sequencing of the collective genetic tapestry present in environmental or clinical samples. This method enables the identification and characterization of zoonotic pathogens within complex microbial communities without any need for previous isolation or culture (Mokili et al., 2012). Metagenome sequencing has become a very useful tool in the study of zoonotic pathogens for the following motives:

Discovery of Novel Pathogens

Metagenome sequencing invented new zoonotic pathogens, which long in the past had been able to go undetected through older methodologies. The sequencing of the total microbial community from samples collected from any wildlife, cattle, or human patients allows the detection of newest viruses, bacteria, or parasites that may harbor zoonotic threats. For example, metagenomics was instrumental in identifying the new COVID SARS-CoV-2, which was identified as a zoonotic pathogen that originated from natural world reservoirs (Zhu et al., 2020). Similarly, metagenomic approaches were used to identify undiscovered zoonotic viruses in bats, rodents, and other flora and fauna species, most of which have the potential to cause future outbreaks (Anthony et al., 2013).

Characterizing Microbial Diversity in Zoonotic Disease Reservoirs

Zoonotic pathogens frequently exist inside diverse microbial ecosystems in animal hosts or environmental reservoirs. Metagenome sequencing offers a complete view of the microbial range present in those ecosystems, supporting researchers apprehend the ecological elements that make contributions to pathogen emergence and transmission. For example, metagenomic research have revealed the wealthy viral range in bat populations, highlighting the position of bats as reservoirs for zoonotic viruses along with Nipah, Hendra, and coronaviruses (Ge et al., 2012). By knowing the microbial groups that coexist with zoonotic pathogens, researchers are expecting to mitigate the chance of zoonotic spillovers.

Studying the Dynamics of Microbiomes During Infections

Metagenome sequencing can further be applied to study how zoonotic pathogens interact with the host-microbiome during contamination. Changes in host-microbiome composition may affect pathogen colonization, host immune responses, and disease effects. For example, metagenomic analyses of the respiration microbiome in people and farm animals have proven how microbial dysbiosis can facilitate the colonization and unfolding of zoonotic pathogens like influenza and *Streptococcus pneumoniae* (Ichinohe et al., 2010). Understanding those dynamics provides insights into the function of the microbiome in zoonotic ailment progression and might inform strategies for microbiome-based interventions.

Bioinformatics Resources to Analyze and Interpret NGS Data

By using the vast amount of data generated by Next-Generation Sequencing, combating infectious diseases requires biologically informative output. It is even easier to detect and monitor zoonotic infections by the computational tools, algorithms, and pipelines provided by the resources of bioinformatics for processing, analyzing, and interpreting such data. The following section elaborates in detail on the major bioinformatics databases and tools used in NGS data interpretation with emphasis on their use in zoonotic pathogen research.

Sequence Alignment and Variant Calling Tools

One of the first steps in NGS data analysis is aligning the raw sequencing reads to a reference genome. Several bioinformatics tools and servers, with various efficiency and accuracy, are available for this resolution. Bowtie2 and BWA (Burrows-Wheeler Aligner) are well-known alignment tools that map brief reads to reference genomes (Langmead & Salzberg, 2012; Li & Durbin, 2009). These tools are highly effective and suitable for large-scale pathogen surveillance studies. Variant calling is done after alignment to find single nucleotide polymorphisms (SNPs), insertions and deletions (indels). Researchers can study the genetic diversity of zoonotic pathogens and follow their evolution by using tools like GATK (Genome Analysis Toolkit) (McKenna et al., 2010), Platypus (Rimmer et al., 2014), VarScan (Koboldt et al., 2009), and FreeBayes (Garrison & Marth, 2012) that are frequently used for SNP and indel discovery.

Metagenomic Data Analysis Pipelines

Metagenomic sequencing is a process of sequencing the entire microbial population, which requires specialized bioinformatics systems to interpret the data sets involved. The most popular tools for metagenomic data analysis are MetaPhlAn and QIIME2, which stands for Quantitative Insights into Microbial Ecology and provides functional annotations and taxonomic characterization of microbial communities (Caporaso et al., 2010; Segata et al., 2012). These tools analyze zoonotic pathogens in their native microbial ecosystems to trace new infections and understand how pathogens interact with the hosts. Another metagenomic classifier that performs effectively is Kraken2, a k-mer-based approach that depends on categorizing pathogens at the species or strain level (Wood et al., 2019).

Phylogenetic and Evolutionary Analysis Tools

Understanding the evolutionary relationships and patterns of animal disease transmission to humans requires a better understanding of phylogenetic analysis. RAxML (Randomized Accelerated Maximum Likelihood) and IQ-TREE are commonly used phylogenetic tools to enable researchers to assemble evolutionary trees from genomic data (Stamatakis, 2014; Nguyen et al., 2015). These tools trace back zoonotic spillover events and track the pathogen evolution through time. Recently, a real-time phylogenetic platform called Nextstrain was extensively used to visualize the unfolding and evolution of emerging zoonotic viruses such as SARS-CoV-2, Ebola, and Zika (Hadfield et al., 2018). The integration of genomic information with metadata like geographic location and sampling date makes Nextstrain present an interactive and dynamic pathogen transmission design.

Functional Annotation and Gene Ontology

Understanding the functional impact of genetic variants in zoonotic pathogens requires functional prediction tools. Prokka is widely used for annotating viral genomes. It quickly identifies non-coding RNA proteins and other genetic material (Seemann, 2014). For eukaryotic bacteria Tools such as InterProScan and BLAST help to locate proteins, motifs, and pathways (Jones et al., 2014). Gene Ontology (GO) and KEGG (Kyoto Encyclopedia of Genes and Genomes) databases provide intuitions into the biological functions and metabolic pathways associated with the identified genes, helping researchers understand the pathogenic mechanisms of zoonotic diseases (Kanehisa et al., 2012).

Data Repositories and Sharing Platforms

Online information archives and repositories play an essential role in zoonotic pathogen research by facilitating data sharing and collaboration. The National Center for Biotechnology Information (NCBI) hosts several databases together with GenBank for nucleotide sequences and SRA (Sequence Read Archive) for raw sequencing data (Leinonen et al., 2011). This repository lets researchers share and get admission to NGS statistics of zoonotic pathogens, facilitating global collaboration in epidemic surveillance and reaction. GISAID is another key platform focusing on sharing influenza and COVID sequences, playing an enormous role in international reaction in the case of viral infections (Shu & McCauley, 2017).

Bioinformatics sources are quite vital in zoonotic pathogen research, which requires interpreting and analyzing the data obtained from NGS. The bioinformatics methods include sequence alignment, variant calling, phylogenetic analysis, and targeted annotation. These tools form the computational framework for understanding the ways of pathogen growth and development, transmission, and many other important mechanisms. Bioinformatics sources will remain essential in establishing timely public health responses and developing well-managed solutions as zoonotic diseases persist in their evolution and emergence.

Future Application NGS-Based Zoonotic Disease

Therefore, Next-Generation Sequencing (NGS) technologies are going to be of prime importance in the emerging era of zoonotic diseases research. Further development of sequences extends the ability of NGS to become an essential tool for real-time monitoring of pathogens with increasing costs reduction and increasing speed and accuracy. It will result in the faster and more efficient recognition of newly identified zoonotic pathogens or during their initial stages of proliferation in human population. Specifically, the diagnostics could be used in the early identification of spillover events by tracking changes in pathogens at different populations of wildlife and livestock, which can naturally precede subsequent outbreaks.

NGS may also fulfill a crucial function in personalized medicinal drug by providing accurate genotypic records on pathogens, which will assist to construct fabricated treatment and vaccination strategies. More precisely, metagenomics will improve the comprehension on pathogen-host-microbiome interplay that will help to decode the mechanisms of the development and transmission of zoonotic diseases. Furthermore, the combination of NGS data with machine learning and artificial intelligence shall create predictive models to predict probable zoonotic diseases based on an analysis of genomic data and environment.

In the future, NGS-based diagnostics could become one of the key roles in the global surveillance systems that will allow researchers and officials of public health to preemptively tackle zoonotic diseases before they develop into a pandemic.

Challenges and Recommendations for Implementation

There are a few challenges in rolling out the resources of computational biology to fight zoonotic diseases with NGS. The most critical challenge is the lack of infrastructure and experience in technical expertise, mainly in limited resource zones where outbreaks of zoonotic diseases frequently happen. This gap makes it difficult to implement real-time application of NGS technologies and bioinformatics tools directly to field settings. In addition, the large amount of data produced by NGS needs powerful computational and storage resources which are difficult to provide in developing nations. Regulation of NGS protocols and bioinformatics pipelines is also another challenge. Inconsistencies in data interpretation due to sequencing platforms and analytical method variations impede harmonized global efforts to track and compare zoonotic pathogens effectively.

In response to these challenges, efforts at capacity building including training in bioinformatics and molecular diagnostics are crucial. The issues of infrastructure deficiencies can also be addressed through scaling up access to the cloud-based computing environment that supplies demanding tools for getting through the analyses. Finally, further development of external cooperation, both at the regional and global levels, and the sharing of data on emerging pathogenic agents, will become an important factor for the improvement of international pathogen surveillance. Introducing high-quality standards and best practices in NGS and adapting them in many areas will contribute to the comparability of outcomes, the identification of risks, and the development of responses to new zoonotic diseases.

Conclusion

Zoonotic disease research is being transformed by the use of computational biology resources, especially bioinformatics and Next-Generation Sequencing. They facilitate quicker pathogen detection, enhance surveillance, and improve our knowledge on the dynamics of pathogen evolution and transmission. Strategic investments in training, interdisciplinary collaboration, and data sharing will enable the seamless integration of new technologies even when such obstacles as the absence of a unified data standard and the need for advanced skills exist. Finally, with new emerging zoonotic diseases as an emerging public health threat around the globe, bioinformatics and NGS are applied to effectively combat them.

References

Anthony, S. J., Epstein, J. H., Murray, K. A., Navarrete-Macias, I., Zambrana-Torrelio, C. M., Solovyov, A., Ojeda-Flores, R., Arrigo, N. C., Islam, A., Ali Khan, S., & Hosseini, P. (2013). A strategy to estimate unknown viral diversity in mammals. *MBio, 4*(5), 1110–1128.

Ardui, S., Ameur, A., Vermeesch, J. R., & Hestand, M. S. (2018). Single molecule real-time (SMRT) sequencing comes of age: Applications and utilities for medical diagnostics. *Nucleic Acids Research, 46*(5), 2159–2168.

Bentley, D. R., Balasubramanian, S., Swerdlow, H. P., Smith, G. P., Milton, J., Brown, C. G., et al. (2008). Accurate whole human genome sequencing using reversible terminator chemistry. *Nature, 456*(7218), 53–59.

Caporaso, J. G., Kuczynski, J., Stombaugh, J., Bittinger, K., Bushman, F. D., Costello, E. K., et al. (2010). QIIME allows analysis of high-throughput community sequencing data. *Nature Methods, 7*(5), 335–336.

Chiara, M., D'Erchia, A. M., Gissi, C., Manzari, C., Parisi, A., Resta, N., et al. (2021). Next generation sequencing of SARS-CoV-2 genomes: Challenges, applications and opportunities. *Briefings in Bioinformatics, 22*(2), 616–630.

Chin, C. S., Sorenson, J., Harris, J. B., Robins, W. P., Charles, R. C., Jean-Charles, R. R., et al. (2011). The origin of the Haitian cholera outbreak strain. *New England Journal of Medicine, 364*(1), 33–42.

Corman, V. M., Eckerle, I., Memish, Z. A., Liljander, A. M., Dijkman, R., Jonsdottir, H., et al. (2016). Link of a ubiquitous human coronavirus to dromedary camels. *Proceedings of the National Academy of Sciences, 113*(35), 9864–9869.

Di Bari, C., Venkateswaran, N., Fastl, C., Gabriël, S., Grace, D., Havelaar, A. H., et al. (2023). The global burden of neglected zoonotic diseases: Current state of evidence. *One Health,* 100595.

Dudas, G., Carvalho, L. M., Rambaut, A., & Bedford, T. (2018). MERS-CoV spillover at the camel-human interface. *eLife, 7,* e31257.

Eid, J., Fehr, A., Gray, J., Luong, K., Lyle, J., Otto, G., et al. (2009). Real-time DNA sequencing from single polymerase molecules. *Science, 323*(5910), 133–138.

Faria, N. R., Quick, J., Claro, I. M., Theze, J., de Jesus, J. G., Giovanetti, M., et al. (2017). Establishment and cryptic transmission of Zika virus in Brazil and the Americas. *Nature, 546*(7658), 406–410.

Garrison, E. P., & Marth, G. T. (2012). *Haplotype-based variant detection from short-read sequencing*. arXiv: Genomics.

Ge, X., Li, Y., Yang, X., Zhang, H., Zhou, P., Zhang, Y., & Shi, Z. (2012). Metagenomic analysis of viruses from bat fecal samples reveals many novel viruses in insectivorous bats in China. *Journal of virology, 86*(8), 4620–4630.

Gire, S. K., Goba, A., Andersen, K. G., Sealfon, R. S., Park, D. J., Kanneh, L., et al. (2014). Genomic surveillance elucidates Ebola virus origin and transmission during the 2014 outbreak. *Science, 345*(6202), 1369–1372.

Goodwin, S., McPherson, J. D., & McCombie, W. R. (2016). Coming of age: Ten years of next-generation sequencing technologies. *Nature Reviews Genetics, 17*(6), 333–351.

Greninger, A. L., Zerr, D. M., Qin, X., Adler, A. L., Sampoleo, R., Kuypers, J. M., Englund, J. A., & Jerome, K. R. (2017). Rapid metagenomic next-generation sequencing during an investigation of hospital-acquired human parainfluenza virus 3 infections. *Journal of Clinical Microbiology, 55*(1), 177–182.

Hadfield, J., Megill, C., Bell, S. M., Huddleston, J., Potter, B., Callender, C., Sagulenko, P., Bedford, T., & Neher, R. A. (2018). Nextstrain: Real-time tracking of pathogen evolution. *Bioinformatics, 34*(23), 4121–4123.

Ichinohe, T., Pang, I. K., & Iwasaki, A. (2010). Influenza virus activates inflammasomes via its intracellular M2 ion channel. *Nature Immunology, 11*(5), 404–410.

Jain, M., Olsen, H. E., Paten, B., & Akeson, M. (2016). The Oxford NanoporeMinION: Delivery of nanopore sequencing to the genomics community. *Genome Biology, 17*, 1–11.

Jones, P., Binns, D., Chang, H. Y., Fraser, M., Li, W., McAnulla, C., McWilliam, H., Maslen, J., Mitchell, A., Nuka, G., Pesseat, S., Quinn, A. F., Sangrador-Vegas, A., Scheremetjew, M., Yong, S. Y., Lopez, R., & Hunter, S. (2014). InterProScan 5: Genome-scale protein function classification. *Bioinformatics, 30*(9), 1236–1240.

Jung Kjær, L., Ward, M. P., Boklund, A. E., Larsen, L. E., Hjulsager, C. K., & Kirkeby, C. T. (2023). Using surveillance data for early warning modelling of highly pathogenic avian influenza in Europe reveals a seasonal shift in transmission, 2016-2022. *Scientific Reports, 13*(1), 15396.

Kanehisa, M., Sato, Y., Kawashima, M., Furumichi, M., & Tanabe, M. (2012). KEGG as a reference resource for gene and protein annotation. *Nucleic Acids Research, 40*(D1), D109–D114.

Koboldt, D. C., Chen, K., Wylie, T., Larson, D. E., McLellan, M. D., Mardis, E. R., Weinstock, G. M., Wilson, R. K., & Ding, L. (2009). VarScan: Variant detection in massively parallel sequencing of individual and pooled samples. *Bioinformatics, 25*, 2283–2285.

Langmead, B., & Salzberg, S. L. (2012). Fast gapped-read alignment with Bowtie 2. *Nature Methods, 9*(4), 357–359.

Larsen, M. V., Cosentino, S., Rasmussen, S., Friis, C., Hasman, H., Marvig, R. L., et al. (2012). Multilocus sequence typing of total-genome-sequenced bacteria. *Journal of Clinical Microbiology, 50*(4), 1355–1361.

Leinonen, R., Sugawara, H., & Shumway, M. (2011). The sequence read archive. *Nucleic Acids Research, 39*(D1), D19–D21.

Li, H., & Durbin, R. (2009). Fast and accurate short read alignment with Burrows-Wheeler transform. *Bioinformatics, 25*(14), 1754–1760.

Liu, L., Li, Y., Li, S., Hu, N., He, Y., Pong, R., et al. (2012). Comparison of next-generation sequencing systems. *BioMed Research International, 2012*(1), 251364.

Loman, N. J., Misra, R. V., Dallman, T. J., Constantinidou, C., Gharbia, S. E., Wain, J., & Pallen, M. J. (2012). Performance comparison of benchtop high-throughput sequencing platforms. *Nature Biotechnology, 30*(5), 434–439.

Maljkovic Berry, I., Melendrez, M. C., Bishop-Lilly, K. A., Rutvisuttinunt, W., Pollett, S., Talundzic, E., Morton, L., & Jarman, R. G., (2020). Next generation sequencing and bioinformatics methodologies for infectious disease research and public health: approaches, applications, and considerations for development of laboratory capacity. *The Journal of infectious diseases, 221*(3), pp.S292–S307.

McKenna, A., Hanna, M., Banks, E., Sivachenko, A., Cibulskis, K., Kernytsky, A., Garimella, K., Altshuler, D., Gabriel, S., Daly, M., & DePristo, M. A. (2010). The Genome Analysis Toolkit: A MapReduce framework for analyzing next-generation DNA sequencing data. *Genome Research, 20*(9), 1297–1303.

Mokili, J. L., Rohwer, F., & Dutilh, B. E. (2012). Metagenomics and future perspectives in virus discovery. *Current Opinion in Virology, 2*(1), 63–77.

Ng, S. B., Turner, E. H., Robertson, P. D., Flygare, S. D., Bigham, A. W., Lee, C., et al. (2009). Targeted capture and massively parallel sequencing of 12 human exomes. *Nature, 461*(7261), 272–276.

Nguyen, L. T., Schmidt, H. A., Von Haeseler, A., & Minh, B. Q. (2015). IQ-TREE: A fast and effective stochastic algorithm for estimating maximum-likelihood phylogenies. *Molecular Biology and Evolution, 32*(1), 268–274.

Pronyk, P. M., de Alwis, R., Rockett, R., Basile, K., Boucher, Y. F., Pang, V., et al. (2023). Advancing pathogen genomics in resource-limited settings. *Cell Genomics.*

Quail, M. A., Smith, M., Coupland, P., Otto, T. D., Harris, S. R., Connor, T. R., et al. (2012). A tale of three next generation sequencing platforms: Comparison of Ion Torrent, Pacific Biosciences and IlluminaMiSeq sequencers. *BMC Genomics, 13*, 1–13.

Rimmer, A., Phan, H., Mathieson, I., Iqbal, Z., Twigg, S. R. F., WGS500, Consortium, Wilkie, A. O. M., McVean, G., & Lunter, G. (2014). Integrating mapping-, assembly- and haplotype-based approaches for calling variants in clinical sequencing applications. *Nature Genetics, 46*(8), 912–918.

Rothberg, J. M., Hinz, W., Rearick, T. M., Schultz, J., Mileski, W., Davey, M., et al. (2011). An integrated semiconductor device enabling non-optical genome sequencing. *Nature, 475*(7356), 348–352.

Seemann, T. (2014). Prokka: Rapid prokaryotic genome annotation. *Bioinformatics, 30*(14), 2068–2069.

Segata, N., Waldron, L., Ballarini, A., Narasimhan, V., Jousson, O., & Huttenhower, C. (2012). Metagenomic microbial community profiling using unique clade-specific marker genes. *Nature Methods, 9*(8), 811–814.

Shu, Y., & McCauley, J. (2017). GISAID: Global initiative on sharing all influenza data–from vision to reality. *Eurosurveillance, 22*(13), 30494.

Sim, S., & Cho, M. (2023). Convergence model of AI and IoT for virus disease control system. *Personal and Ubiquitous Computing, 27*(3), 1209–1219.

Stamatakis, A. (2014). RAxML version 8: A tool for phylogenetic analysis and post-analysis of large phylogenies. *Bioinformatics, 30*(9), 1312–1313.

Suminda, G. G. D., Bhandari, S., Won, Y., Goutam, U., Pulicherla, K. K., Son, Y. O., & Ghosh, M. (2022). High-throughput sequencing technologies in the detection of livestock pathogens, diagnosis, and zoonotic surveillance. *Computational and Structural Biotechnology Journal, 20*, 5378–5392.

Witney, A. A., Cosgrove, C. A., Arnold, A., Hinds, J., Stoker, N. G., & Butcher, P. D. (2016). Clinical use of whole genome sequencing for Mycobacterium tuberculosis. *BMC Medicine, 14*, 1–7.

Wood, D. E., Lu, J., & Langmead, B. (2019). Improved metagenomic analysis with Kraken 2. *Genome Biology, 20*(1), 257.

Zankari, E., Hasman, H., Cosentino, S., Vestergaard, M., Rasmussen, S., Lund, O., et al. (2012). Identification of acquired antimicrobial resistance genes. *Journal of Antimicrobial Chemotherapy, 67*(11), 2640–2644.

Zhu, N., Zhang, D., Wang, W., Li, X., Yang, B., Song, J., et al. (2020). A novel coronavirus from patients with pneumonia in China, 2019. *New England Journal of Medicine, 382*(8), 727–733.

Antimicrobial Resistance

6

Suman Biswas, Ripan Biswas, and Indranil Samanta

Abstract

Antimicrobial resistance is an escalating global health crisis that jeopardizes the efficacy of antibiotics and other antimicrobial agents, rendering previously treatable infections increasingly challenging to manage. Resistant pathogens emerge and spread due to excessive and improper use of antibiotics in health care, farming, and animal rearing, as well as insufficient infection prevention and poor hygiene. Antimicrobial resistance presents substantial public health obstacles, contributing to extended illnesses, escalated healthcare expenditures, and elevated mortality rates. Addressing this challenge necessitates a multifaceted strategy, encompassing the promotion of antimicrobial stewardship, the development of innovative therapeutic interventions, the enhancement of surveillance systems, and the fostering of global cooperation to implement effective policies. Failure to urgently and collectively address this issue risks propelling the world into a post-antibiotic era, wherein routine infections may become potentially life-threatening. This chapter emphasizes the critical need for sustained efforts to address antimicrobial resistance and maintain the effectiveness of modern medical treatments, preferring a One Health approach.

S. Biswas
Department of Avian Sciences, West Bengal University of Animal and Fishery Sciences, Mohanpur, West Bengal, India

R. Biswas
Department of Veterinary Public Health & Epidemiology, West Bengal University of Animal and Fishery Sciences, Kolkata, West Bengal, India

I. Samanta (✉)
Department of Veterinary Microbiology, West Bengal University of Animal and Fishery Sciences, Kolkata, West Bengal, India

R. Biswas et al. (eds.), *Concept of One Health and Zoonoses*,
https://doi.org/10.1007/978-981-95-6627-3_6

Keywords

AMR · Antimicrobial resistance · Antibiotic resistance · One Health

Introduction

Globally, antimicrobial resistance poses a public health concern, influenced by factors impacting human, animal, and environmental health. Many infectious diseases that could formerly be successfully treated with any one of several drug classes have developed resistance to most of these medications, and in some cases, nearly all of them. Antimicrobial resistance is present wherever antimicrobials are utilized, encompassing human and animal populations, agricultural and aquaculture settings, as well as diverse ecological niches such as water, soil, and wildlife habitats. This widespread resistance is exacerbated by sewage contamination, pharmaceutical waste, and agricultural runoff. AMR is an ecological issue that affects human, animal, and environmental health through intricate interactions between different microbe populations. To mitigate health risks at the human-animal-environment interface, a multidisciplinary, multisectoral, and collaborative strategy is required, aligning with the One Health framework.

One Health is the "collaborative effort of multiple health science professions, together with their related disciplines and institutions-working locally, nationally, and globally" to promote the best possible health outcomes for humans, domestic animals, wildlife, plants, and the environment. One Health is an interdisciplinary and comprehensive concept that considers how closely connected the health of people and animals are to each other and to the ecosystem in which they live. The World Health Organization (WHO), the G8, the International Monetary Fund (IMF), and other top regulatory bodies have all identified AMR as a significant twenty-first century global health threat. All these forums agreed that because diverse ecosystems are involved in the acquisition, emergence, and diffusion of AMR, coordinated and interdisciplinary activities are required. One Health case studies on the genesis of AMR and the dynamics of multi-drug-resistant pathogen transmission show that human, animal, and environmental professionals must work together in order to mitigate global AMR.

Use of Antimicrobials in Different Sectors

Certain antimicrobial agents are specifically employed to combat human infections such as tuberculosis (e.g., isoniazid) or other infections that animals are generally not treated (e.g., Cattle with bovine tuberculosis are typically culled instead of treated). Some others, such ionophores and flavophospholipols, are only used in veterinary practice due to their toxicity to humans. The vast majority of antimicrobial classes are utilized for both human and animal applications, encompassing domestic mammals, avian species, farmed fish, shellfish, honeybees, and other organisms. Tetracyclines, streptomycin, and a few other antimicrobials are sometimes employed in horticulture to address and forestall bacterial contamination in fruits like apples and pears (such as "fire blight," which is caused by Erwinia amylovora). Antimicrobials are primarily utilized in human medicine to address clinical infections in individual patients, with occasional prophylactic use in groups or individuals (e.g., to avoid meningococcal disease) or in individuals (e.g., post-surgery). The application of antimicrobials in veterinary medicine varies substantially between companion and food-producing animals. The use of antibiotics in companion animals is much similar as human being; that is, the medications are mostly given to each animal individually to treat clinical infections, with some usage for prophylactic in specific animals, such as recovery from surgery. In contrast, when antimicrobial therapy is necessary for clinically infected food animals, the medications are frequently administered to the whole group via feed or water, for logistical and economic considerations, even if most animals are asymptomatic (e.g., pig pens, broiler and layer flocks) through feed or water for practical and efficient reasons, even when most animals are asymptomatic. However, this is often mischaracterized by contemporary healthcare professionals as "therapeutic" application. In addition, antimicrobials may be employed to treat specific clinical conditions in animals, such as dairy cows afflicted with mastitis, in a manner analogous to their application in human medical contexts.

Metaphylaxis is the widespread use of therapeutic antimicrobial dosages in a cohort of animals at elevated risk of contracting an infection. As an illustration, antimicrobials might be given to groups of calves entering a feedlot via injection, owing to their elevated susceptibility to bovine respiratory disease complex. Within extensive animal populations, the application of antimicrobial prophylaxis is infrequent, typically reserved for the management of severe, highly contagious illnesses such as meningococcal disease. When some animals in a group display symptom of infection, the standard industry practice involves treating the entire group therapeutically. Animals without clinical disease are at increased risk of developing a bacterial infection due to exposure to pathogens, poor hygiene, or overcrowding.

Antimicrobials in food animals are most commonly used to compensate the unhygienic practices, improper housing systems, and inadequate animal health management systems. The long-term, low-dose application of antimicrobials to enhance growth in food animals is a highly contested practice. Antimicrobial growth promoters are frequently utilized for extended durations at subtherapeutic dosages, targeting the entire animal population. This practice accelerates the propagation of resistant bacteria within and among animal populations, as well as their dissemination to humans via food and environmental routes, thereby contributing to antimicrobial resistance. Benefits were previously mostly obtained from the antimicrobials' ability to prevent disease, not from improving feed efficiency or other productivity effects. The World Health Organization (WHO) recommends not to use antimicrobials to promote growth because of these factors, and it has been banned in Europe, United States, and Canada. Recent data from the World Organisation for Animal Health reveal a decrease in the usage of antimicrobials as growth promoters in animals, with 41% of 146 reporting countries implementing this practice compared to 51% of 151 countries in 2012.

Concept and Genesis of Antimicrobial Resistance

In 2006, Gerry Wright and colleagues first used the term "antibiotic resistome." Wright and associates defined the soil resistome as "resistance determinants present in the soil," highlighting that multidrug resistance was more prevalent in environmental populations than previously acknowledged. The resistome has since been described as a "compilation of all the Antibiotic Resistance Genes and their progenitors in both pathogenic and non-pathogenic bacteria." Specifically, it comprised "all ARGs, including those found in pathogenic bacteria, antibiotic-producing organisms, and harmless non-pathogenic bacteria." Concerning the origins of antibiotic resistance genes, the resistome is hypothesized to encompass clinical, environmental, and intrinsic resistance genes, alongside protoresistance genes, which function as a substantial source of ARG progenitors. Further research into the resistome has improved comprehension, encompassing the classification of resistance mechanisms like acquired, intrinsic, silent/cryptic, and protoresistance. On the basis of the resistome concept, key information regarding antimicrobial resistance (AMR) have been published public since 2006 where previous hypotheses were validated, and the following important novel findings were proposed:

(i) AMR has existed for a long time and is present in many.
(ii) The antibiotic resistance genome is intricate.
(iii) ARGs originate and are stored in the environmental resistome.
(iv) The structure and dynamics of the microbial population in the natural environment typically determines the resistome.
(v) Human activities influence the environmental resistome.
(vi) The transmission of ARG is caused by mobile genetic elements.
(vii) ARGs circulate between people, animals, and the natural world.

These research outcomes have enabled novel strategies, such as the "One-Health" paradigm, which is designed to improve understanding of antimicrobial resistance across human, animal, and environmental contexts.

Epidemiology

The manner in which diverse nations have integrated existing scientific knowledge into antimicrobial policy is exemplified by the historical trajectory of initiatives aimed at mitigating the One Health dimensions of antibiotic utilization in the context of food-producing animals. Both the microbiological and toxicological impacts of antimicrobial usage in animals have been the subject of public health concerns since the 1950s.

Indian Perspective

It is common to find antimicrobial compounds in drinking water, stormwater, and groundwater. Additionally, the incapacity of traditional sewage treatment facilities (STPs) to eliminate antimicrobials from wastewater is one of the causes of their prevalence in the environment. Agricultural and livestock operations are among the main sources of antimicrobials in water, in addition to STP effluents. The presence of antibiotic-resistant genes was documented in both the influent and effluent of sewage treatment plants in Delhi and Nagpur, as well as in irrigation canal water that received treated effluent from one of the STPs, according to a study. The research suggested that extensive post-treatment is essential for eliminating antibiotic-resistant genes, thereby rendering the water appropriate for reuse. Furthermore, antibiotic contaminants, including Ampicillin, Fluoroquinolones, and Cephalosporins, were detected in the influents and effluents of sewage treatment plants near the Yamuna River, with Ampicillin exhibiting the highest concentration. These antibiotics enter waterways mostly through their use in the livestock, aquaculture, and agricultural industries; as a result, they eventually find their way into wastewater treatment facilities. Following wastewater treatment, antibiotics can reach the environment either through the treated effluent or through STP sludge used as manure.

Global Perspective

Farmers and veterinarians who provide therapeutic antimicrobial drugs to animals have known for decades that product quality must be maintained by following withdrawal periods. So long as the prescribed withdrawal periods and other labeling regulations were followed, this uncomplicated approach did not significantly hinder the veterinarian's ability to use antimicrobials. This was also straightforward to convey and apply. The implementation of One Health strategies to combat antimicrobial resistance in animals has been impeded due to the complex, unpredictable nature of antimicrobial resistance dynamics and the challenges in assessing, managing, and communicating microbiological risks compared to drug toxicity. Antimicrobials in animal feed, especially those given without a veterinarian prescription for growth promotion, were the initial focus of worries about the risks of antibiotic resistance to human health resulting from their usage in animals. The Swann Committee suggested in 1968 that "feed antibiotics" and "therapeutic antibiotics" be regulated separately, with the latter only being used on a veterinarian's prescription. Following the United Kingdom and other European nations' implementation of these guidelines, the utilization of penicillin, tetracyclines, sulfonamides, and other antimicrobials for therapeutic applications in both humans and animals as growth enhancers was subsequently banned in Europe.

Nonetheless, countries such as the United States and Canada diverged from the European custom of explicitly differentiating between antimicrobials intended for veterinary applications and those used as feed additives for food-producing animals.

The utilization of antimicrobial feed additives presented concerns beyond Europe's borders. Since 1969, various U.S. entities, such as the Food and Drug Administration, the National Academy of Sciences, and the Office of Technology Assessment, have addressed the use of antimicrobials in animals, with a particular focus on their incorporation into animal feeds. Early efforts to revoke authorization for the "sub-therapeutic" use of antimicrobial agents in animal feeds, based on the assertion that antimicrobial-resistant bacteria frequently transmit to humans and cause serious disease, lacked sufficient epidemiological evidence. It was estimated that the sub-therapeutic application of penicillin and tetracycline contributed to an annual range of 1–400 fatalities in the United States due to antimicrobial-resistant salmonellosis. Based on a risk assessment model that analyzed the proportion of these instances attributable to antimicrobial use in livestock feed, the estimated number of annual fatalities in the United States was approximately six.

In the 1990s, renewed worries emerged regarding antimicrobial resistance within human, agricultural, and veterinary sectors. Concerns during the 1990s primarily involved the therapeutic application of fluoroquinolones and the use of avoparcin as a growth promoter, in addition to the development and spread of multidrug-resistant *Salmonella enterica* serovar Typhimurium DT104. Avoparcin, a glycopeptide antimicrobial used as a feed additive in Europe, Australia, and other countries, was selected for its resistance to vancomycin. Concurrently, increasing apprehension regarding resistance to fluoroquinolones in critical human infections highlighted the utilization of these antimicrobials in animal husbandry, especially within the poultry industry, where they were typically administered orally via water. *Fluoroquinolone resistance emerged in Campylobacter jejuni isolated from both poultry and human clinical infections shortly after the drug's introduction for veterinary purposes.* Due to the adverse effects of fluoroquinolones on human health, specifically the development of resistance in *C. jejuni*, the FDA revoked its approval for the use of these drugs in poultry.

In 1986, Sweden initiated the prohibition of antimicrobial growth promoters in food animals, becoming the first European country to do so. In 1995, Denmark prohibited the use of avoparcin. This was followed by bans on other antimicrobial growth promoters by 1999, achieved through a combination of voluntary industry programs and government regulations. As of January 1, 2006, the European Union implemented a comprehensive prohibition on the utilization of all antimicrobial growth promoters. Despite this ban, the EU has permitted the therapeutic application of approved antimicrobials in animals, albeit with specific limitations, including a restriction on the use of third-generation cephalosporins in poultry; this approval encompasses fluoroquinolones and a variety of other significant antiviral medications for human applications.

Impact of AMR on One Health

Antimicrobial resistance genes are found in the microbiomes of humans, animals, and the environment, reflecting the various elements of the One Health framework. The One Health initiative is a cooperative effort that involves various fields at the local, national, and global levels. Its goal is to optimize the health of humans, animals, and the environment through policy, research, education, and practical application. Zoonotic diseases, which impact both human and animal populations, were initially recognized as critical issues. The One Health concept, developed in 2008 following collaboration with the Food and Agriculture Organization, World Organisation for Animal Health, and World Health Organization, prioritized the significance of ecosystems. Infectious diseases, especially those of zoonotic origin, represent the central focus and objectives of One Health strategies. During the 2011 FAO-OIE-WHO tripartite summit, the close relationship between antimicrobial resistance and zoonoses resulted in its designation as a central concern within the One Health agenda. Recently, there has been increased interest in comprehending communication within local ecosystems and in pinpointing the drivers of the worldwide AMR crisis, although the One Health paradigm has traditionally emphasized the interconnections and mutual dependencies among sectors at a local level.

Environmental Health

In the context of various illnesses, environmental factors play a crucial role in transmission, thereby prolonging the cycle of pollution and consumption. Environmental resistance is exacerbated by pollution, poor sanitation, organic fertilizers, and industrial waste, which serve as disease reservoirs. The application of antimicrobials as insecticides in less developed countries is exceeding safety limits for produce. Analysis of produce imported from India to Nepal revealed that antimicrobial pesticide residues were present in 93% of eggplant samples and all tomato and chili samples. Multi-resistant bacteria may evolve as a result of metabolic processes caused by antimicrobial contamination in soil and water habitats. Following manure applications, numerous *Salmonella* serotypes have been found to spread antibiotic resistance in commercial pig farms. The public health infrastructure is strained by the influx of infectious diseases, a consequence of the compromises made in urban development and the associated decline in environmental conditions, specifically sanitation and air quality. For example, the high levels of air pollution in different Indian cities have raised the prevalence of respiratory diseases, which has put pressure on the need for antimicrobials. More and more scientific studies have emphasized how the environment contributes to the increase in AMR; nevertheless, the effects have frequently gone unnoticed by surveillance systems and health regulations.

Animal Health

According to the World Health Organization, the utilization of antimicrobials in animal husbandry primarily serves three purposes: enhancing growth, preventing disease, and providing therapeutic veterinary care. The first two applications are strongly associated with the misuse of antimicrobials in animal populations housed in densely populated environments. It has been observed that low concentration of antimicrobials affects development of resistance in animals. Antibiotic-resistant bacteria (ARBs) and genes (ARGs) may be selected for using antibiotics in animals and can spread to humans. Studies on AMR in animals have shown a favorable relationship between antimicrobial use, resistance profile, and dependence on mobile genetic elements (MGE). It has been observed that the resistance profile of slaughter house workers is influenced by animal resistance profile to antibiotics/antimicrobials. Wildlife animals are also known to spread ARGs, in addition to livestock species. Compared to noncontacted animals, human-contacted wild animals' resistance profiles showed a greater diversity of ARGs. Migratory birds were identified as the primary source of ARGs' environmental spread among wild animals. Furthermore, it is possible that companion animals, such dogs, serve as ARB and ARG reservoirs due to the presence of bacteria harboring developing ARGs in these animals. To mitigate the spread of antimicrobial resistance and uphold the One-Health approach, a comprehensive investigation and assessment of antimicrobial resistance genes in companion animals are crucial, given their close interaction with humans.

Human Health

Animal origin ARB can be spread to human beings through direct contact, food products, or from the environment. Humans and food animals both have similar strains of resistant bacteria, that can be indicative of animal-to-human bacterial spread.

Humans and food animals from North America, Europe, Africa, and Asia have been found to have plasmid-mediated resistance in *Salmonella* sp., *Klebsiella* sp., and E. Coli to colistin, a last-resort class of antimicrobial drug. A recent systematic review indicates that food-producing animals contribute substantially to the incidence of human extra-intestinal extended-spectrum cephalosporin-resistant *E. coli* infections.. A study revealed a Spearman correlation coefficient of 0.94 between the prevalence of amoxicillin resistance in *E. coli* isolated from food animals and humans, strongly suggesting the transmission of antimicrobial-resistant bacteria from food animals to humans. An intervention to decrease antimicrobial use in animals led to a 24% reduction in the incidence of multidrug-resistant diseases in humans. While the study's findings are significant, some researchers argue that the extent of resistance transfer from animals to humans is limited and suggest that reducing antibiotic use in food animals might have adverse health consequences for both humans and animals. In the context of the ongoing discussion about the connection between antimicrobial use in food animals and its effects on human health, it is essential to conduct further research into resistance dynamics and their implications for the health of both humans and animals.

Plant Health

Antimicrobial resistance genes and AMR organisms are ubiquitous in the environment encompassing various plant cultivation methods employing diverse processes. In fact, soil microbes are the source of the great majority of the antibiotic resistance genes found in pathogenic bacteria. During the early 1980s, the potential for genetically modified plants to transfer genes to indigenous bacterial flora, subsequently leading to antimicrobial resistance within horticultural practices, garnered considerable interest. In soil and decomposing plant debris, these genes might persist for long stretches of time. Transfer of these genes into the genomes of native soil bacteria is improbable, as they lack the necessary sequence similarity surrounding the resistance genes to facilitate homologous recombination. In the context of plant transformation, these genes undergo stable integration within the host genome, contrasting with their presence on mobile genetic elements. The introduction of zoonotic pathogens and phytopathogens into plants indicates microbial contamination (such as *E. coli*) has been the subject of attention in more recent times. The plant health is disrupted by field worker, use of contaminated equipment, wildlife infiltration and defecation into fields and irrigation water. Intentionally adding biocontrol agents to crops to prevent disease could harbor genes for antibiotic resistance and act as a reservoir for the spread of those genes. Numerous cross-sectional investigations conducted worldwide have documented the presence of various zoonotic pathogens and bacterial population that are resistant to antibacterial agents on fresh fruits and vegetables. *E. coli* strains found in the nasal and fecal bacteria of sheep that grazed on streptomycin-treated pastures showed a higher incidence of both streptomycin-resistant and multi-drug-resistant strains when compared to sheep grazing on control pastures. Outbreaks of foodborne illnesses have been linked to mangoes and cucumbers contaminated with streptomycin- and tetracycline-resistant strains of *Salmonella* spp. As indicated by the research cited, the complex ecology of antimicrobial resistance is influenced by several factors, including plant-based agriculture, thus presenting a significant public health concern.

Diagnostic Approach to Combat AMR

Antimicrobial resistance has been identified as dynamic and constantly growing, which highlights the significance of state-of-the-art technology to fully investigate its diversity and clarify its dynamics. The use of these sequence-based techniques for AMR research has increased across all One-Health sectors through recently developed whole genome sequencing (WGS) techniques, metagenomics, new databases, and improved diagnostic tools.

Metagenomics

Metagenomics, a sequence-based approach for investigating genomes from a diverse microbial community, is presently employed to understand and elucidate the complexity of antimicrobial resistance across various One Health sectors. Bacteria generally acquire antibiotic resistance genes through horizontal gene transfer. Understanding HGTs is critical as they offer comprehensive insights into the dissemination of AMRs within and across various One Health settings. Metagenomics, a high-throughput DNA sequencing method, enabled the analysis of antibiotic resistance genes. This technique is suitable as it facilitates the utilization of environmental genomic data, bypassing the need for bacterial isolation through culture. Metagenomics assists in characterizing microbial diversity by identifying genes and reconstructing complete microbial community genomes. The ability of this approach to identify antimicrobial resistance genes within a microbial community, alongside its capacity to quantify species abundance, underscores its sensitivity as a key benefit. Currently, metagenomics is utilized to study microbial diversity in both clinical and environmental samples, and it is now considered an alternative to rRNA sequencing.

Certain challenges, however, diminish the effectiveness of metagenomic analysis. One limitation is the diminished sensitivity toward the less abundant microbial populations, which may also include antibiotic resistance genes. Moreover, the specificity in identifying impactful bacterial variants is compromised due to the potential for diverse phenotypic traits among variants. The integration of metagenomic sequencing with functional genomics could potentially mitigate these limitations.

Whole Genome Sequencing

However, it is crucial to reconcile the disparities observed between culture-dependent testing and whole-genome sequencing. Whole-genome sequencing is considered a valuable method for examining antimicrobial resistance and forecasting the resistance patterns of bacterial strains. Categorizing antimicrobial resistance genes and their diverse origins within the One Health framework is crucial due to the complexity of microbial communities found in different environments. Recent investigations utilizing these approaches have revealed associations between mobile genetic elements and the dissemination of antimicrobial resistance across numerous One Health sectors, allowing for the identification of specific ARGs that pose potential health risks. In the future, these advanced methodologies will be indispensable for elucidating the mechanisms by which antimicrobial resistance and antimicrobial resistance genes propagate and are transferred throughout different One Health sectors.

Databases

Within this framework, several online databases are readily available, encompassing ResFinder, AMRFinder, CARD, and ARG-ANNOT, alongside others. In addition to the previously mentioned databases, certain resources are tailored to specific pathogens, including Dream TB for *Mycobacterium tuberculosis* and Kleborate for *Klebsiella* spp. A new compilation of freely accessible online resources for the identification of ARGs in bacterial pathogens has been published. NCBI has also made available a database containing approximately 5000 resistance genes identified globally.

Diagnostics

Early and precise diagnostic methods considerably influence both treatment administration and antibiotic sensitivity. Innovative methods facilitating swift pathogen identification and antibiotic susceptibility assessment have emerged. Several techniques, including rapid immunochromatography, automated time-lapse microscopy, and matrix-assisted laser desorption ionization-mass spectrometry, present potential utility in this domain. Nevertheless, the implementation of these approaches presents challenges, encompassing operational costs due to the high price of reagents, which is particularly prohibitive for researchers in less developed nations. Additionally, these tools necessitate a considerable workforce of trained and proficient personnel. Moreover, molecular biological methods like FilmArray BC-ID and Xpert Carba-R are accessible for the detection of ARGs. PCR/electrospray ionization-mass spectrometry technology offers a marked advancement by enabling the detection of over 500 pathogens and diverse ARGs within hours. These diagnostic tools offer the key advantage of sample applicability, thereby accelerating the process of susceptibility testing and pathogen identification. According to current literature, these techniques are poised to become standard laboratory procedures.

Preventive Approaches

To mitigate the dangers of antimicrobial resistance, WHO, in collaboration with other international bodies and many individual countries, has developed comprehensive action plans. The World Health Organization's Global Action Plan employs a One Health framework to combat antibiotic resistance, recommending that member-states incorporate this strategy when formulating their national action plans (Biswas, 2022) The subsequent subsections delineate the five primary objectives of the WHO Global Action Plan.

Training, Education, and Communication

It is crucial for all individuals to understand the significance of maintaining proper hygiene to mitigate the spread of infections, the necessity of complying with therapeutic guidelines provided by antimicrobial prescribers, and a fundamental understanding of both the benefits and potential dangers associated with antibiotic usage for both personal and communal health. A more profound understanding of the One Health aspects of antimicrobial resistance would be particularly advantageous for companion animal owners, agricultural producers, veterinary professionals, and individuals employed in the food production and related industries. Animal owners should recognize the potential for shared bacterial infections between humans and animals, and they should adhere to veterinary guidance on antibiotic use and disease prevention. The use of antimicrobials should be restricted to the treatment of clinically affected animals. Additionally, farmers should be well-versed in sustainable agricultural practices that minimize or eliminate the need for these substances. They must also be proficient in enhancing overall animal care practices and in mitigating stressful, densely populated, and unsanitary environments that promote disease development in animal and plant populations to reduce the need for antimicrobial treatments. Veterinarians are expected to demonstrate the behaviors, attitudes, and knowledge indicative of appropriate antimicrobial stewardship to protect the health and welfare of their patients, their clients' financial interests, and the health of the public. Human health professionals would also gain from comprehending the One Health dimensions of antimicrobial resistance and the methods employed to curb the dissemination of pathogens, including antibiotic-resistant bacteria, such as infection control, enhanced hygiene, and improved sanitation.

A variety of approaches can be employed to improve awareness and understanding of the One Health considerations related to antimicrobial resistance. These include initiatives in health promotion and protection from public and animal health organizations, educational campaigns aimed at the general public and animal owners, outreach programs targeting farmers, veterinary consultations, industry-specific publications for agricultural producers, and continuing education for veterinary and medical professionals.

Proper Research and Surveillance

Antimicrobial resistance issues necessitate research and surveillance to provide and develop solutions. Surveillance of antimicrobial resistance and the utilization of antimicrobials within human and animal sectors is imperative at regional, national, and global levels to assess the extent, patterns, and health consequences of resistance. Such surveillance should be capable of identifying antimicrobial resistance trends that are of clinical significance in both human and animal populations. Surveillance data should inform educational programs designed to curb antimicrobial resistance, as well as guidelines for antimicrobial use and stewardship efforts. A crucial component of a One Health surveillance strategy for antimicrobial resistance involves the collection of suitable bacterial samples from a range of sources. These sources include specimens obtained from humans, animals, and environmental sites, such as hospitals, long-term care facilities, community environments, veterinary clinics, farms, food supplies, and various environmental locations. In order to enable international comparisons, antimicrobial use surveillance should also be conducted in human, veterinary, and agricultural contexts. This surveillance should also yield national estimates of antimicrobial consumption in human and animal species, along with relevant denominators (such as population size, animal population, and population correction unit). Reports should be issued promptly to facilitate stakeholder utilization, and surveillance data should be integrated across sectors in analysis, interpretation, and presentation. Investigations focusing on particular bacterial species, including pathogens, intestinal commensals, and environmental bacteria, are essential for elucidating the mechanisms of resistance development and dissemination within and between ecological niches. Further research is needed to identify suitable and cost-effective alternatives to antimicrobials for disease prevention, growth enhancement, and improved production efficiency in order to reduce antimicrobial use in animal populations.

Increased Sanitation and Hygiene

It is widely acknowledged that decreasing the spread of antimicrobial resistance in human populations can be achieved through effective infection control practices within healthcare settings, particularly in hospitals. However, the existing infection control strategies are not being fully leveraged within veterinary practices. Within most sectors of food animal production, the implementation of biosecurity and other disease control measures is essential at the farm level, particularly in intensive operations involving swine, ducks, and poultry. In addition to certain farm-level animal disease management programs, national-level initiatives are also employed, particularly to prevent the dissemination of foreign animal diseases within a population. However, the reduction of antimicrobial use in animals, whether for preventative or therapeutic purposes, necessitates the implementation of health management strategies to prevent disease in animals, especially in contexts where high antimicrobial use is critical, a consideration of particular importance within veterinary practice (e.g., coliform mastitis in dairy cattle, PPR in goats).

Essential actions to mitigate human exposure to antimicrobial resistance originating from environmental sources and routes encompass enhancing the safety of food and potable water, particularly in less developed nations, and diminishing environmental contamination caused by the pharmaceutical sector. Effective strategies for reducing or controlling foodborne pathogens at the farm, slaughter, and subsequent processing stages are essential to prevent the transmission of enteric bacteria such as *Salmonella* spp. and *Campylobacter* spp. to humans. Minimizing environmental bacterial contamination necessitates effective sewage treatment and measures to improve the microbiological safety of potable water.

Judicial Use of Antimicrobials

An integrated One Health approach to improve antimicrobial stewardship involves various regulatory, voluntary, and other actions to preserve the effectiveness of antimicrobials across human, animal, and agricultural settings. This approach requires initiatives that promote responsible antimicrobial utilization, coupled with the ongoing assessment of antimicrobial usage and the implementation of systems designed to continuously improve both prescribing practices and antimicrobial utilization. This process can be challenging due to the potential for conflicting interests across various sectors. Generally, national regulatory measures to address antimicrobial resistance are more prevalent in the veterinary and agricultural sectors compared to human health. Regulatory approval for antimicrobials used in animals typically necessitates demonstration of safety and efficacy for human application.

Drug classification constitutes a vital approach in the fight against antimicrobial resistance. This classification provides a framework for risk management strategies designed to prevent and mitigate antibiotic resistance within food animal production. Antimicrobial classes are categorized based on two criteria, with those deemed critically important receiving the highest priority based on supplementary criteria. Third- and fourth-generation cephalosporins and quinolones represent two classes of antimicrobials that are considered to be of the highest priority. The aforementioned antimicrobial classification can be integrated into antimicrobial risk management strategies through various approaches. A well-known example involves the cessation of medically important antimicrobials for growth promotion in animals. Furthermore, classification can be incorporated into formularies or therapy cascade schemes (e.g., first-line, second-line, etc. choices).

Economic Improvement

The WHO Global Action Plan primarily focuses on the human health sector; however, it is also essential to invest in the animal health sector to diminish the reliance on antimicrobials, including the development of additional vaccines and alternative non-antimicrobial disease control strategies.

Collaborative Approaches of National and International Organizations

The containment of antimicrobial resistance has been significantly advanced by One Health initiatives, facilitated through various international bodies. Since the early 1990s, the WHO has convened numerous multidisciplinary, multisectoral expert consultations and advisory groups. Moreover, these bodies have amassed considerable empirical data and expert consensus concerning the impact of antimicrobial usage in animals on human well-being. They have formulated extensive guidelines suitable for all parties involved, encompassing regulatory agencies, the pharmaceutical and animal husbandry sectors, veterinary professionals, agricultural producers, public health organizations, and the general public.

The Food and Agriculture Organization and the World Health Organization manage the global food standards known as Codex Alimentarius. Recognizing the importance of antimicrobials for human and animal well-being, Codex suggested in 2001 that the FAO, OIE, and WHO collaborate to tackle the issue of AMR. An outcome was a series of cooperative workshops facilitated by specialists from the Food and Agriculture Organization, World Health Organization, and World Organisation for Animal Health, focusing on antimicrobial use in non-human animals and antimicrobial resistance, which implicitly embraced a One Health approach.

The European Union has been a key player in global One Health initiatives addressing antimicrobial resistance, particularly in regulating veterinary antimicrobials and implementing integrated surveillance of antimicrobial resistance and usage. The European Food Safety Authority, the European Centre for Disease Prevention and Control, and the European Medicines Agency all operate under the One Health framework. The Committee for Medicinal Products for Veterinary Use, part of the EMA, is responsible for regulating veterinary pharmaceuticals within the European Union. Since 2009, the European Surveillance of Veterinary Antimicrobial Consumption project has overseen the monitoring of veterinary antimicrobial usage in the European Union and the European Economic Area.

Summary and Conclusion

AMR's worldwide appearance and spread make it imperative to advance a coordinated, interdisciplinary One Health strategy to lower its threats to the health of human being, animals, environment, and plants. Public health around the world is adversely affected by the global spread of AMR infections and ARGs spurred on by antibiotic abuse, poor sanitation infrastructure, and insufficient control mechanisms. The FAO-WOAH-WHO commitment and the worldwide evaluation of the One Health approach would support AMR prevention through antibiotic stewardship, political activism, awareness campaigns, and education on antibiotic use.

Reference

Biswas, R., Debnath, C., Bandyopadhyay, S., & Samanta, I. (2022). One health approaches adapted in low resource settings to address antimicrobial resistance. *Science in One Health, 1*, 100011.

Zoonotic Pathogens as Agents of Bioterrorism

7

Prasanna Vadhana A and Ripan Biswas

Abstract

Zoonotic pathogens, which are infectious agents transmitted from animal reservoirs to human hosts, represent a significant threat as potential bioterrorism agent. The natural presence of these pathogens in animal populations, their capacity to induce severe illness in humans, and their potential for rapid transmission render them attractive options for malicious individuals seeking to cause harm. Several zoonotic pathogens, including *Bacillus anthracis*, *Yersinia pestis*, Ebola virus, and avian influenza viruses, have a history of causing severe outbreaks and could potentially be utilized as bioweapons to generate fear, disrupt communities, and strain healthcare resources. Moreover, the dual-use potential of biotechnology exacerbates the risk, as advancements in genetic engineering could bolster the virulence, transmissibility, or resistance of zoonotic pathogens.

Keywords

Bioterrorism · Bioweapon · zoonotic bioweapons

Prasanna Vadhana A (✉)
Department of Veterinary Microbiology, College of Veterinary and Animal Sciences, Rani Lakshmi Bai Central Agricultural University, Jhansi, Uttar Pradesh, India

R. Biswas
Department of Veterinary Public Health and Epidemiology, West Bengal University of Animal and Fishery Sciences, Kolkata, West Bengal, India

R. Biswas et al. (eds.), *Concept of One Health and Zoonoses*,
https://doi.org/10.1007/978-981-95-6627-3_7

Introduction

Biological weapons are devices or agents designed to intentionally spread disease-causing organisms or toxins through various means such as aerosols, food, water, or insect vectors. Bioterrorism involves the deliberate release or threat of release of biological agents—including viruses, bacteria, fungi, or their toxins—to inflict disease or death on humans, livestock, or crops, aiming to create fear, disrupt societies, and manipulate governments. Zoonotic pathogens, which naturally transmit between animals and humans, pose a serious threat when weaponized for bioterrorism. Their high infectivity, virulence, and ability to withstand environmental conditions make them prime candidates for deliberate dissemination. The impact of zoonotic bioweapons extends beyond public health to include significant damage to food security, livestock industries, and ecosystems. Throughout history, biological agents have been employed in warfare and terrorism, illustrating their potential for large-scale disruption and devastation. Healthcare professionals in human and veterinary sectors need to be aware of the essentials of bioterrorism and biological weapons, as these may be used as part of a terrorist attack in any part of the world. Therefore, these sectors should work under one health line to tackle them more effectively. Thus, healthcare professionals need to be trained and prepared in case of a potentially catastrophic event, where quick action and decision-making may potentially save lives. Bioterrorism can have widespread and severe consequences across multiple sectors of society. The intentional release of biological agents can lead to widespread disease outbreaks, resulting in high morbidity and mortality, overwhelming healthcare systems, and causing fear-induced panic that exacerbates social instability. Livestock populations may be severely affected, leading to food shortages, economic losses due to loss of life, poor quality animal products, and potential transmission of zoonotic diseases to humans. The spread of zoonotic pathogens to wildlife could lead to the persistence of infection in nature, as these infections are difficult to control in wildlife. The introduction of pathogens into ecosystems can disrupt biodiversity, damage natural resources, and result in long-term ecological consequences. Additionally, the economic burden of bioterrorism is immense, including healthcare costs, loss of agricultural productivity, trade restrictions, and the expenses associated with containment, decontamination, and recovery efforts.

Biological agents can be disseminated using various methods, making them highly adaptable to terrorist activities. The major routes include aerosol dissemination of fine particles, allowing inhalation and rapid spread, contaminating water sources and distribution systems to affect large populations, contaminating food production or distribution chains, using vectors, and contaminating surfaces or objects leading to fomite transmission.

History of Bioterrorism

Period	Agent used
Fourteenth century of BC	Diseased rams (possibly infected with tularemia) to their enemies.
In the sixth century of B.C.	Assyrians poisoned enemy wells with rye ergot (*Claviceps purpurea*, a fungus), and Solon used the purgative herb hellebore during the siege of Krissa.
In the fourth century BC	Greek historian Herodotus relates that Scythian archers used to infect their arrows by dipping them in a mixture of rotten cadavers of human blood and manure.
1155	Roman Emperor Frederick Barbarossa is said to have dumped human cadavers down wells.
1346	Plague was transmitted to the citizens of Caffa by the hurling of diseased cadavers into the besieged city.
1422	Lithuanian army used to hurl the manure made of infected victims into the town of Carolstein.
1495	Spanish are used to mix wine with blood of leprosy to sell to their French foes in Naples, Italy.
1650	National armed forces of the republic of Poland used to fire saliva from rabid dogs towards their enemies.
1710	Russian troops catapult human bodies of plague victims into Swedish cities.
1763	British distributed blankets from smallpox patients to Native Americans.
1797	Napoleonic armies flooded the plains around Mantua (Italy), to augment the spread of malaria.
American Civil War in 1863	Confederates used to sell cloth from yellow fever and smallpox patients to Union troops.
World War I	Germans transported horses and cattle inoculated with disease-producing bacteria, such as *Bacillus anthracis* and *Pseudomonas pseudomallei* and spread cholera in Italy and plague in St. Petersburg in Russia.
World War II	Japan used plague and anthrax as bioweapon.

Classification of Bioterrorism Agents

In 1998, Carus defined "Bioterrorism is the threat of use of biological agents by individuals or groups motivated by political, religious, ecological, social or for other ideological objectives to inculcate fear or cause illness or death in order to achieve their objectives."

The Centre for Disease Control and Prevention (CDC) classified bioterrorism agents into three categories—A, B, and C—using the risk-matrix analysis process to assess potential biological threat agents. The criteria to meet the bioterrorism agent are given below:

- Ability of the agent to be easily disseminated.
- Capability to be transmitted from person to person.
- Potential to cause significant morbidity and mortality.
- Capacity to cause public panic and social disruption.

Category A (Highest Threat)

These are the highest priority agents, represent organisms that pose a risk to national security because they can be easily disseminated or transmitted person-to-person; cause high mortality, with potential for major public health impact; might cause public panic and social disruption; and require special action for public health preparedness. Tables 7.1 and 7.2, and 7.3 list examples of category A, B, and C agents, respectively.

Category B (Moderate Threat)

These agents cause lower mortality rates compared with category A agents, have relative ease of dissemination, and require special resources for surveillance and diagnostics. Table 7.2 gives a list of examples of category B agents.

Table 7.1 Details of Category-A agents as per CDC

Type of Category-A agents	Name of the pathogens	Possible route of dissemination	Disease
Bacterial	*Bacillus anthracis*	Aerosol/water/soil	Anthrax
	Yersinia pestis	Droplet/vector	Plague
	Francisella tularensis	Droplet/vector	Tularemia
Viruses	**Arenaviridae** (a) Lassa fever virus (b) Machupo virus (c) Junin virus (d) Guanarito virus (e) Sabia virus	Aerosol/food	Lassa fever
	Bunyaviridae (a) CCHF virus (b) Rift Valley Fever virus	Droplet/Vectors	CCHF Rift Valley Fever
	Filoviridae (a) Ebola virus (b) Marburg virus	Droplet	Ebola hemorrhagic fever Marburg Disease
	Flaviviridae (a) Yellow fever virus, (b) Omsk hemorrhagic fever virus (c) Kyasanur Forest disease virus	Vectors	Yellow fever Omsk hemorrhagic fever Kyasanur Forest disease
	Poxviridae (a) Variola major virus	Aerosol/Droplet	Smallpox
Toxins	Botulinum *Clostridium botulinum* (Anderson, 2012)	Food	Botulism

Table 7.2 Details of Category-B agents as per CDC

Type of Category-A agents	Name of the pathogens	Possible route of dissemination	Disease
Bacterial	*Brucella* species	Food/water/soil	Brucellosis
	Salmonella spp.	Food/water	Salmonellosis
	Escherichia coli 0157:H7	Food/water	Hemorrhagic colitis, hemolytic-uremic syndrome (HUS)
	Shigella spp.	Food/water	Shigellosis
	Burkholderia mallei	Aerosol/droplet	Glanders
	Burkholderia pseudomallei	Aerosol/droplet	Pseudo Glanders
	Vibrio cholerae	Food and water	Cholera
Viruses	**Togaviridae** Western equine encephalitis virus (WEE) Eastern equine encephalitis virus (EEE) Venezuelan equine encephalitis virus (VEE)	Vectors Vectors Vectors	Encephalitis Encephalitis Encephalitis
Other agents	*Cryptosporidium parvum*	Food/water	CCHF Rift Valley fever
	Chlamydia psittaci	Aerosol	Ebola hemorrhagic fever Marburg disease
	Rickettsia prowazekii	Vectors	Yellow fever Omsk hemorrhagic fever Kyasanur Forest disease
Toxins	Epsilon toxin *(Clostridium perfringens)*	Food	Intoxication
	Enterotoxin-B *Staphylococcus* spp. (Poli, 2011)	Food	Intoxication

Category C (Emerging Threats)

These agents have the potential for future bioterrorism threats due to their availability, ease of production, and adaptability (van Aken and Hammond, 2003).

Role of Animals in Bioterrorism Events

Animals play a crucial role as sentinels in detecting and forecasting bioterrorism threats. Due to their heightened sensitivity to certain biological agents and their frequent exposure to environmental hazards, animals often manifest symptoms before humans, providing an early warning system. Surveillance of domestic animals, wildlife, and livestock can assist in identifying potential bioterrorism events and mitigating their impact. Studies indicate that certain bioterrorism agents affect

Table 7.3 Details of Category-C agents as per CDC

Type of Category-C agents	Name of the pathogens	Possible route of dissemination	Disease
Viruses	*Nipah virus*		Nipah viral disease
	Hantaviruses	Aerosol	Hemorrhagic fever with renal syndrome Hantavirus pulmonary syndrome
	Yellow fever virus		Yellow fever
	Tick-borne encephalitis complex	Vectors or food	Encephalitis
	Influenza viruses	Droplet	Influenza/ bird flu
	Coronaviruses	Droplet	SARS/COVID-19/MERS
	Any emerging pathogens	–	
Bacterial	Any emerging pathogen	–	

animals before humans, allowing for early detection. For example, in the 1979 Sverdlovsk anthrax outbreak, livestock deaths occurred before human cases, suggesting that monitoring animal populations could have provided an early alert. Similarly, cats infected with *Yersinia pestis* (plague) in experimental conditions showed symptoms within 1–2 days, which could be an early indicator of an outbreak before human cases emerge. Additionally, animals can serve as markers for ongoing exposure risks. The persistence of pathogens in soil, water, or air can be monitored through active surveillance of livestock, pets, or wildlife. For instance, outbreaks of *Francisella tularensis* (tularemia) in livestock have been linked to contaminated environmental sources, highlighting the role of animals in detecting bioterrorism-related exposure. The geographic distribution of sick or dead animals may also provide clues to the spread of an intentionally released pathogen. Furthermore, mobile animal populations such as migratory birds, shipped livestock, and international pet trade animals can aid in tracking and forecasting the spread of bioterrorism agents. The spread of West Nile virus through birds demonstrated how monitoring animal movements can help predict and contain outbreaks (Galloway et., 2015).

Mitigation of Bioterrorism

Although naturally occurring disease outbreaks present substantial difficulties, the threat of bioterrorism introduces additional complexities stemming from intentional malevolence and the potential for engineered pathogens with heightened virulence or resistance. Effective preparedness for the threat of bioterrorism necessitates a multifaceted approach. This includes establishing robust surveillance systems capable of promptly identifying unusual disease patterns, developing rapid diagnostic capabilities, maintaining stockpiles of essential medications and vaccines, and implementing well-defined response plans that involve close collaboration among public health agencies, healthcare providers, and law enforcement authorities. Collaborative efforts among nations and the exchange of pertinent information are essential for the successful prevention and alleviation of bioterrorist threats. Combating bioterrorism necessitates a multifaceted strategy encompassing international cooperation, a robust public health system, advanced surveillance capabilities, and rapid response mechanisms. Essential strategies comprise:

Proactive Measures and Mitigation Bolstering global biosecurity protocols, augmenting laboratory safety measures, and regulating access to hazardous pathogens are essential for deterring bioterrorism. Public health infrastructure must be equipped to promptly identify and address disease outbreaks.

Surveillance Systems and Information Sharing The early identification of bioterrorism risks is contingent upon the implementation of comprehensive surveillance systems and the exchange of intelligence among nations. Advancements in biotechnology and data analysis capabilities can enhance the capacity to recognize and monitor potential threats.

Research and Development Developing effective countermeasures through research on vaccines, therapeutics, and diagnostic technologies for potential bioterrorism agents is vital. However, advancements in biotechnology must be carefully balanced with ethical considerations to mitigate the risks of dual-use applications.

Promoting Public Awareness and Professional Education Educating the populace and medical professionals about the risks of bioterrorism and the corresponding response procedures can help mitigate panic and facilitate a coordinated effort during an incident.

Concerted International Collaboration Bioterrorism is a global phenomenon that is not constrained by national borders. International frameworks, such as the Biological Weapons Convention, and cooperative actions among countries are essential for mitigating this threat.

One Health approach: Acknowledging the interdependence among human, animal, and environmental health domains is essential for effectively addressing zoonotic diseases that could be vulnerable to bioterrorist exploitation.

Over all, addressing the threat of bioterrorism requires a comprehensive, coordinated, and collaborative approach. This involves using advances in science, technology, and international cooperation to protect global health and security.

References

Anderson, P. (2012). Bioterrorism: Toxins as weapons. *Journal of Pharmacy Practice, 25*(2), 121. SAGE Publishing. https://doi.org/10.1177/0897190012442351

Galloway, S. E., Petzing, S. R., & Young, C. G. (2015). Reassessing biological threats: Implications for cooperative mitigation strategies [Review of reassessing biological threats: Implications for cooperative mitigation strategies]. *Frontiers in Public Health, 3*, 251. Frontiers Media. https://doi.org/10.3389/fpubh.2015.00251

Poli, M. (2011). Toxins: Overview and general principles. In *Encyclopedia of bioterrorism defense* (p. 1). https://doi.org/10.1002/0471686786.ebd0122.pub2

van Aken, J., & Hammond, E. (2003). Genetic engineering and biological weapons. *EMBO Reports, 4*, S57–S60. Springer. https://doi.org/10.1038/sj.embor.embor860

Anthrax

8

Ripan Biswas and Indranil Samanta

Abstract

Anthrax, a zoonotic infectious disease stemming from the bacterium Bacillus anthracis, continues to pose a significant threat to animal and human populations. This chapter delivers a comprehensive overview of anthrax, encompassing its etiology, epidemiology, clinical manifestations, diagnosis, treatment, and prevention. The chapter focuses on the life cycle of *Bacillus anthracis*, highlighting its capacity to produce robust spores that enhance its environmental resilience. It also presents various modes of anthrax transmission, direct contact with diseased animals or their derivatives, inhalation of spores, and consumption of contaminated foodstuffs. This chapter also defines the potential misuse of anthrax as a bioweapon, underlining the necessity for inclusive surveillance and preparedness measures. Additionally, the text explores preventive strategies, encompassing vaccination programs for livestock and high-risk populations, as well as approaches for environmental decontamination and post-exposure prophylaxis.

Keywords

Anthrax · Welder's anthrax · Zoonotic anthrax · Woolsorters' disease · Milzbrand · Cumberland disease · Charbon · Malignant carbuncle · Malignant pustule · Splenic fever

R. Biswas (✉)
Department of Veterinary Public Health and Epidemiology, West Bengal University of Animal and Fishery Sciences, Kolkata, West Bengal, India

I. Samanta
Department of Veterinary Microbiology, West Bengal University of Animal and Fishery Sciences, Kolkata, West Bengal, India

R. Biswas et al. (eds.), *Concept of One Health and Zoonoses*,
https://doi.org/10.1007/978-981-95-6627-3_8

History

Anthrax showed up for the first time way back in 1491 B.C. evident in the early literatures of Mesopotamia and the *Book of Genesis*. The typical symptoms of anthrax were described in the Old Testament mentioning the time of the fifth and sixth Egyptian plague. The disease anthrax is mentioned in the initial written records of the Greeks, Romans, and Hindus. In 29 B.C., Roman poet, Virgil also mentioned about this disease in his third Georgic (poems on agriculture and animal husbandry). In 1613, the notorious "Black Bane" occurred in Europe resulting in more than 60,000 deaths in humans and cattle, and it became the first European pandemic. In 1850, the first microbiological breakthrough was made by Pierre Rayer who described "filiform bodies" from the blood of a sheep that may have died due to anthrax. In 1863, Casimir-Joseph Davaine proposed that the "corpuscles" were the etiological factor of anthrax, demonstrating its transmissibility to cattle, sheep, horses, mice, and guinea pigs via subcutaneous inoculation of contaminated blood. Tiegel and Klebs documented in 1864 that the anthrax pathogen's ability to cause infection was eliminated when blood contaminated with anthrax was filtered through a porous clay device (bacterial filter). When Robert Koch was posted in Wollstein, Germany, as district medical officer, he began to investigate a major health problem as 528 people and 56,000 farm animals died in 4 years and shown that the disease can be transmitted by inoculating uninfected animals with the blood from infected animals. He inoculated blood from a sheep that had died of suspected anthrax into a mouse, and after performing the autopsy, he found rod-shaped microscopic creatures in the blood, lymph nodes, and spleen. In 1876, Robert Koch decisively evidenced that anthrax was the first disease, caused by microbial (bacterial) agent. In another experiment, Robert Koch isolated anthrax organisms from sheep skin lesions and cultivated a pure culture by growing the bacilli on ox eye aqueous humor, subsequently injecting the bacilli into healthy sheep. He studied those cultures and described how the bacilli multiplied and formed round spores inside themselves when the environment was not good, especially without oxygen. This proved to be a self-protective mechanism. All these experiments played a pivotal role toward the development of "Koch's postulates" and "Germ theory" of the disease. In 1881, Louis Pasteur proved that buried anthrax-infected carcasses were the primary source of new infections in an area. He also confirmed Robert Koch's discovery that earthworms could transport anthrax spores from the carcasses to the soil's surface. In another investigation, he defined that chickens were immune because of its high body temperature but by lowering the temperature to 37 °C chickens became susceptible to anthrax. Louis Pasteur inoculated heated anthrax bacilli into 25 sheep, and all those sheep survived except one pregnant one, which died with another problem, but all the sheep died within 2 days after being challenged with virulent anthrax bacilli. Thus, Louis Pasteur discovered the first live bacterial vaccine against anthrax and confirmed the causative agent was an "organism" not any "toxin." Wool sorters' disease: Between 1846 and 1877, this ailment surfaced in Bradford, specifically among those employed as sorters of alpaca and mohair, suggesting a significant occupational hazard. In 1879, John Henry Bell's research provided conclusive evidence that woolsorters' disease was, in fact, caused by anthrax. In 1979–1980, Zimbabwe had the biggest anthrax outbreak ever, with 9400 cases on the skin and 182 people dying. Due to the civil war, the population consumed and managed anthrax-infected animals because of food shortages.

Etiological Agent and Ecology

The name anthrax comes from the Greek word "anthrakis," which translates to coal, a reference to the dark skin sores that develop in the disease's cutaneous form. Anthrax is an infection caused by the bacterium *Bacillus anthracis*, which is part of the Bacillaceae family. This bacterium is characterized as Gram-positive, spore-forming, and encapsulated, non-motile, rod-shaped, and its dimensions range with a width ranging from 1 to 1.2 μm and a length between 3 and 5 μm. It thrives in both oxygen-rich and low-oxygen environments. Though it is an aerobic bacterium but due to its ability to form spores, it can persist in soil for many years even in environments lacking oxygen. Spores remain unaffected under adverse settings and able to withstand extreme temperatures, high pressure, pH changes, chemical interactions, lack of nutrients, and even exposure to ultraviolet light. When spores are in their dormant phase, they transition into the active vegetative form of bacteria upon receiving signals to germinate. It can be grown readily on nutrient agar, though 5–7% sheep and horse blood agar are the medium of choice. *B. anthracis* colonies, when incubated overnight at 37 °C, display a grey to grey-white color and measure between 0.3 and 0.5 mm in diameter. These colonies are non-hemolytic, have a moist surface resembling ground glass, and are distinctly sticky when touched with an inoculating loop. The growth extends back toward the parent colony in a prominent curl, all oriented in the same direction, which is known as "medusa head" (a female figure of Greek mythology with living venomous snakes in place of hair) appearance. The methionine and thiamine are essential for growth of the bacteria in addition to multiple amino acids as the sources of nitrogen. The organisms hydrolyze casein medium containing adenine, thiamine, uracil, glucose, tryptophan, glycine, and cysteine. Sporulation requires the presence of oxygen and is most favorable at temperatures ranging from 25 to 30 °C, with a preferred growth pH of 7.4. Germination of spores needs fresh nutrients and aerobic conditions. The bacterial capsule consists of γ-linked poly-D-glutamic acid, which imparts a mucoid look to the colony.

The virulence of this organism depends on the formation of a capsule. The capsule is inherently non-toxic and does not provoke an immune reaction from the host. Consequently, the organism escapes the phagocytic action and afterward the disease is manifested by anthrax toxin. Biochemically, *B. anthracis* ferments sugars producing acids and hydrolyze starch.

The genome of *B. anthracis* is highly enriched with A + T content while only 35% of G + C content resulting in higher buoyant density of DNA and lower melting temperature. The genome is tripartite and consists of single circular chromosomal DNA with two virulence plasmids, pXO1 (182 kb) and pXO2 (96 kb) encoding toxins and capsule-producing genes, respectively. The *pag*, *lef*, and *cya* genes on the pXO1 plasmid are organized to code for protective antigen (PA), lethal factor (LF), and oedema factor (EF), respectively. When *B. anthracis* grows in conditions with increased CO2 concentrations, the expression of its toxin genes is amplified. The pXO1 plasmid is equipped with a trans-acting *atxA* gene, which oversees the regulation of genes found on both pXO1 (*pag*, *lef*, and *cya*) and pXO2. The pXO2 plasmid has the genes that encode the cap operon accountable for the production of a polyglutamate capsule, which authorizes the pathogen to escape the host immune response by shielding itself from phagocytosis. Two virulence plasmids, pXO1 (182 kb) and pXO2 (96 kb), are involved in the encoding of genes that lead to the production of toxins and a capsule, respectively.

Ames, Sterne, and Vollum are three well-characterized strains out of 89 strains of *B. anthracis*, known till date. Ames is highly virulent as it is having both plasmid pXO1 and pXO2. In 1981, a cow in Texas, USA, was found to be infected with anthrax, leading to the isolation of the Ames strain. This particular strain is prevalent in both the United States and the United Kingdom. The geographic origin of Sterne strain is Canada, and this strain is having toxigenic properties with no virulence. Sterne strain contains only pXO1 plasmid but has no capsule-producing plasmid (pXO2). Vollum is a low virulent strain isolated from an infected cow in Oxford and is commonly detected in Zimbabwe, United Kingdom, and Spain. Unlike Sterne, the Pasteur strain contains the pXO2 plasmid but lacks pXO1.

Host Range and Susceptibility

Anthrax is mainly a disease-affecting herbivores, with cattle and sheep being the most commonly impacted, while goats and horses are less frequently affected. Humans are somewhat in between these animals and those with greater resistance, such as pigs, dogs, and cats, though cases in pigs and elephants are seldom reported. Some occurrences of anthrax have been documented in dogs, carnivorous animals as well as wild animals associated with scavenging of anthrax infected carcasses. Amphibians, reptiles, and avian species are naturally resistant because of its high body temperature while some reports are there in ostriches, crows, canaries, and ducks in low grade.

Epidemiology

Anthrax is a zoonotic disease that primarily affects animals and can later be transmitted to humans, either directly or indirectly. The CDC has classified *B. anthracis* as a high-priority Category A agent because of its ability to spread easily, its high mortality rates, its potential to cause widespread outbreaks, and the necessity for specific preparedness measures. Although the disease has been identified for centuries, it has not been well-known methodically that how the healthy herbivores acquire the disease while they graze. The disease is primarily observed in hot and humid region around the world and has lately moved into cooler areas as a result of the movement of infected animals and the spread of contaminated animal products. The sporulated form of anthrax organism is generally shed by the infected or dead animals which act as the main source of the disease for other animals as well as soil contamination. *B. anthracis* is apparently non-invasive to the susceptible host's body but it is to be believed that a cut wound or lesion is required to initiate the infection. In connection with higher incidence rate and dry hot condition, animals are forced to feed on dry, spiky grass close to the ground, causing some oro-gastrointestinal damage. Meats, hides, hair, wool, and bones are capable of being transported over long distances to aid various industries, crafts, and to be used as ingredients in animal feed. Cross-contamination may have occurred during the shipping and handling of the products. Factors such as rainfall and temperature significantly impact the incidence of anthrax cases. Soil characteristics such as a pH level greater than 6.5, abundant organic material, adequate moisture, and temperatures exceeding 15.5 °C foster the transformation of *B. anthracis* spores into vegetative bacteria, which eventually results in an increased production of spores. *B. anthracis* spores can enter the human body through three primary routes: inhalation into the lungs, penetration through skin lesions, or ingestion via the gastrointestinal tract. Once inside, these spores can transform into their vegetative state. The ability of these spores to remain viable in soil for extended periods is facilitated by high soil pH and elevated calcium levels. As a result, the presence of alkaline pH, increased calcium concentration, organic matter in the soil, moisture levels, and the mentioned ambient temperature are pivotal in influencing the occurrence of a major outbreak in a particular geographic location. While grazing in dry, dusty areas where an animal previously succumbed to anthrax, animals breathe in the spores. The inhaled spores remain un-germinated in the lungs for even months. Deliberate release of anthrax spores can develop disease also. Biting flies, predominantly Tabanus, Stomoxys, and Hippobosca species are considered to be important mechanical vectors to transmit the disease from infected to healthy one. Anthrax may also be transmitted by the scavenger animals. Inadequate reporting and insufficient surveillance system of unexpected livestock death is a very common picture throughout the world. According to the global disease prevalence data, reported to World Organization for Animal Health (WOAH), the epidemics in livestock are more common in certain nations across Africa, Asia, the Americas, and southern Europe. In many other countries, the disease occurs sporadically. Several anthrax outbreaks have been recorded in animals from nearly most of the countries throughout the world. Ecological, social, political, economic, and cultural factors continue to influence the process of occurring anthrax worldwide.

Virulence Factors and Toxins

The toxins and the capsule are considered as the chief virulence factors of *B. anthracis*. The poly-D-glutamic acid capsule allows the bacteria to evade the host's immune defenses by obstructing the macrophages' ability to engulf and process the bacteria. The exotoxin complex is comprised of jointly acting proteins, namely, Protective Antigens (PA, 83 kDa), Lethal Factor (LF, 87 kDa), and Oedema Factor (EF, 89 kDa) are synthesized during the bacterial growth's log phase. In the formation of toxins, the protective antigen (PA) functions as the binding domain (B), whereas the lethal factor (LF) and edema factor (EF) serve as the active domains (A). These components independently combine to produce lethal toxin (PA + LF = LT) and edema toxin (PA + EF = ET), respectively. The toxins normally bind with the two receptors of anthrax toxin on host cells CMG2 (capillary morphogenesis gene 2) and TEM8 (tumor endothelial marker 8), aiming to penetrate the endosome. Once the 83 kDa form of PA (PA83) attaches to receptors on the surface of the host cell, it is cleaved by furin-like protease enzymes, resulting in the formation of a smaller, active version known as PA63. Through the process of endocytosis, the toxin-receptor complex is internalized and transported to the early endosome. Once PA is internalized from the cell surface, it experiences a change in structure due to the low pH within endosomes, resulting in the creation of a pore in the cellular membrane that functions as a translocation channel. This channel facilitates the movement of LF and EF into the cytosol, in which the host cell is disrupted by enzymatic actions. Once the LF and EF toxins have entered the late endosomal compartment, EF stays linked to the late endosome membranes, while LF is released into the cytoplasm. Lethal factor (LF) is a zinc-dependent metalloprotease that disrupts the MAP kinase or MEK family, leading to impaired cell division, differentiation, and stress response, which eventually results in apoptosis. Edema Factor (EF) is an adenylate cyclase that relies on calmodulin (CaM) to function. It triggers the excessive formation of cyclic-AMP (adenosine monophosphate) (cAMP), leading to cellular water loss and chloride ion depletion, which in turn results in extracellular edema. Additionally, it indirectly reduces the oxidative burst and the phagocytic function of neutrophils.

Pathogenesis

Upon entering a mammalian host, spores encounter a nutrient-rich environment that stimulates their germination into vegetative encapsulated bacteria, which then begin to produce toxins. This process is also influenced by specific germination factors unique to the host. At the initial site of entry, spores infect macrophages, transform into vegetative cells, and proliferate within the tissues, initiating anthrax toxin production within 3 h of germination. These toxins interfere with cell signaling pathways, as previously discussed, to suppress the host's innate immune responses during the early stages of infection and to trigger vascular collapse in later stages. The toxins effectively disable the innate immune response, allowing the bacteria to replicate and spread to regional lymph nodes, the bloodstream, and other tissues.

Episode after access of spores by inhalation: Germination begins either en route to or upon reaching the tracheobronchial lymph nodes. At this point, the vegetative cells are released from the phagocytes, allowing them to multiply. Then it spread through efferent lymph ducts to blood circulation and tissues. Primarily, flu-like symptoms like cold, mild respiratory discomfort, shortness of breath, nausea are observed, followed by severe respiratory collapse resulting in hemorrhagic mediastinitis and pulmonary edema.

Episode after access through cutaneous tissue: This type of infection is started with a small papule in the skin which looks like a site insect bite. Within 24–48 h, the small papule becomes large and turns into a painless ulcer characterized by necrotic center with raised edges. After 7–8 days, the lesion turns into a black "eschar" surrounded by oedema. Permanent scar mark is seen after curing the disease at the place of infection. Sometimes this form of infection can be systemic in nature when bacteremia occurs.

Episode after access by ingestion: Gastrointestinal anthrax can be divided into two types based on the lesions it causes: abdominal and oropharyngeal. Ileum and cecum part of the intestine are mostly affected resulting in blood vomiting, colic, diarrhea, loss of appetite, and finally development of massive ascites. Conversely, lesions from oropharyngeal anthrax appear in the mouth, leading to a sore throat, swallowing difficulties, and neck swelling caused by edema and cervical lymphadenopathy.

Clinical Manifestations in Animals

The period for anthrax to manifest, regardless of the infection pathway, spans from 36 to 72 h. The initial indication in a group of animals is the unexpected death of one or more members. When animals are under stress, they may exhibit a rise in body temperature and swelling in the area of the submandibular fossa. Skin lesions are typically not observed in cases of animal anthrax.

Herbivores: Among the herbivores, cattle are more susceptible group of animals within the class of mammalian group. Because of its grazing habit, cattle pull the grass from the soil with roots unlike sheep and goats which mostly browse on plants. After an incubation period ranging from 3 to 5 days, the disease can manifest in either a peracute or acute form, influenced by the concentration of bacterial cells involved. Peracute form starts with staggering walk and recumbency without showing any other sign. The diseased animal dies within a few minutes after convulsions or spasms. The acute form may be observed with high fever (>105 °F), restlessness, labored breathing, edematous swelling in the neck or chest region, and finally death within 12–14 h after appearance of these symptoms. Animals may appear to have dropped dead in their places with oozing of tarry blood from mouth, nose, and anus. Sometimes ulcerative lesions may be shown in the tongue and oral mucosal layer of the oral cavity.

Equine: Generally, this group of animals suffer from acute form of disease appearing the symptoms of high temperature (>105 °F), shortness of breath, bluish skin discoloration, abdominal pain, diarrhea with traces of blood, localized swelling with dead tissue at the center, and bloody discharge from natural openings.

Swine: Pigs show mild sub-acute to chronic form of disease, characterized by local swelling due to edema in the head and chest areas, fever accompanied by swollen lymph nodes, inflammation of the throat, and black necrotizing lesions on the skin and mucous membranes. Intestinal connection can result in anorexia, icterus, vomiting, diarrhea (which may be bloody) or constipation. Historically, pharyngeal anthrax is more common in pigs in earlier times due to feeding of waste contained with bone fragments or other sharp objects that cause lesions, but recently, outbreaks have predominantly been of the intestinal variety. Pigs that have recovered and show no symptoms might still have active lesions in their tonsils and cervical lymph nodes when they are slaughtered.

Carnivores: Carnivores, specifically wild scavengers, are comparatively resistant to infection with *B. anthracis* and chronic infection may occur. Typically, following an incubation period of 3–5 days, dogs, cats, and other wild carnivores usually bear a resemblance to the disease in pigs, with gastrointestinal and/or pharyngeal signs. Inflammation and swelling in the neck, accompanied by puffiness in the lips, chin, and tongue, can be detected. These animals succumb quickly after the first signs of illness, experiencing bleeding from their natural orifices and failing to recover.

Birds: Anthrax has been reported in poultry, pigeons, and eagles but occurrences are very rare. Several reports of anthrax have been recorded in ostriches. Acute septicemic form of disease is seen in birds with sudden death occurring soon after the clinical signs appear.

Clinical Manifestations in Humans

Human anthrax has been classified into two types, agricultural or non-industrial and industrial. Agricultural anthrax happens when veterinarians, butchers, slaughterhouse workers, and even ranchers, accidentally or during slaughter, come into contact with tissues from animals that are infected or deceased. Conversely, industrial anthrax affects those working in the processing of animal products like bones, hides, and wool that are contaminated. According to the CDC, 65% cases are industrial in nature. On the basis of route of entry, the disease is found in four forms. Any of these forms can develop life-menacing anthrax septicemia or anthrax meningitis.

Cutaneous anthrax: 95% of anthrax cases are known to be cutaneous form of infection which is mainly acquired through skin following an insect bite or a small cut lesion. Lesions are usually occurred on exposed regions like hands, wrists, neck, and face and often commence with itchiness. The time frame for the onset of symptoms in cutaneous anthrax can vary from 1 to 20 days. After getting contact with the spores, a small papule (popular stage) appears in 1–2 days. A ring of vesicle (vesicular stage) or blister develops around the papule on third to fourth day of infection. No pus is formed and the lesion is not painful if there is no secondary bacterial infection although in the nearby lymph nodes, one might notice tenderness and swelling. The blister becomes hemorrhagic and by 5–7 days it starts to dry to form a depressed black scab (eschar stage) which is known as "malignant pastule." The eschar starts to resolve by tenth day. It takes nearly 40–42 days to resolve the eschar and is not accelerated by the treatment. However, the organisms can sometimes spread and cause serious illnesses including septicemia and meningitis.

Gastrointestinal anthrax: This represents the second-most frequently occurring variety that generally develops after consuming contaminated undercooked meat and animal products. After penetrating the intestinal epithelium, the spores are germinated and develop ulcerative lesions resulting in hemorrhages, obstruction or perforation. Ileum and colon parts are commonly affected whereas oropharyngeal anthrax is distinguished by clinical signs of local orientation. A slight fever, a general feeling of discomfort, nausea, loss of appetite, and diarrhea might be observed. Rapid onset may present with blood vomiting, severe abdominal cramps, increased fluid accumulation in the abdomen, and bloody diarrhea, potentially resulting in shock, coma, and death.

Inhalation anthrax: It is the deadliest but rare form. Anthrax occurs when spores, often linked to industrial environments or biological warfare, are inhaled. Initially, the symptoms are not specific but it can develop slowly. During the early stage, individuals may experience symptoms like malaise, fever, tiredness, weakness, and a dry cough that is sometimes accompanied by mild chest pain. As time progresses, the symptoms become more complex, but the prodromal stage concludes with the onset of severe shortness of breath, rapid heartbeat, bluish skin, life-threatening blood infection, and shock within a day or two. At the acute phase, the fatality rate has been observed to be as high as 90–100% among patients, irrespective of the treatment administered.

Injectional anthrax: In 2009/2010, Injectional anthrax, a newly identified form of the disease, surfaced due to the injection of heroin laced with anthrax spores. This type of anthrax has been reported from Germany, Denmark, France, United Kingdom, and Scotland.

Anthrax meningitis: Anthrax meningitis can be developed from the other four forms of disease. A typical form of meningoencephalitis is seen after a prodromal period of 1–6 days when the bacteria or exotoxin can spread to the brain, spinal cord, and meninges through the bloodstream or lymphatic system. Patients rapidly develop comatose conditions and die within 24 h. Blood is often found in the cerebrospinal fluid. This condition is clinically characterized by the presence of symptoms including fever, headache, neck stiffness, agitation, and delirium.

Anthrax sepsis: The condition arises when *B. anthracis* spreads through the lymphatic and blood systems from an initial infection in the skin, gastrointestinal tract, or lungs. Symptoms include a high fever, severe toxemia, and shock, which can ultimately lead to death.

Welder's anthrax: It is a newly identified occupational anthrax, characterized by severe pneumonia, reported by CDC during the time of 1994–2020. In total, there have been seven reported cases, with six of them being welders and one being a metal worker. Four of the patients worked in Texas and three in Louisiana. Researchers have suggested that metal fumes, particularly those containing iron, might provide a conducive environment for bacterial proliferation, potentially increasing the bacteria's ability to attach to lung tissues or disrupting the lung's immune defenses through oxidative stress.

Anthrax as Bioweapon

Bioweapon has been considered as a biological agent which is relatively inexpensive to produce, readily accessible, easy to store, and induces widespread panic and fear compared to the actual physical harm. The CDC has designated *Bacillus anthracis* as a category A organism. Anthrax spores were utilized as a biological weapon during both World War I and World War II. In 2001, *B. anthracis* spores were mailed to various prominent individuals in the United States, resulting in 22 people being exposed when the spores were enclosed in envelopes. Post mortem lesions in animals: The common post mortem lesions are noticed in animals; most of the cases rigor mortis do not develop, oozing of dark tarry color blood from natural orifices, ulcerative hemorrhage in mucus and serous membrane, enlargement of lymph nodes and spleen, degeneration of liver and kidneys, bloating and fast decomposition of carcass.

Post Mortem Lesions in Humans

Hemorrhagic thoracic lymphadenitis and hemorrhagic mediastinitis are commonly observed.

Regarding Biosafety Level

For diagnostic tests, BSL-2 facility is appropriate but for production of organisms, organism activities related with aerosol BSL-3 should be used. For tasks such as bacterial counts and sterility tests involving liquid cultures, it is crucial to carry out these activities in biological safety cabinets within laboratories that meet, as closely as possible, the standards of basic BSL3, or at a minimum, BSL2 laboratories (Tables 8.1 and 8.2).

Table 8.1 Samples collection from human

Type of disease	Type of sample	Purpose
Cutaneous anthrax	Vesicular fluid	McFadyean
Inhalation anthrax	Blood	McFadyean
	CSF	McFadyean
	Nasal swab	Culture on blood agar
Gastrointestinal anthrax	Blood	McFadyean
	Ascitic fluid	McFadyean
	Peritoneal fluid	McFadyean
Anthrax meningitis	CSF	McFadyean
	Blood	McFadyean

Table 8.2 Collection of samples from animals suspected of having died from anthrax

Condition of carcass	Type of sample	Purpose
Fresh carcass	Blood drawn from a vein, or a combination of blood and fluid from a body cavity if scavengers have accessed it, or a piece clipped from the ear tip	McFadyean and culture
Putrefied carcass	A sample or swab of tissue rich in blood vessels, such as from the nostril, eye socket, or any other bloody material. Soil found beneath the head and tail	Culture on blood agar
Very old carcass	Swab of nostril and eye sockets. Soil from where body fluids are supposed to have fallen	Culture on blood agar

Preparedness for Collection of Samples

- Before collecting the samples, wear the apron and double covered gloves.
- Carry disinfectant with a concentration of 10,000 ppm hypochlorite, along with disposal bags and hand-washing supplies. Ensure any existing cuts, scrapes, or wounds are covered before donning PPE.
- After gathering samples, clean or wipe your gloved hands with a 10% hypochlorite solution and remove the outer gloves.
- Place used PPE in disposal bags, making sure to separate items that can be autoclaved from those that cannot.
- The inner gloves should be the last to be discarded. Dispose of sharps in a designated sharps container.

Transportation of Collected Samples

- Once the samples have been collected in a sterile container without contamination, the containers should be cleaned with a 10,000-ppm hypochlorite solution.
- After removing the outer gloves, they should be placed in the secondary double-bagged container.
- This container must then be sealed and stored in a high-quality cool box with a lid.
- The secondary and the outer container should be labeled with hazard indication (Table 8.3).

Prevention and Control Strategies

Control measures are intended to break the cycle of infection.

- Rapid identification and treatment of infected animals should be started.
- As anthrax is a notifiable disease, local veterinary personals should report to the higher authority immediately.
- Effective control of anthrax can be achieved through sufficient disease monitoring, thorough outbreak investigation, and the utilization of historical data.

Table 8.3 Diagnosis

Microscopy	Fresh specimens from an untreated animal, McFadyean (polychrome methylene blue) reaction, revealed a large number of the capsulated bacilli with dark blue rods with square ends are encased in a pink capsule. The capsule of anthrax bacilli, which is cultivated on nutrient agar or broth, is not stained by Gram's and Giemsa stains though it can be used as a differential stain.
Culture	After incubating overnight at 35–37 °C on nutrient agar or 5–7% sheep or horse blood agar, the colonies develop to a size of 2–3 mm in diameter. They are characterized by their irregular shape, raised profile, dull and opaque appearance, and a greyish-white color, resembling a "medusa head." for the isolation of B. anthracis from samples that are contaminated or suspected, blood agar media with polymyxin-B, lysozyme, EDTA, and thallous acetate is employed.
Animal inoculation	When 0.1 mL of a 1 in 100 dilution from a 24-h broth culture is injected intraperitoneally into mice or guinea pigs, the animal typically succumbs within 48–72 h. Using McFadyean's methods, smears taken from the heart blood and spleen show the presence of characteristic anthrax bacilli.
Ascoli's test	This test is designed to detect the presence of a thermostable anthrax antigen in tissue extracts. Developed by Ascoli in 1911, the thermo precipitation test involves a ring of precipitation forming when the boiled tissue extract in a test tube comes into contact with anthrax antiserum. The process involves chopping the tissue in saline, boiling it for 5 min, straining it, and then placing it over anti-anthrax serum in a narrow tube. When the tissue contains the anthrax antigen, a precipitation ring will form at the boundary between the two liquids within 5 min at room temperature.
ELISA	At present, this method is recognized as the leading serological technique using microtitre plates that are coated with the protective antigen (PA) and lethal factor (LF) of the anthrax toxin. The toxin antigens are genuinely unique to B. anthracis. The presence of IgG against PA influences both the host's immune response and the reaction to the anthrax vaccine.
PCR	PCR-based molecular confirmation is gaining popularity as an independent method for various types of specimens and is now widely accessible worldwide through commercial kits.
ENIA	The European Nanoparticle Immuno Assay (ENIA) is an extremely sensitive immunoassay developed with European nanoparticles to detect PA in serum. It has been reported to be 100 times more sensitive than ELISA.

- Prophylaxis, vaccination, and quarantine practice must be followed.
- It is necessary to limit access to areas believed to be sources, such as feed or pastures.
- Sustainable vaccination of animals and proper disposal of infected carcass can control the spreading of disease effectively.
- To prevent birds and scavengers from spreading spores and bacilli, the carcass should be enclosed with a fence and covered.
- Disinfection of affected premises and materials must be done.

Incineration is the most effective method to ensure the destruction of the bacteria as well as spores. Although the deep burial (6 ft deep) method is not reliable but may be an alternative along with the disinfectants.

Recognition of anthrax spots by analyzing the ecology and epidemiology of the disease can prevent the natural occurrence of the disease.

Essential steps for people include rapidly determining the source of an animal outbreak and identifying individuals who have been exposed to it.

Veterinary Vaccines: In 1937, Sterne developed a live spore vaccine, obtained from a rough strain (34F2) of Bacillus anthracis, suspended in glycerol with saponin as adjuvant. In Italy, Pasteur strain vaccine is used for goats and horses. In Italy, an attenuated Bacillus anthracis strain, named "Carbosap," is used for vaccination of ovine and bovine. In endemic areas, annual vaccination is recommended before the season when generally outbreaks occur. Microencapsulated Bacillus anthracis Sterne strain 34F2 spores are used for formulation of oral vaccine which can be used on wildlife populations.

Vaccine in humans: In 2008, the guidelines for administering human anthrax vaccines were updated to change the method from subcutaneous to intramuscular. The revised schedule involves five doses given at 0 and 4 weeks, followed by additional doses at 6, 12, and 18 months, with yearly boosters thereafter. When medically warranted, such as for patients with coagulation issues or those on anticoagulant medications, the subcutaneous route for administering AVA (Anthrax Vaccine Adsorbed) remains a feasible alternative. This information sheet includes both methods of administration.

Anthrax Vaccine Adsorbed: The FDA has approved only one human anthrax vaccine, which is made from culture filtrates of a toxigenic but non-virulent, non-encapsulated mutant strain called V770-NP1-R. This AVA vaccine incorporates aluminum hydroxide, which serves to adsorb the Protective Antigen (PA) and acts as an adjuvant.

Live vaccines: In May 1881, Louis Pasteur was the first to employ live-attenuated bacilli. Both Russia and China utilize live attenuated strains in their vaccines for humans. In Russia, live spores from the attenuated strain STI-1 are used, while in China, the attenuated strain A16R is employed.

Treatment for Animals

Penicillin is a commonly used drug for *B. anthracis* infection in animals although streptomycin gives synergistic effect along with penicillin. Some countries do not allow to treat the infected animals.

Treatment for Humans

Penicillin G continues to be the primary treatment option, with doxycycline and ciprofloxacin now acknowledged as the leading alternatives for managing naturally occurring anthrax. The World Health Organization suggests the use of intramuscular procaine penicillin, oral amoxicillin, or penicillin V to manage mild and uncomplicated cases of cutaneous anthrax. For cases of cutaneous anthrax accompanied by significant swelling, intravenous penicillin G is advised. Antibiotics are only effective against the vegetative form and do not destroy the spore form. That is why, in case of inhalation anthrax, 60 days of antibiotics therapy are recommended as there might be a chance of remaining the spores in dormant form.

References

Burke, R. L., Kronmann, K. C., Daniels, C. C., et al. (2012). A review of zoonotic disease surveillance supported by the armed forces health surveillance Center. Zoonoses and Public Health, 59(3), 164–175. https://doi.org/10.1111/j.1863-2378.2011.01440.x

de Perio, M. A., Hendricks, K. A., Dowell, C. H., et al. (2022). Welder's anthrax: A review of an occupational disease. Pathogens, 11(4), 402. https://doi.org/10.3390/pathogens11040402. Published 2022 Mar 26.

Doganay, M., & Demiraslan, H. (2015). Human anthrax as a re-emerging disease. Recent Patents on Anti-Infective Drug Discovery, 10(1), 10–29. https://doi.org/10.2174/1574891x10666150408162354

Goel, A. K. (2015). Anthrax: A disease of biowarfare and public health importance. World Journal of Clinical Cases, 3(1), 20–33. https://doi.org/10.12998/wjcc.v3.i1.20

Islam, M. S., Hossain, M. J., Mikolon, A., et al. (2013). Risk practices for animal and human anthrax in Bangladesh: An exploratory study. Infection Ecology & Epidemiology, 3, 10.3402/iee.v3i0.21356. https://doi.org/10.3402/iee.v3i0.21356. Published 2013 Nov 27.

Tang, S., Moayeri, M., Chen, Z., et al. (2009). Detection of anthrax toxin by an ultrasensitive immunoassay using europium nanoparticles. Clinical and Vaccine Immunology, 16(3), 408–413. https://doi.org/10.1128/CVI.00412-08

WHO. (2008). Anthrax in humans and animals (4th ed.). World Health Organization.

Zasada, A. A. (2018). Injectional anthrax in human: A new face of the old disease. Advances in Clinical and Experimental Medicine, 27(4), 553–558. https://doi.org/10.17219/acem/68380

Part II

Bacterial Zoonoses

Brucellosis

9

Maninder Singh

Abstract

Brucellosis, a widespread zoonotic disease caused by bacteria of the genus *Brucella*, remains a significant public health concern globally. This chapter provides a concise overview of brucellosis, encompassing its etiology, transmission, clinical manifestations, diagnosis, and control. The disease primarily affects livestock animals, such as cattle, sheep, goats, and pigs, but can be transmitted to humans through contact with infected animals or consumption of contaminated animal products, particularly unpasteurized milk and dairy products. Human brucellosis presents with a wide range of nonspecific symptoms, including fever, fatigue, joint pain, and sweating, making diagnosis challenging. The chapter further highlights the importance of laboratory testing for accurate diagnosis and the role of antibiotics in treatment. Control and prevention strategies, including animal vaccination and surveillance programs, are vital for mitigating the impact of brucellosis on both human and animal health.

Keywords

Brucellosis · Bang's disease · Undulant fever · In humans—Undulant fever, Malta fever, Mediterranean fever, Rock fever of Gibraltar fever · In animals: Bang's disease, contagious abortion, epizootic abortion, Ram's epididymitis (*B. ovis* infection in sheep)

M. Singh (✉)
Department of Veterinary Public Health and Epidemiology, College of Veterinary Science, Guru Angad Dev Veterinary and Animal Sciences University, Bathinda, India

R. Biswas et al. (eds.), *Concept of One Health and Zoonoses*,
https://doi.org/10.1007/978-981-95-6627-3_9

Brucellosis is a bacterial disease of livestock caused by the Gram-negative bacteria of genus *Brucella* and is known for huge economic losses to the livestock industry. In Latin America, the annual losses due to bovine brucellosis are estimated to be $600 million. In India, average losses are $6.8 per cattle, $18.2 per buffalo, $0.7 per sheep, $0.5 per goat, and $0.6 per pig. The disease is a classical zoonosis and has significant public health impact. Annually 500,000 human cases of brucellosis are reported worldwide.

History

- Although the brucellosis is known to exist since Roman era, the cause was only identified in late nineteenth century when David Bruce was assigned the work to identify the cause of mysterious disease leading to death in British soldiers posted at Malta island. In 1887, David Bruce successfully isolated a bacterium from the spleen of dead soldiers and named it *Micrococcus melitensis*. However, David Bruce was not able to identify the source of infection to soldiers.
- In 1905, it was Themistocles Zammit who isolated the similar bacterium from raw milk of goats and identified that consumption of unpasteurized milk from Maltese goats to be the source of infection to soldiers. Based on his findings, Goat's milk was banned, and the outbreak of mysterious disease ended.
- Alongside, in 1897, Bernhard Bang isolated the similar bacterium from cattle abortion and named the bacterium as Bang's bacillus.
- In 1918, American microbiologist Alice Evans identified the relationship between Malta fever and Bang's disease and suggested to pasteurize the milk.

Etiology

Brucella is a Gram-negative coccobacilli bacterium belonging to family *Brucellaceae*, class alphaproteobacterial, and phylum proteobacteria. The bacterium is facultative intracellular, non-motile, non-capsulated, non-spore forming, devoid of flagella, devoid of plasmids, and modified Ziehl Neelsen positive. On the basis of host preference, *Brucella* genus has six classical species viz., *Brucella abortus* (cattle and buffalo), *B. melitensis* (sheep and goat), *B. suis* (pigs and boars), *B. canis* (dogs), *B. ovis* (sheep and goat), and *B. neotomae* (desert wood rat). Apart from these classical six species, new species have been isolated viz., *B. ceti* (dolphins), *B. pinnipedialis* (seals), *B. microti* (wild voles), and *B. inopinata* (human breast implant).

Being a Gram-negative bacterium, *Brucella* has lipopolysaccharide (LPS) in its cell wall. On the basis of nature of O-polysaccharide chain of LPS, species are categorized as smooth species and rough species. *B. abortus*, *B. melitensis,* and *B. suis* are further divided into biovars. *B. abortus* strains are microaerophilic in nature and require 5–10% CO_2 for primary isolation from field samples (Boral et al., 2009).

Table 9.1 List of *Brucella* species and their different attributes

S. no.	Brucella species	Natural host	Colony type	Zoonoses
1.	B. abortus (bv. 1–6, 7, 9)	Cattle	Smooth	++
2.	B. melitensis (bv. 1–3)	Sheep and goat	Smooth	+++
3.	B suis biovar:			
	1–3	Pig	Smooth	++
	2	Wild boar, hare	Smooth	+
	4	Reindeer, caribou	Smooth	++
	5	Rodent	Smooth	–
4.	B. ovis	Sheep	Rough	–
5.	B. neotomae	Desert wood rat	Smooth	+
6.	B. canis	Dog	Rough	+
7.	B. ceti	Dolphins	Smooth	+
8.	B. pinnipedialis	Seals	Smooth	+
9.	B. microti	Voles	Smooth	?
10.	B. inopinata	Human	Smooth	++

Brucella species, their natural hosts, type of colony, etc. are depicted in the following Table 9.1.

World Health Organization has categorized *Brucella* in risk group 3 of infectious agents having high individual severity but limited community transmission, and it is advised to use biosafety level 3 for the handling and manipulation of clinical samples for *Brucella* isolation. The agent has also been used a bioterrorism agent in the history by American army (*B. suis*) and is classified as category B agent (Seleem et al., 2010).

Epidemiology

Brucellosis is one of the most widely prevalent zoonosis globally. The disease is reported throughout the globe except the countries where bovine brucellosis in domestic animals has been eradicated. Nevertheless, the disease is emerging and re-emerging in new areas due to spill over from wild ruminants to domestic ruminants or uncommon host species, e.g., the infection of *Brucella abortus* in sheep and goats leading to secretion in mammary glands is reported. The wild animals as reservoirs of *B. abortus* reported are bison and elk in Greater Yellowstone area in the United States. Bovine brucellosis is enzootic in majority of countries of the world especially developing countries. Brucellosis is endemic in countries of Western Asia, India, Middle East, Africa, Southern Europe, and Southern Americas. Countries which are free of bovine brucellosis in domestic cattle are from Northern America, Oceania, and Europe regions and include Australia, Canada, Cyprus, Denmark, Finland, Netherland, Norway, New Zealand, Sweden, and the United Kingdom (Khurana et al., 2021). The infection of *Brucella abortus* in cattle and yaks is reported. Dogs can be infected by *B. abortus*, *B. melitensis,* and *B. suis*.

The epidemiology of human brucellosis depends upon the disease in domestic ruminants of the region. Therefore, the incidence of human brucellosis is higher in the animal brucellosis endemic areas. Among *Brucella* species, the most pathogenic to humans is *Brucella melitensis* followed by *B. abortus* and *B. suis*. Human brucellosis by *B. canis* is reported. *Brucella* species isolated from marine mammals also exhibit zoonotic potential.

Transmission

In animals, *Brucella* has the predilection for reproductive tract and mammary glands. Consequently, the bacterium leads to late-term abortions in animals. The abortion secretions, excretions, placenta, and aborted fetus carry high load of bacteria and contaminate the environment on the farm leading to transmission of infection to other healthy animals through contaminated feed and water. Aerosolization of bacteria under contaminated environment may happen. Calves from infected cows may have latent infections. Bulls infected with brucellosis shed bacteria in their semen and may transmit the infection through artificial insemination. Regarding transmission by natural service/venereal transmission, there are variations among species, and venereal transmission is reported in dogs and pig. Venereal transmission through natural service in cattle is rare while transmission through artificial insemination is well known when the semen is deposited in uterus of the animal. In case of cattle, another important route of transmission is by consumption of raw milk of infected animals by sucking calves.

In humans: the infection is generally occupational associated. Veterinarians and farm house workers on contact with secretions and excretions of abortion materials may get infected; slaughter house workers may also get infected through aerosols or direct contact with viscera. In general public, the most important route of transmission is consumption of unpasteurized milk and milk products such as butter, cheese, and whey; among cheese, consumption of soft cheese is the most significant. The consumption of improperly cooked meat and organs such as liver and spleen is another important source of infection. In veterinarians, accidental inoculation of live vaccines such as *B. abortus* S19 and *B. melitensis* Rev 1 may lead to disease. Among laboratory workers, aerosols having *Brucella* organism is most significant factor. Although the infectious dose of *Brucella* varies with the route of inoculation and immune system of the host, the infection requires 10–100 organisms to infect by respiratory route. Venereal transmission in humans is reported through intrauterine infection and breast feeding. Sexual transmission is also reported.

Pathogenesis

On exposure through oral or nasal route, the bacterium passes through digestive or respiratory tract mucous membrane and is internalized by the phagocytes such as macrophages, dendritic cell, and neutrophils transporting the bacterium to the local lymph nodes. Through the lymphatic system, the pathogen spreads across the different organs of the body. Organs of reticuloendothelial system (lungs, liver, spleen, and bone marrow) are the most preferred sites of the bacterium. Apart from reticuloendothelial sites, organism has predilection for placental trophoblasts in gravid uterus and mammary glands. In male animals, genital organs are infected with continuous or intermittent shedding of bacteria in the semen. In humans, brucellosis is a systemic infection and any organ can be infected with a predilection to joints and liver.

Although *Brucella* is devoid of classical virulence factors, such as plasmid, flagella, and exotoxins, it has various regulation systems which ensure its replication within the phagocytic cells. After internalization in the target cells, the bacterium transits via a phagosome; the phagosome should fuse with lysosome to form phagolysosme where the pathogens are destroyed. However, *Brucella* prevents the fusion of phagosome with lysosome by using various virulence factors, such as periplasmic cyclic β-1,2-glucans and Type IV secretion system (named *VirB*), and reaches its final destination—Endoplasmic Reticulum (ER). ER is the replicating niche of the bacterium within the cell which is probably because ER is responsible for production and transportation of cellular molecules. Intracellular replicating pathogens often parasitize ER to gain the access to store of cellular molecules. The bacterium has lipopolysaccharide which favors the early replication of bacterium after internalization and prevents the generation of innate and adaptive immunity. *Brucella* has modified its pathogen-associated molecular patterns (PAMP) to defy the innate immune response (González-Espinoza et al., 2021).

Brucella has tropism for trophoblast cells in gravid uterus owing to presence of erythritol sugar and steroid hormones (progesterone). Erythritol is preferentially used as carbon and energy source by *Brucella.* However, *B. abortus* S19 vaccine strain has no preference to erythritol and still causes abortion in pregnant animals indicating the involvement of other factors. As *Brucella* multiplies in the placenta extensively, it leads to placentitis and abortion. Young animals may get infected but show clinical manifestations only on pregnancy.

Clinical Signs

In animals, the clinical manifestations are mainly reproductive failures. Pregnant animals abort in the last trimester due to placentitis, and bacterium is shed in a large number in placenta. Disease is self-limiting in cattle, and the abortion occurs once or twice in infected animals, although shedding of pathogen in the subsequent parturitions continue leading to carrier state in animal. Other reproductive problems include still birth, retention of placenta, and decrease in milk yield. Abortions are reported in all domestic species including cattle, buffalo, sheep, goat, bitches, and pigs. In male animals, orchitis and epididymitis are the manifestations which may resolve with time but with continuous or intermittent shedding of the bacteria in the semen. In many cases, joints are swollen, especially the knee joint. In dogs, the infection is associated with pronounce bacteremia, and pathogen is shed in all body secretions including urine and saliva. In horses, the infection has been associated with skin abscesses leading to poll evil and fistulous withers.

Humans are the accidental hosts of *Brucella,* and infection in humans may change from acute to chronic. Acute cases are associated with the invasion of the bacteria with subsequent dissemination in the body. The clinical signs are undulant fever with fever rising in the evening, malaise, fatigue, joint pain, back pain, and irritability. As the symptoms are not conclusive for a disease, the disease is often not diagnosed and is associated with pyrexia of unknown origin (PUO). The symptoms worsen as the bacteria involve other organs especially liver and spleen leading to hepatomegaly and splenomegaly. Other organs involved are brain and heart. The mortality rate ranges from 2% to 5% due to endocarditis. The condition is treatable in humans. Abortions in females are unusual but may occur due to placentitis in first or second trimester; the abortions are associated with bacteremia (erythritol sugar is absent in human placenta) (Mantur et al., 2007).

Diagnosis

There are various approaches for diagnosis of brucellosis but for clinical purposes, the majority of cases are diagnosed by serological tests. The gold standard for diagnosis is isolation of bacterium.

Diagnosis in animals: Diagnosis in animals, especially livestock is required to achieve two objectives: (a) surveillance of disease in an area and (b) clinical diagnosis of disease. To achieve surveillance of disease in an area, milk ring test is used to detect brucellosis-infected herds followed by individual animal screening by Rose Bengal plate test and ELISA. The clinical diagnosis of disease is generally required in abortion cases involving isolation of bacterium. Both approaches in brief are detailed below.

Surveillance of brucellosis is done to determine the absence/presence of disease in an area, and if present, prevalence of disease which will indicate the relative priority of disease in the study area. It involves, first, application of milk ring test (MRT; also called Abortus Bang ring test) on the pooled milk samples collected from dairy herds. The test is based on the interaction of anti-*Brucella* antibodies (these antibodies are of lipophilic nature and thus bind with the fat globules present in milk) with colored antigen leading to colored ring formation at top in the positive herds. The test is mainly used in cattle and buffalo and is not recommended for sheep and goats due to their smaller fat globules size. The herds positive in MRT are further subjected to individual animal testing by Rose Bengal plate test (RBPT). RBPT is an agglutination test based on the interaction between *Brucella abortus* S99 (a smooths strain) and anti-*Brucella* antibodies present in serum. As the antigen is a smooth strain, the antigen is used to detect infection of *B. melitensis* and *B. suis* in sheep/goats and pigs, respectively, although the test procedure for these species has not been standardized. The test is not useful to detect infection of rough strains such as *B. ovis* and *B. canis*. For these species, test based on rough LPS is employed.

Furthermore, RBPT is a very sensitive test in the vaccinated animals and may show positive results in the adult vaccinated animals. Animals positive in RBPT are confirmed by complement fixation test. Fluorescent Polarization assay is also an OIE approved test to detect disease in animals; indirect ELISAs may be used to identify herds or animals positive for brucellosis. As each of these serological tests have certain limitations such as cross-reaction with other bacteria, it is advised to apply various tests to confirm the diagnosis on the basis of serological tests.

The isolation of bacterium, abortion materials, vaginal discharge, milk, urine, and supramammary lymph node (collected during post mortem) are used. The media used is trypticase soy agar, Columbia agar, blood agar, serum dextrose agar, glycerol dextrose agar, and potato dextrose agar. The antibiotic supplements such as Farrell's medium are added in the basal media. For primary isolation of *B. abortus*, incubation in microaerophilc environment (5–10%CO_2) is required (WOAH, 2022). *Brucella* is stained best by modified Ziehl Neelsen. Biochemical tests (CO_2 requirement, H_2S production, growth in presence of basic fuchsin, and thionin dyes and agglutination with A and M monospecific antisera) are used to detect *Brucella* species and the biovar of *B. abortus*, *B. melitensis,* and *B. suis*.

Note: Serum agglutination test (SAT) is another widely used test for serological diagnosis of brucellosis. However, World Organization of Animal Health has stopped recommending this test owing to high non-specific positive reactions. Nevertheless, the test is still used in various countries due to its ability to enumerate the titre of the antibodies in the serum. The test is based on agglutination between antigen and antibodies present in the diluted serum sample.

Diagnosis in humans: Diagnosis in humans is widely done on the basis of serological diagnosis by RBPT and SAT. SAT is further used to monitor the efficacy of the treatment as the effective treatment reduces the antibody titre spontaneously. Furthermore, extension of SAT, called 2-Mercaptoetahnol test, is used to monitor the titre of IgM and IgG antibodies (2-Mercaptoetahnol cuts the disulphide bonds of IgM reducing their agglutination power). Active brucellosis infection is associated with high IgG titre. Blood samples from the active infections are used for isolation by Castaneda method.

Nucleic acid-based assays: Although *Brucella* is a very homogenous genus with more than 90% similarity among *Brucella* species, the species and biovar-level identification has been achieved by nucleic-acid-based assays. Genus-level identification methods have been developed on bcsp gene or 16S rRNA gene. First species-level PCR used for identification was AMOS-PCR developed by Bricker and Halling (1991) identifying *B. abortus* (biovar 1,2, and 4), *B. melitensis*, *B. ovis,* and *B. suis*. Further development led to Bruce-ladder PCR which was able to identify all biovars of *B. abortus*, *B. abortus S19*, *B. pinnipedialis,* and *B. ceti*. Further modification to detect *B. microti* and differentiate *B. suis* and *B. canis* have been developed (Brcuel-ladder v2.0). Real-time PCR methods have been employed in many studies. For epidemiological investigations and subtyping Multilocus Variable Number tandem Repeat typing (MLVA) and Multilocus Sequence typing MLST techniques are being used.

Control and Prevention

At farm level in animals: In animals, brucellosis is not treatable, therefore control and prevention is based on strong biosecurity measures. As the disease may enter in the herd through a purchased new asymptomatic carrier animal, all purchased animals should be tested for brucellosis. To prevent transmission in an already infected herd, the infected animal should be separated from other animals during the calving period; it is advised to use calving pen in infected farms.

At large population level in animals: in the endemic areas, the only way to control brucellosis in large populations is vaccination. Vaccination reduces the secretion of pathogen during abortion and parturition, thus limiting the chances of transmission in the herd. For cattle, *B. abortus* S19 vaccine was developed and licensed in 1941 which has successfully controlled the disease in many countries. The vaccine has live attenuated smooth strain of B. abortus S19. The vaccine is used in 4–8 months female calf (calfhood vaccination) as the usage in adults is associated with persistent antibodies leading to false positive in serological tests. Also the vaccine is not used in male calves and bulls because of localization in testicles leading to shedding in the semen. B. abortus S19 is pathogenic to humans and accidental inoculation in veterinarians invariably leads to disease. The efficacy of vaccine is reported to be 70%. However, disease in vaccinated individuals living in high exposure environment may occur indicating the requirements of good hygienic measures in vaccinated farms.

As the vaccination with *B. abortus* S19 is associated with persistent serological reactions, rough strain-based vaccine *B. abortus* RB51 (live attenuated) has been approved for use in the United States and other countries. The vaccine is not used in adult cattle due to its abortifacient potential. The other disadvantage is that owing to its rough LPS nature the infection of RB51 in humans is not detectable by routine serological tests. Moreover, RB51 is rifampicin resistant, a drug of choice for treatment of brucellosis in humans. In sheep and goats, *B. melitensis* Rev1 is the prescribed vaccine used in female lambs and kids.

To circumvent the disadvantages associated with the above vaccines, research on vaccines to identify safe and effective candidates are continuing. Recombinant proteins like Omp16, Adk, SecB, and subunit vaccines with proteins BP26, Omp25, L7/L12 have been evaluated to be safer but with lower protection titer.

In humans: In humans, control and prevention is mainly oriented around reducing the exposure to the bacteria during high-risk events such as abortion. Veterinarians and farmers should use protective clothing during handling of such cases. Consumption of raw/unpasteurized milk should be avoided. The disease is treatable in humans with rifampicin (600/900 mg per day) and doxycycline (200 mg per day) as drug of choices. Relapses are reported. Live vaccines B. abortus 19-BA and B. abortus 104 M are being used in former Soviet Union and China, respectively.

References

Boral, R., Singh, M., & Singh, D. K. (2009). Status and strategies for control of brucellosis—A review. *Indian Journal of Animal Sciences, 79*(12), 1191–1199.

González-Espinoza, G., Arce-Gorvel, V., Mémet, S., & Gorvel, J. P. (2021). *Brucella*: Reservoirs and niches in animals and humans. *Pathogens (Basel, Switzerland), 10*(2), 186. https://doi.org/10.3390/pathogens10020186

Khurana, S. K., Sehrawat, A., Tiwari, R., Prasad, M., Gulati, B., Shabbir, M. Z., Chhabra, R., Karthik, K., Patel, S. K., Pathak, M., Iqbal Yatoo, M., Gupta, V. K., Dhama, K., Sah, R., & Chaicumpa, W. (2021). Bovine brucellosis – A comprehensive review. *The Veterinary Quarterly, 41*(1), 61–88. https://doi.org/10.1080/01652176.2020.1868616

Mantur, B. G., Amarnath, S. K., & Shinde, R. S. (2007). Review of clinical and laboratory features of human brucellosis. *Indian Journal of Medical Microbiology, 25*(3), 188–202. https://doi.org/10.4103/0255-0857.34758

Seleem, M. N., Boyle, S. M., & Sriranganathan, N. (2010). Brucellosis: A re-emerging zoonosis. *Veterinary Microbiology, 140*(3–4), 392–398. https://doi.org/10.1016/j.vetmic.2009.06.021

WOAH (World Organization for Animal Health). (2022). Brucellosis infection with *B. abortus, B. melitensis* and *B. suis. Terrestrial manual.* (https://www.woah.org/en/what-we-do/standards/codes-and-manuals/). Assessed on 01.07.2022.

References

[illegible]

Tuberculosis

10

Ripan Biswas and Indranil Samanta

Abstract

Tuberculosis (TB), an infectious disease caused by *Mycobacterium tuberculosis*, poses a significant global public health challenge. This chapter provides an overview of both human and zoonotic tuberculosis (zTB) and highlights their interconnectedness. Human TB is commonly transmitted through airborne respiratory droplets, whereas zoonotic TB, primarily caused by *Mycobacterium bovis,* is transmitted through direct interaction with infected animals or by consuming animal products that are contaminated. The symptoms of human and zoonotic TB are often similar, complicating diagnosis. This chapter highlights the significance of distinguishing between human tuberculosis and its zoonotic counterparts to enable effective treatment and control measures. Additionally, it focuses on the necessity of adopting a One Health approach that integrates efforts to monitor and manage health issues in both humans and animals are essential to tackle the interconnected nature of these diseases and reduce their effects on human and animal populations.

Keywords

TB · Zoonotic TB · Tuberculosis · Zoonotic tuberculosis · Consumption · Scrofula · The King's touch · Phthisis · The white plague and the captain of all these men of death

---

R. Biswas (✉)
Department of Veterinary Public Health and Epidemiology, West Bengal University of Animal and Fishery Sciences, Kolkata, West Bengal, India

I. Samanta
Department of Veterinary Microbiology, West Bengal University of Animal and Fishery Sciences, Kolkata, West Bengal, India

R. Biswas et al. (eds.), *Concept of One Health and Zoonoses*,
https://doi.org/10.1007/978-981-95-6627-3_10

History

Tuberculosis or the signs and symptoms resembling tuberculosis have been described in countless civilizations because of the vintage time. The description has been found in Greek, Chinese, and Arabic literature which provide us a clue for its existence on this planet for the remaining (?) 150 million years ago. In mummies from the Egyptian pre-dynastic and Peruvian pre-Colombian times, researchers have discovered distinctive vertebral lesions linked to tuberculosis. In Ancient Greece, tuberculosis was well known as Phtisis, and Hippocrates identified Phtisis as a fatal disease, particularly affecting younger adults, marked by characteristic lung damage. In Turkey, the first proof of human tuberculosis was observed in a 500-year-old skull with a bone lesion. In 174 AD, Clarissimus Galen, acting as the personal doctor to the Roman Emperor Marcus Aurelius, recognized the contagious characteristics of tuberculosis. Isocrates remarked on the contagious nature of tuberculosis, while Aristotle explored how "king's evil" could spread among animals like cattle and swine. During Roman times, Celso documented extrapulmonary conditions like scrofula, Pott's disease, and tuberculosis lupus. The sickness used to be recognized as "king's evil" in England and France, and it used to be appreciably believed that folks affected should be recovered after royal contact, and this exercise used to be endured in France up to 1825. Girolamo Fracastorius (1483–1553) suggested that "particles" could be transmitted through both direct and indirect human contact. In 1720, the English physician Benjamin Marten published *A New Theory of Consumption*, where he suggested that tuberculosis was caused by an infectious agent. In the eighteenth century in Western Europe, tuberculosis became an epidemic with high mortality among young people. For this reason, TB was also called "the robber of youth." In 1779, Sir Percivall Pott, a British surgeon defined as "Pott's disease" manifested with vertebral collapse and spinal cord paralysis. In the eighteenth century, the term "white plague" was coined to describe the severe pale appearance of individuals suffering from tuberculosis. Due to its epidemic reach across Europe and North America, tuberculosis was dubbed the "Captain of All These Men of Death," as it was responsible for one in every four fatalities. In 1834, the term "tuberculosis" was first introduced by Johann Lukas Schonlein. He derived this name from the Latin word "tubercula," which translates to "a small lump," a feature present in every form of the disease. Jean Antoine Villemin (1827–1892), a doctor serving in the French army, made a remarkable contribution by proving that tuberculosis could be passed from humans to animals and also among animals. On March 24, 1882, during a session of the Berlin Society of Physiology, Robert Koch revealed his groundbreaking discovery of the microorganism responsible for pulmonary tuberculosis, which he termed the "tuberkel virus." Later, in 1889, Theobald Smith identified tuberculosis affecting cattle.

Etiological Agent and Ecology

In both humans and animals, tuberculosis is triggered by an organism that belongs to the Mycobacterium tuberculosis complex (MTBC). This group is made up of several genetically related Mycobacterium species, each of which has the potential to cause illness in one or more hosts. Throughout their history, the species comprising the Mycobacterium tuberculosis complex (MTBC) have been subject to a variety of taxonomic and nomenclatural revisions. The *Mycobacterium tuberculosis* complex (MBTC) encompasses several species, including *Mycobacterium tuberculosis* (Mtb), *Mycobacterium bovis*, *Mycobacterium bovis* bacilli Calmette–Guerin (BCG), *Mycobacterium africanum*, *Mycobacterium pinnipedii*, and *Mycobacterium microti*. *Mycobacterium bovis* (*M. bovis*) is responsible for causing bovine tuberculosis (bTB), while *Mycobacterium tuberculosis* (*M. tuberculosis*) is the pathogen behind human tuberculosis (TB). A new member of MTBC has been reported around the world that is *Mycobacterium orygis* (*M. orygis*). The point of origin has been specified for Southeast Asia, which includes the countries of India, Pakistan, Bangladesh, and Nepal. These pathogens have a 16S rRNA sequence that is identical, with their nucleotide sequences showing up to 99.9% similarity. The *Mycobacterium tuberculosis* complex is characterized as a gram-positive, acid-fast bacillus (AFB) that is non-motile, does not possess a capsule, and is non-spore forming. It thrives in aerobic environments, is catalase-negative, relies on oxidative metabolism, and is found intracellularly, classifying it within the Mycobacteriaceae family. In 1896, Mycobacterium was referred to as a "fungus bacterium" because it can develop mold-like layers when grown in liquid environments. The ideal temperature range for the growth of tubercle bacilli is 35–37 °C, whereas they are unable to grow at temperatures of 25 °C or 41 °C.

Being Gram-positive, this group of bacteria does not take up the Gram-stain dyes because the walls of the organisms are enriched with mycolic acid. Similar cell walls are present in the genera *Corynebacterium, Nocardia,* and *Rhodococcus* and the genera *Gordomia*, *Tsukamurella*, *Nocardia,* and *Rhodococcus* are closely related to this complex. The mycobacterial cell wall is predominantly made up of mycolic acid, which comprises 70–90 carbon atoms, in addition to arabinogalactan and a peptidoglycan layer. Together, these components form what is known as the mycolyl-arabinogalactan-peptidoglycan (mAGP) complex.

Among the various lipids are lipoarabinomannan (LAM), phosphatidylinositol mannoside (PIM), sulfated glycophospholipid (SL), trehalose dimycolate (TDM), and dimycocerosate phthiocerol (DIM). The thick, waxy coating contributed to both the resistance to acids and the water-repellent properties. Mycolic acid, a waxy coating, is instrumental in rendering bacteria resistant or impermeable to a wide range of disinfectants, standard laboratory stains, antibiotics, and physical harm. It has also the ability to develop serpentine structures which are known as "cords," and the cord factor is mainly associated with virulent strains of *M. tuberculosis*. Due to its thick waxy outer coating this group of organisms are also referred as "ducks of the microbial world." It has a slow generation time of approximately 20–36 h as the uptake of nutrients is restricted by the lipid layer. The bacilli can survive for 20–30 h in sputum and can be killed at 80 °C for 15–20 min. Two hours of direct sunlight were sufficient to destroy the bacteria. It is sensitive to UV radiation, formaldehyde, and glutaraldehyde.

Classification of Mycobacterium

Order: Actinomycetales Class: Actinomycetes Family: *Mycobacteriaceae* Genus: *Mycobacterium*	
Mycobacterium tuberculosis complex (MBTC)	Nontuberculous mycobacteria (MTM)
Mycobacterium tuberculosis *Mycobacterium bovis*, *Mycobacterium bovis* bacilli Calmette–Guerin (BCG) *Mycobacterium africanum* *Mycobacterium pinnipedii* *Mycobacterium microti* *Mycobacterium orygis* *Mycobacterium mungi* *Mycobacterium canettii*	Non-cultivable *Mycobacterium leprae*, *Mycobacterium lepromatosis*
	Non-chromogens slow growing *Mycobacterium avium* complex (MAC), *Mycobacterium hemophilum*
	Non-photochromogens slow growing *Mycobacterium kansasii, Mycobacterium marinum*
	Scoto-chromogens slow growing *Mycobacterium gordonae, Mycobacterium xenopi*
	Rapid growing *Mycobacterium chelonae, mycobacterium fortuitum*

Host Susceptibility

All MBTC members cause tuberculosis, a chronic granulomatous disease that affects humans and many other animals.

Mycobacterium species	Susceptible host
M. tuberculosis	Humans
M. canetti	Humans
M. africanum	Humans
M. bovis	Bovine, goats, and humans
M. bovis var BCG	Vaccination purpose
M. caprae	Goats
M. microti	Rodents, immunocompromised human patient
M. pinnipedii	Seals
M. mungi	Mongoose *(Mungos mungo)*
M. orygis	Waterbucks, antelopes, and gazelles

Although *M. bovis* is recognized as the primary zoonotic organism, instances of *M. caprae*, *M. orygis*, *M. pinnipedii*, and *M. microti* have also been documented in humans. *Mycobacterium tuberculosis* is widely acknowledged as the leading cause of tuberculosis in humans across the globe. However, a significant body of evidence suggests that the incidence of *Mycobacterium bovis*, which leads to bovine tuberculosis, may be underestimated in humans as a contributor to zoonotic tuberculosis.

Epidemiology

Before the emergence of COVID-19, tuberculosis was the leading cause of death from a single infectious agent, outpacing AIDS. In 2020, tuberculosis (TB) was identified as the 13th most common cause of death worldwide. According to the World Health Organization (WHO), the disease affected 10 million people, leading to 1.2 million deaths. Furthermore, an additional 208,000 fatalities were linked to TB-AIDS syndrome. HIV and tuberculosis form a perilous alliance, with each disease accelerating the other's progression. In 2020, approximately 215,000 people died from tuberculosis associated with HIV. In the WHO African Region, the prevalence of TB linked to HIV was the highest, with 85% of the patients having their HIV test results recorded. Across the globe, men represent the most at-risk group, followed by women and children, with numbers reaching 5.8 million, 3.2 million, and 1.0 million, respectively. The data indicate that individuals aged 15 and older experience more hardship than younger ones. Based on data from the WHO in 2020, 86% of the newly reported TB cases are concentrated in 30 countries identified as having a high burden of tuberculosis. Eight countries accounted for two-thirds of the world's tuberculosis cases, with India leading at 27%. China was next with 9%, followed by Indonesia at 8%. The Philippines reported 6%, while Pakistan contributed 5%. Nigeria and Bangladesh each accounted for 4%, and South Africa reported 3%. Additionally, the European region and the WHO region of the Americas reported 6% and 3% of the global cases, respectively. Globally, latent tuberculosis infection is estimated to affect around 1.7 billion people, which represents 23% of the world's population.

AIDS and other immunosuppressed patients, including intravenous drug users, alcoholics, and people living in very crowded conditions, such as prisons or those in densely populated urban areas such as slums, are especially vulnerable to tuberculosis. The most common source of infection for human tuberculosis is sputum of patients suffering from pulmonary tuberculosis (PTB). Urine, pus, pleural, peritoneal, and pericardial fluids, feces, and gastric contents in human type of infection may also act as a source of infection for the new human individual. Bovine tuberculosis (bTB), also known as zoonotic tuberculosis (zTB), can spread to humans and other animals through exhaled droplets, saliva, milk, urine, feces (stemming from both intestinal lesions and ingested sputum from pulmonary lesions), semen, vaginal and uterine discharges, as well as secretions from open peripheral lymph nodes. Tuberculosis risk factors include malnutrition, widespread poverty, rapid urbanization, dense populations, a higher prevalence of diabetes, and significant air pollution. These factors are particularly relevant in developing countries where malnutrition is often widespread.

The connection between tuberculosis (TB) and poverty is profound, with factors linked to poverty—such as malnutrition, overcrowding, inadequate ventilation, and poor sanitation heightening the risk of TB infection and the chance of it developing into a clinical condition. This interconnection creates a harmful cycle where poverty and TB reinforce each other; TB reduces an individual's ability to work and increases the financial burden of treatment, thereby deepening their poverty. TB is a health and social problem in the context of both clinical and epidemiological entities. Overcrowding, poor housing, low sunlight (absence of UV rays) and ventilation, large families, unawareness, and ignorance of the mode of transmission are predisposing factors for TB.

Mode of Transmission

Inhaling droplet nuclei measuring 1–5 μm and consuming milk, dairy items, and meat that are contaminated are common transmission routes. Droplet transmission is usually limited to a distance of 30–60 cm between the source and the susceptible host. Because of their tiny size, droplet nuclei have the ability to linger in the air from a few minutes to several hours. Almost all transmissions occur indoors.

Zoonotic Tuberculosis (zTB)

Bovine TB is found globally, with the exception of Antarctica and certain regions such as the Caribbean islands, parts of South America, and Australia, where it has been eliminated through rigorous testing and slaughter protocols. Tuberculosis poses a threat to wild animals, farm animals, and humans alike. In developing regions, the close interactions among these groups are considered to significantly elevate the likelihood of the disease spreading.

Bovine TB's epidemiology is shaped by a variety of risk factors, including genetic, behavioral, biological, and environmental influences, which play a crucial role in the disease's transmission, infection establishment, and symptom expression. At the herd level, various risk factors such as the size and type of the herd, the extent of the farm, farm management practices, interactions with adjacent herds, and cattle movement are crucial for the disease's occurrence. According to the WHO, 1.4% of new cases are zoonotic tuberculosis (TB). While cattle serve as the primary hosts for *M. bovis*, buffalo also play a crucial role as a maintenance and reservoir host, facilitating the transmission of the disease between species. Animals such as goats, sheep, pigs, horses, camels, and household pets can be susceptible to *M. bovis* infection. Wild boars, possums, badgers, and deer are among the wildlife species recognized as reservoir hosts. Zebu cattle are thought to have a higher resistance to tuberculosis compared to European cattle. Nevertheless, the infection primarily affects the respiratory system, with the bacteria targeting the lungs for infection development. Cattle that are infected are likely a source of infection because they release a large number of organisms into the environment through droplet nuclei. During artificial insemination, bulls that are infected can spread the disease through semen that has become contaminated.

Reverse Zoonoses or Zooanthroponoses

It refers to any pathogen that is normally present in humans and can be transmitted to other vertebrates. In Africa, there have been reports of reverse transmission involving *M. bovis*, *M. tuberculosis*, and *M. africanum*. In India, the MANU strain of *M. tuberculosis* is more frequently observed in cattle compared to *M. bovis*. *Mycobacterium tuberculosis* is frequent in human clinical cases of TB; however, unlike *M. bovis*, it is generally regarded as nonzoonotic. Interestingly, the concept has recently been improved by research citations. There are various reports by different researchers showing its zoonotic potential in zoo animals, dogs, monkeys, and many other species. Although the agent does not induce progressive tuberculosis in animals, the bacillus can remain in their tissues for some time, particularly in the lymph nodes. This persistence sensitizes the animals to mammalian tuberculin, thereby complicating the diagnostic process.

Pathogenesis and Immunity

This pathogen mainly targets host macrophages, which are key cells in both innate and adaptive immunity. In pulmonary tuberculosis, the pathogen passes through the airways and reaches the lower portion of the lung alveolar space. It interacts with alveolar epithelial cells, macrophages, and dendritic cells, but most preferably interacts with alveolar macrophages and other dendritic cells.

Aerosolization: The tuberculosis pathophysiology cycle both starts and concludes with aerosolization. Infected individuals generate these aerosols via intense breathing actions such as coughing, sneezing, shouting, or singing (human). The organism is then able to stay alive in the air and become airborne. Some of these droplets, which are smaller than 5 μm and hold 1–3 bacilli, can attain the alveolar sacs after inhalation.

Phagocytosis: Bacilli interact with alveolar macrophages, dendritic cells, and monocytes, although alveolar macrophages (dust cells) play a leading role. The bacilli attach to the dust cells through various receptors, including mannose, scavenger, complement (CR1, CR3, CR4), Fc, and surfactant protein receptors (SPR). The mannose receptor functions as a pathogen-recognition receptor, influencing the movement of molecules, the presentation of antigens, the differentiation of macrophages, and the regulation of inflammatory responses. The interaction between mannose receptors and Grb2 triggers the Rac/Pak/Cdc-42 signaling pathway, which aids in the absorption of mycobacteria. The Rac/Pak/Cdc-42 pathway recruits the Src homology 2 (SH2) domain, which includes protein tyrosine phosphatase 1

(SHP-1). Protein tyrosine phosphatase 1 (SHP-1) restricts the activity of phosphatidylinositol 3-phosphate (PI3P), a trafficking phospholipid that prevents phagosome-lysosome fusion. In contrast, mycobacteria secrete secretory acid phosphatase (SapM), which eliminates phosphatidylinositol 3-phosphate (PI3P) from phagosomes. Moreover, PI3P functions as a docking molecule that interacts with the proteins found within lysosomes. The Rac/Pak/Cdc-42/SHP-1/PI3P signaling pathway promotes the proliferation of mycobacteria inside alveolar macrophages.

Additionally, bacilli feature a unique collection of cell wall lipids that contribute to their defense against macrophages. Within the lipid layer, the presence of sulfated glycolipids (SL) and trehalose dimycolate (TDM) serves to prevent the fusion of lysosomes with phagosomes. Dimycocerosate phthiocerol (DIM) prevents acidification and increases the permeability of phagosomes. Macrophages phagocytose bacilli and attempt to destroy them; however, mycobacteria inhibit the fusion of phagosomes with lysosomes, thereby evading destruction, initiating replication, and releasing DNA, RNA, proteases, and lipids. Concurrently, dendritic cells make their way to the lymph nodes to engage T-helper cells. In parallel, macrophages secrete cytokines and vascular endothelial growth factor (VEGF), which initiate angiogenesis and promote the increased vascularization of lesions. Cytokines activate the innate immune system by recruiting natural killer (NK) cells, dendritic cells (DC), neutrophils, and macrophages to perform a range of functions. T-helper cells assist in directing TH1, regulatory T cells (Tregs), and B cells to the germinal center, where granulation is initiated. This granuloma acts as a detention center for bacteria and prevents their systemic spread. Subsequently, immunocompromising inhibits the granuloma from enclosing the bacterial cells. As a result, bacilli spread and multiply at various sites and show clinical manifestations. From this point on, the infection can spread to the new susceptible host by following the same aerosolization pathway. Therefore, we can conclude that *Mycobacterium* phagocytosis recruits a cascade of events that result in either effective control of the infection, followed by latent tuberculosis, or the development of active disease. Before the advent of antibiotics, patients were commonly treated with cod-liver oil and exposure to sunlight, both of which serve as sources of 25-hydroxyvitamin-D, known for its role in modulating the immune system. Cathelecidin is responsible for breaking down microbial membranes within the phagolysosomes of macrophages. The cathelicidin antimicrobial peptide (CAMP) is a type of polypeptide predominantly located in the lysosomes of macrophages and polymorphonuclear leukocytes (PMNs). The synthesis of cathelecidin is entirely contingent upon adequate levels of 1,25-hydroxyvitamin D. Several studies have linked vitamin D deficiency to a heightened vulnerability to developing active tuberculosis.

Clinical Manifestations in Bovine

Bovine tuberculosis (bTB) is characterized by a gradual progression, with symptoms that may take months or even years to appear. The incubation period varies, starting from as little as 2 months and potentially lasting several years.

Most bovines do not develop clinical manifestations, but if clinical signs appear, they will be extremely variable and nonspecific. As the disease advances, animals experience weakness, loss of strength, and ultimately death due to underlying toxemia. Bovines often exhibit clinical symptoms such as weight loss and emaciation, decreased appetite, weakness, a low-grade and inconsistent fever, swollen lymph nodes, and respiratory issues characterized by a moist, sporadic cough that tends to worsen in the morning, during cold weather, or after physical exertion, leading to shortness of breath. Involvement of the gastrointestinal tract can lead to alternating episodes of diarrhea and constipation. The superficial lymph node swells and may occasionally burst. Swelling of deeper lymph nodes can unpredictably obstruct blood vessels, airways, or the digestive system. Eye diseases, recurrent miscarriages, and infertility have been noted due to lesions in the uterus.

Clinical Manifestations in Humans

Classically, two types of infections are commonly observed in humans, i.e., primary and secondary tuberculosis represent two different stages of the disease. Primary tuberculosis is characterized by a newly acquired infection that occurs when the immune system is unable to contain or prevent the initial infection, and it is frequently observed in individuals with compromised immune systems. During this period, infected people can spread the disease to other susceptible humans by producing infected aerosols.

Interestingly, if the immune system fails to eliminate the bacteria, then the bacteria are enclosed in the granuloma and patients remain asymptomatic. This stage is called latent tuberculosis infection (LTBI), in which bacteria form protective biofilms in necrotic tissue. Subsequently, secondary tuberculosis begins as the immunocompromised condition permits the bacteria to become reactive within the granuloma, resulting in pulmonary tuberculosis (PTB), extrapulmonary tuberculosis (EPTB), and miliary tuberculosis.

Latent tuberculosis infection most frequently leads to pulmonary tuberculosis, which is marked by symptoms such as coughing, coughing up blood, losing weight, diminished appetite, sweating at night, lack of desire to eat, and fever.

Extrapulmonary tuberculosis (EPTB) occurs when the infection extends to areas such as the lymph nodes, pleura, gastrointestinal tract, genitourinary system, and bones.

Miliary tuberculosis is marked by the widespread distribution of granulomas, leading to the formation of tuberculomas throughout the body. The term "miliary" refers to numerous small pulmonary nodules scattered through lung-like millet seeds. Miliary tuberculosis represents less than 2% of all tuberculosis cases and can comprise up to 20% of extrapulmonary tuberculosis cases in adults with normal immune function.

Pleural Tuberculosis
In countries with a high prevalence of tuberculosis, pleural tuberculosis accounts for up to 30% of all extrapulmonary tuberculosis (EPTB) cases. The clinical manifestations remain the same as pulmonary tuberculosis.

Abdominal Tuberculosis
Approximately 11% of all cases of extrapulmonary tuberculosis (EPTB) are abdominal. The most affected site of abdominal tuberculosis is the ileocecal region, then comes the ascending colon, followed by the jejunum, appendix, duodenum, stomach, esophagus, sigmoid colon, and rectum.

CNS Tuberculosis
It is a severe and often fatal form of extrapulmonary tuberculosis (EPTB) that mainly occurs in young children. Tuberculosis, meningitis, and intracranial tuberculoma are common forms of brain TB.

Bone and Joint Tuberculosis
Nearly 10–15% of all extrapulmonary tuberculosis (EPTB) cases are bone and joint tuberculosis.

Urogenital Tuberculosis (UG-TB)
It represents 15% of all cases of extrapulmonary tuberculosis and accounts for 3–4% of all pulmonary tuberculosis cases. Its incidence is 20 times higher in individuals who have received kidney transplants compared to the general population. Generally, the spread of infection descends from the kidneys to other urogenital organs.

Scrofula or Tubercular Lymphadenitis or Lymph Node Tuberculosis
From the earliest times, lymph node tuberculosis has been called Scrofula or King's evil. Cervical lymphadenitis represents the most frequent extrapulmonary manifestation of TB, as the infection bypasses the tonsils, adenoids, and sinonasal/osteomyelitis of the ethmoid bone. Fine-needle aspiration cytology (FNAC) is an effective diagnostic screening tool.

TB-AIDS Syndemic

Among individuals living with HIV, tuberculosis (TB) ranks as a predominant opportunistic infection. It is recommended that all patients with TB and suspected persons should be offered HIV testing and counseling. All PLHIV (People Living with HIV) should be tested for tuberculosis for all available opportunities.

Multidrug-Resistant Tuberculosis (MDR-TB)

The ongoing issue of drug-resistant tuberculosis remains a significant public health challenge. The resistance to isoniazid and rifampicin, the two most potent first-line medications, is particularly alarming. When a tubercle organism is resistant to isoniazid and rifampicin, the most potent drug for tuberculosis and the development of tuberculosis is called MDR-TB. MDR-TB and RR-TB (rifampicin resistant tuberculosis) require treatment with second-line drugs. According to the "Global tuberculosis report 2021" by WHO, the worldwide incidence of MDR-TB and RR-TB has remained consistent over the past decade, accounting for 3–4% of all tuberculosis cases.

Extensively Drug-Resistant Tuberculosis (XDR TB)

XDR-TB is a form of tuberculosis that shows resistance to rifampicin, any fluoroquinolone drug, and at least one of the medications bedaquiline or linezolid.

Post-Mortem Lesions in Bovine

The lesion is marked by the formation of granulomas where the bacteria are located. These granulomas are generally yellowish-white or grayish-white, with a caseous, caseocalcareous, calcified nature, and are frequently encapsulated. Some tubercles are so small that they might go unnoticed by the unaided eye unless the tissue is sliced into sections. In cattle, tubercles are found in lymph nodes of the head and thorax. Tubercles are also seen in the lungs, spleen, liver, and surfaces of body cavities.

Regarding Biosafety Level

Contact with laboratory-generated aerosols is the most significant hazard. Given that the infectious dose of M. tuberculosis is remarkably low ($ID50 < 10$ bacilli), it is essential to regard clinical specimens from individuals suspected or diagnosed with tuberculosis as potentially infectious, thereby requiring strict adherence to safety protocols. Furthermore, accidental needle sticks are recognized as potential hazards. The WHO biosafety guidelines specify that any handling of samples suspected to contain MDR *M. tuberculosis* (discussed later in this chapter) should be carried out in biosafety level 3 (BSL-3) facilities.

Diagnosis of Tuberculosis in Cattle

Microscopy acid-fast staining	Ziehl-Neelsen (ZN) or fluorescent staining methods are employed to identify acid-fast bacilli (AFB) in clinical samples. Microscopic examination and acid-fast staining are the simplest, quickest, and most cost-effective diagnostic techniques, though they are the least sensitive. Under the microscope, acid-fast bacteria are seen as red, whereas non-acid-fast bacteria or tissues appear blue or green.
Culture	Mycobacterium can be isolated by cultivating the bacteria on a medium that contains eggs, known as the Lowenstein-Jensen (LJ) medium.

Tuberculin test (delayed hypersensitivity test)	The tuberculin test is the conventional method used to identify bovine tuberculosis. The process involves injecting bovine tuberculin purified protein derivative (PPD) into the skin and then observing for any swelling, indicative of delayed hypersensitivity, at the injection site after 72 h. The tuberculin test is commonly administered on the mid-neck, although it can alternatively be conducted on the caudal fold of the tail. It is intriguing to note that the skin on the neck showed a higher sensitivity to tuberculin than the skin on the caudal fold. Delayed hypersensitivity may take between 3 and 6 weeks to appear after an infection, potentially leading to a false negative result if the infection is recent. In nationwide campaigns aimed at eradicating the disease, it is recommended to administer doses of purified bovine tuberculin protein derivatives, with a maximum of 5000 IU. The volume of each injection should be limited to 0.2 mL. There are four techniques for conducting the tuberculin test. The PPD of bovine tuberculin is derived from the M. bovis strain AN 5, with a concentration of 1 mg PPD per ml, which is equivalent to 2000 tuberculin units (TU) per 0.1 mL, and it is preserved using 0.5% phenol. Single intradermal test (SID): 72 h following the injection, the thickness of the skinfold was assessed. Stormont test/double intradermal test: In alignment with the single I/D test procedure, a second tuberculin dose was administered at the identical location (same dosage and method) precisely 1 week following the initial injection. Twenty-four hours following the administration of the second dose, if the area exhibited heat, pain, swelling, and an increase in thickness of at least 5 mm, the reaction was identified as positive for tuberculosis. While this test is extremely precise and reliable, it necessitates three separate appointments. Short thermal test: In the procedure, 4 mL of diluted tuberculin was administered subcutaneously into the neck region of the cattle, ensuring that the animals maintained their normal body temperature throughout. The temperature was recorded again at 4, 6, and 8 h post-injection, and if it exceeds 104 °C, the case might be considered positive. Comparative test: In this test, the procedure mirrored that of the single I/D test, but avian and mammalian tuberculins were administered at the same time, positioned 12 cm apart on the same side of the neck, with one placed above the other. After 72 h, the test results are evaluated, and the reaction with the greater swelling identifies the organism that caused the sensitization.
ELISA	ELISA is advantageous due to its ease of use, yet its sensitivity is often compromised by the slow and uneven maturation of the humoral immune response in cattle. When complex antigens like tuberculin or M. bovis culture filtrates are utilized in cattle, the specificity is also low. The USDA has granted a license for this test in the United States for certain species, including elephants and nonhuman primates, while it has also been approved for use with badgers in the United Kingdom. Nonetheless, ELISA is not advised for use with animals.
PCR	To assess the severity of infection in cattle with bovine tuberculosis, real-time PCR was employed and compared to IFN gamma mRNA levels in blood culture. RT-PCR serves as an additional effective diagnostic tool for detecting bovine tuberculosis in cattle.
Gamma-interferon assay	The assay functions by measuring the IFN-γ released from sensitized lymphocytes when they are exposed to the specific antigen PPD-tuberculin over an incubation period of 16–24 h.

Lymphocyte proliferation assay	This laboratory test evaluates how peripheral blood lymphocytes respond to tuberculin PPD (PPD-B) and PPD derived from Mycobacterium avium (PPD-A). This procedure can be executed with either whole blood or lymphocytes that have been extracted from peripheral blood samples.

Diagnosis of Tuberculosis in Human

The most common specimens for diagnosing tuberculosis are sputum, bronchial washing, and early-morning gastric aspirates, and tissue biopsies are used for microscopy and culture. The CSF, urine, and other fluids were centrifuged for examination.

Test	Assay principle	Use
Imaging techniques		
Chest X-ray	Imaging of the lungs	Active TB disease screening
Microscopy		
Conventional sputum smear microscopy	Direct visualization of mycobacteria by using light microscopy	Active TB disease diagnosis
LED (light-emitting-diode) fluorescence smear microscopy	Direct visualization of mycobacteria using fluorescence microscopy	Active TB disease diagnosis
Culture-based techniques		
Liquid culture with DST (drug susceptibility testing)	Mycobacterial culture on liquid media	Active TB disease diagnosis Drug resistance
Antigen detection techniques		
LAM (lipoarabinomannan assay) lateral flow assay	Antigen detection	Active TB disease diagnosis in HIV-positive individuals
Molecular techniques (nucleic acid amplification tests-NAAT)		
Xpert MTB/RIF	NAAT (qPCR)	Active TB disease diagnosis Drug resistance (rifampicin)
First-line LPA (line probe assay)	NAAT (LPA)	Diagnosis of active tuberculosis disease Resistance to drugs (isoniazid and rifampicin)
Second-line LPA	NAAT (LPA)	Drug resistance (fluoroquinolones and second-line injectable drugs)
Loop mediated isothermal amplification assay	NAAT (LAMP)	Active TB disease diagnosis
TST (tuberculin skin tests)/Mantoux test	An intradermal injection of five tuberculin units (5 TU) of purified protein derivative (PPD) was administered, and the results were assessed 48–72 h later	Latent tuberculosis infection (LTBI)

Test	Assay principle	Use
IGRA (interferon gamma release assay)	Tests conducted in vitro assessed cell-mediated immune responses by evaluating the release of IFNγ from T cells after they were stimulated with antigens encoded by RD1	Latent tuberculosis infection (LTBI)

Prevention and Control Strategies

Human tuberculosis is preventable by setting the following points:

- Early case detection and effective therapy for infectious individuals are essential in the community.
- Identification of active cases from the community by involving a deliberate search, often on a house-to-house basis or in workplaces.
- Reducing overcrowding decreases the risk of infection.
- A medical intervention aimed at preventing tuberculosis infection from developing into active TB serves as a prophylactic treatment for TB.
- Giving the Bacillus Calmette-Guerin (BCG) vaccine to children can help protect them, especially from severe forms of tuberculosis.
- It is advisable to consume pasteurized milk and thoroughly cooked meat.

The WHO's proposed strategies for eradicating tuberculosis:

Comprehensive Patient Care and Prevention

- Early detection of tuberculosis involves universal testing for drug susceptibility and systematic screening of contacts and high-risk populations.
- Treatment for all individuals with TB, including those with drug-resistant strains, along with patient support.
- Joint TB/HIV initiatives and management of coexisting conditions.
- Preventive treatment for high-risk individuals and vaccination against tuberculosis.

Comprehensive Policies and Supportive Systems

- Political commitment with ample resources dedicated to tuberculosis care and prevention.
- Active participation of communities, civil society organizations, and both public and private healthcare providers.
- Implementation of universal health coverage policies and regulatory frameworks for case notification, vital statistics, the quality and rational use of medications, and infection control.
- Social protection, poverty alleviation, and measures addressing other determinants of tuberculosis.

Enhanced Research and Innovation

- The discovery, creation, and swift adoption of novel tools, interventions, and strategies.
- Research aimed at optimizing implementation, maximizing impact, and fostering innovations.

Preventive and Control Measures for Cattle

- Regular monitoring during slaughter, culling of diseased animals, limiting the movement of affected livestock, and eliminating animals that test positive for TB.
- Avoiding overcrowding in animals.
- Provide good hygienic conditions in the shed.
- Abattoir surveillance and postmortem confirmation.

Treatment for Humans

Once the disease is confirmed the treatment regimen will be decided on the basis of the followings:

Result of sputum smear.
History of previous anti-TB treatment.
The disease was ruled out categorically as pulmonary or extrapulmonary.
Anti-tuberculosis drugs:

<table>
<tr><th>First-line drugs</th><th colspan="2">Second-line drugs</th><th>Third-line drugs</th></tr>
<tr><td rowspan="3">Isoniazid (H)
Rifampicin (R)
Pyrazinamide (Z)
Ethambutol (E)
Streptomycin (S)</td><td>Kanamycin (Km)
Amikacin (Am)
Capreomycin (Cm)</td><td>Injectables</td><td rowspan="3">Clofazimine (Clf)
Bedaquiline (Bdq)
Delamanid
Linezolid (Lzd)
Amoxyclav (Amx/Clv)
Clarithromycin (CLr)
Meropenem (Mem)
Imipenem/Cilastatin (Ipm/Cs)
High dose—Isoniazid (H)
Thiacetazone (Thz)</td></tr>
<tr><td>Levofloxacin (Lfx)
Moxifloxacin (Mfx)
Ofloxacin (Ofx)</td><td>Fluoroquinolones</td></tr>
<tr><td>Ethionamide (Eto)
Prothionamide (Pto)
Cycloserine (Cs)
Terizidone (Trz)
Para Aminosalicylic acid (PAS)</td><td>Oral bacteriostatics</td></tr>
</table>

Directly Observed Therapy (DOT) is a case management tool designed to ensure that patients follow their treatment plans. The system involves a qualified healthcare professional or another trained individual watching a patient take each dose of anti-TB medication and recording it. DOT can help prevent the development of drug resistance, the failure of treatment, or the recurrence of illness after the treatment period ends.

Treatment for Animals

Antibiotics have been employed in the treatment of tuberculosis in various animals, notably those that are domesticated or housed in animal settings. The most effective first-line antituberculous treatments include isoniazid (H), rifampicin (R), pyrazinamide (Z), ethambutol (E), streptomycin (S), ethambutol (EMB), and streptomycin (SM). Capreomycin (CAP), ethionamide (ETH), cycloserine, and thioacetazone are classified as second-line medications. In most cases, cattle that react positively to tests are culled, although some nations might initially implement test-and-segregation strategies before transitioning to a test-and-slaughter approach. However, treatment is generally not recommended for cattle.

Bibliography

Abrahams, K. A., & Besra, G. S. (2018, February). Mycobacterial cell wall biosynthesis: A multifaceted antibiotic target. *Parasitology, 145*(2), 116–133. https://doi.org/10.1017/S0031182016002377. Epub 2016 Dec 15. PMID: 27976597; PMCID: PMC5964476.

Golden, M. P., & Vikram, H. R. (2005, November 1). Extrapulmonary tuberculosis: An overview. *American Family Physician, 72*(9), 1761–1768.

Heemskerk, D., Caws, M., Marais, B., & Farrar, J. (2015). *Tuberculosis in adults and children.* Springer.

Jeremiah, C., Petersen, E., Nantanda, R., Mungai, B. N., Migliori, G. B., Amanullah, F., Lungu, P., Ntoumi, F., Kumarasamy, N., Maeurer, M., & Zumla, A. (2022, March). The WHO global tuberculosis 2021 report – Not so good news and turning the tide back to End TB. *International Journal of Infectious Diseases, 20*, S1201-9712(22)00149-7. https://doi.org/10.1016/j.ijid.2022.03.011. Epub ahead of print. PMID: 35321845; PMCID: PMC8934249.

Natarajan, A., Beena, P. M., Devnikar, A. V., & Mali, S. (2020, July). A systemic review on tuberculosis. *The Indian Journal of Tuberculosis, 67*(3), 295–311. https://doi.org/10.1016/j.ijtb.2020.02.005. Epub 2020 Feb 28.

Olea-Popelka, F., Muwonge, A., Perera, A., Dean, A. S., Mumford, E., Erlacher-Vindel, E., Forcella, S., Silk, B. J., Ditiu, L., El Idrissi, A., Raviglione, M., Cosivi, O., LoBue, P., & Fujiwara, P. I.. (2017, January). Zoonotic tuberculosis in human beings caused by Mycobacterium bovis—A call for action. *The Lancet Infectious Diseases, 17*(1), e21–e25. https://doi.org/10.1016/S1473-3099(16)30139-6. Epub 2016 Sept 30.

Pai, M., Behr, M. A., Dowdy, D., Dheda, K., Divangahi, M., Boehme, C. C., Ginsberg, A., Swaminathan, S., Spigelman, M., Getahun, H., Menzies, D., & Raviglione, M. (2016, October 27). Tuberculosis. *Nature Reviews Disease Primers, 2*, 16076. https://doi.org/10.1038/nrdp.2016.76

Leptospirosis

11

Ripan Biswas and Indranil Samanta

Abstract

Leptospirosis, an extensively occurring zoonotic disease instigated by pathogenic spirochetes belonging to the *Leptospira* genus, constitutes a major health hazard for both humans and animals on a global scale. This chapter presents a comprehensive overview of leptospirosis, covering its epidemiology, transmission dynamics, clinical features, diagnostic approaches, and preventive strategies. Humans can contract leptospirosis through either direct or indirect contact with the urine of infected animals, such as rodents, livestock, and wildlife. Environmental conditions, such as flooding and inadequate sanitation, are essential determinants of disease transmission. The clinical presentation of human leptospirosis involves a broad spectrum, from mild, influenza-like symptoms to severe, potentially life-threatening complications, such as Weil's disease and pulmonary hemorrhage syndrome. This chapter emphasizes the diagnostic intricacies of leptospirosis owing to its nonspecific symptoms and the necessity of laboratory testing for definitive diagnosis. This chapter also highlights the crucial role of thorough disease monitoring systems, specific rodent control measures, and preventive approaches, such as vaccination initiatives and the use of personal protective gear, in effectively reducing and managing the spread of leptospirosis among both human and animal populations.

R. Biswas (✉)
Department of Veterinary Public Health and Epidemiology, West Bengal University of Animal and Fishery Sciences, Kolkata, West Bengal, India

I. Samanta
Department of Veterinary Microbiology, West Bengal University of Animal and Fishery Sciences, Kolkata, West Bengal, India

R. Biswas et al. (eds.), *Concept of One Health and Zoonoses*,
https://doi.org/10.1007/978-981-95-6627-3_11

Keywords

Leptospirosis · Weil's syndrome · Cane-Cutter's fever · Swamp fever · Stuttgart disease · Mud fever · Akiyami · Autumn fever · Swineherd's disease · Hemorrhagic jaundice · Rice-field fever · Canicola fever · Red water of calves

History

Leptospirosis is a bacterial infection that impacts both humans and animals, prevalent across the globe, particularly in developing regions, and presents significant public health issues at the convergence of animal, human, and environmental systems. In 1886, Adolph Weil was the pioneer in identifying leptospirosis in a patient from Heidelberg who exhibited symptoms such as jaundice, fever, an enlarged spleen, kidney failure, and conjunctivitis. This condition was linked to outdoor jobs that required interaction with water. Leptospirosis was documented as an occupational hazard of rice harvesting as "rice field jaundice" in ancient China, and the Japanese term akiyami, meaning autumn fever. In 1915, Inada and Ido discovered both spirochetes and specific antibodies present in the blood of Japanese miners suffering from infectious jaundice.

In that same year, two teams of German doctors, Uhlenhuth, Fromme, and Hubener, Reiter, discovered the organism in the blood of guinea pigs that had been injected with the blood of soldiers infected during the trench warfare of World War I. In 1907, Stimson was the first to identify the organism responsible for the disease by employing a silver staining technique. He detected the spirochaetes in the kidneys of a patient who had reportedly succumbed to yellow fever. The organism was named *Spirochaeta interrogans* because of its characteristic question-mark shape. In 1917, canine leptospiral disease was recognized, and subsequently, the discovery was made that rats serve as a source of infection for humans. Subsequently, leptospiral infections in livestock have been documented.

Etiological Agent and Ecology

Leptospirosis is an infection caused by multiple species of the *Leptospira* genus, which are spirochetes classified under the Leptospiraceae family and the Spirochaetales order. In the period before 1989, the genus *Leptospira* was categorized into two distinct species: *L. interrogans*, which encompassed all pathogenic strains, and *L. biflexa*, which included the non-pathogenic saprophytic strains that were isolated from environmental sources. For ease of classification, serovars with antigenic similarities were grouped into serogroups. There are over 250 serovars of *L. interrogans* and more than 60 serovars of *L. biflexa*. Each strain or serovar of these species is determined by agglutination after cross-absorption with the homologous antigen surface lipopolysaccharide (LPS) of the bacteria.

The serological classification system has been superseded by a genotypic classification system based on DNA. The genus Leptospira has been reclassified into 21 species, but this DNA-based categorization poses challenges for clinical microbiologists, as it clearly conflicts with the long-established system previously used. Clinical cases have identified several species, including *L. interrogans*, *L. borgpetersenii*, *L. alstonii*, *L. inadai*, *L. kirschneri*, *L. licerasiae*, *L. noguchi*, *L. alexanderi*, *L. fainei*, *L. santarosai*, *L. terpstrae*, *L. weilii*, and *L. wolffii*.

It is slender, tightly coiled, obligatory, aerobic, and highly motile. The organisms measure approximately 0.25 μm by 6–25 μm and are capable of passing through filters with a pore size of 0.45 μm. The axial filaments are connected to the membrane's inner surface, and their periodic contraction results in the spiral bacteria's rotary motion and movement. The organism's lipopolysaccharides share a composition with other gram-negative bacteria, yet they can be stained with a carbol fuchsin counter stain. Initially, leptospires were observed using silver staining techniques, and the Warthin-Starry stain was widely employed for histological analysis. To examine these bacteria, techniques such as silver impregnation, staining, immune peroxidase staining, and immunofluorescence were utilized. Because of its narrow dimensions, this organism cannot be observed using a light microscope. Consequently, dark-field or phase-contrast microscopy of wet samples is essential for the direct visualization of spirochetes, as these bacteria do not effectively absorb the stain. The bacteria can be cultivated in artificial media that contain 10% rabbit serum or 1% bovine serum albumin, supplemented with vitamins B1 and B12, long-chain fatty acids, and ammonium salts. The ideal conditions for the organism's growth are temperatures between 28 °C and 30 °C and a pH range of 6.8–7.4. Ellinghausen-McCullough-Johnson-Harris (EMJH) medium, along with Korthof's and Fletcher's media, are among the commonly utilized media. The bacteria exhibit a slow rate of growth, necessitating the disposal of the culture only after 13 weeks post-inoculation. This growth is linked to optimal oxygen levels and is referred to as Dinger's ring or disk. The organism can be preserved for extended periods by repeatedly sub-culturing it in semisolid agar with hemoglobin or by using lyophilization or storage at −70 °C. Bacteria are unable to survive in conditions that are dry, hot, acidic, or treated with basic disinfectants.

Host Range and Susceptibility

Both animals and humans can serve as either primary or incidental hosts. Leptospirosis persists through the continuous occupation of the proximal renal tubules in animals that act as carriers. An animal that is infected may not show any symptoms and can excrete the organism in its urine throughout its lifetime. While dogs and pigs serve as temporary carriers for a few months, rodents typically remain carriers for their entire lives. As a result, rodents are regarded as the primary source of infection. Leptospira serovars are typically sustained in one or more mammalian or marsupial hosts, which may or may not exhibit clinical symptoms. Dogs serve as reservoir hosts for the serovar Canicola, while cattle are the main hosts for the serovar Hardjo. However, this serovar is also sustained in sheep, goats, red deer, and wapiti. Serovar Bratislava and Pomona are maintained in pigs. Sometimes, horses act as maintenance hosts for Bratislava. Rodents and insectivores act as reservoir hosts for a range of serovars, which include those belonging to the serogroups *Icterohaemorrhagiae*, *Grippotyphosa*, and *Sejroe*. Rats play a crucial role as hosts for the serovars *Icterohaemorrhagiae* and *Copenhageni*, both of which are part of the *Icterohaemorrhagiae* serogroup. Pathogenic strains of Leptospira are also carried by various domestic animals and numerous wild creatures, including wild boars, raccoons, and skunks.

Epidemiology

Pathogenic strains of *Leptospira* are distributed worldwide, except Antarctica, and are found mostly in warm, humid environments, such as tropical, subtropical, and temperate zones. Numerous outbreaks have been recorded worldwide after floods, heavy rainfall, and cyclones. Owing to climate change, the frequency of natural calamities is increasing, which is also helping to develop the disease. Therefore, heavy rain, flooding, and high temperatures are factors that influence the development of idle conditions for the disease. Several outbreaks have been recorded from different water bodies, such as swimming pools, fountain drinking water, lakes, rivers, canals, ponds, water rafting, and other similar activities.

The disease is primarily spread through the urine of rodents or infected animals and through contact with skin that has been broken or injured. The disease can also be transmitted through contaminated food or water by ingestion, as the organisms enter the host through the abraded mucosa and conjunctiva. In a population of natural maintenance hosts, infections persist through both vertical and horizontal transmission methods. Aerosol infection can occur via inhalation through the mucous membrane of the respiratory tract.

Humans are regarded as incidental hosts, contracting infections through contact with the urine of infected carrier mammals, either directly or through the contamination of soil, water, or food. Human-to-human transmission is possible by sexual intercourse, breast feeding, and the trans-placental route, but is rarely seen. Although the disease occurs in both sexes, it is predominantly recorded in young adult males from urban and rural areas. It is a very important occupational zoonotic disease in sewer workers, agricultural personnel, veterinarians, veterinary nurses, animal handlers, kennel attendants, dairy farmers, slaughterhouse workers, butchers, mine laborers, construction workers, military personnel, and fishermen. In urban areas, slam dwellers remain at a high risk due to poor sanitation. In rural settings, workers in rice fields and those in sugarcane plantations are highly susceptible to infections.

Pathogenesis

While the exact processes of pathogenesis are not fully understood, it is evident that bacteria invade the host's body through the conjunctiva, minor cuts, mucous membranes, and the walls of the genital tract. Pathogenic Leptospira then rapidly invades the bloodstream or via lymphatics and spreads to all tissues. Corkscrew-like movement is seen in the viscous barrier and connective tissues, and this type of movement helps them become highly invasive in nature. After binding to the host cell, cytokines, such as IL-6, IL-10, and TNF-α, are released along with some antimicrobial peptides to limit the invasive damage caused by the pathogens. By escaping the phagocytosis process, organisms survive and grow exponentially in the bloodstream and tissues with a doubling time of approximately 8 h. Phagocytosis occurs in macrophages in the presence of specific antibodies and complements. The majority of anti-leptospiral antibodies target the organism's LPS, as these LPSs are antigenically active and distinct for each serogroup. The mechanisms of leptospirosis can be divided into two primary categories: the direct effects caused by Leptospira bacteria and the host's immune reaction to the infection. After reaching a higher concentration, the organisms damage the endothelium and tissues because of the endotoxins secreted by the pathogen. The primary lesion manifests in the endothelium of small blood vessels, necrosis of renal tubules, liver and lung damage, causing organ ischemia, meningitis, myositis, and placentitis. Antigens are detected by toll-like receptors and NOD-like receptors within the host's body. Immunity to leptospirosis is primarily humoral. The organisms gather in the host's proximal convoluted tubules, leading to colonization. From this point on, the continuation of the disease and release of the bacteria into the environment occurs through urination. Frequent and unplanned injury to the lining of small blood vessels is regarded as a significant pathological factor in leptospirosis. Research has shown that these pathogens persist and multiply within phagolysosomes.

Clinical Manifestations in Animals

The incubation period varied between 7 and 12 days. Leptospirosis can manifest in three different forms: (i) an acute or severe manifestation of illness resulting from septicemia, marked by indications of endotoxemia, such as hemorrhaging, hepatitis, nephritis, and meningitis; (ii) a subacute or moderately severe condition characterized by nephritis, hepatitis, agalactia, and meningitis; and (iii) chronic form of disease characterized by abortion, still birth, and infertility.

Cattle: During this septicemic phase, extensive intravascular hemolysis may occur because of hemolysin, and hemoglobinuria may be recorded more commonly in calves than in adults. Localization of the pathogen occurs in the kidney if the animal successfully passes the septicemic phase. Vascular damage was also found in the kidney, along with anemic anoxia and hemoglobinuric nephrosis. Calves often exhibit acute or septicemic forms of the disease, which are marked by symptoms such as fever, loss of appetite, eye inflammation, and diarrhea. In more severe instances, they may also experience jaundice, blood in urine, hemolytic anemia, pneumonia, or neurological signs like lack of coordination, excessive salivation, and muscle stiffness.

In abortions of adult cattle, stillbirths sometimes give birth to weak calves with neonatal mortality. Abortion can occur at any stage of pregnancy from 4 months onward in cattle. Abortion in pregnant animals may occur due to invasion of the placenta, followed by degeneration of the placenta by pathogens. Retaining the placenta may be observed in abortive cows, and infertility can be a consequence. In certain instances, there may be a rapid onset of agalactia or a reduction in milk production. The milk becomes thick, yellowish, and tinged with blood, showing minimal signs of inflammation in the mammary glands.

Dogs: The significance of diseases in dogs is considerable because they have the potential to pass infections to humans. The primary clinical signs are generally nonspecific and may include pyrexia, depression, anorexia, stiffness, myalgia, shivering, and weakness. In acute cases, renal failure or jaundice in dogs may be observed. Symptoms include vomiting, dehydration, hepatitis, and death. They excrete organisms in the urine for very long period even after antibiotic therapy. Canicola is the most prevalent serovar, but Icterohaemorrhagiae can lead to severe and potentially fatal infections. Chronic infections may either be asymptomatic or manifest as an unexplained fever.

Pigs: Illnesses in pigs might be more important than those affecting cattle or goats. Pigs often reside in enclosures for extended durations.

Disease in Human

Incubation Period: 2–26 days (generally 7–12 days).

In the mild form, the disease may present as an influenza-like illness with headaches and myalgia. Weil syndrome is the term used for severe leptospirosis, which is marked by symptoms such as jaundice, kidney dysfunction, and a tendency to bleed.

Leptospirosis can occur in two forms: anicteric and icteric.

	Anicteric		Icteric	
	First phage	Second phage	First phage	Second phage
	3–7 days	0–30 days	3–7 days	0–30 days
	Septicemic	Immune	Septicemic	Immune
Clinical findings	Fever, myalgia, headache, conjunctival suffusion (red eye), abdominal pain, vomiting	Meningitis, uveitis, rash, and fever. In cases of leptospirosis, the central nervous system (CNS) is primarily impacted by aseptic meningitis. Complications such as optic neuritis, uveitis, iridocyclitis, chorioretinitis, and peripheral neuropathy tend to be more common during the immune phase	High fever, jaundice, hemorrhages, renal failure and multi organ failure (Weil's syndrome is usually caused by *L. icterohaemorrhagiae*)	

	Anicteric		Icteric	
	First phage	Second phage	First phage	Second phage
	3–7 days	0–30 days	3–7 days	0–30 days
	Septicemic	Immune	Septicemic	Immune
Status	Blood and CSF	Urine	Blood and CSF	Urine

The infection is frequently misdiagnosed because its symptoms are not specific. Individuals suffering from leptospirosis may often be incorrectly diagnosed with conditions such as malaria, dengue hemorrhagic fever, or viral hepatitis.

Diagnosis of Leptospirosis

During the initial 7–10 days, bacteria can be detected in the blood and cerebrospinal fluid, and after this period, the microorganism may appear in fresh urine. Both direct and indirect laboratory techniques are employed to diagnose leptospirosis in humans and animals. The direct method targets the identification of an organism or its components, whereas the indirect method is concerned with detecting specific antibodies.

Direct examination for Leptospira	
Darkfield microscopic examination (DFM)	Dark field microscopic examination can be performed using blood, urine, cerebrospinal fluid, and dialysate fluid. Under a dark field microscope, the organism can be directly observed for the presence of long, motile, coiled leptospires. The materials under examination may be centrifuged to concentrate the Leptospira cells in the deposit. More likely, the preparations can be stained using silver impregnation technique. For applying immunofluorescent technique on routine basis, preparation of conjugates in the laboratory for the specific purpose is recommended against the commercial conjugates. Additional newer techniques for demonstrating leptospires in tissue materials and urine include detection of leptospiral antigens using monoclonal antibodies. It is a qualitative test for initial screening of suspected cases of leptospirosis. Serum is the choice of material for examination under DFM. Test serum should be examined immediately or 3–4 h after the collection. Serum should not be refrigerated or should not be transported on ice, because ice-cold condition will affect the motility of the leptospires.
Culture isolation	Isolation of the organism can be attributed either in liquid medium or by animal inoculation. The EMJH medium, developed by Ellinghausen, McCullough, Johnson, and Harris, is employed to isolate organisms from urine, materials from abortions, uterine discharges, and tissues obtained during autopsy. It may take up to 6 weeks for organisms to develop. Korthof's and Fletcher's medium are also suitable for use. It is important to collect samples for culture before starting antibiotic treatment. Cultures should be incubated at 30 °C and monitored regularly over a span of 4–6 months. Use DFM to observe the growth of leptospira.

PCR	The PCR technique is capable of identifying DNA in the bloodstream from as early as 5–10 days after the disease begins and can continue to do so up to the 15th day. It also facilitates the detection of leptospires in blood samples that are culture-negative if the patient has been administered an effective antimicrobial treatment but has not yet cleared nonviable organisms. Additionally, in urine samples, the PCR method can differentiate between a carrier state and a shedding state
Indirect methods/serological test	
MAT	The Martin and Pettit test, also referred to as the gold standard for diagnosing leptospirosis, involves using live Leptospira cells as the antigen. In seroprevalence research, a titre of 1:100 or higher suggests a current or previous infection. A suspected case is confirmed by detecting a fourfold rise in the titre through paired acute and convalescent serum samples. However, chronic cases may exhibit lower titres, and similar issues arise with host-adapted strains like Hardjo in cattle. Although the test is both sensitive and specific, its application is somewhat restricted in many laboratories due to challenges in maintaining Leptospira strains. Despite its widespread use and numerous efforts by researchers and organizations to standardize it, achieving consistent results across different laboratories remains challenging.
ELISA	A range of techniques has been devised to identify anti-leptospiral antibodies, incorporating various antigen preparations, assay protocols, and platforms, including both plate and dipstick tests. The LEPTO Dipstick assay represents a novel diagnostic approach for leptospirosis, utilizing a broadly reactive antigen to predominantly detect IgM antibodies.

Treatment

Tetracyclines, Penicillins, streptomycin, and dihydrostreptomycin are used to treat leptospirosis. Fluid therapy, blood transfusions, and respiratory support are among the supportive treatments that may be administered, along with other necessary care. Penicillins or doxycycline traditionally have been found the choice of treatment for humans and dogs with leptospirosis. Ceftriaxone and cefotaxime are also having the efficacy in human leptospirosis. Intermittent hemodialysis or continuous renal replacement therapy (CRRT) is recommended in dogs with severe anuric leptospirosis.

Prevention and Control

Completely eliminating the disease seems unlikely because numerous rodent species serve as infection reservoirs. Therefore, effective leptospirosis prevention strategies are based on understanding its epidemiology and transmission methods. The disease can be avoided by reducing exposure and implementing protective measures, vaccination, and chemoprophylaxis either before or after exposure.

Vaccines are administered to animals to safeguard both the animals themselves and the humans who interact with them. Vaccination plays a crucial role in programs aimed at controlling and eradicating diseases. Vaccines for leptospirosis are available and commonly used in dogs, pigs, and cattle. These commercial vaccines, which are whole-cell products, are accessible worldwide for cattle, pigs, and dogs. In household pets like dogs, vaccinating against the bacteria not only protects the animal but also the family members.

(a) Rodent control Denying access to food by proper disposal of garbage and storage of food in rat-proof containers and houses is the most effective way to remove the rodents from domestic environment.
(b) Control measures in animals: Maintaining good hygiene in the animal sheds/houses is the most effective way to control the spread of infection among animals and from animals to man. Effluents from the animal houses should be properly disposed of. Identification and isolation of infected animals and treating them separately is recommended.
(c) Control of infection in man: Human vaccination contributes to a certain level of defense against infections, but it is currently available only in countries like China and Cuba. It is important to inform people about how infections spread and the need to take precautions to avoid contact with environments contaminated by the urine, tissues, or blood of rodents or carrier animals. Scientists and other individuals involved in such high-risk activities should wear protective gear and follow standard hygiene practices. There is an urgent need for carrying out epidemiological study of leptosiprosis in man and animals so that a proper control program can be worked out.

To prevent human leptospirosis, it is essential for high-risk groups such as sewer workers and farmers to practice occupational hygiene, which includes wearing waterproof shoes and gloves.

Bibliography

Bharti, A. R., Nally, J. E., Ricaldi, J. N., et al. (2003). Leptospirosis: A zoonotic disease of global importance. *The Lancet Infectious Diseases, 3*(12), 757–771. https://doi.org/10.1016/s1473-3099(03)00830-2

Di Azevedo, M. I. N., & Lilenbaum, W. (2021). An overview on the molecular diagnosis of animal leptospirosis. *Letters in Applied Microbiology, 72*(5), 496–508. https://doi.org/10.1111/lam.13442

Levett, P. N. (2001). Leptospirosis. *Clinical Microbiology Reviews, 14*(2), 296–326. https://doi.org/10.1128/CMR.14.2.296-326.2001

Rajapakse, S. (2022, January). Leptospirosis: Clinical aspects. *Clinical Medicine (London, England), 22*(1), 14–17. https://doi.org/10.7861/clinmed.2021-0784. PMID: 35078790; PMCID: PMC8813018.

Samrot, A. V., Sean, T. C., Bhavya, K. S., et al. (2021). Leptospiral infection, pathogenesis and its diagnosis—a review. *Pathogens, 10*(2), 145. https://doi.org/10.3390/pathogens10020145. Published 1 Feb 2021.

Listeriosis

12

Ajay Pratap Singh

Abstract

Listeriosis is recognised as lesser-known foodborne illness opportunistic foodborne infection caused by *Listeria monocytogenes* in humans. The disease is manifested in one of the several clinical forms: gastroenteritis, meningitis, neuro-encephalitis, chorioamnionitis, abortions, febrile gastroenteritis, and neonatal infections. The disease is relatively rare in occurrence, however, in immunocompromised patients, the mortality rate can go up to 20–30%. The patients commonly affected are expectant mothers, new-borns, the elderly people with co-morbidities, or people with impaired/compromised immune system. Although the infection can also take place in normal individuals. In addition to humans, *L. monocytogenes* is associated with invasive disease in many different animal species, including farm ruminants (cattle, sheep, and goats). Only some strains of *L. monocytogens* (more common) and *L. ivanovii* are pathogenic and have been associated with diseases.

Keywords

Listeriosis · Circling disease · Silage disease · Listerial abortion

A. P. Singh (✉)
Department of Veterinary Microbiology, College of Veterinary Science & Animal Husbandry, U.P. Pandit Deen Dayal Upadhyaya Pashu Chikitsa Vigyan Vishwavidyalaya Evam Go-Anusandhan Sansthan, Mathura, Uttar Pradesh, India

R. Biswas et al. (eds.), *Concept of One Health and Zoonoses*,
https://doi.org/10.1007/978-981-95-6627-3_12

History

In 1891, *Listeria monocytogenes* were possibly identified from tissue sections of patients in Germany. *L monocytogenes* was isolated by Hülphers in Sweden in 1910 from a rabbit's necrotic liver, and he assigned the scientific name *Bacillus hepatis*. *Listeria monocytogenes* isolation from spinal fluid of meningitis patients was first documented in a study from 1917. In 1924, Murray, Webb, and Swan made the first distinction between a septicaemic condition affecting rabbits and guinea pigs that was characterised by massive mononuclear leucocytosis and an isolated tiny gram-positive bacillus, which they named *Bacterium monocytogenes*. J. H. Harvey Pirie in 1927, advocated renaming the genus to *Listerella* in honour of Josepf Lister. Jahn had proposed a similar homonym for the Mycetozoan in 1906, Pirie suggested rechristening of genus *Listerella* as *Listeria* in 1940. Nyfeldt discovered *Listeria monocytogenes* for the first time in humans in 1929, and in the same year, Gill characterised the circling disease, an *L. monocytogenes*-related ailment in sheep. The first incidence of neonatal listeriosis was reported in 1936, while the first instance of adult listeriosis was recorded in 1929. It wasn't until 1953 that the first instance of foodborne listeriosis was identified, when the mother of twin stillbirths was found to have consumed raw milk from a cow with listerial mastitis. Since Murray's discovery of *Listeria monocytogenes* until 1981, listeriosis has largely been regarded as a disease of veterinary interest, with sporadic human infections being linked to close contact with infected animals. The first significant attention to *Listeria monocytogenes* as a human foodborne pathogen came from a Canadian listeriosis incident in 1981, which was linked to consumption of frozen cabbage. Several listeriosis outbreaks were recorded in North America and Europe in the late 1980s and early 1990s from Ready-to-Eat milk and meat products establishing it as mainstream human pathogen.

Etiology

Currently, the genus Listeria has 26 identified species and six subspecies, and it may be classified into two separate clades based on either strictly or loosely related to *L. monocytogenes*. Before 1985, five sensu stricto species were found, including *L. monocytogenes*, *L. innocua*, *L. ivanovii*, *L. seeligeri*, and *L. welshimeri*. This clade is of significance to public health since it comprises species known to cause human and animal diseases. In 2010, *L. marthii* was classified as the sixth sensu stricto species. Recently, three novel species (*L. farberi*, *L. immobilis*, *L. cossartiae*) were grouped into the sensu stricto clade (Carlin et al., 2021).

Listeria sensu stricto	Listeria sensu lato
L. monocytogenes *L. innocua* *L. ivanovii* subsp. *ivanovii* and subsp. *londoniensis* *L. seeligeri* *L. marthii* *L. welshimeri* *L. farberi* *L. immobilis* *L. cossartiae* subsp. *cossartiae* and subsp. *cayugensis*	*L. grayi* subsp. *grayi* and subsp. *murrayi* *L. fleischmannii* subsp. *fleischmannii* and subsp. *coloradonensis* *L. floridensis* *L. aquatica* *L. costaricensis* *L. goaensis* *L. thailandensis* *L. valentina* *L. newyorkensis* *L. cornellensis* *L. rocourtiae* *L. weihenstephanensis* *L. grandensis* *L. riparia* *L. booriae* *L. portnoyi* *L. rustica*

Listeria monocytogenes is a bacterium that may infect both humans and animals, and *L. ivanovii* is precisely pathogenic to sheep and cattle. *L. ivanovii* and *L. seeligeri* have occasionally been connected to human infections. The majority of recorded cases of human infection are caused by *L. monocytogenes*, making it by far the most significant and prevalent. *Listeria monocytogenes* is a significant foodborne pathogen that can result in life-threatening disease.

Listeria Monocytogenes

Listeria monocytogenes are small Gram-positive rod, 1–2 µm in length and 0.5 µm wide, with bacteria often lying parallel to each other in palisades and thread-like forms present in some cultures. Growth occurs between 3 °C and 45 °C but the optimum temperature is 30–37 °C. The organism multiplies readily in aerobic or microaerophilic conditions at pH values as high as 9.6. Growth is absent or scanty in complete anaerobic conditions and multiplication is inhibited by pH values lower than 5.6. Although they are facultative anaerobes, the growth is enhanced by 5–10% CO_2. They are catalase-positive, oxidase-negative, hydrolyse aesculin, tolerate 10% salt (NaCl), and will not grow on MacConkey agar. They are widely spread in nature and have been isolated from both sick and healthy animals as well as from the soil, water, silage, and sewage. Based on somatic (O) and flagellar (H) antigens, 13 serotypes of *L. monocytogenes* have been described and divided into four phylogenetic subdivisions, lineages I to IV (Yin et al., 2019). Lineage-I includes serotypes 1/2b, 3b, 4b, 4d, 4e, and 7; lineage II includes serotypes 1/2a, 1/2c, 3a, and 3c; lineage III and IV include serotypes 4a, 4ab, 4c, and atypical 4b. The 14th serotype of *L. monocytogenes* Serovar 4 h was recently reported.

Epidemiology

Reservoirs

Listeria monocytogenes is a common environmental pathogen that thrives in decaying plant materials, soil, and water as a saprophyte. Additionally, *L. monocytogenes* has been discovered in effluent, sewage, silage, and slaughterhouse waste. Domesticated ruminants most likely play a significant part in the maintenance of *Listeria* spp. in the rural environment, through a continual faecal-oral enrichment cycle. The bacteria may be excreted in the milk of infected goats, cows, and ewes or shed through faeces, making cattle, sheep, goats, and poultry possible reservoirs. The pathogen can be recovered from the faeces of both pregnant women and healthy people and the human gastrointestinal tract acts as a significant reservoir for the infection. While 2–10% of the general population are asymptomatic carriers, the percentage of people whose faecal samples have tested positive for *L. monocytogenes* ranges from a low 0.5% to as high as 29%. *L. monocytogenes* in the faeces is not always correlated with the signs of illness, although it can cause epidemics in the population through secondary transmission. A small percentage of asymptomatic, healthy participants had the bacterium found in their faeces. Several species of birds are susceptible to *Listeria* infection, including chickens, turkeys, geese, ducks, pigeons, parrots, wood grouse, snowy owls, eagles, and canaries. By contaminating the environment and litter during poultry production, chickens act as primary carriers and reservoirs of *Listeria* and contribute significantly to its spread.

Associated Serovars

Of the 98% of human listeriosis outbreaks, only four *L. monocytogenes* serotypes—1/2a, 1/2b, 1/2c, and 4b—have been identified. Serotype 1/2a strains have been found in a wide range of foods all over the world and account for more than 50% of the *L. monocytogenes* isolates found in food and the environment, whereas serotype 4b is primarily responsible for listeriosis in humans.

Incidence

Listeria monocytogenes is rarely associated with human-related infections with sporadic occurrence of human listeriosis cases. Food contamination is thought to be the main means of transmission and may account for up to 99% of cases. Although listeriosis is a relatively rare foodborne illness, its incidence in humans ranges between 0.1 and 1.1 cases/million with an average case fatality rate close to 30%. There were 2493 cases, including 499 deaths, due to foodborne listeriosis in the United States during 1996–1997. The individuals with underlying disease condition leading to immunosuppression such as hematologic malignancies, solid tumours, organ transplantation (specially renal), AIDS, treatment with tumour necrosis factor-alpha antagonists, diabetes mellitus, end-stage renal disease (including haemodialysis and peritoneal dialysis), iron overload, collagen-vascular diseases, other serious chronic non-malignant diseases, liver disease and alcoholism or undergoing glucocorticoid therapy are at increased risk of *Listeria monocytogenes* infection. Among all the foodborne pathogens, *L. monocytogenes* had the second-highest case fatality rate (21%) with 90.5% hospitalisation rate. The incidence of listeriosis in males and females is approximately equal.

Transmission

While the modes of transmission for *L. monocytogenes* can include vertical (mother to child), zoonotic (animal to man), and nosocomial (hospital acquired), it is generally considered that most cases of human listeriosis involve foodborne transmission.

Animal to Human Transmission

Foodborne Transmission

Human listeriosis outbreaks are frequently linked to food-borne illness. Although a wide range of foods may be contaminated with *L. monocytogenes*, cases of listeriosis seem to be mostly linked to ready-to-eat (RTE) food products, because of their capacity to withstand high salt concentrations, survive at relatively low pHs, and multiply at refrigerated temperatures. The majority of human-associated outbreaks (85%) are attributed to foodborne *Listeria* infections. The earliest reported human incidences directly linked consumption of dairy products, with cow milk being identified as a source of the listeriosis (Henderson et al., 2020). Microbiological studies have revealed that *L. monoytogenes* can be found in raw milk samples (2.5–6% samples). Specific incidents of listeriosis or serious outbreaks of this disease might arise from consuming milk and dairy products contaminated with *L. monocytogenes*. The most frequent sources of listeriosis among all dairy products are soft cheeses and unpasteurised milk. Additionally, the bacteria create biofilms on the equipment's surfaces, which are highly resistant to sanitation procedures and can serve as a persistent source of post-pasteurisation contamination.

Listeriosis outbreaks are more frequently linked to frozen produce, such as broccoli, cauliflower, peas, onions, peppers, sweet potatoes, maize, blueberries, cranberries, peaches, raspberries, and strawberries. They are also more frequently linked to chopped celery, pre-packaged salad mixes, asparagus soup, and other similar foods. Furthermore, new food sources such as cantaloupe melons, stone fruits like nectarines, peaches, plums, have been linked to a number of human Listeria outbreaks. Vegetable contamination can happen after harvest, possibly from contaminated soil, or it might happen when the veggies are being sold. Agricultural land may become contaminated by manure that contains a high level of pathogens due to faecal shedding of bacteria. The survival of bacteria in soil is significantly influenced by the soil pH, microbiota, and moisture level.

Prepared, ready-to-eat meat and meat products, such as cooked, ready-to-eat sausages, sliced meats, chilled pâtés, or meat spreads, have become more popular in developed nations. *L. monocytogenes* is frequently found in meat products. *L. monocytogenes* may contaminate ready-to-eat meat products at a number of points, including the raw material, during the manufacturing process, or by contact with contaminated unprocessed raw materials, dirty surfaces, or people. Food processing equipment like slicers, spiral freezers, packaging equipment, and conveyor belts are frequently contaminated with *L. monocytogenes*. Additionally, contamination of meat products is possible during postharvest processing (heat treatment, smoking, fermentation process).

L. monocytogenes, having the demonstrable ability to grow in the presence of high salt concentration, is of particular significance to the seafood industry. Several varieties of seafood, including raw shrimp, lobsters, crabs, mussels, fish, and ready-to-eat smoked fish, have been found to possess *L. monocytogenes*. *L. monocytogenes* is occasionally found in raw seafood, with levels varying from 1% to 10%, while 3–10% incidences have been reported in RTE seafood items. The likelihood of increased contamination during processing appears to be higher in the seafood sectors as a result of the contaminated water usage during processing. Humans may carry *L. monocytogenes* and can contaminate the surface of processing plants.

Zoonotic

Veterinarians, farm workers, abattoir workers are at increased risk of contracting the disease. *Listeria* can enter the bloodstream through superficial skin sores, conjunctiva, mucous membranes, or by ingesting food or drink that has been contaminated with saliva, faeces, nasal secretions, or aborted animal tissues. Cases of cutaneous listeriosis are typically caused by occupational exposure to infectious material from bovine abortion, which typically affects farm workers or veterinarians after an obstetrical procedure. The new-born may exhibit skin lesions as a result of prenatal infection (*granulomatosis infantiseptica*).

Vertical

L. monocytogenes in maternal blood stream can make direct contact with placental cells. *L. monocytogenes* can cross the placental barrier using cell-to-cell spread to travel through placental cells to reach the foetus. Additional pathways for *L. monocytogenes* entrance into the placenta could be provided by bacterial translocation in maternal immune cells.

Animal-to-Animal Transmission

Feeding of spoiled silage to ruminants may serve as major source of *Listeria* infection. *Listeria* grow rapidly in improperly stored silage and cause herd epidemics. Rabbits and rodents are more likely to contract listeriosis than dogs and cats. It is believed that contaminated food is the cause of the infection in rodents and pets. Consuming food and water contaminated with saliva, faeces, nasal secretions, and aborted animal material can spread listeriosis. Rodents and carrier animals can also transmit the infection.

Pathogenesis

Both *L. monocytogenes* and *L. ivanovii* are facultative intracellular parasites that can infiltrate a range of normally non-phagocytic cells, including endothelial, epithelial, and hepatocyte cells, as well as precise phagocytic cells like macrophages. Pathogenic listeriae primarily enter the body through the intestine, where they penetrate the lining of the digestive system and initiate the infections. The number of organisms consumed, the host's susceptibility, and the virulence of the particular strain are few of the variables that affect the invasion of intestinal tissue. The ability of *L. monocyogenes* to successfully cross intestinal barrier is essential for its systemic infection (blood-brain and placental). The three possible routes for cross-intestinal translocation are as follows.

1. *M cell transcytosis:* Microfold (M) cells are primarily found in the follicle-associated epithelium that covers the surfaces of intestinal lymphoid tissues like Peyer's patches and isolated lymphoid follicles. M cells are actively involved in generalised transport of many enteric pathogens including *L. monocytogenes* from the intestinal lumen into the subepithelial region by a transepithelial transport system known as transcytosis.
2. *InlA-mediated uptake: L. monocytogenes* internalin A (InlA) and internalin B (InlB) specifically interact to the E-cadhrin receptor located on the basolateral surface of the host cell. InlA facilitates the internalisation of luminal *L. monocytogenes* by transcytosis in the form of a membrane-bound vacuole (Drolia & Bhunia, 2019). *L. monocytogenes* escapes the vacuole by secreting the pore-forming toxin listeriolysin O (LLO) and two phospholipases [broad specific phosphatidylcholine phospholipase C (PC-PLC), and phosphatidylinositol-specific phospholipase C (PI-PLC)] to enter the cytoplasm. Within cytoplasm, a bacterial protein ActA, promotes actin polymerisation in the host cell and gives *L. monocytogenes* the ability to travel across the cytosol and finally force admission into nearby cells, hence facilitating its spreads from cell to cell, as most of the strains are non-motile at 37 °C.
3. *Alternate Listeria adhesion protein (LAP) mediated translocation:* Pathogenic *Listeria* species secrete Listeria adhesion protein (LAP), a common alcohol acetaldehyde dehydrogenase protein. Once secreted, the LAP interacts with the heat shock protein 60 (Hsp60) receptor present in apical side of the host cell. NF-B signalling is initiated by LAP binding to Hsp60, which then activates myosin light chain kinase (MLCK). Following activation of MLCK, the main TJ proteins (claudin-1 and occludin) and AJ protein (E-cad) are redistributed within cells, which led to the increased intestinal permeability and the disruption of the intestinal epithelial barrier. Additionally, in the early infection phase (24–48 h), LAP-mediated epithelial cytokine production (TNF-∞ and IL-6) enhances epi-

thelial barrier dysfunction and *L. monocytogenes* translocation, while evading innate immune defences (Ryan, 2016).

Following intestinal translocation, the liver is believed to be their initial target organ. Listeriae multiply rapidly in the liver until the infection is self-limited by a cell-mediated immune response. Listeriae growth in the liver may cause a protracted low-level bacteraemia, which could progress to invasion of the secondary target organs (brain and the gravid uterus). *L. monocytogenes* is more easily able to enter the uterine environment when the maternal blood is in direct touch with the syncytiotrophoblasts placental cells and extra-villous cytotrophoblasts. Furthermore, the movement of *L. monocytogenes* within the immune cells in a concealed manner may promote its conspicuous uterine spread.

Hematogenous spread from the digestive tract allows the bacterium to enter the central nervous system (CNS). The choroid plexus epithelium gives *L. monocytogenes* access to the CNS and results in a meningitis infection. On the other hand, *L. monocytogenes* can enter the brain parenchyma through the cerebral capillary endothelium, and *L. monocytogenes* infected macrophages can enter the middle cerebral artery and pass-through the endothelial cells, causing cerebritis, which results in the development of brain abscesses. According to a recent animal study, the infection can propagate along the axons and cause more lesions by moving inside the trigeminal nerve's axons once the bacteria have entered the CNS through the peripheral nervous system.

Clinical Symptom

Human Listeriosis

Febrile Gastrointestinal Infections Consumption of food with exceptionally high levels of *L. monocytogenes* ($>10^7$/g) results in a non-invasive form of listeriosis that manifests gastroenteritis with vomiting and diarrhoea, usually self-resolving. Typical signs and symptoms associated with febrile listerial gastroenteritis include chills, fever, diarrhoea, headache, abdominal pain and cramps, nausea, vomiting, fatigue, joint and muscle pain, and myalgia.

Perinatal Infections Listeriosis primarily affects women in their third trimester of pregnancy and is characterised by a moderate "flu-like" disease with symptoms like fever, chills, malaise, arthralgia, back pain, and diarrhoea. In many cases, the infection is undetectable or subclinical, yet it can still cause premature birth, foetal death, and stillbirth (Barikbin et al., 2016). *Listeria* infection during the first trimester may cause a sudden abortion. In later stages of pregnancy, the outcome could be a stillbirth or a severely ill new-born. There is little evidence that listeriosis increases the likelihood of complications in future pregnancies. It is rarely severe or life-threatening to the mother.

Neonatal Infection Listeriosis can manifest in new-borns in early-onset or late-onset forms. *Granulomatosis infantisepticum* is an early-onset condition that manifests in new-borns less than 7 days old and is characterised by miliary micro abscesses and granulomas, which most frequently affect the skin, liver, and spleen. Premature birth, respiratory distress, and circulatory failure are the hallmarks of early-onset listeriosis. The majority of early-onset cases have sepsis, and roughly 20% have meningitis. Late-onset neonatal illness appears 10–30 days after delivery in mothers who are asymptomatic. Most of the time, neonates with late-onset are healthy and full-term births. Neonatal listeriosis, which often appears one to several weeks after delivery and is characterised by extensive granulomas, late-onset meningitis, and early-onset pneumonia and/or sepsis in neonates, who survive listeriosis with mortality rate typically in the range of 20–30%.

Septicaemic Infections/Meningoencephalitis or Sepsis The most frequent clinical manifestations of listeriosis are sepsis and meningitis. Seizures, tremor, ataxia, nuchal stiffness, and high fever are all typical signs of listeriosis of the CNS. Encephalitis involving the brainstem is the most prevalent form of non-meningitis CNS listeriosis (rhombencephalitis).

Occupational Cutaneous Infections Widespread pustular rashes on the hands and arms can result from direct contact with infected materials, accompanied by headache, myalgia, and fever. Hematogenous dispersion in immunocompromised hosts with invasive illness may result in skin lesions and associated regional lymphangitis and adenopathy. Skin listeriosis could become self-limiting; however, symptoms could linger for longer than a month without antibiotics.

Glandular Listeriosis Resembles infectious mononucleosis with swelling of the salivary glands and nuchal lymph nodes.

Typhoid Listeriosis Characterised by high fever and is particularly frequent in immunocompromised individuals.

Atypical Listeriosis Rare cases of have been described with symptoms such as endocarditis, purulent (mononuclear) pleural exudates, pneumonia, urethritis, and abscesses.

Animal Listeriosis

L. monocytogenes mostly causes encephalitis and uterine infections in ruminants. Clinically ill animals can shed *L. monocytogenes* in their faeces, but healthy animals can also carry the infection latently. Neurological symptoms as circling, excessive salivation, and unilateral facial paralysis mark the encephalitic form of animal listeriosis. Ingestion of the bacterium results in the septicemic/visceral form in young animals, which mostly affects the gastrointestinal tract. Intestinal epithelial cells and specialised epithelial cells covering Peyer's patches are major predilection site for bacteria. The occult infection may develop into bacteraemia and culminate into fatal septicaemia.

Neonatal septicaemia or late-term abortions are characteristics of uterine infections. Some *Listeria* sp. infections can cause foetal infection, which can result in stillbirths and neonatal deaths. Accidental inoculation of *L. monocytogenes* into the eye, especially during silage feeding may results in keratitis and eye infections in ruminants. Animals exhibiting CN VII dysfunction (Facial Nerve dysfunction), exposure keratitis may develop due to loss of palpebral movement. The udder is not easily invaded by *L. monocytogenes;* hence, mastitis is uncommon. If mastitis occurs, the animal may have prolonged shedding of the bacteria in the milk.

L. monocytogenes seldom cause disease in dogs or cats. Pets infected with *L. monocytogenes* often exhibit only moderate gastrointestinal symptoms, such as diarrhoea and vomiting. Occasionally, pets may witness more severe symptoms, including fever, muscle discomfort, breathing issues, preterm delivery, and even death. Dogs and cats can contract listeriosis after eating contaminated pet food without showing any clinical signs. More often, these carrier animals excrete *L. monocytogenes* in their faeces, and possible transmission to human contacts.

Diagnosis

Isolation and Identification

L. monocytogenes is readily isolated from clinical specimens obtained from normally sterile sites such as cerebrospinal fluid, blood, and amniotic fluid.

Pre-enrichment In most instances, directly plating of suspected material onto conventional media fails to grow *L. monocytogenes*. Hence, for selective isolation of *Listeria monocytogenes* from faeces need cold enrichment of samples at 4 °C for several weeks followed by selective enrichment and plating at higher temperatures (30 °C). Because this lengthy procedure is not amenable for routine testing of food or biologic specimens, a shortened enrichment procedure comprising of pre-enrichment culture in a liquid medium such as buffered *Listeria* Enrichment Broth (BLEB) supplemented with acriflavine (10 mg/mL) and nalidixic acid (40 mg/mL) and cycloheximide (50 mg/mL) and incubation at 30 °C for 24 h is more popularly used for routine isolation (U.S. Food and Drug Administration, 2017).

Selective Isolation *L. monocytogenes* grows on a wide variety of non-selective plating media (e.g., blood agar, chocolate agar, tryptic soy agar, and brain heart infusion agar). Addition of acriflavine, lithium chloride, nalidixic acid, nitrofurazone, potassium tellurite, and potassium thiocyanate may be required for selective isolation of *L. monocytogenes* from food and contaminated samples. The pre-enriched samples are inoculated onto Esculin-based *Listeria* selective agars viz. Oxford agar containing Polymyxin B sulfate Acriflavine Ceftazidine as selective agent, Polymyxin Acriflavin Lithium-chloride Ceftazidime Esculin Mannitol (PALCAM) agar, Lithium Chloride-Phenylethanol-Moxalactam (LPM) Medium, Modified Oxford Agar (MOX) containing colistin methane sulfonate (1% w/v), and Sodium (or ammonium) moxalactam or Chromogenic agars (RAPID'L.mono™, CHROMagar™ Listeria). On PALCAM agar, *Listeria* colonies appear grey-green, are approximately 2 mm in diameter, and have black, sunken centres. They also have a black halo against a cherry red background Oxford agar was useful for the isolation of *L. monocytogenes* from milk and dairy products.

Cultural and Staining Characteristics

Small round transparent colonies appear on the blood agar after 24 hours of incubation. They become greyish-white with a diameter of 0.5–2 mm after 48 h of incubation. *Listeria monocytogenes* and the non-pathogenic *L. seelegeri* produce a narrow zone whereas *L. ivanovii* produces a wide zone of β-haemolysis. Sometimes haemolysis can only be noted under the colonies. Smooth and rough colony forms may be seen. Gram-stain reveals the presence of Gram-positive rods. Primary cultures are often pleomorphic and may be confused with *Corynebacterium* and *Streptococcus* species. The organisms easily decolourise and may be mistaken for Gram-negative rods. All the *Listeria* species hydrolyse aesculin. *L. monocytogenes* has a typical tumbling motility in a 2–4 h broth culture incubated at 25 °C when examined by the hanging drop method. When grown on semi-solid motility medium or gelatine, they have a characteristic umbrella-shaped growth on the subsurface. Modified CAMP tests using *Staphylocccus aureus* and *Rhodococcus equi* help distinguish the β-haemolytic *Listeria* species.

Animal Inoculation As many strains of *L. monocytogenes* and *L. ivanovii* can be non-pathogenic, laboratory animals can be used to assess the virulence of this agent.

1. *Anton conjunctivitis test:* inoculation into the conjunctiva of a rabbit or guinea pig. *L. monocytogenes* will cause a purulent keratoconjunctivitis within 24–36 h after inoculation.
2. *Immunocompromised Mouse Pathogenicity Test:* Intraperitoneal inoculation of carrageenan treated mice with a broth culture of either *L. monocytogenes* or *L. ivanovii*. Virulent *Listeria* strain causes death within 5 days with necrotic lesions present in the liver.

Treatment

Antibiotic treatment is seldom required in most people with intestinal illness. In human patients with systemic infection, a variety of antibiotics like penicillin, ampicillin, and amoxicillin have been used extensively, however, ampicillin alone or in combination with gentamicin remains the treatment of choice (Temple & Nahata,

2000). Treatment with trimethoprim/sulfamethoxazole (TMP/SMX) offers an alternative in patients with known allergic to penicillin. Vancomycin has been indicated in listerial meningitis and endocarditis. Other antibiotics used include erythromycin, meropenem, linezolid, and rifampin.

Control Measures

Control Measures at Food Production System Feedstuffs and low-quality silages are major sources of Listeria contamination, which is linked to listeriosis outbreaks in animals. The growth and survival of Listeria in silage materials can be prevented by proper storage and quick pH acidification (lower than 4.5). By properly disinfecting soils, animal ingestion of soil-contaminated pastures may be minimised. Birds, mice, and rats should not be allowed access because they can contaminate feed, feed bunkers, water, and other surfaces on farms. Damages of silage-making facilities inflicted by birds or other vermin can significantly permit oxygen penetration through silages and resulting increase in bacterial growth. Agricultural farm contamination can be minimised by proper manure management and storage.

Control Measures at Food Processing System Handling areas and equipment for raw milk and pasteurised products must be separated. Unauthorised personnel, such as truck drivers, receivers, and raw product handlers from the processing area, must be prohibited from entering the clean area. It is necessary to put in place an internal monitoring system for cleaning, sanitising of the environment that includes routine cleaning and inspecting plant floors and drains. Avoid producing aerosols during processing, especially in storage and packaging areas. Keep track of daily cleaning, sanitising, and maintenance of equipment to avoid post-processing contamination. The employees must be adequately trained and advised in personal hygiene and health, good manufacturing practices, and plant sanitation practises. A newer approach to inhibit growth of *L. monocytogenes* on RTE foods is addition of bacteriocins [cationic antimicrobial peptides (CAMPs) produced by bacteria that inhibit the growth of a broad spectrum of pathogens]. Nisin and pediocin are FDA-approved bacteriocins to be used in a variety of food products.

Control Measures at Household

Avoid High-risk Foods Pregnant women, geriatric persons, and immunocompromised patients should avoid foods more likely to be contaminated by *Listeria monocytogenes*, such as unpasteurised dairy products, soft cheeses, raw vegetables, prepared deli meats and salads, refrigerated meat spreads, and smoked seafood (World Health Organization, 2004). Freshly pasteurised milk and yoghurt, washed and prepared vegetables and salads, and cooked foods to be consumed immediately.

Prepare Food Safely Keep raw meat and poultry separate from other foods. Before eating raw vegetables and fruits, thoroughly wash them under running water. Cook all animal-origin foods thoroughly, including beef, pork, poultry, fish, and eggs. Wash all kitchen surfaces, cutting boards, and utensils in hot, soapy water after finishing. Separate uncooked meats from vegetables, cooked foods, and ready-to-eat foods. Cook meat, poultry, and egg dishes until the centre temperature reaches 160 °F. After handling raw foods, wash your hands, knives, and cutting boards. Perishable and ready-to-eat foods should be consumed as soon as possible. Recently, a team of researchers have developed a probiotic based on bioengineered strain of *Lactobacillus*, which could prevent infections in high-risk persons by blocking the pathway translocation of *Listeria monocytogenes* from the intestinal wall cells into the bloodstream.

References

Barikbin, P., Sallmon, H., Hüseman, D., Sarioglu, N., Weichert, A., von Weizsäcker, K., Bührer, C., & Koehne, P. (2016). Clinical, laboratory, and placental findings in perinatal listeriosis. *Fetal and Pediatric Pathology, 35*(5), 307–314.

Carlin, C. R., Liao, J., Weller, D., Guo, X., Orsi, R., & Wiedmann, M. (2021). *Listeria cossartiae* sp. nov., *Listeria immobilis* sp. nov., *Listeria portnoyi* sp. nov. and *Listeria rustica* sp. nov., isolated from agricultural water and natural environments. *International Journal of Systematic and Evolutionary Microbiology, 71*(5), 004795.

Drolia, R., & Bhunia, A. K. (2019). Crossing the intestinal barrier via listeria adhesion protein and internalin A. *Trends in Microbiology, 27*(5), 408–425.

Henderson, L. O., Erazo Flores, B. J., Skeens, J., Kent, D., Murphy, S. I., Wiedmann, M., & Guariglia-Oropeza, V. (2020). Nevertheless, she resisted–role of the environment on *Listeria monocytogenes* sensitivity to nisin treatment in a laboratory cheese model. *Frontiers in Microbiology, 11*, 635.

Ryan, V. E. (2016). *Recombinant Listeria adhesion protein expressing probiotics protect against Listeria monocytogenes infection in animal models*. Master Thesis, Purdue University, West Lafayette, Indiana Open Access Theses. 894. https://docs.lib.purdue.edu/open_access_theses/894

Temple, M. E., & Nahata, M. C. (2000). Treatment of listeriosis. *Annals of Pharmacotherapy, 34*(5), 656–661.

U.S. Food and Drug Administration. (2017). Chapter 10: Detection of Listeria monocytogenes in foods and environmental samples, and enumeration of Listeria monocytogenes in foods. In *Bacteriological analytical manual*. Available at: https://www.fdagov/food/laboratory-methods-food/bam-detection-and-enumerationlisteria-monocytogenes

World Health Organization. (2004). *Risk assessment of Listeria monocytogenes in ready-to-eat foods: Technical report*. Food and Agriculture Organization.

Yin, Y., Yao, H., Doijad, S., Kong, S., Shen, Y., Cai, X., Tan, W., Wang, Y., Feng, Y., Ling, Z., & Wang, G. (2019). A hybrid sub-lineage of *Listeria monocytogenes* comprising hypervirulent isolates. *Nature Communications, 10*(1), 1–16.

Plague

13

Ripan Biswas and Indranil Samanta

Abstract

Plague is a zoonotic illness triggered by the bacterium *Yersinia pestis*, has precipitated severe pandemic outbreaks across history, and remains a pressing public health concern at present. This chapter offers a concise overview of plague, encompassing its epidemiology, transmission pathways, clinical presentations, and control strategies. Plague predominantly impacts rodent population, with human infections arising from bites by fleas that are carriers of the disease. In humans, the disease appears in three primary forms: bubonic, septicemic, and pneumonic. The bubonic variant, which is the most common, is marked by the enlargement of lymph nodes. Septicemic plague develops when the infection disseminates to the circulatory system, whereas pneumonic plague impacts the respiratory system and may be transmitted directly between individuals. The chapter emphasizes the critical role of timely identification and antibiotic therapy in the effective management of plague. The text moreover highlights the necessity of coordinated monitoring frameworks, rodent population control interventions, and public health education initiatives to preempt and manage plague outbreaks effectively.

Keywords

Anthrax · Black death · Pestis minor · Septicemic plague · The great plague "Ring a Ring o' Rosie"—a plague rhyme · Pneumonic plague · Bubonic plague

R. Biswas (✉)
Department of Veterinary Public Health and Epidemiology, West Bengal University of Animal and Fishery Sciences, Kolkata, West Bengal, India

I. Samanta
Department of Veterinary Microbiology, West Bengal University of Animal and Fishery Sciences, Kolkata, West Bengal, India

R. Biswas et al. (eds.), *Concept of One Health and Zoonoses*,
https://doi.org/10.1007/978-981-95-6627-3_13

History

Plague ranks among the earliest recorded infectious diseases and has profoundly influenced the political history, art, and religious beliefs of humanity. The bacterium *Yersinia pestis*, which is responsible for the disease, was named after Alexandre Yersin, a bacteriologist of French-Swiss origin. It is believed that plague has aggrieved human population from thousands of years. Through the use of molecular tools, scientists have discovered the genome of the plague bacterium in two skeletons from the Bronze Age, which are about 3800 years old. The first major plague epidemic recorded in history took place in China in 224 BC. Three major plague pandemics have been documented in sixth century, fourteenth century, and nineteenth century.

In the sixth century, the Justinian Plague, named after the Byzantine ruler Justinian I, is believed to have originated in either India or China. It reached Constantinople, the Roman Empire's capital, around 541–542 AD. This devastating pandemic lasted for two centuries, from 541 to 750 AD, and unfolded in at least 18 waves, claiming the lives of over 100 million individuals. The Mediterranean basin was primarily affected, and the plague is thought to have significantly contributed to the Roman Empire's decline.

The second pandemic, known as the "Black Death" or the great plague, arose in the fourteenth century. It is believed to have originated in central Asia, sparking epidemics in China and India. The disease then spread along caravan routes to the Near and Middle East, eventually reaching Messina and Sicily via Genoese ships that transported rats infested with fleas. This devastating bubonic plague, dubbed the "Black Death," resulted in the deaths of at least one-third of Europe's population, exceeding 25 million people during the fourteenth century.

The third plague pandemic began around 1855 in Yunnan Province, located in southwest China, although outbreaks had been occurring since 1772, eventually spreading to Taiwan and Hong Kong. From 1910 to 1920, the disease spread to Japan, India, Australia, and both North and South America through maritime routes. In India alone, around 20 million people perished over the two decades from 1898 to 1918. It has been suggested that the death toll from plagues throughout the entirety of recorded history could likely reach 200 million. Amid the recent pandemic, researchers pinpointed a bacterium as the cause and established that the plague spreads through bites from fleas carrying the infection.

In 1894, Alexandre Yersin was the first to identify the plague bacillus in Hong Kong by analyzing the enlarged lymph nodes, or buboes, of a patient suffering from the plague. In 1898, French scientist Paul-Louis Simond was tasked with examining a plague epidemic in Bombay, now Mumbai, India. During his investigation, he discovered the presence of the plague bacillus in the tissues of rats that had died. He subsequently theorized that the organism was spread from rat to rat and from rats to humans through the agency of rat fleas. In the years after 1900, the worldwide dissemination of the plague was limited by enforcing rules that managed rat populations in port areas and mandated the inspection and rat-proofing of vessels.

In 1967, India experienced a significant decline in the plague mortality rate, dropping from 183 to 1.8 per population, and eventually reaching zero. The last recorded case was in the Mulbagal area of Kolar district in 1966. In the years 1983 and 1984, there were occasional reports of suspected human cases from the regions of Himachal Pradesh and Karnataka. The National Centre for Disease Control's plague surveillance unit, formerly known as the National Institute of Communicable Diseases, identified and documented cases of sylvatic plague in the area where Karnataka, Andhra Pradesh, and Tamil Nadu meet. In 1994, the discontinuation of the plague surveillance and control unit led to the reappearance of both bubonic and pneumonic plague in the Beed district of Maharashtra and the city of Surat in Gujarat. In 2002, Hatkoti in Shimla, Himachal Pradesh, experienced a localized outbreak of pneumonic plague, which was followed by another outbreak in 2004 in Dangud, Uttarkashi, Uttarakhand.

Etiological Agent and Ecology

Plague is a notable zoonotic disease of bacterial origin, triggered by *Yersinia pestis*, a species within the *Enterobacteriaceae* family. *Y. pestis* is a Gram-negative, aerobic, pleomorphic, non-motile, non-sporulating, facultative intracellular bacillus or coccobacillus that multiplies in macrophages which exhibit bipolar staining property (hair pin appearance) with Wright, Giemsa or Wayson stains. *Y. pestis* is characterized by three virulent biovars—Antiqua, Medievalis, and Orientalis—and one avirulent biovar, Microtus, all of which are zoonotic by nature. The organism can grow at wide range of temperature, i.e., 4–40 °C but the optimum growth is seen at 28–30 °C within 24–72 h. The bacteria thrive best at a pH of 7.4. *Y. pestis* is susceptible to disinfectants such as 1% sodium hypochlorite, 2% glutaraldehyde, formaldehyde, 70% ethanol, and those based on iodine and phenol. The organism can be inactivated by exposure to autoclave and hot air oven.

Host Range and Susceptibility

Plague predominantly affects rodents and lagomorphs, but humans may accidentally contract the disease through the bite of an infected rat flea or by coming into direct contact with an infected host, which could be either a human or an animal. The infection is maintained in nature by one or more maintenance or enzootic host like wild rodents. The organism has been documented in more than 200 rodent species, with notable presence in prairie dogs, ground squirrels, antelope ground squirrels, chipmunks, rats, wood rats, mice, Siberian marmots, voles, jerboas, and some gerbils. In felids, house cat, wild cats and mountain lions are susceptible to plague. Carnivores can be infected after the ingestion of contaminated rodents. Camels are highly susceptible to the disease if exposed to dead- infected rodent, contaminated carcasses, or excrement. The disease has been also reported from deer, dogs, goats, coyotes, foxes, badger, skunks, and primates. A recent study has determined that soil serves as a potential habitat for the plague organism, capable of surviving for months to several years alongside wild rodents. Another study documented, plague organism in protozoa, including amoebae.

Epidemiology

The natural reservoir of plague organism is wild rodents. Other rodents and mammals also act the reservoir which has been discussed in the previous section. The disease is maintained in nature by using a cycle between wild rodents or logomorphs and fleas. Intermittently, the reservoir animals experience epizootics with the influence of different environmental and biological factors which increase the chances to transmit the disease to other species of animals. Human can get the infection from the tissue of wild reservoirs directly or by infected fleas. Sometimes domesticated animals can act as the "channel" which can carry the organism from the wildlife and transmit into human. In urban areas, especially rats are capable to cause epidemic in human population. The oriental rat flea, *Xenopsylla cheopis*, is the primary biological vector responsible for the widespread transmission of the disease. Although other fleas, such as *Ctenocephalides* spp. (dog and cat fleas) and *Pulex irritans* (human flea), can also carry the plague organism, they are significantly less effective as vectors compared to *Xenopsylla cheopis*. Both sexes of flea can bite and transmit the disease.

A rat flea has the ability to consume up to 0.0005 mL of blood, which might contain as many as 5000 plague bacilli. Then the bacilli multiply in the gut of rat flea and block the proventriculus of the rat flea. Due to persistent colonization of bacteria, biofilm is formed in the proventriculus which blocks the passage of blood through the gut of the flea. Such flea is called "blocked flea." Blocked flea transmits the bacteria to the susceptible host by regurgitation at flea bite site. This blockage increases the tendency of biting rate as the flea starves to death.

In certain countries, ticks are considered potential mechanical carriers. While direct transmission between animals or humans is possible, its significance depends on the specific clinical manifestation of the disease. Although theoretically possible, the transmission of bubonic plague from one person to another appears to be either extremely rare or non-existent. Different routes of transmission of plague to human is listed below:

- Bite of rat flea infected plague organism
- Direct contact with the infected draining fluid of buboes through broken skin and mucous membrane
- Direct contact with infected animals including its bites or scratches
- Inhaling droplets emitted by humans or animals infected with pneumonic plague (within a 2-m distance) is most common in humans in densely populated and poorly ventilated environments
- Bacterial ingestion through contaminated food (e.g., undercooked or spoiled meat)
- Fomites transmission is possible within a short time

The spread of this disease is influenced by environmental conditions, human activities, low-income levels, inadequate hygiene, and poor sanitation. Environmental factors include season, humidity, temperature, rainfall, habitat, and human dwellings. The enzootic plague affecting rodents is found in remote, sparsely inhabited regions, such as the semi-arid or arid grassland steppes of Central Asia and the savannahs of Eastern Africa. Enzootic transmission usually results in a low likelihood of incidental human infection, with such cases being both rare and sporadic. Alternatively, the epizootic plague swiftly spread among vulnerable rodents through the rat flea, which then passed the infection on to humans. Four plague foci have been recognized throughout the world-.

Name of the foci	Plague sites	Host	Vector
North American plague foci	Boarder area of Canada and Mexico, Northeastern Arizona, New Mexico, southern Colorado, Pacific Coast	Burrowing ground squirrels, prairie dogs, wood rats, antelope ground squirrels, deer mice	Ground squirrel fleas, wood rat fleas, prairie dog fleas, and chipmunk fleas
South American plague foci	Brazil, Bolivia, Peru, Ecuador Paraguay, Argentina, and Venezuela	Commensal rats, cotton rats, wild cavies, rice rats, field mice, cane mice, and domesticated guinea pigs	*X. cheopis*, *Polygenis* and *Pleochaetis* spp., *Pulex irritans*
African plague foci	In the regions of East and Southern Africa, including nations like the Democratic Republic of the Congo, Uganda, Kenya, Tanzania, Zambia, Zimbabwe, Mozambique, Botswana, South Africa, Namibia, and Angola, along with Madagascar, an island situated in the Indian Ocean	Gerbils, aquatic rodents, several species of grass mice, multimammate mice, and human-associated rats	*Xenopsylla* and *Dinopsyllus* spp.
Asian plague foci	Regions such as Iran, Kurdistan, Transcaucasia, the surrounding areas of the Caspian Sea, and the southeastern plains of Russia and Kazakhstan	Gerbils	*Xenopsylla* and *Nosopsyllus* spp.
	Central Asia encompasses regions such as the mountainous areas of Kazakhstan, northeastern China, Mongolia, Manchuria, and Transbaikalia	Marmots	*Oropsylla*, *Rhadinopsylla*, and Citellophilusspp
	Northern-Central China, Mongolia, the Central Asian plains and steppes	Ground squirrels	Citellophilus and Neopsyllaspp
	India	Gerbil, Tateraindica, Mellardiameltada, various field mice, palm squirrels, *Bandicotabengalensis*, *Bandicotaindica*, *R. rattus*, and *R. norvegicus*	*X. cheopis* and *Xenopsylla astia*
	Myanmar, Indonesia and Vietnam	The black rat, *R. rattus* subsp. *diardii*, and the Polynesian rat, *R. norvegicus*, shrew	X. cheopis, Stivaliuscognatus

Plague naturally occurs in various parts of the world. Recently, the majority of cases have been reported in Africa (80%), followed by Asia (15%), with the remaining cases coming from the Americas, Madagascar, Tanzania, the Democratic Republic of Congo, Vietnam, Mozambique, Namibia, and Peru. People of all ages and genders are equally vulnerable to the plague. Activities related to certain occupations, such as hunting, grazing, farming, harvesting, and construction, create opportunities for direct contact between humans and fleas.

Virulence Factors and Toxins

This organism possesses various virulence factors, among which the plasminogen activator (Pla), yersinia outer proteins (Yops), soluble V antigen, and fraction 1 (F1) envelope glycoprotein antigen are noteworthy. The plasminogen activator (Pla), a protease located on the bacterial surface, aids the bacilli in surviving within the flea's gut. Yersinia outer proteins (Yops) and the soluble V antigen facilitate survival within macrophages. The fraction 1 (F1) envelope glycoprotein antigen impedes the phagocytic process when opsonizing antibodies are absent.

Pathogenesis

Y. pestis loses its capsular layer during the time of its growth in the gut of the rat flea. From the blocked flea, some bacilli are regurgitated into the new host through abraded skin or mucous membrane. Most of the organisms are phagocytosed and killed by the neutrophils but few organisms enter the macrophages. In the macrophages, bacilli are not killed rather they get a protective environment to synthesize their virulence factors. Then the organisms destroy the macrophages and come out in extracellular environment where again phagocytosis is inhibited by the activity of *Yersinia* outer membrane proteins (YopH and YopE). Finally, *Y. pestis* organisms spread into regional lymph nodes via lymphatic system. Affected lymph nodes become edematous, congested at the primary stage of the disease with minimal inflammatory infiltrates and vascular injury. All these characteristics facilitate to develop "Black Buboes" which contain huge number of plague organisms. The bacilli can also spread to blood circulation within few minutes after biting by infected flea. *Y. pestis* has the ability to infect any organ, but it frequently results in multifocal liver necrosis, widespread interstitial myocarditis accompanied by cardiac enlargement, and extensive hemorrhagic necrosis of the spleen. Primary septicemia is seen due to multiplication of bacteria in blood circulation in absence of any "bubo," but secondary septicemia is developed when infection come into the bloodstream from primary lymphadenitic and pulmonary infection after failing of local defensive mechanisms.

Primary pneumonic plague occurs from the infected droplets of a person or animal. Secondary pneumonia occurs as a result of the hematogenous spread linked to bubonic or septicemic plague.

Clinical Manifestations

The incubation period of plague is generally 2–7 days but it may be shorter for primary pneumonic plague, i.e., 24 h. Animals can experience three types of plague, which are also found in humans: bubonic plague, pneumonic plague, and septicemic plague. This disease is a rapidly developing febrile disease. Inappropriate and untreated bubonic plague often develop pneumonic and septicemic plague.

Clinical Manifestations in Animals

Cats: Bubonic plague is commonly seen in cat with pyrexia, lethargy, anorexia, and swelling by means of pain of lymph node that may or may not form abscess followed by draining.

In septicemic plague, cats will show lethargy, fever, and anorexia with no obvious swelled or abscessed lymph nodes. The cat may show the signs of sepsis including diarrhea, vomiting, tachycardia, cold extremities, ocular discharges, dehydration, weight loss pale or brick-red mucous membranes, irregular pulse, and respiratory distress syndrome.

Cats suffering from bubonic or septicemic plague may develop pneumonic plague, which is marked by breathing difficulties such as shortness of breath and coughing up blood. Neurological signs also may be noted such as incoordination.

Dogs: Dogs are less susceptible than cats. Fever, tiredness, swelling of the submandibular lymph nodes, mouth sores, and coughing may be the only signs of a subclinical infection.

Rodents: In rodents, the disease is variable from mild to fatal. High mortality is seen during the time of epizootics in rodents and lagomorphs.

Clinical Manifestations in Humans

Humans typically encounter three types of plague: bubonic, septicemic, and pneumonic.

Bubonic Plague Usually, the incubation period of bubonic plague is 2–6 days, but it may vary sometimes. Primarily patient shows signs like headache, fever (100–104 °F), chills, muscle ache, and joint pain. Tenderness of the regional or proximal lymph nodes to the site of inoculation of the bacilli is noticed just after the onset of the previous signs. Inguinal and femoral lymph nodes are commonly affected, and afterward axillary lymph nodes are also involved. All affected lymph nodes swell and become tender. This condition of lymph nodes is known as "bubo." The bubo may be presented with surrounding edema and desquamation. If treated with appropriate antimicrobials, bubos can disappear within 2–5 days. If not addressed, the swelling and pain may persist for a week or more, possibly resulting in a toxic fever with symptoms such as a rapid heartbeat, stupor, anxiety, confusion, and, less frequently, seizures, and disorientation. A mild version of the bubonic plague is referred to as pestis minor.

Septicemic Plague The time it takes for septicemic plague to develop after exposure is between 1 and 4 days. This form of plague is developed due to excessive endotoxemia and spread of infection. Primary septicemic plague appears without lymphadenitis, and at the initial stage, the disease may not be ruled out without blood culture. Acute fever, chills, weakness, and malaise are the cardinal signs accompanied by digestive issues like nausea, vomiting, diarrhea, and abdominal pain. Due to the involvement of multiple organs, systemic inflammatory response syndrome (SIRS) and disseminated intravascular coagulation (DIC) may be developed. In advance stages, development of thrombi may be found in tip of the ears and digits followed by gangrene. Acute renal failure and stage of unconsciousness is seen as pre-terminal actions. Untreated septicemic plague is generally fatal.

Pneumonic Plague (Human-to-Human Transmission) Pneumonic plague is the most severe type of plague, with an incubation period ranging from 1 to 7 days. However, in most instances, symptoms manifest within 3–4 days following exposure. Most of the time, the disease starts with fever, shivering, headache, muscle pain, weakness, and respiratory discomfort. Starting from the second day, symptoms such as a productive cough, chest discomfort, sputum production, worsening chest pain, rapid breathing leading to difficulty in breathing, hemoptysis, cardiopulmonary insufficiency, cyanosis, and circulatory collapse are observed in this condition. All these signs are distinguished as primary plague pneumonia, while in secondary plague pneumonia the involvement of all lobes of lungs is observed followed by cyanosis. In secondary pneumonic form, sputum production is slight and sputum become thick.

Some other plague syndromes also may be observed such as meningeal plague and pharyngeal plague.

Meningeal Plague/Plague Meningitis Meningitis is an abnormal demonstration of plague. The onset of meningitis is often postponed because of improper or inadequate use of antibiotics for primary plague. Meningeal plague, also known as plague meningitis, occurs when Y. pestis bacteria spread through the bloodstream, leading to symptoms such as fever, headache, changes in mental state, and the presence of polymorphonuclear leukocytes in the cerebrospinal fluid.

Pharyngeal Plague/Plague Pharyngitis This rare syndrome is marked by symptoms such as fever, a sore throat, and swelling of the cervical lymph nodes. Pharyngeal involvement arises from contact with individuals suffering from pneumonic plague, the presence of cervical or sub-mandibular buboes, or the ingestion of raw contaminated meat.

Plague as Bioweapon

Plague bacilli is considered one of the priority pathogens to be used as bioweapon because of its rapid spread and high fatal rate capability. The pathogen has been classified as category "A" agent by CDC. Two highly heinous occurrence of plague as bioweapon are "Siege of caffa" and "Japanese Unit 731."

Siege of Caffa In 1346, during the siege of Caffa, the Mongol (Tartar) forces besieging the well-defended Genoese seaport, now known as Feodosia in Ukraine, were struck by a plague epidemic. The disease spread to the inhabitants of Caffa when the attackers catapulted infected corpses into the city. Consequently, Italians fled from Caffa, carrying the plague to Mediterranean ports.

Japanese Unit 731 During World War II and the Second Sino-Japanese War, Unit 731 of the Imperial Japanese Army conducted experiments, developed, and deployed biological weapons on the population in the northeastern region of China, particularly in Manchuria.

Diagnosis

Accurate epidemiological documentation and thorough physical examination are crucial for the prompt diagnosis of plague. Due to its potential for aerosol transmission, the organism must be managed under containment conditions appropriate for Class-3 pathogens. Routine samples can be managed safely within BSL-2 settings, but BSL-3 precautions are required for the manipulation of isolates to adequately safeguard laboratory personnel. The presence of plague is confirmed by identifying the bacilli in the fluid from buboes or skin lesions, as well as in sputum and blood smears. In all variations of the disease, blood culture results can occasionally be positive. Suitable specimen (Bubonic plague-aspirated pus or fluid, Septicemic plague-blood, pneumonic plague-sputum and blood, Meningeal plague-CSF, and Postmortem-splenic tissue) is to be collected for obtaining appropriate result in different forms of plague.

Microscopy	Exudate or sputum smears are treated with Gram's stain, Wright stain, Giemsa stain, Wayson's stain, or Methylene blue stain for staining. The identification of bipolar-staining, oval-shaped, gram-negative bacteria with a "safety pin" appearance allows for a rapid preliminary diagnosis of plague Fluorescence in situ hybridization (FISH) can also be implied Confirming the presence of *Y. pestis* can be achieved by using immune-specific staining to demonstrate the FI capsular antigen
Culture	The samples can be grown in blood agar, MacConkey agar, nutrient agar, and ghee broth at 27 °C. On blood agar, the colonies appear dark brown, on MacConkey agar colonies are colorless and in ghee broth, a characteristic stalactite growth is produced
Animal inoculation	When exudates are injected subcutaneously into guinea pigs or mice, the animals typically succumb within 2–5 days. Bacilli can be observed in a blood smear

Dipstick test	FI glycoprotein can be spotted by using monoclonal antibodies. It is rapid test and gives reliable result within 15 min
CFT	Complement fixation test (CFT) can be done by using the antibodies to FI antigen in patient's serum
ELISA	Method using Protein A-peroxidase to identify anti-F1 antibodies across various species
PCR	Traditional polymerase chain reactions (PCR) targeting the *pla*, *caf1*, *inv*, and *yopM* genes reduce the diagnostic time to 3–4 h, while real-time PCR completes the process in just 2 h

Prevention and Control Strategies

- Mass awareness program should be initiated among the people and advising them to avoid flea bites and not to handle animal carcasses.
- In endemic areas, quick reporting facilitates to stop exposure of other animals and humans.
- People should not be in contact with the body fluids and tissues of any suspected individual.
- Standard precautions should be taken during collecting and handling the specimens.
- Adequate control measures should be taken for fleas and rodents.
- A proper flea management plan should be tailored specifically for both dogs and cats. It is important to prevent them from consuming tissues from other infected animals.
- Vaccinations: During epizootics, endangered black-footed ferrets were vaccinated due to their high susceptibility to plague. Bait vaccine was also used to prevent plague in ferrets and prairie dogs.

The WHO advises against vaccination for the general population, recommending it only for those at high risk, such as laboratory staff frequently exposed to contamination and healthcare professionals.

Killed vaccine is given subcutaneously at 1–3 months interval for two doses and followed by a third dose after 6 months. While vaccination offers considerable defense against the bubonic plague, it does not provide the same level of protection against the pneumonic plague. This immunity is effective for a period of no longer than 6 months.

Treatment

Treatment should be initiated as soon as plague is suspected. Gentamicin and fluoroquinolone are the drugs of choice though CDC has recommended a series of antimicrobials to treat plague successfully.

Antimicrobials	Doses		Route	Type of plague
Ciprofloxacin	Adults	400 mg at 8 h interval	IV	Pharyngeal plague Septicemic plague Bubonic plague Pneumonic plague
		750 mg at 12 h interval	PO	
	Children	10 mg/kg at 8–12 h interval	IV	
		15 mg/kg at 8–12 h interval	PO	
Levofloxacin	Adults	750 mg at 24 h interval	IV/ PO	Pneumonic plague Bubonic plague Septicemic plague Pharyngeal plague
	Children	Weight < 50 kg: 8 mg/kg at 12 h interval (maximum 250 mg/dose) Weight ≥ 50 kg: 500–750 mg at 24 h interval	IV/ PO	
Gentamicin	Adults	5 mg/kg at 24 h interval	IV/ IM	Pneumonic plague Bubonic plague Septicemic plague Pharyngeal plague
	Children	7.5 mg/kg at 24 h interval	IV/ IM	
Streptomicin	Adults	1 g at 12 h interval	IV/ IM	Pneumonic plague Bubonic plague Pharyngeal plague Septicemic plague
	Children	15 mg/kg at 12 h interval (maximum 1 g/ dose)	IV/ IM	
Doxycycline	Adults	200 mg loading dose, then 100 mg at 12 h interval	IV/ PO	Pharyngeal plague Bubonic plague
	Children	Weight < 45 kg: 4.4 mg/kg loading dose, then 2.2 mg/kg at 12 h interval (maximum 100 mg/dose) Weight ≥ 45 kg: same as adult dose	IV/ PO	

Bibliography

Ansari, I., Grier, G., & Byers, M. (2020). Deliberate release: Plague – A review. *Journal of Biosafety and Biosecurity, 2*(1), 10–22. https://doi.org/10.1016/j.jobb.2020.02.001

Barbieri, R., Signoli, M., Chevé, D., et al. (2020). Yersinia pestis: The natural history of plague. *Clinical Microbiology Reviews, 34*(1), e00044-19. https://doi.org/10.1128/CMR.00044-19. Published 2020 Dec 9.

Butler, T. (2009). Plague into the 21st century. *Clinical Infectious Diseases, 49*(5), 736–742. https://doi.org/10.1086/604718

Demeure, C. E., Dussurget, O., Mas Fiol, G., Le Guern, A. S., Savin, C., & Pizarro-Cerdá, J. (2019). Yersinia pestis and plague: an updated view on evolution, virulence determinants, immune subversion, vaccination, and diagnostics. *Genes and Immunity, 20*(5), 357–370. https://doi.org/10.1038/s41435-019-0065-0

Sebbane, F., & Lemaître, N. (2021). Antibiotic therapy of plague: A review. *Biomolecules, 11*(5), 724. https://doi.org/10.3390/biom11050724. Published 2021 May 12.

Yang, R. (2017). Plague: Recognition, treatment, and prevention. *Journal of Clinical Microbiology, 56*(1), e01519-17. https://doi.org/10.1128/JCM.01519-17. Published 2017 Dec 26.

Glanders

14

Ripan Biswas and Indranil Samanta

Abstract

Glanders is a zoonotic disease caused by the bacterium *Burkholderia mallei*, which predominantly affects horses but can also be transmitted to humans. This chapter presents an overview of glanders, encompassing its transmission, clinical presentation, diagnosis, and control. Humans typically become infected through direct exposure to infected animals or contaminated environments. Glanders in humans can present in diverse clinical manifestations, encompassing localized skin infections, respiratory complications, and systemic sepsis. Diagnosis can be difficult owing to the non-specific character of symptoms and laboratory verification is crucial. This chapter highlights the importance of prompt diagnosis and treatment with appropriate antibiotics. This chapter also emphasizes the imperative for implementing control measures within equine populations, such as surveillance, testing, and quarantine protocols, to mitigate the transmission of glanders to both animal and human hosts.

Keywords

Glanders · Farcy · Equine zoonoses · Malleus · Droes · Equinia · Malleus

R. Biswas (✉)
Department of Veterinary Public Health and Epidemiology, West Bengal University of Animal and Fishery Sciences, Kolkata, West Bengal, India

I. Samanta
Department of Veterinary Microbiology, West Bengal University of Animal and Fishery Sciences, Kolkata, West Bengal, India

R. Biswas et al. (eds.), *Concept of One Health and Zoonoses*,
https://doi.org/10.1007/978-981-95-6627-3_14

History

Glanders is an extremely infectious and deadly zoonotic illness that mainly impacts horses, mules, and donkeys. During 450–425 BC, the disease was first mentioned by Hippocrates in the Greek literatures. Aristotle called this disease as *malleus,* which means "malignant sickness" and during 400–500 AD the disease was described by the Romans. In the sixteenth century, William Shakespeare alluded to the disease in his comedy "The Taming of the Shrew," and a century later, Alexandre Dumas made mention of it in his renowned work *The Three Musketeers*. In 1664, Sollysel in France recorded the infectious characteristics of the disease. Until the beginning of nineteenth century, the zoonotic potential was not discovered. In 1882, the organism was first isolated by Loeffler and Schutz from the liver of an infected horse in Germany, with Bouchard, Charrin, and Capitan making similar discoveries later that year. In 1890, Helman, Kalning, and Pearson discovered mallien from the cultures of the organism for diagnostic purpose. The organism has undergone multiple reclassifications, being placed in the genera *Bacillus, Corynebacterium, Mycobacterium, Loefflerella, Pfeifferella, Malleomyces, Actinobacillus,* and *Pseudomonas* at different times. In 1992, on the basis of 16S ribosomal ribonucleic acid gene sequences, DNA homology, fatty acid composition, cellular lipid, and phenotypic characteristics the genus was proposed as *Burkholderia,* which is still in use. Given the lethal and transmissible characteristics of the disease affecting both humans and soliped animals, the organism is considered a potential candidate for use in biological warfare. The Centers for Disease Control and Prevention (CDC) has identified this bacterium as a potential category B biological warfare agent.

Etiological Agent and Ecology

Glanders is an infectious disease that affects both solipeds and humans. It is caused by the Gram-negative bacterium *Burkholderia mallei*, which belongs to the order Burkholderiales, the family Burkholderiaceae, and the genus *Burkholderia. Burkholderia mallei, once referred to as Pseudomonas mallei*, is a Gram-negative bacterium that does not form spores, lacks motility, flagella, and a capsule, and is an aerobic pathogen that must live inside mammalian cells. The bacillus measures between 0.3 and 0.8 μm in width and 2 to 5 μm in length, and it is only faintly stained when using simple stains. The uneven staining property of *B. mallei* occurs due to the presence of lipoid granules. *Burkholderia pseudomallei*, the pathogen responsible for melioidosis, is closely related to and appears to have evolved from *B. mallei.* The organism is sensitive to the external environment, and it is inactivated within 24 h by the exposure to direct sunlight and can be killed by using the most common disinfectants like sodium hypochlorite (500 ppm available chlorine) and other chlorine compounds, chlorine, potassium permanganate, formalin, copper sulphate, 70% ethanol, and 2% glutaraldehyde. Exposure to a temperature of 550 °C for 10 min can destroy it. In moist and humid environments, the organism can survive for 3 to 5 weeks, while in contaminated stables, such as those with bedding and manure, it can persist for approximately 6 weeks. In tap water, it remains viable for at least a month. Electron microscopy has verified the existence of a capsule-like structure. This capsule, made up of neutral carbohydrates, serves to shield the cell from adverse environmental conditions. Identifying the organism in tissue samples is challenging, as they can appear with a beaded pattern. The organism's appearance in culture media changes based on the culture's age and the medium type. Older cultures exhibit pleomorphism. When cultivated at 37 °C for a period of 48 h on nutrient, blood, and MacConkey agar media, the organism forms colonies that are smooth, creamy, and viscid in appearance. It can be cultivated as both an aerobic and facultative anaerobe when nitrogen is present.

Host Range and Susceptibility

The natural reservoirs of the Glanders are the soliped animals. Horses, mules, and donkeys are the major susceptible host though cases have been reported from domesticated cats, tigers, lions, leopards, and other carnivorous animals. Small ruminant like sheep and goat may be infected with Glanders but cattle and pigs were reported to be resistant. Birds are highly resistant to the infection. Guinea pig, hamsters, and mice are susceptible which institutes the basis of "Strauss reaction," a diagnostic test. The disease is having zoonotic potentiality and proceeds as occupational threat for veterinarians, stable workers, animal workers, farriers, jockey, cops, and horse handlers. Untreated infection is generally fatal.

Epidemiology

Glanders is prevalent in regions such as Asia, the Middle East, Africa, and Central and South America. However, the disease has been successfully eliminated in countries like Great Britain, Denmark, Switzerland, Sweden, Portugal, Germany, France, Spain, Australia, Japan, South Africa, Canada, and the United States. Due to some recent outbreaks, the infection is considered as reemerging infection. During last three decades, the infection has been reported from Brazil, Mongolia, Iraq, Ethiopia, Turkey, Iran, the United Arab Emirates, former Soviet Union, Bahrain, Cameroon, Curacao, and India. Many Indian states such as Maharashtra (2006), Uttar Pradesh (2006, 2007, 2010, and 2011), Punjab (2007), Uttarakhand (2007), Himachal Pradesh (2007 and 2010), Haryana (2007), and Chhattisgarh (2009 and 2010) reported Glanders from equines.

The primary means of infection appears to be the consumption of food or water contaminated by secretions from the respiratory system or infected skin sores of carrier animals. Subclinical and chronic infection can shed the organism constantly as well as intermittently in food and water. The disease also can be transmitted through inhalation via contaminated aerosols. Reports indicate that stallions can transmit venereal diseases to mares, and vertical transmission is also possible. Mechanical transmission is possible by flies which act as vectors. The infection is facilitated by proximity of animals and different stress conditions. Humans can contract infections through contact with sick animals, contaminated objects, tissues, or bacterial cultures. Besides contact with animals, there have also been cases of transmission between humans.

Pathogenesis

The effective infection of host cells relies on the strategic use of numerous virulence factors and mechanisms, which manipulate various host processes and pathways. These approaches are pivotal in determining pathogen virulence and present novel avenues for the development of vaccines and therapeutic solutions. *Burkholderia mallei* utilizes a type III secretion system to escape from the phagosome into the cytosol, where it can multiply and employ actin-based motility to navigate within the cell, as the bacteria are non-motile in vivo. Additionally, *B. mallei* can trigger interactions between cells that result in the creation of multinucleated giant cells (MNGCs). These cells are believed to offer a protective environment and metabolic resources for the bacteria. However, there is limited understanding of the bacterial or host factors necessary for the formation of MNGCs. The presence of a type VI secretion system has been identified as essential for the effective spread both within and between cells, as well as for the fusion of host cells. Nevertheless, it is crucial to obtain comprehensive insights into the specific molecular pathways that drive the virulence of *B. mallei*. After breaching the mucosal barrier, *Burkholderia mallei* navigates through the lymphatic system and ultimately arrives at the lymphatic nodes. After multiplying in the lymph nodes, it then circulates through the bloodstream. In cases of nasal glanders, granulomas and ulcers appear in various tissues, and nodules along with severe ulcers are found on the mucous membranes of the nasal passages. In pulmonary glanders, one can detect nodular lesions beneath the pleura, as well as dispersed miliary granulomatous nodules in the lungs, characterized by a caseonecrotic center. In cases of acute glanders, there is a presence of severe bronchopneumonia and swelling in the lung lobes. Additionally, serous fluid might accumulate in the para-nasal sinuses, trachea, and nasal cavity.

Glanders as a Bioterrorism Weapon

CDC has scheduled this bacterium as latent agent for biological warfare under category B. Throughout the US Civil War, the Anglo-Boer War in South Africa, and World War I, there were instances of mass euthanasia or death of horses and other equines. During World War I, German agents deliberately employed *Burkholderia mallei*. During World War II, Japanese forces deployed the organism against horses, civilians, and prisoners. In the 1980s, the former Soviet Union deployed this organism as a weapon in a limited manner against the Mujahideen in Afghanistan.

Clinical Manifestations in Animals

The incubation period for glanders can span from a few days to several months, with 2–3 weeks being the most common duration. In animals, the disease can appear in three distinct forms: pulmonary, nasal, and cutaneous. The progression of glanders can be either acute or chronic.

The acute form of glanders encompasses pulmonary, nasal, and cutaneous types. It is marked by symptoms such as a high fever, coughing, nasal discharge, and the presence of ulcers and nodules in the upper respiratory tract and lungs. Ultimately, death results from septicemia and anoxic anoxia. Chronic form also involves three types of glanders-.

Nasal Glanders Deep ulcers and nodules commonly develop in the nasal cavity, especially on the nasal septum. Many of these nodules get ruptured and a thick mucopurulent, sticky yellowish green discharge occurs from the nostrils. The discharge can be abundant, affecting one or both sides, and may become bloody. The exudate is extremely contagious. The ruptured nodules transform into deep ulcers with jagged, elevated, and inflamed edges. These ulcers take a long time to heal, eventually forming star-shaped scars. An infection in the nasal area can extend to affect the lower parts of the respiratory system.

Pulmonary Glanders This form of glanders is often observed in the majority of clinical cases and is typically found alongside another form of glanders. The infection is characterized by formation of round, grayish or white, firm, and encapsulated nodules throughout the lung tissue with fever or febrile episodes and progressive debilitation.

Cutaneous Glanders (Farcy) When the skin becomes infected, it involves the superficial lymphatic vessels and nodes. These lymph vessels become thickened and appear as firm cords beneath the skin, known as "farcy pipes." The nodules often rupture and yellowish exudates discharge from the deep ulcers which recover slowly.

In the cutaneous form known as 'farcy,' the lymph nodes become swollen, and nodular abscesses ranging from 0.5 to 2.5 cm in size, referred to as 'buds,' form. These abscesses ulcerate and release a yellow, oily pus. Additionally, dry ulcers may appear. Occasionally, pyogranulomatous nodules are observed in the liver and spleen.

Occult Glanders It is seen in chronic or subclinical or asymptomatic infected horse, and this may function as a potential source of infection because of the persistent or sporadic discharge of bacteria.

Clinical Manifestations in Human

The acute version of the disease is characterized by an incubation period ranging from 1 to 14 days, whereas the chronic version can have an incubation period extending up to 12 weeks. In humans, potential indicators include localized skin infections, ongoing skin issues, acute pulmonary disease, and bacteremia. Additionally, individuals may experience low-grade fevers, especially from afternoon to evening, along with general malaise, fatigue, headaches, muscle pain such as backaches, swollen lymph nodes, and chest pain. Cases of human glanders have also reported symptoms such as excessive tearing, sensitivity to light, and diarrhea. Nonetheless, when localized infections begin to spread, they can result in infections affecting the lungs, the bloodstream, or various tissues.

Nasal Glanders Nasal glanders is identified by the swelling and inflammation of the nasal area, which is subsequently accompanied by a profuse nasal discharge. The infection has the potential to invade the nasal septum and bone tissue, resulting in ulcerative lesions and fistulae. Additionally, it can extend to the lower respiratory tract, where it manifests as bronchitis, leading to coughing and the production of mucopurulent sputum.

Pulmonary Glanders This type of human glanders is typically marked by conditions such as pneumonia, lung abscesses, pleuritis, and pleural effusion, which manifest through symptoms like coughing, difficulty breathing, chest discomfort, mucopurulent sputum, fatigue, high fever (over 102 °F), chills, headaches, muscle aches, lymphangitis, sore throat, pleuritic chest pain, rapid breathing, shortness of breath, discharge, and gastrointestinal symptoms. Symptoms such as dizziness, chills, muscle pain, nausea, night sweats, intense headaches, rapid heartbeat, weight loss, and mucosal eruptions are reported without specific details.

Cutaneous Glanders It is manifested by appearing as a single blister and gradually develops ulcer at the site of entry which becomes hemorrhagic. This localized infection is developed at the site of entry of the organisms. Inflammation may reach up to regional lymph glands and cause lymphangitis.

Regarding Biosafety Level

Laboratory procedures should be performed in a biosafety level (BSL)-3 facility or in a BSL-2 environment with BSL-3 precautions, which involve wearing personal protective equipment (PPE) and using respiratory protection. When handling samples and culture isolates, it is essential to wear gloves and personal protective equipment within a biosafety cabinet. Any procedures that might create aerosols, such as the preparation of standard inoculums for identification systems, should be carried out in a BSL-2 cabinet or under BSL-3 safety protocols.

Diagnosis

To isolate *B. mallei*, one can collect pus or discharge from open sores, the lungs, abscesses, nasal passages, or materials obtained during a necropsy. The serum (2–5 ml), nasal swab, and lymph node biopsy can be collected in sterile vial for laboratory testing by maintaining the cold chain. Commonly, these samples are contaminated with *Pseudomonas* and *Pasteurella* which make difficulties in isolation process.

Microscopy: *B. mallei* is visible under a microscope. A smear is made from pus or fluid and then stained using methylene blue or Gram stain. The bacteria appear as Gram-negative rods with rounded ends.

Culture: It is preferable to conduct isolation from lesions that are both unopened and uncontaminated. Bacteria are aerobic or facultative anaerobic, and it can be grown on glycerol agar at 37°for 72 h. Growth is notably less vigorous on plain nutrient agar, and it is sparse on gelatin.

Animal inoculation test (Strauss reaction): In a male guinea pig, the suspected material is injected into the peritoneal cavity. If the test is positive, the animal develops severe localized peritonitis, orchitis, and inflammation of the scrotal sac, known as the Strauss reaction. However, this reaction is not exclusive to glanders, as it can also be triggered by other organisms.

Immunological test/ Mallein test: In 1890, Mallein introduced a technique for identifying *B. mallei* in equines. This method involves using a purified protein derivative (PPD) derived from *B. mallei* that has been subjected to heat treatment. It is a recommended procedure in international trade for detecting chronic infections, carriers, or latent cases. This test, which assesses allergic hypersensitivity, is frequently employed alongside the complement fixation test (CFT). False positive reaction may be shown due to cross reaction with *Streptococcus equi*. This test is not recommended as diagnostic tool in human.

Indtradermo-palpebral mallein test: To test suspected horses, 0.1 ml of mallein is administered into the lower eyelid, and after 48 h, a positive reaction is indicated by swelling of the eyelid, blepharospasm, and severe purulent conjunctivitis.

Opthalamic mallein test: To test horses suspected of infection, a few drops of mallein are placed in their eyes. If the horse has a positive reaction, it will show signs of swelling around the face and eyelids, along with a discharge from the eyes. Despite its use, this method is not as dependable as the intradermo-palpebral test.

Sub cutaneous mallein test: An injection of 2.5 ml of diluted mallein is given at the midpoint of a horse's neck. It is essential that the horse's body temperature is below 102 °F before the procedure. A positive test result is characterized by a marked increase in temperature, surpassing 104 °F, within the initial 15 h, along with painful swelling at the injection site. This test is generally avoided due to its potential to disrupt serological diagnostic processes.

Serological Test

The Complement Fixation Test (CFT): Mallein and CFT are tests recommended by WOAH for the international trade of horses. This test has an accuracy rate of 90–95%, and positive results can be detected within a week of infection, as well as in cases of chronic infection.

ELISA: Avidin–biotin dot ELISA was standardized but not validated.

Immunoblot assay: This is the best evaluated serological test but it cannot differentiate glanders from melioidosis.

Rose Bengal Plate Agglutination Test: The antigen consists of a bacterial suspension that has been inactivated by heat and is tinted with Rose Bengal. This test is exclusively conducted in Russia on horses and other animals. It shows a high degree of cross-reactivity with brucellosis.

Molecular Test: PCR, along with real-time and multiplex PCR, serves as swift and trustworthy techniques for detecting glanders.

Prevention and Control Strategies

- Positive animal cases must be identified very fast followed by euthanasia and proper disposal (burning or burying) of carcass to be done quickly to ensure successful prevention and control.
- Extreme precautionary measures should be followed during handling of morbid animals and contaminated fomites. PPE should be used. Carcass should not be opened.
- Bedding materials, rope, utensils, and residual feed are recommended to burn or bury.
- Traditional disinfectants that can be utilized both inside and outside the farm include iodine, mercuric chloride in alcohol, potassium permanganate, 1% sodium hypochlorite, 70% ethanol, and 2% glutaraldehyde.
- Awareness program should be taken for the people who are engaged with equines.
- There is no vaccine available for animals or human.

Treatment for Animals

This disease requires mandatory reporting. Under the Glanders and Farcy Act of 1899, any animals found to be infected must be put down, and their remains must be disposed of in a manner that ensures safety. Test and slaughter policy should be employed in every susceptible case. To date, no treatment with veterinary drugs is competent to cure the infection.

Treatment for Human

Though human glanders is very rare, but it has been reported that Sulfadiazine is effective against glanders. Before the advent of antibiotic treatments, the majority of human cases of glanders resulted in death, with over 90% of those infected not surviving the illness. Because *B. mallei* is a facultative intracellular pathogen, aminoglycosides and other antibiotics are unable to enter the host cell. Intensive intravenous therapy is administered using imipenem, meropenem, or ceftazidime, and may optionally include trimethoprim-sulfamethoxazole (TMP-SMX). The intravenous administration should be maintained for at least 10 days, and it might be extended based on how severe the illness is. The duration of oral treatment is influenced by the disease's severity and can last anywhere from 12 weeks to 12 months. TMP-SMX, optionally combined with doxycycline, is the recommended medication for oral therapy.

Bibliography

Centers for Disease Control and Prevention (CDC). http://www.cdc.gov/glanders/

Dvorak, G. D., & Spickler, A. R. (2008). Glanders. *Journal of the American Veterinary Medical Association, 233*(4), 570–577. https://doi.org/10.2460/javma.233.4.570

Khan, I., Wieler, L. H., Melzer, F., et al. (2013). Glanders in animals: A review on epidemiology, clinical presentation, diagnosis and countermeasures. *Transboundary and Emerging Diseases, 60*(3), 204–221. https://doi.org/10.1111/j.1865-1682.2012.01342.x

Kovalev, G. K. (1971). Sap (obzor) [Glanders (review)]. *Zhurnal Mikrobiologii, Epidemiologii, i Immunobiologii, 48*(1), 63–70.

Lehavi, O., Aizenstien, O., Katz, L. H., & Hourvitz, A. (2002). *Harefuah, 141*(Spec 88–91), 119.

Mariappan, V., Vellasamy, K. M., Anpalagar, R. R., et al. (2022). One health surveillance approaches for melioidosis and glanders: The Malaysian perspective. *Frontiers in Veterinary Science, 9*, 1056723. Published 2022 Dec 15. https://doi.org/10.3389/fvets.2022.1056723

Van Zandt, K. E., Greer, M. T., & Gelhaus, H. C. (2013). Glanders: An overview of infection in humans. *Orphanet Journal of Rare Diseases, 8*, 131. Published 2013 Sept 3. https://doi.org/10.1186/1750-1172-8-131

Wittig, M. B., Wohlsein, P., Hagen, R. M., et al. (2006). Ein Ubersichtsreferat zur Rotzerkrankung [Glanders--a comprehensive review]. *Deutsche Tierärztliche Wochenschrift, 113*(9), 323–330.

World Organisation for Animal Health. (2019). *Manual of diagnostic tests and vaccines for terrestrial animals*. OIE.

World Organization for Animal Health [OIE] . Manual of diagnostic tests and vaccines for terrestrial animals [online]. : OIE; 2013. Glanders. Available at: http://www.oie.int/fileadmin/Home/eng/Health_standards/tahm/2.05.11_GLANDERS_FINAL.pdf

Salmonellosis

15

Ajay Pratap Singh

Abstract

With 1.3 billion incidents of the illness and more than 3 million fatalities annually, salmonellosis is a global public health concern. Salmonellosis is a leading cause of bacterial intestinal infection in people and animals alike. Salmonellosis is the second most often reported gastrointestinal infection and a leading source of foodborne outbreaks. More than 2500 Salmonella serovars have been discovered; the majority of them have been linked to human infections, although only a small subset of these serovars is significant for public health.

Keywords

Salmonella · Food-borne infection · Enteric fever · Paratyphoid fever

History

In 1839, Soholerin made the initial discovery of the bacterium, which was later isolated from a typhoid fever patient by Karl Eberth in 1880. In 1885, Daniel Elmer Salmon and Theobald Smith discovered related bacteria in pigs and gave it the name *Bacillus cholerasuis*. Lignieres first suggested the name *Salmonella* for the genus in 1900, and it was officially adopted in 1934 to honour the American veterinary bacteriologist Daniel E. Salmon. Following Gartner's isolation of the organism in 1888, the link between *Salmonella* and human gastroenteritis was established. In 1892, Loeffler isolated *S.* Typhimurium from an infected mouse. Rattger made the

A. P. Singh (✉)
Department of Veterinary Microbiology, College of Veterinary Science & Animal Husbandry, U.P. Pandit Deen Dayal Upadhyaya Pashu Chikitsa Vigyan Vishwavidyalaya Evam Go-Anusandhan Sansthan, Mathura, Uttar Pradesh, India

R. Biswas et al. (eds.), *Concept of One Health and Zoonoses*,
https://doi.org/10.1007/978-981-95-6627-3_15

discovery of *S.* Pullorum. Achard and Bensaud discovered an organism in 1896 that they named *Bacillus paratyphique*; however, according to Boycott (1911), this organism was actually *S. schottmulleri*. Schottmuller in 1900 showed that two distinct types of paratyphoid bacilli existed; these were afterward renamed as *S.* Paratyphi A and B.

Etiology

Being a member of the *Enterobacteriaceae* family, *Salmonella* spp. typically has gram-negative, rod-shaped morphology. *Salmonella* is facultative anaerobic, non-capsulated, non-spore forming, exhibit motility using peritrichous flagella (except *S.* Pullorum and *S.* Gallinarum). While 37 °C is the optimum growth temperature, growth occurs in a wide range of temperatures (5–47 °C) and pH levels (4.0–9.0), with an ideal pH of 7.0. Generally, they ferment glucose, mannitol, arabinose, maltose, dulcitol, sorbitol, mucate, trehalose, xylose, mannose, and rhamnose with the production of acid with or without gas while lactose, sucrose, raffinose, salicin, or adonitol are not fermented. Salmonella decarboxylate the amino acid lysine, ornithine, and arginine, but not glutamic acid. *S.* Typhi can be distinguished from *S.* Paratyphi by lacking the enzymes ornithine and lysine decarboxylase. Oxidase, urease, Beta-galactosidase, phenylalanine deaminase, DNase, o-nitrophenyl-β-D-galactopyranoside (ONPG) test, Indole, and Voges-Proskauer test are all negative for the genus *Salmonella*. Except *S.* Paratyphi A, *S.* Cholerasui, *S.* Typhi, *S.* Suis, and *S.* Sendai most of the members of genus *Salmonella*, reduce nitrates to nitrites and positive for catalase, methyl red and produce hydrogen sulphide (H_2S).

The genus *Salmonella* has two species: *S. enterica*, the type species, and *S. bongori*. Furthermore, there are six different subspecies of *S. enterica*. (*S. enterica* subspp. *enterica*; *S. enterica* subspp. *salamae*; *S. enterica* subspp. *arizonae*; *S. enterica* subspp. *diarizonae*; *S. enterica* subspp. *houtenae*; and *S. enterica* subspp. *indica*). Out of 2659 currently recognised *Salmonella* serovars, 1547 serovars (59%) of the them belong to *S. enterica* subspp. (*S. enterica* subspp. *enterica*), which is responsible for roughly 99% of infections in animals and humans. Serotypes in *S. enterica* subspecies (*S. enterica* subspp. *salamae*), (S. *enterica* subspp. *arizonae*), (*S. enterica* subspp. *diarizonae*), (*S. enterica* subspp. *houtenae*), (*S. enterica* subspp. *indica*), and *S. bongori* are usually isolated from cold-blooded animals and the environment but rarely from humans.

Epidemiology

Reservoirs

Many sick, recovering, and sub-clinical human carriers have *Salmonella* bacteria colonising their gastrointestinal tracts. The faecal shedding of salmonella contaminates the environment and may enter human food chain through contaminated drinking water and food. Domestic animals such as cattle, sheep, goats, pigs, horses, chinchillas, minks, chimpanzees, rabbits, and guinea pigs serve as significant salmonella reservoirs. Both cats and dogs have previously been linked to salmonellosis and can act as asymptomatic carriers. Pigs are another important source of *Salmonella* species, however, most infections in pigs are asymptomatic. *Salmonella* spp. is commonly isolated from domestic chickens, turkeys, guinea fowl, pheasants, quails, sparrows, parrots, ducks, geese, canaries, bull-finches, ring-necked pigeons, pea-fowls, ostriches, emus, rhea, cassowaries, and kiwis. Wild birds are also a known reservoir for *salmonella*, most frequently found in passerines, mainly in bull finches and greenfinches, gulls, waterfowl, birds of prey, and doves. Hedgehogs, wild geckos, wild griffon vultures, wild rodents, and reptiles are other animals that may have harbouring Salmonella and pose a risk for transmission. A variety of insects, including beetles, that are frequently found in close proximity to domestic animals have been reported to carry *Salmonella*. *Salmonella* may naturally accumulate in cockroaches.

Transmission

Humans

People can get *salmonella* enteric fever by direct contact, ingesting contaminated food or drink, or being in close proximity to an infected person. Contact with farm animals or exposure to polluted surroundings are the two most frequent ways that non-Typhoidal *Salmonella* is transmitted to people. Close contact with both wild animals and pets has apparently been connected to human cases of salmonellosis. *Salmonella* spp. is a naturally occurring microbiota in the gut or gallbladder of animals. Transportation, mixing or crowding, food scarcity, parturition, exposure to cold, concurrent viral or parasite illness, rapid change of diet, or overfeeding can all contribute to an increase in Salmonella spp. shedding burden to the environment. *Salmonella* infection enters the human food chain via foods of animal origin. Workers on poultry and animal farms may mechanically carry *Salmonella* on their hands, feet, or clothing, which can cause infection in both humans and animals. Faecal contamination of pork, beef, and poultry meat with *Salmonella* spp. may occur at production environment or at slaughterhouses. Nevertheless, water and excrement may also infect raw veggies. Houseflies can also spread *Salmonella* spp. in poultry meat and eggs.

Animals

It is commonly known that *Salmonella* may be found in milk, faeces, urine, parturition fluid, and membranes, and that transmission can happen when infected food or water is consumed. Livestock feeds are commonly contaminated and are thought to be potential causes of salmonellosis in food animals, particularly those that contain bone meal, fish meal, and meat meal. Other animals may get an infection from contamination in the agricultural environment. Faeces with *Salmonella* can contaminate pasture, plants, soil, food, and water. The infection is transmitted by wildlife, which can also serve as a vector for transmission to domestic animals. Dogs are known as intermittent shedders and often function as asymptomatic carriers of *Salmonella* spp. Pigs may get the disease while they are still in the reproductive stage, from their feed, from other farms' animals, from the surroundings, or through the bite of a vector (insect, flea).

Poultry
Hatcheries, brooding, and raising units are frequently contaminated by the *salmonella* pathogen. The lateral spread of the organism from infected breeding flocks to their offspring and even to chicks hatched from eggs laid by infected descendants can occur by ovarian transmission. *Salmonella* on eggshells has the potential to break through the shell during incubation. *Salmonella* can horizontally spread by direct bird-to-bird contact, as well as consumption of contaminated food, drink, waste, or workers, equipment, or the environment. Spread within a flock can also occur as a result of cannibalism by ill birds, scratching in infected soil or litter, skin wounds, or eating contaminated eggs.

Incidence
In underdeveloped nations, salmonellosis is one of the most often documented bacterial food- and water-borne infections. *Salmonella* Typhi infection ranks tenth, but non-typhoidal *S. enterica* is the fourth most significant cause of preventable diseases. *Salmonella* infection is more likely to affect young children under the age of 5, the elderly, and those who have immunosuppression. Around 90 million diarrhoea-related illnesses are thought to be caused annually by *Salmonella* spp., with 85% of those cases being related to food. In South Asia, sub-Saharan Africa, East Asia, and the Pacific, where the illness is widespread, there are an estimated 200 million to over 1 billion cases of salmonellosis each year. Sub-Saharan Africa has an estimated incidence rate of 227 per 100,000 people per year, whereas enteric fever (typhoid) is the most widespread bacterial systemic infection in South Asia, with an estimated incidence rate of over 100 per 100,000 people per year and around 75,000 fatalities. For the global burden of Salmonella-induced gastroenteritis, South East Asia ranks third as a super area.

The three most often found serovars isolated from people are *S.* Weltevreden, *S.* Rissen, and *S.* Anatum. With prevalence rates of 24.7%, 23.5%, and 6.2%, respectively, *S.* Enteritidis, *S.* Typhimurium, *S.* Newport, and *S.* Heidelberg are among the most prevalent serovars in Europe and America and are becoming more significant in Southeast Asia, Africa, and the Western Pacific. Together, they account for roughly 60% of human cases. Several Asian nations, including China, India, Vietnam, Pakistan, and Indonesia, have enteric fever incidence rates that are more than 100 cases per 100,000 people per year. The greatest incidence rates are seen in Pakistan and India, with 451.7 cases and 214.2 cases per 100,000 people, respectively. With 93.8 million cases and 155,000 fatalities annually, gastroenteritis is the most prevalent *Salmonella* infection globally. *S.* Enteriditis was the most frequently isolated serotype associated with NTS infections worldwide (65%). The next two clinical isolates, *S.* Typhimurium and *S.* Newport, each contributed 12% and 4% of the isolates.

Zoonotic Potential

S. enterica subspp. *enterica* is responsible for the majority of the isolates that cause sickness in humans and other animals. A few serovars, including Salmonella serovars Typhi, Paratyphi, and Hirschfeldii, are human pathogens. They have no large animal reservoirs and are mostly spread from human to person. The other Salmonella serovars, sometimes known as nontyphoidal Salmonella, are zoonotic or may become so in the future. *Salmonella* serovar Typhimurium and *Salmonella* serovar Enteritidis are the most prevalent serovars that infect people globally. Newport, Javiana, Muenchen, Bareilly, Montevideo, Heidelberg, Saintpaul, and Infantis are only a few of the serovars that have been identified from sick people.

Pathogenesis

The infective dosage is between 10^6 and 10^8 cells, however, even a dose of 10 cells can cause salmonellosis in certain persons. Pathogenic salmonellae consumed in food survive passage past the stomach acid barrier and enter the intestine. In order to get inside the host cell, Salmonella triggers its own phagocytosis. The first step in the invasion process is the bacterium's first binding to specialized receptors on the surface of the epithelial cell (*S.* Typhimurium utilizes the giant non-fimbrial adhesin SiiE to bind the transmembrane mucin MUC1,*S.* Typhi exploits the Cystic Fibrosis Transmembrane Conductance Regulator channel). Salmonella injects its effectors through the intestinal epithelial cell membrane into the cytoplasm via type III secretion systems (multi-channel proteins) expressed in Salmonella Pathogenicity Islands.

The bacterial effectors subsequently activate the signalling pathway and cause the host cell's actin cytoskeleton to be reconstructed, resulting in the outward expansion or "ruffling" of the epithelial cell membrane, allowing the organisms to be engulfed (pinocytosis). When epithelial cells are invaded, proinflammatory cytokines such IL-1, IL-6, IL-8, TNF-2, IFN-U, MCP-1, and GM-CSF are released, which triggers an inflammatory response. The initial inflammatory reaction produces diarrhoea and may result in ulceration and mucosal damage. Following invasion of the epithelium, the organisms multiply intracellularly before being absorbed by the reticuloendothelial cells and moving on to the mesenteric lymph nodes and eventually the entire body through the systemic circulation. The reticuloendothelial system limits and regulates the organism's ability to spread. The bacteria can disseminate from the intestines to cause systemic disease infecting the liver, spleen, gallbladder, bones, meninges, and other organs.

Clinical Symptom

Human Salmonellosis

Typhoidal salmonellosis (Enteric fever): These are severe systemic types of salmonellosis. Following an incubation phase of 10 to 14 days, the symptoms appear. Enteric fevers are sometimes preceded by gastroenteritis, which normally resolves before the beginning of systemic illness. There are no particular signs or symptoms for enteric fevers, however, they do include fever, anorexia, headaches, myalgias, and constipation. During the illness, enteric fever has a distinct fever pattern, beginning with a low-grade fever (> 37.5 °C to 38.2 °C) and progressing to a high-grade fever (> 38.2 °C to 41.5 °C) in the second week. Children are more prone to experience diarrhoea, whilst immunocompromised people are more likely to experience constipation. In addition to fever, people with the infection may also experience muscle pain, heart palpitations, hepatomegaly, splenomegaly. In 1–30% of patients, "Rose spots," a blanching erythematous maculopapular rash with lesions 2–4 mm in diameter that commonly appear on the belly and chest but can also occur on the back, arms, and legs, have been recorded. Occasionally, neonatal infection can be caused by enteric fever and may complicate pregnancy. Fever might last for a month or longer if the patient is not treated. If appropriate antibiotic treatment is not given, enteric fevers, which are serious diseases, might be deadly. In endemic areas, roughly 15% of infected people have gastrointestinal problems such as pancreatitis, hepatitis, and cholecystitis. One of the most serious gastrointestinal problems that can emerge from the rupture of Peyer's patches and cause bloody diarrhoea is gastrointestinal haemorrhage. Other enteric fever consequences include typhoid encephalopathy, mental instability, and potentially fatal pneumonia.

Non-typhoidal Salmonellosis (Gastroenteritis) Nontyphoidal Salmonella is a foodborne illness that is mostly spread by commercially manufactured food that has been contaminated with animal faeces. It is caused by all nontyphoid serotypes of the Salmonella genus, with the exception of *S.* typhi and *S.* Paratyphi A, B, and C. Food poisoning is most typically caused by serotypes *S.* Typhimurium, *S.* enteritidis, *S.* newport, and *S.* heidelberg, however, *S.* Cholerasuis and *S.* Dublin can also cause diarrhoea. The duration of incubation for Salmonella gastroenteritis (food poisoning) is determined by the number of bacteria consumed. Symptoms generally appear from 6 to 48 h after consuming contaminated food or drink. NTS is classified as non-invasive or invasive. The vast majority of non-invasive NTS begin as gastrointestinal self-limiting infections. The non-invasive NTS infection is characterised by gastroenteritis or "stomach flu," an inflammatory illness of the gastrointestinal system accompanied by symptoms such as stomach discomfort, non-bloody diarrhoea, vomiting, nausea, abdominal cramps, and lack of appetite. Although invasive salmonellosis is less prevalent, it is a more serious condition that can include meningitides, sepsis, septic aortic aneurysms, septic arthritis, headaches, and muscle or articular pain.

Animal Salmonellosis

Cattle Most affected calves are between the ages of 2 and 6 weeks. Septicaemia in newborn calves can be fatal in 6 to 12 h. Calves become dull and depressed, fail to suckle and accompanied by diarrhoea which can be fatal. Calves have higher rectal temperature and grey pasty faeces that include fresh blood and mucous. Older calves may experience watery, foul-smelling diarrhoea with mucosal casts, which is sometimes associated with an infection with S. typhimurium and causes further dehydration and an emaciated appearance. Those calves that survive the disease episode grow slowly and remain stunted. S. Dublin may cause articular infections and dry gangrene of the extremities (ears, limbs, and tail), infection of the neck vertebrae, which causes paralysis in all four legs and leads to recumbency in two to four-month-old calves. Abortion, preterm delivery, stillbirth, or the birth of weak non-viable calves occurs in pregnant adult animals (Underwood et al., 2015). Abortion happens between the ages of 5 and 8 months of pregnancy, with placental retention. Systemic illness is unusual in *S.* Dublin, but it is more prevalent in *S.* Typhimurium, where abortion might occur after septicaemia and enteritis.

Dogs Two or more serotypes of *Salmonella* may colonise large intestine and mesenteric lymph nodes. Animals that are young and pregnant seem to be particularly vulnerable to salmonellae. Clinical salmonellosis in dogs is characterised by hyperpyrexia (104°–106 ° F), anorexia, intermittent diarrhoea, bloody stools, vomiting, weight loss, cough, nasal haemorrhages, hyperirritability, incoordination, partial posterior paralysis, blindness, "running fits and barking", depression, and lochia. The acute phase of the illness might last 4 to 10 days and mortality rate rarely cross 10%. It should be emphasised, however, that most salmonellae-infected dogs are asymptomatic and are long term, sporadic faecal shedders of the pathogen(s) and may serve as sources of salmonellosis for human beings.

Diagnosis

Bacterial Isolation The isolation of *Salmonella enterica* and S. Typhi from faeces, urine, blood, and bone marrow is required for the conclusive diagnosis of enteric fever. The blood culture is positive in 80% or more of untreated patients, although the isolation rate for patients receiving antibiotics is as low as 40%. Bone marrow culture is recommended over blood culture as the bone marrow aspirate as the detection rate is as high as >80%, especially when the patient has already undergone antimicrobial medication. Salmonella enterica may be isolated from faeces in up to 30% of typhoid fever patients and 20% of urine samples.

For the transport of specimen, non-selective pre-enrichment medium like buffered peptone water is used. For initial enrichment of Salmonella, Modified Semisolid Rappaport Vassiliadis (MSRV), Rappaport-Vassiliadis Salmonella (RVS) broth, Muller-Kauffmann Tetrathionate Novobiocin (MKTTn) broth, and Leifson's selenite F (SF) is used as a selective enrichment medium for rapid detection of motile

Salmonella spp. in faeces and food products. For examination of contaminated materials such as stools, sewage, foods, etc., the use of selective media is recommended, e.g., Kristensen, Lester, & Jurgens brilliant green agar (Beef extract, Peptone Sodium chloride, Phenol-red solution), Leifson's desoxycholate-citrate agar (Peptone, Beef Infusion Lactose, Sodium citrate Sodium desoxycholate Iron ammonium citrate:, Wilson & Blair bismuth-sulfite agar (Glucose Disodium phosphate Ferrous sulfate Bismuth indicator Sodium sulfite Brilliant green) and SS agar (Beef extract, Peptone, Lactose, Bile salts Sodium citrate, Sodium thiosulfate, Iron citrate, Neutral red, Brilliant green).

Serologic Assays Salmonellae are Gram-negative, flagellated, facultative anaerobic bacilli possessing three major antigens: H or flagellar antigen, O or somatic antigen, and Vi (capsular) antigen (possessed by only a few serovars). The different species are serotyped according to these three different antigens.

1. H (flagellar) antigen may occur in either or both of two forms, phase 1 and phase 2. There are over 1800 known serovars which current classification considers being separate species. The organisms tend to change from one phase to the other.
2. O (somatic) antigens occur on the surface of the outer membrane and are determined by specific sugar sequences on the cell surface.
3. Vi (capsular) antigen is a superficial antigen overlying the O antigen (additional surface antigen). It is present in a few serovars, the most important being Salmonella Typhi, but also present in *Salmonella* Paratyphi C and *Salmonella* Dublin.

After identifying the species, serotyping is conducted on a fresh, pure culture of *Salmonella* isolated on a non-selective agar medium such TSA (Tryptic Soy Agar). To begin, use polyvalent O antiserum to test the isolate. The vast majority (almost 98%) of *Salmonella* found in warm-blooded animals have an O antigen matching the agglutinins present in OMA, OMB, and OMC sera. Agglutination with one of these three groups indicates that the isolate is positive for that group. To identify the O antigen(s), individual monovalent O antisera are used. When a strain fails to agglutinate the OMA, OMB, or OMC polyvalent sera, it is advised to test it with VI serum.

Antigen Detection

Assays to detect bacterial antigens such as Vi, O9, and Hd in urine have been used as diagnostic methods for enteric fever but with variable results.

Antibody Detection

Serologic reactions to O, H, and Vi antigens are often observed in the sera of persons with probable enteric fever by the Widal test in typhoid fever. By the conclusion of the first week of sickness, agglutinating antibodies against O, H, and Vi antigens had generally developed. A positive Widal test needs both acute- and convalescent-phase blood samples obtained around 10 days apart; a fourfold rise in antibody titer is required for a positive result. A threshold titer of 200 for O agglutination or 100 for H agglutination would properly diagnose 74% of patients with blood culture positivity. The results are not conclusive and should be taken with caution because titers can be raised in response to a variety of different illnesses. Single serum specimens with high titers from people living in endemic disease regions have minimal diagnostic value. ELISAs are more sensitive than the Widal "O" and "H" antigen-based tests and have been used to evaluate the normal antibody response to LPS, flagella, Vi capsular polysaccharide, or outer membrane protein antigens during enteric fever. Typhi dot is an immunodot test that detects specific IgM and IgG antibodies against a 50-kDa Salmonella serovar Typhi outer membrane protein (OMP).

Molecular Assays

Salmonella serovars Typhi and Paratyphi A have been detected mostly in blood using nucleic acid amplification techniques such as conventional PCR and real-time PCR. TrueNAT™ Salmonella Chip-based Real-Time PCR is available for qualitative detection and identification of Salmonella in human blood.

Treatment

Nontyphoid salmonellosis does not usually necessitate hospitalisation, and only fluid treatment is required in situations of severe diarrhoea. Antibiotic treatment is seldom required and recommended in septicaemic, enteric fever, and localised infection syndromes. Treatment for general salmonellosis includes restoring fluid loss by oral and intravenous means, as well as reducing discomfort, nausea, and vomiting. Antibiotics should be used to treat typhoid and enteric fevers.

Control Measures

The primary cause of typhoidal salmonellosis is the faeces of humans contaminating food and water. Therefore, the key to lowering salmonellosis sickness is appropriate food and water cleanliness, pasteurisation of milk and other dairy products, correct handling and cooking of food, and excluding human faeces from the food production chain. Because poultry and cattle are primary reservoirs for human infection, lowering the amount of salmonellae burden in these species would significantly reduce nontyphoidal salmonellosis in humans. Strict feeding management, all-in-all-out production, breeding control, sanitary locks, bug and rodent control, and the eradication of carrier animals are some of the methods used for minimising infection of cattle, pig, and poultry farms. All employees who handle food in slaughterhouses, food processing facilities, and restaurants should receive hygienic practises training. Food should also be cooked and stored properly in food processing facilities, restaurants, and homes. Cross-contamination of animal carcasses should be minimised.

Only two immunisation types—inactive and live attenuated—are now acknowledged as being efficacious against enteric fever. The Ty21a vaccine is a live, attenuated, oral immunisation based on the attenuated Salmonella serovar Typhi strain Ty21a, whereas the parenteral Vi vaccine is based on the capsular polysaccharide antigen of Salmonella serovar Typhi strain Vi. The Ty21a and Vi polysaccharides' overall protective effectiveness has been assessed to be 48% and 55%, respectively (Guzman et al., 2006).

References

Guzman, C. A., Borsutzky, S., Griot-Wenk, M., Metcalfe, I. C., Pearman, J., Collioud, A., Favre, D., & Dietrich, G. (2006). Vaccines against typhoid fever. *Vaccine, 24*(18), 3804–3811.

Underwood, W. J., Blauwiekel, R., Delano, M. L., Gillesby, R., Mischler, S. A., & Schoell, A. (2015). Biology and diseases of ruminants (sheep, goats, and cattle). In *Laboratory animal medicine* (pp. 623–694). Academic.

Campylobacteriosis

16

Ajay Pratap Singh

Abstract

Campylobacter is an important foodborne enteropathogen responsible for causing acute gastroenteritis in humans, often with symptoms like watery or bloody diarrhoea, abdominal pain, headache, fever, chills, and dysentery. This condition is referred to as Campylobacteriosis. *Campylobacter* species are a common cause of bacterial foodborne and waterborne infections worldwide. It can be found in gastrointestinal tracts of animals and birds, making it relatively easy for contamination to occur in the food chain, especially through undercooked poultry. Infections are more common in children, the elderly, and individuals with weakened immune systems, such as immunocompromised individuals. The symptoms of *Campylobacter* infection can vary from mild to severe and may include diarrhoea, vomiting, fever, and abdominal pain. In some cases, the diarrhoea may be bloody.

While *Campylobacter* is primarily associated with gastrointestinal symptoms, it has been reported to cause various systemic infections, including septic thrombophlebitis, endocarditis, neonatal sepsis, pneumonia, and bloodstream infections. These are relatively rare but can occur, especially in vulnerable individuals. There is a significant association between *Campylobacter* infection and neurological complications. Guillain–Barré syndrome (GBS), Miller Fisher syndrome (MFS), and severe demyelinating neuropathy have been linked to *Campylobacter* infection. This can result in serious nerve damage and muscle weakness. *Campylobacter* has been associated with conditions such as colorectal cancer, Barrett's oesophagus, Reiter's arthritis, and irritable bowel syndrome. However, the precise mechanisms of these associations are not yet fully understood. In

A. P. Singh (✉)
Department of Veterinary Microbiology, College of Veterinary Science & Animal Husbandry, U.P. Pandit Deen Dayal Upadhyaya Pashu Chikitsa Vigyan Vishwavidyalaya Evam Go-Anusandhan Sansthan, Mathura, Uttar Pradesh, India

R. Biswas et al. (eds.), *Concept of One Health and Zoonoses*,
https://doi.org/10.1007/978-981-95-6627-3_16

addition to gastrointestinal and systemic infections, *Campylobacter* has been reported to cause extraintestinal infections such as brain abscesses, meningitis, lung infections, bacteraemia, and reactive arthritis. *Campylobacter* species have also been reported to be associated with extraintestinal infections such as brain abscesses, meningitis, lung infections, bacteraemia, and reactive arthritis.

Keywords

Campylobacter · Guillain–Barré syndrome · Miller Fisher syndrome · Vibrionic abortion and Vibrionic enteritis · Avian vibrionic hepatitis · Campylobacter enteritis

History

The initial identification of bacteria resembling Campylobacter dates back to 1886, when Theodor Escherich, a German paediatrician, observed spiral-shaped microorganisms in the intestines of infants afflicted with "cholera infantum", a condition now understood as a type of gastroenteritis. Decades later, on February 2, 1906, British veterinarians John McFadyean and Stewart Stockman made a significant breakthrough by identifying "large numbers of peculiar organisms" in uterine mucus samples from a pregnant ewe, using Loeffler's blue staining technique. By 1913, similar bacteria were detected in aborted bovine foetuses, and by 1927, they were successfully isolated from the faeces of diarrhoeic cattle. In 1931, the work of Jones and colleagues established a link between microaerophilic vibrios and bovine dysentery, leading to the provisional classification of the organism as *Vibrio jejuni.* The first documented foodborne outbreak attributed to Campylobacter occurred in 1938 in Illinois, where contaminated milk caused illness in a large number of individuals, highlighting the pathogen's potential for foodborne transmission. In 1944, Doyle isolated another vibrio species from diarrhoeic pigs, naming it *Vibrio coli.* Although Escherich's early observations laid the groundwork for recognizing these spiral bacteria, it took advancements in microbiology and laboratory techniques to fully elucidate the characteristics of Campylobacter species. The term Campylobacter was formally proposed by Sebald and Vernon in 1963 to differentiate these organisms from Vibrio species. The name originates from the Greek word "kampylos", meaning "curved", reflecting the bacteria's distinctive morphology.

Aetiology

The Campylobacter genus is classified within the family Campylobacteraceae, order Campylobacterales, class Epsilonproteobacteria, and phylum Proteobacteria. It encompasses 27 species, 9 subspecies, and 3 biovars. Members of this genus are characterized as small, Gram-negative, spirally curved rods, measuring 0.2–0.9 μm in width and 0.5–5 µm in length. They are non-spore-forming and typically motile, possessing a single polar flagellum at one or both ends of the cell. These bacteria are primarily microaerophilic, relying on a respiratory-type metabolism, and thrive in environments with 3–6% oxygen. Their growth is further enhanced by the presence of 2–10% carbon dioxide. Most species are mesophilic, with optimal growth temperatures ranging from 25 °C to 45.5 °C. However, species pathogenic to humans, such as *Campylobacter jejuni*, exhibit optimal growth at around 42 °C.

Several Campylobacter species are significant human enteric pathogens, responsible for causing gastrointestinal illnesses. The primary pathogenic species include *Campylobacter jejuni* subsp. *jejuni* (Cjj), *Campylobacter jejuni* subsp. *doylei* (Cjd), *Campylobacter coli*, and *Campylobacter fetus*. In addition to these major pathogens, certain species are considered minor human pathogens due to their association with fewer infections. These include *Campylobacter concisus*, *Campylobacter upsaliensis*, *Campylobacter lari*, and *Campylobacter hyointestinalis*. Furthermore, some Campylobacter species are primarily veterinary pathogens but can also pose risks to human health. Notable examples include *Campylobacter fetus* subsp. *venerealis* (Cfv) and *Campylobacter fetus* subsp. *fetus* (Cff).

Campylobacter species are classified into various serotypes based on different antigenic properties. Two common serotyping schemes are utilized:

1. *Lior Serotypes*: This classification is based on the heat-stable antigenic properties of the bacteria. It divides *Campylobacter* into distinct serotypes, and over 100 Lior serotypes have been reported. While this system is valuable for epidemiological tracking and differentiation, only a subset of these serotypes has clinical significance.
2. *Penner Serotypes*: The Penner serotyping system is based on the heat-labile (O) antigen of *Campylobacter*. It is a more extensive scheme, with over 600 reported serotypes. As with Lior serotypes, not all Penner serotypes are of clinical importance.

Epidemiology

Members of the Campylobacter genus often exist as commensal organisms within the gastrointestinal (GI) tracts of diverse hosts, including domestic pets, livestock, and wild animals. They are also commonly detected in contaminated food. Numerous species within this genus have been linked to human illnesses, including gastroenteritis, inflammatory bowel disease (IBD), and periodontitis. Among these, *Campylobacter jejuni* and *Campylobacter coli* are the most prominent species associated with global foodborne disease outbreaks (Silva et al., 2011). Notably, these two species typically colonize the intestinal tracts of domestic and wild animals without causing any apparent harm to their hosts.

Host Susceptibility

Numerous animals, including cattle, sheep, swine, cats, dogs, and poultry, serve as natural hosts for Campylobacter. Among these, dogs and cats can act as carriers for multiple Campylobacter species. Wild animals, particularly migratory birds, also play a significant role as reservoirs for these bacteria, facilitating their spread into the environment and enabling transmission to other animals and humans.

Campylobacter species associated with human and animal infections

Campylobacter species	Disease in humans	Disease in animals	Animal reservoir
Campylobacter coli	Gastroenteritis, acute cholecystitis, meningitis, bacteraemia, and sepsis	Gastroenteritis, infectious hepatitis	Pig, poultry, cattle, sheep, goat, monkey, and dog
Campylobacter fetus subsp. *fetus*	Gastroenteritis, septicaemia, abortion, meningitis, epidural and brain abscesses, endocarditis, and peritonitis	Bovine and ovine spontaneous abortion	Cattle, horse, sheep, reptiles (pet turtle), and kangaroo
Campylobacter fetus subsp. *venerealis*	Septicaemia	Bovine infectious infertility	Cattle, sheep
Campylobacter curvus	Gastroenteritis, ulcerative colitis, Barrett's esophagitis, periodontitis, and bronchial and liver abscesses	–	Human and dog
Campylobacter gracilis	Periodontitis, empyema, abscesses, and IBD	–	–
Campylobacter hyointestinalis subsp. *hyointestinalis*	Gastroenteritis	Porcine and bovine enteritis	Cattle, deer, pig, hamster
Campylobacter jejuni subsp. *doylei*	Gastroenteritis and septicaemia	–	–
Campylobacter jejuni subsp. jejuni	Gastroenteritis, IBD, coeliac disease, reactive arthritis, Guillain–Barré syndrome, septicaemia, meningitis, urinary tract infection, abortion, proctitis, and septicaemia	Spontaneous abortion, avian hepatitis, and gastroenteritis	Poultry, cattle, pig, sheep, dog, cat, ostrich, wild birds, mink, rabbit, insects, and water
Campylobacter lari subsp. *Concheus*	Gastroenteritis	–	Shellfish
Campylobacter lari subsp. *lari*	Gastroenteritis and septicaemia	Avian gastroenteritis	Poultry, wild birds, dog, shellfish, and horse

Campylobacter species	Disease in humans	Disease in animals	Animal reservoir
Campylobacter Pelopidas	Gastroenteritis	–	Shellfish
Campylobacter sputorum bv. Sputorum	Gastroenteritis, abscesses	Spontaneous abortion	Human, cattle, pig, and sheep
Campylobacter rectus	Gastroenteritis, IBD, periodontitis, empyema thoracis, and vertebral abscess	–	Human and dog

Transmission

Animal-to-Human Transmission

Direct animal-to-human transmission of Campylobacter typically involves close contact with infected animals or their faeces. This route of transmission is particularly relevant for individuals who work with livestock, poultry, or domestic pets, as these animals are common reservoirs for Campylobacter species, especially *Campylobacter jejuni* and *Campylobacter coli*. Farmers, veterinarians, and slaughterhouse workers are at increased risk of Campylobacter infection due to direct contact with infected animals. Handling poultry, cattle, or pigs without proper hygiene measures can lead to the ingestion of the bacteria, often through contaminated hands or surfaces. Dogs, cats, and other household pets can carry Campylobacter without showing symptoms. Direct contact with infected pets, particularly through handling faeces or cleaning litter boxes, can result in transmission if proper handwashing is not practised. Activities such as petting zoos or farm visits can expose individuals, especially children, to Campylobacter.

Food-Borne Transmission

Food-borne transmission is the most common route of human infection. Food-producing animals, such as poultry, cattle, and pigs, serve as the primary reservoirs for Campylobacter. These animals often carry the bacteria in their gastrointestinal tracts without showing signs of illness. During slaughter and processing, Campylobacter can spread from the intestines to meat products through faecal contamination, leading to the presence of the bacteria in raw or undercooked meat. Consumption of contaminated meat products, particularly undercooked poultry, is a leading cause of Campylobacter-related illnesses (Myintzaw et al., 2021). Poultry is estimated to account for 50–70% of human infections, as Campylobacter jejuni, the most prevalent species, frequently colonizes the intestines of birds (Je et al., 2023). Beef and pork products can also harbour the bacteria, though to a lesser extent. Cross-contamination in kitchens, where raw meat comes into contact with other foods or surfaces, further amplifies the risk of infection. Unpasteurized bovine milk and milk products are common vehicles for *Campylobacter*-associated foodborne infection transmission. Environmental contamination also plays a critical role in the transmission of Campylobacter. The bacteria can be shed into the environment through animal faeces, contaminating soil, water, and crops. Irrigation of crops with contaminated water or the use of manure as fertilizer can introduce Campylobacter into fresh agriculture produce and finally in human food chain. Postharvest factors involved in the contamination of fruits and vegetables include improper hygiene practices of the equipment, utensils, and handlers; use of contaminated water to wash the foods; irregular storage (temperature and physical environment), improper packaging; and cross-contamination at the retail.

Water-Borne Transmission

Besides food, water may serve as a significant environmental reservoir for *Campylobacter*. Human or animal faeces containing these microorganisms are the primary contamination source for water resources. Consumption of contaminated drinking water is a significant cause of campylobacteriosis outbreaks.

Incidence

Campylobacteriosis is a prevalent bacterial infection with a significant global impact on public health. The incidence of Campylobacteriosis varies across regions and is influenced by multiple factors, including hygiene practices, food safety measures, and surveillance programmes. In developed countries, Campylobacteriosis is a well-documented public health concern. The annual incidence of this infection ranges from 4.4 to 9.3 cases per 1000 population. Notably, *Campylobacter* has earned its reputation as the most frequent pathogen associated with foodborne outbreaks in several developed nations, including Singapore, Australia, Canada, New Zealand, and the European Union (Kaakoush et al., 2015). *Campylobacter* species are recognized as a leading bacterial cause of zoonotic disease on a global scale. In Europe, approximately 1% of the human population is affected by *Campylobacter* infections each year. In the United States, the incidence is estimated at 13 cases per 100,000 persons annually. Within the developed world, *Campylobacter jejuni* is the primary culprit behind most human *Campylobacter* infections. In developing countries, the prevalence of *Campylobacter* infection is interestingly lower when compared to developed nations largely due to lower reported incidence in developing countries in the absence of robust national surveillance programmes.

Pathogenesis

The initial interaction between *Campylobacter* and the host's tissues occurs within the intestine. The GI tract is equipped with various defence mechanisms to prevent the invasion of pathogens, including the mucous layer and intestinal epithelial cells. The location of infection can vary between different hosts. In humans, *Campylobacter* infections predominantly occur in the small intestine. However, in poultry, the primary site of colonization is often the caecum. The presence of polar flagella and the spiral cell shape in *Campylobacter* play a crucial role in its movement through the mucus layer in the host's gastrointestinal tract. Once *Campylobacter* bacteria have successfully crossed the mucous layer in the gastrointestinal tract, they encounter the intestinal epithelial cells. Adhesion is the process by which bacterial adhesins (cell surface structures) interact with specific receptors on the host's epithelial cells. The adhesion process is complex and involves specific interactions between the bacterial cell surface adhesins and the host cell receptors. These interactions are typically irreversible and specific, meaning *Campylobacter* binds tightly to the host's epithelial cells. When *Campylobacter* adheres to the host epithelial cells, it triggers a signalling process within the host cell. As part of this signalling cascade, the GTPases are activated which in turn induce changes in the host cell's actin cytoskeleton. This reorganization of the actin network leads to the formation of filopodia and lamellipodia, which are cellular structures that can engulf the adherent *Campylobacter* cells. The reorganization of the actin cytoskeleton and the formation of these cellular structures facilitate the internalization of *Campylobacter* cells into the host epithelial cell.

Once internalized by the host epithelial cell, *Campylobacter* resides within a membrane-bound compartment. This compartment provides a degree of protection for the bacterium, shielding it from the host's immune defences, such as antibodies and phagocytes. *Campylobacter* can survive and replicate within this intracellular compartment. By doing so, it can persist for extended periods of time inside the epithelial cell. *Campylobacter* can remain inside the host cell until conditions become favourable for it to induce a cytotoxic response. This response may involve various bacterial virulence factors, such as toxins or other molecules, which can damage the host cell. *Campylobacter* produces a tripartite toxin known as cytolethal distending toxin (CDT). The CDT is translocated into the host cell cytoplasm towards the nucleus, where it is internalized, and causes cell cycle arrest in the G2/M transition phase through blocking of CDC2 kinase involved in the entry into mitosis leading to cell distension and apoptotic cell death.

Clinical Symptoms

Humans

Human *C. fetus* subspp *fetus* infections exhibit a spectrum of clinical signs and symptoms. These presentations can range from an acute diarrhoeal illness to systemic illness. The clinical manifestations vary depending on the localization of the disseminated pathogen.

Septicaemia Campylobacteriosis most often occurs in the form of gastroenteritis or enteritis only. Symptoms diseases are diverse and include diarrhoea, often with blood in the stool, nausea, vomiting, abdominal pain in the umbilical area, and fever. Disease develops within 1–7 days of infection and is often preceded by symptoms such as headaches, muscle pain, or fever. Symptoms of campylobacteriosis usually disappear spontaneously after 2–7 days; however, cases of chronic diarrhoea are also known. Approximately 30% of patients may experience only flu-like symptoms.

Neurological Infections Infections can result in neurological complications, including meningitis, meningoencephalitis, subdural empyema, or brain abscesses. A previous infection with *C. jejuni* may occur in some people and cause neurological complications such as Guillain–Barré syndrome or Miller Fisher syndrome.

1. *Guillain–Barré Syndrome (GBS)* is an autoimmune disorder of the peripheral nervous system, characterized by the immune system mistakenly attacking the body's own nerve cells. One of the known triggers for GBS is an infection with the bacterium *Campylobacter jejuni*. In GBS, molecular mimicry plays a crucial role in the pathogenesis, where antibodies produced by the host against *Campylobacter* cross-react with gangliosides on peripheral nerve tissue, causing demyelination and damage to axonal nerves. GBS can lead to severe muscle weakness, paralysis, and, in some cases, respiratory failure.
2. *Miller Fisher Syndrome:* In addition to GBS, *Campylobacter* infection has been associated with the development of Miller Fisher syndrome (MFS). MFS is considered a variation of GBS and is characterized by a triad of symptoms: ataxia (loss of coordination), ophthalmoplegia (paralysis or weakness of the eye muscles), and areflexia (absent or reduced reflexes). Like GBS, MFS often develops following exposure to an infectious pathogen. *Campylobacter jejuni* is a common pathogen associated with MFS, further emphasizing its role in these neurological disorders.

Maternal and Perinatal Infections Pregnant women can also experience *C. fetus* infections throughout pregnancy, from early stages to full-term birth. Clinical signs in pregnant mothers typically include fever, sometimes accompanied by diarrhoea. Spontaneous abortions, without additional clinical signs, have also been reported. In cases where living infants are born, many suffer from *C. fetus* sepsis, often leading to meningitis. In a study of 14 cases of infant *C. fetus* sepsis, 9 resulted in a fatal outcome, underscoring the severity of neonatal infections. Perinatal infections are most commonly associated with a confirmed *C. fetus* infection in the mother. *C. fetus* subspp. *Venerealis* is primarily regarded as a pathogen affecting cattle and is not typically associated with human infections. However, isolated cases of *C. fetus* subspp *venerealis* have involved isolates from vaginal discharges, mirroring bovine infections where this subspecies colonizes the genital tract.

Reactive Arthritis *Campylobacter* infection may also lead to reactive arthritis, a spondyloarthropathic disorder. Reactive arthritis is characterized by the inflammation of various body parts, including joints and other tissues. Individuals affected by reactive arthritis typically experience joint pain, swelling, and stiffness, along with other symptoms such as eye inflammation and urinary problems. *Campylobacter*-induced reactive arthritis is considered a result of the immune system's response to the infection, where the bacterium may trigger an aberrant immune response that targets the joints and other tissues.

Vascular Pathology Some cases involve vascular pathology, including mycotic aneurysms, endocarditis, vasculitis, thrombophlebitis, or pericarditis.

Animals

*Campylobacter*iosis in animals can produce a spectrum of clinical scenarios, influenced by factors such as host immunity, bacterial virulence, gene expression, and more. Infected animals may present with various clinical signs. The most common clinical manifestations in animals include abdominal pain, fever, diarrhoea, blood in faeces, and the presence of inflammatory cells in faeces. These signs reflect the inflammatory nature of the infection by *C. jejuni* subsp *jejuni*.

- *Dogs and Cats*: *C. jejuni* (subspecies *jejuni*) is known to cause acute diarrhoea in dogs and cats. Diarrhoea in dogs, especially those under six months old, can last 5–15 days, with watery to bloody and sometimes bile-stained stools. Cats under six months old may also experience bloody diarrhoea. Some infected cats may remain asymptomatic. Additionally, *C. jejuni* (subspecies *jejuni*) has been isolated in the vaginal discharge from late-pregnancy abortions in dogs.
- *Cattle and Sheep*: *Campylobacter* spp, including *C. jejuni, C. fetus, C. hyointestinalis*, and *C. sputorum*, can cause enteritis and abortion in cattle and sheep. However, the role of *Campylobacter*spp in enteric disease of ruminants is still controversial, as clinically normal cattle can carry substantial numbers of *Campylobacter*. Faecal shedding in cattle can lead to milk and beef contamination.
- *Birds*: Birds, including intensively farmed poultry, have high infection and carrier rates of *Campylobacter* spp, especially *C. jejuni*. In broilers, the pathogen can colonize palatine lymphoid tissues and the crop, facilitating rapid transmission. An unidentified *Campylobacter* species has been linked to necrotizing hepatitis in chickens. *C. jejuni* has been isolated from the small intestines of clinically ill birds, with symptoms including hepatitis, lethargy, loss of appetite, weight loss, and yellow diarrhoea. Wild birds commonly carry the organism but rarely manifest disease.
- *Exotic Pets*: *Campylobacter* gastrointestinal disease has been reported in exotic pets such as ferrets, mink, primates, hamsters, guinea pigs, mice, and rats. Clinical signs often include mucoid, watery, bile-streaked diarrhoea, anorexia, vomiting, and fever. While prolonged infections can occur, they are generally self-limiting with mild clinical signs.

Diagnosis

The foundation of diagnosing Campylobacteriosis is the isolation of the pathogen from clinical material. Bacteriological tests typically involve direct microscopic examination, isolation, and identification of microorganisms. Stool samples are preferred sample type for intestinal infections. It should be collected during the acute phase of the disease into a sterile vessel or a transport medium containing activated charcoal (e.g., Amies, Stuart, or Cary-Blair medium). Rectal swab is an acceptable alternative for stool sample. In cases of generalized infection, blood or cerebrospinal fluid must be collected.

I. **Culture and Isolation**: *Campylobacter* are cultured for 1–5 days under microaerophilic conditions at 42 °C. The choice of culture medium depends on the sample type, with selective media often used, enriched with 5% sheep or horse blood. Majority of selective media are supplemented with vancomycin and cefoperazone for inhibiting other enteric bacteria. Common culture media used for isolation of *Campylobacter* species are Campylobacter blood-free selective medium (CBFS) (ideal for isolating *C. jejuni* and *C. coli*), charcoal cefoperazone desoxycholate agar (CCDA), and Karmali Agar. Routine blood-based agar medium is Skirrow Agar, Preston Agar (used for specific sample types), Butzler Agar, and Campy cefex agar. Enrichment Broths may enhance the yield of culture, especially where low organism numbers are expected.

II. **Preliminary Identification Tests**: Upon growth of microorganisms on media, preliminary identification tests are conducted such as Gram's staining, cytochrome, and oxidase test. For further identification of colonies showing Gram-negative bacilli, morphology, positive motility, and cytochrome oxidase production are screened on Columbia agar medium with 5% sheep blood. In some countries, a filtration method is employed for isolating *Campylobacter* bacilli from stool samples. This method selectively passes *Campylobacter* bacilli through a membrane filter while retaining other bacteria. The grown colonies are further identified.

III. **Blood Cultures**: *Campylobacter* species are occasionally isolated from blood cultures, primarily *C. fetus*, *C. jejuni*, and *C. upsaliensis*.

IV. **Direct Detection in Stool Samples**: Direct detection methods such as Gram's staining, phase contrast, and dark-field microscopy may be used.

V. **Stool Antigen Tests:** Commercial stool antigen tests like *CAMPYLOBACTER* QUIK CHEK™, Premier™ CAMPY enzyme immunoassay and RIDASCREEN® *Campylobacter* have been developed. *Campylobacter* stool antigen tests are diagnostic assays used to detect the presence of bacteria in stool samples. The tests are based on immunoassays for the detection of specific antigens (proteins) associated with *Campylobacter*.

VI. **Molecular Approaches**: Nucleic acid amplification tests (NAATs), including polymerase chain reaction (PCR) and other molecular-based methods, have been employed for direct detection, species-level identification, and typing of *Campylobacters* directly from stool samples. The BD Max™ Enteric Bacterial Panel, Luminex® xTAG® Gastrointestinal Pathogen Panel (GPP), Luminex® Verigene® Enteric Pathogens (EP), and Qiagen QIAstat-Dx® Gastrointestinal Panel are all molecular-based assays used for the detection of various pathogens, including *Campylobacter*, from stool samples. These panels offer the advantage of detecting multiple pathogens simultaneously in a single test, which can be particularly beneficial in diagnosing and managing gastrointestinal infections.

VII. **Serodiagnosis:** *Campylobacter* antigens such as flagellin, lipopolysaccharide, and other protein antigens are targeted for serodiagnosis. An enzyme-linked immunosorbent assay (ELISA) that utilizes purified *Campylobacter* antigens, namely, OMP18 and P39, is used for serodiagnostic test designed for the detection of antibodies specific to *Campylobacter* in the serum or plasma of patients.

Zoonoses

Campylobacter species are notorious zoonotic agents that have the potential to serve as reservoirs for antimicrobial-resistant genes. Campylobacters, including *C. jejuni* and *C. coli*, are significant zoonotic pathogens that infect about 1% of Western Europe's population annually. While they play a role in animal infections, their impact on food-borne human infections can lead to long-term health issues like reactive arthritis and Guillain–Barré syndrome. Campylobacters are widely present in nature, domestic animals, and the environment. They are commonly found on various raw foods, with poultry being a significant source.

Treatment and Prevention Strategy

1. *Biosecurity Measures*: Implementing strict biosecurity measures in poultry farms is crucial to prevent *Campylobacter* colonization. These measures aim to minimize the introduction and spread of the bacteria within the poultry flock. Biosecurity measures include controlling access to farms, maintaining good hygiene practices, and preventing contact with wild birds and other potential carriers of the bacteria.
2. *Hygiene Practices*: Hygiene practices at the primary production level are essential in reducing *Campylobacter* contamination in poultry. These practices include boot dip, hygiene barrier, handwashing, and equipment sanitization.
3. *Reduction of Intestinal Colonization*: Reducing *Campylobacter* colonization in poultry is a highly effective strategy to lower the bacterial load in these animals. This may involve optimizing poultry feed, managing the gut microbiota, and using probiotics to limit *Campylobacter*'s ability to colonize the intestinal tract.
4. *Vaccination*: While no commercial vaccine for *Campylobacter* in poultry has been developed yet, research is ongoing in this area. Vaccination is a potential strategy to reduce colonization in poultry and subsequently decrease the risk of transmission to humans. Studies have shown that specific antibodies reactive against *Campylobacter* jejuni can significantly reduce bacterial loads in poultry. Developing an effective and practical vaccine is a promising avenue for future control.

5. *Food Safety Measures*: Ensuring proper food safety practices in food processing, distribution, and preparation is essential to prevent the transmission of *Campylobacter* to consumers. This includes thorough cooking of poultry, proper storage, and cross-contamination prevention.
6. *Surveillance and Monitoring*: Regular testing and monitoring of poultry flocks for *Campylobacter* contamination can help identify potential issues early and implement corrective actions.

References

Je, H. J., Singh, S., Kim, D. W., Hur, H. S., Kim, A. L., Seo, E. J., & Koo, O. K. (2023). Systematic review and meta-analysis of Campylobacter species contamination in poultry, meat, and processing environments in South Korea. *Microorganisms, 11*(11), 2722. https://doi.org/10.3390/microorganisms11112722

Kaakoush, N. O., Castaño-Rodríguez, N., Mitchell, H. M., & Man, S. M. (2015). Global epidemiology of Campylobacter infection. *Clinical Microbiology Reviews, 28*, 687. https://doi.org/10.1128/cmr.00006-15

Myintzaw, P., Jaiswal, A. K., & Jaiswal, S. (2021). A review on Campylobacteriosis associated with poultry meat consumption. *Food Reviews International, 39*(4), 2107–2121. https://doi.org/10.1080/87559129.2021.1942487

Silva, J., Leite, D., Fernandes, M., Mena, C., Gibbs, P. A., & Teixeira, P. (2011). Campylobacter spp. as a foodborne pathogen: A review. *Frontiers in Microbiology, 2*, 200. https://doi.org/10.3389/fmicb.2011.00200

Vibriosis

17

Ajay Pratap Singh and Ambika Arun

Abstract

Vibrio spp. represents a group of common bacteria characterized by their curved rod or comma like shape and Gram-negative nature. They are naturally found in various aquatic environments, including freshwater, estuaries, and the vast expanses of the sea. These bacteria play a pivotal role in a spectrum of infections collectively referred to as vibriosis. The severity of these infections can span a wide range, from mild ailments to life-threatening conditions. Marine environments serve as the favored habitat for *Vibrio* bacteria, and infections often arise from contact with contaminated water or the consumption of contaminated seafood. Although over a 100 distinct *Vibrio* species have been described, it is a selected group of approximately 12 that cause infections in humans. *Vibrio cholerae*, a notable member of this bacterial family, is responsible for the devastating disease cholera. Several non-cholera *Vibrio* spp. orchestrate onset of disease recognize as "Vibriosis". Some species, such *as V. cholerae* non-O1 and *V. mimicus*, have demonstrated the ability to adapt and survive in freshwater environments. In addition to their adaptability, there are halophilic vibrios, which are specifically adapted to saline environments. The sodium dependence is a key characteristic of halophilic vibrios, reflecting their specialization for salt-rich environments.

Keywords

Vibrio parahaemolyticus · Kanagawa phenomenon · Gastroenteritis

A. P. Singh (✉) · A. Arun
Department of Veterinary Microbiology, College of Veterinary Science & Animal Husbandry, U.P. Pandit Deen Dayal Upadhyaya Pashu Chikitsa Vigyan Vishwavidyalaya Evam Go-Anusandhan Sansthan, Mathura, Uttar Pradesh, India

R. Biswas et al. (eds.), *Concept of One Health and Zoonoses*,
https://doi.org/10.1007/978-981-95-6627-3_17

History

The journey of *Vibrio* bacteria, from their early taxonomical descriptions by O. F. Müller in the late eighteenth century to their recognition as significant pathogens underscores the critical evolution of our understanding of waterborne diseases and global epidemiology. Müller's work, spanning during 1773 to 1786, marked the description of eight species within the genus *Vibrio*, notably acknowledging the spirilli form nature of three of them. It was Pacini's pioneering work in 1854, isolating and identifying microorganisms, which he aptly named "vibrions," from cholera patients, that brought attention to this genus. However, an interesting transition occurred in the nomenclature of the *Vibrio* genus. Initially, *Vibrio* Müller, as designated in 1773, was associated with a zoological genus. In contrast, the bacterial genus was assigned the name *Vibrio* by Pacini in 1854. This taxonomical distinction marked a pivotal moment in the classification of these microorganisms. The term "vibrio" itself, derived from Latin, conveys the essence of quivering, reflecting the characteristic motility of these microorganisms. The cholera bacillus, *Vibrio cholerae*, was first isolated by Robert Koch in 1882 during an epidemic of cholera in India and Egypt (Colwell, 2006). The history of Vibriosis is rooted in the recognition of this bacterial disease in cattle, dating back to 1909 in England. This early understanding of Vibriosis laid the foundation for subsequent investigations into the causative organism. In 1918, Smith, in the United States, not only isolated the responsible pathogen but also bestowed upon it the name *Vibrio fetus*. For decades, this pathogen was primarily associated with natural infections in domestic animals. Until the year 1950, no other pathogenic species of *Vibrio* had been officially recorded. It was in that year when Fujino made a significant discovery by isolating *Vibrio parahaemolyticus* from a severe outbreak of food poisoning in Osaka, Japan. This discovery marked the emergence of *V. parahaemolyticus* as a recognized human pathogen, and it has since become a well-documented cause of gastrointestinal infections associated with the consumption of contaminated seafood. In 1976, Hollis in the United States reported a new species called *V. vulnificus*. Unlike other *Vibrio* species, *V. vulnificus* stands out because it can cause infections outside the intestines, and its biochemistry is different from the rest of the members under the genus. Notably, in 1979, the researchers observed that *V. vulnificus* infections could lead to primary sepsis (bloodstream infection) and wound infections, which earned the pathogen its name—"vulni" for wound and "ficus" for to make. Further, in 1995, a pathogenic strain of *V. parahaemolyticus* emerged and triggered the first pandemic in the history of this species. The impact of this strain was profound, as a decade later, it caused one of the world's most severe diarrhea outbreaks in Chile, affecting over 10,000 individuals.

Etiology

Vibrio belong to the Gram-negative bacterial family called *Vibrionaceae*. They are distinctive in their appearance, resembling curved rods or commas. *Vibrio* cells typically take on a curved rod shape, measuring about 0.5 μm in width and 1.5–3.0 μm in length. They can exist as single cells or in a diploid form, generating distinctive S-shaped or spiral structures. Vibrios are highly motile organisms with one to three polar flagella at one end often enclosed in a protective sheath. *Vibrio* are facultative anaerobes with the capability of both fermentative and respiratory metabolism, allowing them to adapt to a variety of environmental conditions with or without the presence of oxygen. *Vibrio* has two chromosomes. This is an unusual feature in the bacterial world. Each of these chromosomes has its own distinct and independent origin of replication. Remarkably, these two chromosomes have been conserved together over time within the *Vibrio* genus, contributing to the genetic diversity and adaptability of these bacteria. Within the genus *Vibrio*, there are several species associated with human disease. Among the 100 currently recognized species within the genus *Vibrio*, 12 have been linked to various human illnesses. Those of prime medical concern are *V. cholerae*, *V. parahaemolyticus,* and *V. vulnificus*. Other species implicated as opportunistic pathogens are *V. fluvialis*, *V. furnissii*, *V. hollisae*, and *V. mimicus*. *V. alginolyticus*, and *V. cincinnatiensis, V. metschnikovii*, and *V. navarrensis*.

Vibrio cholerae: Over 70 different serotypes identified for *V. cholerae*, the serotype O1 stands out as the etiologic agent of cholera. The two biovars, "**classical**" and "**El Tor**", within serotype O1 exhibit differences in sensitivity to antibiotics, susceptibility to specific phages, and their ability to agglutinate chicken erythrocytes. The global prevalence of the El Tor biovar underscores its significance in cholera epidemiology.

Vibrio vulnificus: Three distinct biotypes of *V. vulnificus* have so far been described. Biotype 1 was originally described as "lactose-positive" *Vibrio*. In recent studies, approximately 85% of clinical specimens associated with human infections have been reported to be lactose positives (Phillips & Satchell, 2017). Strains belonging to biotype 2 are a major source of infections in eels. Biotype 3was associated with seafood-mediated bacteremia and wound infections in a total of 62 patients, which had been in contact with pond-cultivated fish in Israel.

Epidemiology

The pandemic strain of *V. cholerae* (serogroup O1, cholera-toxin+) primarily colonizes the human intestinal tract, whereas other serotypes of *V. cholerae* (*V. cholerae* non-O1) can be found in various locations. *V. parahaemolyticus* is mainly associated with gastroenteritis, while *V. vulnificus* is linked to wound infections and septicemia. *Vibrio* species are routinely isolated from various sources, including blood samples, wounds, eyes, ears, and gallbladders. Infections with *Vibrio* spp. often result from the ingestion of untreated water or contaminated fish or shellfish. Raw or undercooked seafood, particularly fish and shellfish, is a common source of food-related gastroenteritis caused by *Vibrio* spp. Consumers of raw or poorly cooked shellfish, such as oysters, mussels, clams, shrimp, and crabs, are at risk of *Vibrio* infections. Contact with contaminated water poses a risk for wound infections, potentially through direct immersion, splashing, or entry through open wounds.

The salinity and temperature of aquatic environments significantly influence the abundance of *Vibrio* species. Salinity between 5% and 25% and water temperatures above 15 °C are associated with variations in *Vibrio* abundance. Environmental temperature and salinity changes, whether short term or long term, impact the concentration of *Vibrio* in water and seafood. The environmental sensitivity can influence infection rates. *Vibrio* infections have been on the rise worldwide, and this trend is attributed to the warming of ocean waters. Ongoing climate change is expected to exacerbate this issue, making it crucial to monitor and adapt to the changing environmental conditions to mitigate the spread of *Vibrio* infections.

While there is no clear evidence that animals in cholera-endemic countries act as significant reservoirs for *V. cholerae* O1, non-O1 serotypes of *V. cholerae* have been isolated from domestic animals, waterfowl, and various wildlife in non-endemic cholera regions. Aquatic birds are known carriers of *V. cholerae*, facilitating its dissemination in areas not endemic for cholera.

Pathogenesis

Key factors contributing to pathogenesis of *V.cholerae*, *V. vulnificus*, and *V. parahaemolyticus* include the *Vibrio* pathogenicity island (VPI), the role of specific toxins, adherence mechanisms, and virulence factors. *V. cholerae* can only cause cholera when it carries the *Vibrio* pathogenicity island (VPI). The island contains the toxin co-regulated pilus (TCP) gene cluster and the CTX bacteriophage, which supplies the cholera toxin gene. Cholera toxin exhibits an AB5 structure, comprising a single catalytic A subunit (CTA) and five B subunits (CTB). The B subunits facilitate binding to host cell receptors, while the A subunit possesses the catalytic activity. CTB subunits bind specifically to GM1 ganglioside receptors on the host cell surface, enabling toxin internalization. After binding to GM1 ganglioside receptors, CT is endocytosed into the host cell. Within the host cell, CTA undergoes proteolytic activation, separating it from CTB. The activated CTA enters the host cell cytoplasm. Activated CTA acts as an adenylate cyclase, catalyzing the conversion of ATP to cyclic AMP (cAMP). The elevation of cAMP levels causes activation of protein kinase A (PKA), which disrupts cystic fibrosis transmembrane conductance regulator (CFTR) chloride channel. This disruption results in the efflux of chloride ions and excessive secretion of sodium and water into the intestinal lumen. The disruption of ion transport and the subsequent massive fluid secretion into the intestine lead to the hallmark symptom of cholera, i.e., profuse watery diarrhea (Pal, 2014). Apart from enterotoxins, *Vibrio* species possess an array of virulence factors that enhance their ability to cause infection including cytolysins (causing cell lysis), collagenase (contributing to the destruction of tissue), siderophores (involved in scavenging iron), hemolysins (responsible for degrading erythrocyte and cellular membranes), capsular polysaccharide (antiphagocytic factor), pili and surface proteins (facilitating adherence and attachment), and flagella-mediated movement (associated with adherence process). Some *Vibrio* species maintain type 3 secretion system (T3SS), enabling the direct translocation of toxic effector proteins from the bacterium into the host cell. Adherence is a prerequisite for *Vibrio* colonization and infection of surfaces, including the gastrointestinal tract and wounds. Factors contributing to adherence include the presence of flagella and hemagglutinins.

Vibrio parahaemolyticus produces several important virulence factors among which thermostable direct hemolysin (TDH) and TDH-related hemolysin (TRH) are important ones (Letchumanan et al., 2014). The presence of TDH and TRH is strongly correlated with the pathogenicity of *Vibrio* parahemolyticus. Strains containing these hemolysins are more likely to cause illness in humans. TDH is known to cause beta hemolysis of human erythrocytes when cultured on an agar medium. This specific reaction is referred to as the **Kanagawa phenomenon**. Apart from these, phospholipase A and lysophospholipase are involved in the breakdown of phospholipids, which are essential components of cell membranes.

Clinical Symptoms

In Humans

Acute Gastroenteritis Among the *Vibrio* species, five have been associated with causing diarrhea in humans. One of the most well-known is *V. cholerae*, recognized as the primary agent responsible for cholera. *V. parahaemolyticus* is another *Vibrio* species with a well-documented role in causing acute gastroenteritis. More recently, *V. fluvialis*, *V. hollisae*, and *V. mimicus* have also emerged as contributors to diarrheal diseases. *V. furnissii* has been isolated from a few individuals with diarrhea, although there is currently insufficient evidence to establish it as a direct cause of diarrhea.

The primary symptoms of *Vibrio parahaemolyticus* gastroenteritis are watery diarrhea and abdominal cramps along with nausea, vomiting and fever. Headache and chills may also be present. The incubation period is between 4 and 96 h (mean 15 h) and the median duration of the illness is 2.5 days but may last for as long as 7 days.

Septicemia

Primary septicemia is the major form of infection with *V. vulnificus* involves a rapidly progressing septicemia with few gastrointestinal signs. Such infections have a reported incubation period from 7 h to several days. The most frequent symptoms are fever, chills, nausea and cardiovascular hypotension. This disease has a high fatality rate of up to 60% in reported cases, and is in most cases affecting predisposed persons.

Cutaneous Infection

These are not foodborne in the strict sense as they usually occur in connection with puncture wounds after handling of raw seafood or trauma and exposure to saline environments. Wound infections range from mild self-limiting lesions to rapidly progressing erythema, cellulitis, and necrosis, occasionally developing to secondary septicemia. Wound infections have a reported mean incubation period of 12 h. Severe wound infections with this bacterium often require extensive surgery of affected tissue or even amputations, and have reported fatality rates ranging from 7% to 25%.

Extraintestinal Infections

Vibrio species have been frequently isolated from extraintestinal infections, including wounds, eyes, ears, and gallbladders. Notably, they are less commonly reported in cases of meningitis, pneumonia, or infections of the reproductive organs or urinary tract.

Organism	Disease associations
V. cholerae O1, O139	Cholera>gastroenteritis
V. cholerae non-O1, O139	Gastroenteritis>wound infections>sepsis
V. parahaemolyticus	Gastroenteritis> > wound infections
V. fluvialis	Gastroenteritis> > otitis media>sepsis, peritonitis, biliary disease
V. furnissii	Sepsis (limited data)
V. hollisae	Gastroenteritis>sepsis
V.. mimicus	Gastroenteritis> > sepsis
V. alginolyticus	Otitis media>wounds infections> > sepsis
V. cincinnatiensis	Meningitis (single case)
V. vulnificus	Wound infections>gastroenteritis>sepsis
V. metschnikovii	Wound infection>pneumonia, sepsis (limited data)
V. harveyi (V. carchariae)	Wound infection>sepsis (limited data)
V. navarrensis	Sepsis>wound infections>gastroenteritis

Disease Spectrum of *Vibrio* and *Vibrio*-Like Organisms Causing Human Infections

In Animals

While most animals do not develop illness from cholera, dogs and cats can become infected if exposed to a high infectious dose of the bacteria. Animals can be exposed to *V. cholerae* through ingestion of contaminated food or water. It may occur from exposure to feces from infected animals or people or ingestion of contaminated water, food, or raw shellfish. Animals infected with *V. cholerae* may develop one or more of the following clinical signs. Large amounts of watery diarrhea, vomiting, dehydration, lethargy, anorexia, and weight loss. *Vibrio* infections are also common in open wounds in animals. These infections can be caused by a variety of species, including *V. vulnificus* and *V. parahaemolyticus*. *Vibrio* infections can be induced experimentally in laboratory animals producing diarrhea. Intestinal disease was also induced in monkeys and kittens fed cell-free filtrates from broth cultures.

Diagnosis

Collection of Samples

When dealing with extraintestinal specimens such as blood, wound, tissue, and the like, the standard procedures for collecting and processing these specimens should be adhered to. No specialized or unique procedures are typically required for the isolation of *Vibrio* from such specimens. For stool specimens, it is recommended to collect them promptly, ideally within the initial 24 h of the onset of illness, and prior to the administration of any antimicrobial agents to the patient. This approach helps to ensure the effective isolation and analysis of *Vibrio* bacteria from stool samples.

Isolation

Marine agar is a valuable culture medium used for the isolation of *Vibrio*. It is particularly well-suited for *Vibrio* species that have a specific requirement for higher levels of sodium ions (Na+) than what is typically found in regular bacteriological media. TCBS (Thiosulfate-Citrate-Bile Salts-Sucrose) Agar serves as another medium, particularly for the isolation and identification of *V. cholerae* and *V. parahaemolyticus* from diarrheal stool specimens. This agar is commonly used as a primary medium for isolating vibrios from clinical samples and environmental sources. It is worth emphasizing that TCBS agar may not support the growth of all *Vibrio* species. Some *Vibrio* species may exhibit poor growth or fail to grow entirely on this specific medium. When isolating *Vibrio* from marine and environmental samples, enrichment cultures are commonly employed. These cultures are typically incubated at room temperature, which ranges from 18 °C to 22 °C. This temperature range is suitable for the growth and enrichment of *Vibrio* species from these samples. In the process of isolating *Vibrio* from seawater samples, a common procedure involves filtering samples of seawater, typically ranging from 5 to 300 mL in volume, through nitrocellulose filters with pore sizes of either 0.22 or 0.45 μm. These filters are subsequently placed onto selective agar media for further analysis and isolation of *Vibrio*. This method allows for the concentration and growth of *Vibrio* bacteria from the seawater samples.

Direct Testing of Stool Sample Detecting specific stool antigens allows for the direct testing of stool samples for *V. cholerae* O1. A commercial latex agglutination test containing *V. cholerae* O1 sero factor "a" monoclonal antibody has been employed to detect the presence of O antigen in the liquid stool.

Kanagawa Test It is also known as the **Kanagawa phenomenon**. It is a diagnostic method used to identify pathogenic strains of *V. parahaemolyticus*. The test detects the presence of a specific exotoxin called Thermostable Direct Hemolysin (TDH). The bacterium is cultured on a special high-salt medium called Wagatsuma agar, which is supplemented with human or rabbit red blood cells. A clear zone of beta-hemolysis forms around the bacterial colonies indicates the strain produces TDH and is considered pathogenic while non-pathogenic environmental isolates does not produces haemolysis.

Molecular Detection

Different methods for detecting *Vibrio* species have proven effective across diverse sources, each carrying its own set of advantages and limitations. These approaches encompass nucleic acid-based techniques like PCR, loop-mediated isothermal amplification, microarrays, biosensor-based methods, as well as immunological methods such as enzyme-linked immunosorbent assay (ELISA) and lateral flow immunoassays (Loo et al., 2022). In addition to these commonly employed techniques, several modern detection methods are currently in development to enhance the efficiency of *Vibrio* detection. Spot dye-based sensor is a simple and efficient kit based on color change for a positive result. It uses selective media to promote *Vibrio* growth, followed by the addition of a dye-based sensor to detect active *Vibrio*. Additionally, a Cas12a-assisted rapid isothermal detection (CARID) system was developed for detecting toxigenic *V. cholerae* serogroups O1 and O139. CRISPR-Cas12a assisted PCR have been developed for the detection of *V. parahaemolyticus* using a specifically designed crRNA guiding the Cas12a enzyme to cleave target DNA, producing fluorescent signals for easy visualization. Another approach involved the use of a rapid nucleic acid lateral flow biosensor (NALFB) to detect *V. cholerae* by hybridizing with fluorescein-labeled single-stranded amplicons, producing visible results in minutes with high specificity and sensitivity. A lateral flow dipstick combined with loop-mediated isothermal amplification (LAMP) have also been developed targeting the rpoX gene.

Treatment

Vibriosis encompasses a range of infectious diseases caused by *Vibrio* species, with *V. cholerae* being a notable pathogen. Effective treatment strategies are essential in managing these infections. Standard treatment for toxigenic *V. cholera* includes the following:

1. *Oral Rehydration Therapy (ORT)*: The cornerstone of cholera treatment is ORT. It involves the administration of a solution containing sodium, chloride, potassium, and glucose in water. ORT effectively counters the massive fluid loss associated with cholera, preventing life-threatening dehydration.
2. *Oral Antibiotics*: In addition to ORT, appropriate oral antibiotics are administered to treat toxigenic *V. cholerae* infections. Commonly used antibiotics include doxycycline, azithromycin, and tetracycline. These antibiotics target the pathogen and help reduce the duration and severity of the disease.

Emerging Treatment Strategies

1. Lytic bacteriophages: The emergence of antimicrobial resistance among *Vibrio* species has led to a renewed interest in lytic bacteriophages as potential therapeutic agents. Bacteriophages are viruses that specifically infect and destroy bacteria. Applying specific lytic bacteriophages may offer an alternative treatment approach to combat *Vibrio* infections. Their precision in targeting pathogenic *Vibrio* strains is a promising avenue for research.
2. Antimicrobial Peptides: Antimicrobial peptides are naturally occurring molecules with the ability to inhibit the growth of bacteria, including *Vibrio* species. These peptides are being explored as potential treatment options due to their broad-spectrum antimicrobial activity. They may prove useful in cases of *Vibrio* infections where traditional antibiotics are ineffective due to resistance.

Prevention and Control

Management of Post-harvesting Sea-Food Items Harvesting areas identified as at risk could be subject to closure during periods when the seawater temperature reaches a level that correlates with high numbers of *Vibrio*.

Transportation of Sea Food Items Additional temperature controls are imposed on the transport of shellfish harvested from state waters that have been confirmed as an original source of product associated with two or more cases of illness due to *Vibrio*. The length of time allowed from harvest to refrigeration varies with the temperature of the water in the harvesting area; e.g., when this exceeds 28.9 °C, the time limit is 10 h. For areas that have not had such an association with *Vibrio* infection, standard controls apply, with time to refrigeration varying between 20 and 48 h, depending on the season or air temperature.

Depuration Technologies Depuration processes are crucial for ensuring the safety of shellfish for human consumption. Shellfish, like clams, mussels, and oysters, can filter and accumulate bacteria and other microorganisms from their surrounding water. Depuration methods help in reducing the microbial load in these shellfish.

Natural depuration, also known as relay, involves transferring shellfish to cleaner waters where they naturally purge themselves of contaminants. On the other hand, land-based depuration facilities use controlled environments like depuration ponds or tanks. In these settings, shellfish are kept in clean water, and the conditions are managed to optimize the depuration process.

Heat Treatment *Vibrio* sp. is relatively sensitive to heat treatment and temperatures above 45 °C have been shown to kill the microorganism. The average reported decimal reduction time for 52 strains kept at 47 °C was 78 s, and heating to 50 °C for 10 min was reported to be sufficient to reduce the number to a non-detectable level.

Irradiation Ionizing irradiation is used as a food preservation method to enhance the safety and extend the shelf life of certain foods. Irradiation is a food safety technology that involves exposing food to ionizing radiation to control or eliminate bacteria, parasites, viruses, and other pathogens. A dose of 1 kilogray (kGy) has been shown to be lethal for *Vibrio* bacteria.

Vaccination There are currently three oral cholera vaccines that are licensed by the World Health Organization and available for global use: Dukoral, Shanchol, and Euvichol. Shanchol is the first low-cost oral vaccine prequalified by the World Health Organization for international use. There is currently no approved vaccine available against *V. vulnificus* and *V. parahaemolyticus*.

References

Colwell, R. R. (2006). A global and historical perspective of the genus vibrio. In *The biology of vibrios* (pp. 1–11).

Letchumanan, V., Chan, K. G., & Lee, L. H. (2014). Vibrio parahaemolyticus: A review on the pathogenesis, prevalence, and advance molecular identification techniques. *Frontiers in Microbiology, 5*, 705.

Loo, K. Y., Law, J. W. F., Tan, L. T. H., Pusparajah, P., Letchumanan, V., & Lee, L. H. (2022). Diagnostic techniques for rapid detection of vibrio species. *Aquaculture, 561*, 738628.

Pal, P. (2014). Role of cholera toxin in vibrio cholerae infection in humans-a review. *International Letters of Natural Sciences, 22*.

Phillips, K. E., & Satchell, K. J. (2017). Vibrio vulnificus: From oyster colonist to human pathogen. *PLoS Pathogens, 13*(1), e1006053.

Further Reading

Faruque, S. M., & Nair, G. B. (Eds.). (2008). *Vibrio cholerae: Genomics and molecular biology*. Caister Academic.

Pukatzki, S., & Almagro-Moreno, S. (2023). *Vibrio spp. infections*. Springer.

Selection and Application of Methods for the Detection and Enumeration of Human-Pathogenic Halophilic Vibrio Spp. in Seafood Food and Agriculture Organization of the United Nations 2017

Escherichia coli Infections

18

Ajay Pratap Singh and Prachi Singh

Abstract

Escherichia coli, is a ubiquitous bacterium with a significant role in the world of microbiology. It has garnered attention as a model organism, a vital component of the human microbiome, and a subject of concern due to its pathogenic strains. *E. coli* is the most prevalent commensal-type species of the family *Enterobacteriaceae* (Genus *Escherichia*), inhabiting the gastrointestinal tracts of warm-blooded animals and humans. The classification of *E. coli* based on pathogenicity is a critical distinction in understanding its impact on human and animal health. *E. coli* can be broadly categorized into two main groups: non-pathogenic and pathogenic strains, each with distinct characteristics and implications for host organisms.

Non-pathogenic *E. coli* primarily reside in the gastrointestinal tract of humans and animals and are considered the major commensal organisms. The commensal has co-evolved with their hosts and play an essential role in maintaining the gut microbiome. They help digest complex carbohydrates and maintain a balanced microbial community ('eubiosis'), which is crucial for digestion and overall physical and mental health. The non-pathogenic strains compete with potential pathogens for nutritional resources and host cellular receptors, helping to prevent the colonization and overgrowth of gut pathogens. Non-pathogenic *E. coli* strains are typically harmless and do not cause disease in the host. They remain confined to the lumen of the intestine, causing no significant harm to the host organism. Pathogenic *E. coli* strains, on the other hand, are associated with a range of diseases in both humans and animals. The strains have acquired various virulence factors, which enable them to colonize host tissues and cause infections. Some

A. P. Singh (✉) · P. Singh
Department of Veterinary Microbiology, College of Veterinary Science & Animal Husbandry, U.P. Pandit Deen Dayal Upadhyaya Pashu Chikitsa Vigyan Vishwavidyalaya Evam Go-Anusandhan Sansthan, Mathura, Uttar Pradesh, India

R. Biswas et al. (eds.), *Concept of One Health and Zoonoses*,
https://doi.org/10.1007/978-981-95-6627-3_18

notable diseases caused by pathogenic *E. coli include urinary* tract infections (UTI) caused by uropathogenic *E. coli* (UPEC), gastroenteritis caused by enteropathogenic *E. coli (EPEC),* enterotoxigenic *E. coli (ETEC)* and haemolytic uremic syndrome (HUS). In animals, pathogenic *E. coli* strains are linked to various diseases, including diarrhoea in farm animals, oedema disease in pigs, dysentery in calves and lambs, and urinary tract infections in dogs and cats. Additionally, they can cause mastitis in bovines.

Keywords

Enteric pathogenic *Escherichia coli* · *E. coli* pathotypes · *Colibacillosis* · *White Scours*

History

In 1885, German paediatrician Theodor Escherich made a groundbreaking discovery when he first isolated *E. coli* from the intestinal tract of infants. At that time, he named this bacterium 'Bacterium coli commune'. Escherich's work focused on studying the characteristics of this bacterium, particularly its pathogenic potential, including its ability to cause infections outside the intestine. In 1919, Castellani and Chalmers renamed the bacterium, defining it under the genus *Escherichia*, with *E. coli being* designated as the type species. This renaming marked an important taxonomical clarification, placing the bacterium under its now-familiar name. While earlier reports suggested that *E. coli* was non-pathogenic and primarily acted as commensals in the intestinal tract, Bray's work in 1945 challenged this notion. Bray identified an association between *E. coli* and severe outbreaks of infantile diarrhoea, suggesting that this bacterium could indeed be pathogenic in certain contexts. In 2004, Kaper and his colleagues expanded our understanding of *E. coli* by describing a wide range of clinical conditions associated with different pathovars of *E. coli.* This highlighted the diverse pathogenic capabilities of *E. coli* strains, leading to a more nuanced classification of pathogenic variants.

In 1973, Milkman initiated a quantitative study of population genetics involving *E. coli.* This research focused on variations in the electrophoretic mobility of enzymes isolated from various *E. coli* strains, shedding light on the genetic diversity within this bacterium. In 1991, Albert and his co-workers made an interesting discovery. They recovered five diarrhoeagenic *E. coli* isolates from children in Bangladesh, initially naming them 'Hafnia alvei'. This discovery highlighted the diversity of *E. coli* and its potential to cause disease. In 2003, Huys made an important contribution to our understanding of *E. coli* by discovering that this bacterium shares common genotypic and phenotypic characteristics with enteropathogenic *E. coli (EPEC).* This led to the renaming of *E. coli* strains with these characteristics as '*Escherichia albertii*'.

Classification

The genus *Escherichia* comprises a group of Gram-negative, rod-shaped bacteria belonging to the family *Enterobacteriaceae*. The most well-known and extensively studied species within this genus is *Escherichia coli*. They are classified under Order: Enterobacterales Family: Enterobacteriaceae**.** *E. coli* strains can be categorized into various groups based on their genetic, functional, and pathogenic characteristics. These strains exhibit a wide range of behaviours, from commensal organisms in the human gut to highly pathogenic variants that can cause severe diseases.

1. **Commensal *E. coli* (Non-Pathogenic):** These strains are a part of the normal gut microbiota in humans and other animals. They generally do not cause disease and have adapted to life in the intestinal tract. They play a role in digestion and nutrient absorption.
2. **Pathogenic *E. coli Strains*:** Pathogenic *E. coli strains* are those that have the ability to cause a variety of diseases in humans and animals. These pathogenic strains often possess specific virulence factors that enable them to colonize host tissues, evade the host immune response, and cause disease.
3. **Environmental *E. coli:*** Some strains of *E. coli are* found in environmental samples, such as water and soil. They are not typically associated with human or animal infections, but can serve as indicators of faecal contamination.
4. **Avian Pathogenic *E. coli (*APEC):** APEC strains cause infections in poultry, leading to conditions like colibacillosis. These strains are of concern in the poultry industry.
5. **Porcine Pathogenic *E. coli (*PPEC):** PPEC strains are associated with swine diseases, including post-weaning diarrhoea and oedema disease in pigs.
6. **Canine and Feline Pathogenic *E. coli:*** These strains can cause infections in dogs and cats, including urinary tract infections and genital tract infections.
7. **Bovine Mastitis *E. coli:*** Some *E. coli* strains are responsible for mastitis in cows, a condition that affects milk production and quality.
8. **Environmental and Laboratory *E. coli*:** Various *E. coli* strains have been adapted for use in biotechnology, genetic engineering, and laboratory research. These strains are often chosen for their ease of cultivation and genetic tractability.
9. **Probiotic *E. coli:*** Researchers have explored the use of certain non-pathogenic *E. coli* strain*s* as probiotics to promote gut health.

Pathotyping This is a method of classification based on the pathogenicity and virulence factors associated with different strains. Pathotyping helps to identify and categorize *E. coli* strains according to their ability to cause specific types of infections and the virulence factors they possess. Here are some common pathotypes of *E. coli:*

- **Enteropathogenic *E. coli* (EPEC):** These strains can cause gastroenteritis in humans, particularly in infants and young children. They are known for their ability to adhere to and damage the intestinal epithelium.
- **Enterotoxigenic *E. coli* (ETEC):** ETEC strains produce toxins that lead to traveller's diarrhoea and cause fluid and electrolyte imbalances in the intestines.
- **Enterohaemorrhagic *E. coli* (EHEC):** EHEC strains produce Intimin and Shiga toxins and are associated with severe foodborne illnesses, including outbreaks of haemolytic uremic syndrome (HUS).
- **Uropathogenic *E. coli* (UPEC):** These strains are responsible for the majority of urinary tract infections (UTIs) in humans. They possess virulence factors that allow them to adhere to and infect the urinary tract.
- **Enteroinvasive *E. coli* (EIEC):** EIEC strains are capable of invading the intestinal epithelium and causing inflammatory diarrhoea, resembling Shigella.
- **Enteroaggregative *E. coli* (EAEC):** EAEC strains are associated with persistent diarrhoea, particularly in children and immunocompromised individuals.

Phylogenetic Typing One common way to categorize *E. coli* strains is by phylogenetic groups. These groups are based on genetic and evolutionary relationships and can help researchers understand the diversity and characteristics of different *E. coli* strains.

The six main phylogenetic groups of *E. coli* are:

1. **Phylogenetic Group A:** This group includes most commensal strains and some pathogenic strains. It is often associated with strains that colonize the human intestine without causing disease.
2. **Phylogenetic Group B1:** Strains in this group are also primarily commensal and are often associated with the human gut. They typically do not possess many virulence factors and are less likely to cause disease.
3. **Phylogenetic Group B2:** This group is known for including many extraintestinal pathogenic *E. coli* (ExPEC) strains. ExPEC strains can cause infections outside of the gastrointestinal tract, such as urinary tract infections and sepsis. They often possess a wide range of virulence factors.
4. **Phylogenetic Group D:** This group is less commonly encountered, but includes some pathogenic strains associated with diarrhoea and intestinal infections.
5. **Phylogenetic Group E:** Group E is often associated with atypical *E. coli* strains that do not fit neatly into the other categories. Some atypical strains that cause disease can be found in this group.

Serotyping This is one of the most widely accepted methods for characterizing *E. coli*. It is based on the identification of specific antigens, including the somatic polysaccharide (O antigen), capsular polysaccharide (K antigen), and flagellar protein (H antigen). This method helps classify *E. coli into* different serotypes, with complete serotyping typically performed in specialized referral laboratories. Based on surface antigens i.e., somatic(O), flagellar(H), and capsular(K) antigens, *E. coli* is differentiated into 190 serogroups, out of which after combination of O and H antigen they are basically assigned into relevant categories e.g. O157:H7.

Biotyping This method relies on selective biochemical tests to identify and characterize *E. coli* based on their metabolic properties. It helps to differentiate between different biotypes of *E. coli.*

R Typing R typing is used to differentiate *E. coli* based on their antibiotic resistance profiles, particularly those mediated by plasmids. However, this method has limitations because plasmids can be easily lost or transferred.

Phage Typing Phage typing involves the use of bacteriophages to subdivide *E. coli into* different groups. This method has been used since 1945, but is less common today due to the limited availability of specific bacteriophages and specialized laboratory requirements.

Colicin Typing Colicin typing assesses the ability of *E. coli* to produce colicins, which are antimicrobial proteins that can inhibit the growth of other bacteria. Different colicin types are used to differentiate strains.

Epidemiology

Enterotoxigenic *Escherichia coli* (ETEC)

Enterotoxigenic *Escherichia coli* (ETEC) is a type of bacteria that causes gastrointestinal infections in humans. These infections are commonly associated with symptoms such as diarrhoea and are often a result of consuming contaminated food or water. ETEC is a significant cause of traveller's diarrhoea and is a public health concern. ETEC is one of the most common causes of bacterial diarrhoea in the world, responsible for millions of cases each year (Al-Abri et al., 2005). It is particularly prevalent in developing countries and in areas with inadequate sanitation and hygiene practices. It affects a significant number of tourists and travellers visiting regions with poor sanitation and exposed to contaminated food and water. ETEC infections can affect people of all age groups, but they are particularly dangerous for infants and young children. Children in developing countries are more susceptible to severe ETEC infections.

Enteropathogenic *Escherichia coli* (EPEC)

Enteropathogenic Escherichia coli (EPEC) are associated with gastrointestinal infections. The transmission of EPEC occurs through a faecal–oral route involving contaminated surfaces, weaning fluids, and human carriers. While outbreaks among adults related to the ingestion of contaminated food and water are infrequent, no environmental source for these outbreaks has been pinpointed. The recognized O serogroups in classical EPEC encompass O55, O86, O111, O114, O119, O127, and O142, while the prevailing H antigens linked to EPEC are H6 and H2. EPEC outbreaks are known to occur, especially in developing countries, and often target young children, particularly those in the early period of weaning. EPEC outbreaks can happen sporadically, and they are more common in regions with poor sanitation and limited access to clean water. These conditions can facilitate the spread of the bacteria. EPEC is commonly associated with paediatric diarrhoea, particularly in children under the age of two. Typical Enteropathogenic *E. coli* (tEPEC) are rarely detected in animals. However, atypical Enteropathogenic *E. coli* (aEPEC) strains have been identified in both diarrhoeic and healthy animals. This suggests that EPEC strains are more adaptable and can colonize the intestines of animals without necessarily causing disease. Their presence in healthy animals raises questions about their potential role as asymptomatic carriers and the dynamics of atypical EPEC infections in animal populations.

Enterohaemorrhagic/Shigatoxigenic/Verotoxigenic *Escherichia coli*

The presence of EHEC/STEC, bacteria responsible for causing foodborne illnesses, has been extensively investigated in various animal species, revealing the zoonotic nature of these infections. Apart from cattle, which are the primary natural reservoir, these pathogens have been identified in the faeces of dairy buffaloes, sheep, pigs, birds, and even fish. This indicates that a wide range of animals can carry EHEC/STEC. *E. coli* O157*:H7* was first identified and reported by Konowalchuk in 1977. In 1982, it gained significant attention due to two outbreaks of foodborne illness. These outbreaks were characterized by cases of bloody diarrhoea and other severe symptoms, which prompted investigations into the source of contamination. *E. coli* O157:H7 is particularly concerning because it can cause severe complications. Haemolytic uraemic syndrome (HUS) and thrombotic thrombocytopenic purpura (TTP) are rare but life-threatening conditions that can result from *E. coli* O157:H7 infection, especially in children. In addition to these severe conditions, it can also cause non-bloody diarrhoea. The transmission of *E. coli* O157:H7 can occur through various routes, including consumption of contaminated food and water. Sources of infection have been associated with unpasteurized apple cider, undercooked ground beef, and other meat products especially those which are undercooked like hamburger, *kabab,* etc. This strain is known for its low infectious dose, meaning that even a small number of bacteria can cause illness. *E. coli* O157:H7 has the ability to withstand extreme environmental stress, which can make it challenging to eliminate. It can persist in various environments, contributing to the risk of contamination. *E. coli* O157:H7 is a concern in both developed and developing countries. Outbreaks and cases have been reported worldwide, and it has led to significant public health efforts to reduce its prevalence and improve food safety. *E. coli* O104:H4 is a rare and highly virulent shiga toxin-producing strain that caused a severe 2011 foodborne outbreak in Germany and France, which was traced to contaminated fenugreek sprouts (Bilinski et al., 2012).

Enteroaggregative *Escherichia coli* (EAEC)

EAEC strain is known for its association with gastrointestinal infections, particularly diarrhoea. EAEC is primarily associated with diarrhoeal illness in infants and young children, particularly in developing countries with limited access to clean water and sanitation facilities. The bacteria adhere to and colonize the intestinal mucosa, leading to prolonged episodes of diarrhoea. EAEC has also been isolated from patients suffering from traveller's diarrhoea. Traveller's diarrhoea is a common condition that affects people when they visit countries with different sanitation and food safety standards. Contaminated food and water sources can transmit EAEC, leading to gastrointestinal symptoms in travellers. In developed countries, EAEC is typically associated with sporadic cases of diarrhoeal disease. EAEC is known for its ability to form characteristic 'stacked-brick' biofilms on the surface of intestinal epithelial cells, allowing it to adhere and persist in the gut. The bacteria produce a range of virulence factors, including adhesins, toxins, and other proteins that contribute to its pathogenicity. No correlation has been established between EAEC strains isolated from humans and various animal species. This suggests that animals are unlikely to serve as a reservoir for the typical human pathogenic EAEC.

Enteroadherent Invasive *Escherichia coli* (EIEC)

EIEC, unlike some other forms of *E. coli,* has a relatively high infectious dose and causes gastrointestinal infections in humans. EIEC primarily spreads through direct person-to-person contact and the ingestion of contaminated food and water. It is often associated with poor sanitation and hygiene practices. EIEC is known for its ability to invade and disrupt the epithelial cells lining the human intestine. This invasion leads to symptoms such as diarrhoea, abdominal pain, and sometimes fever, which can be severe.

Diffusely Adherent *Escherichia coli* (DAEC)

DAEC is a category of *E. coli* that has been isolated from individuals with diarrhoea, but its role in causing diarrhoea remains controversial and not fully understood. While some studies have isolated DAEC strains from individuals suffering from diarrhoea, the direct link between DAEC and diarrhoea remains inconclusive. DAEC strains were initially isolated in children with diarrhoea in Mayanmar and Bangladesh in 1992. Some research indicates a significantly higher prevalence of DAEC in adults with diarrhea (around 18.8%) compared to asymptomatic adults (4.2%). Prevalence differs geographically, with studies in Asia showing high overall DAEC rates (22.8%).

Pathogenesis

Escherichia coli is a versatile bacterium that can have both commensal and pathogenic strains. The pathogenesis of *E. coli* involves various mechanisms that allow pathogenic strains to cause disease in humans and animals. Crucial mechanisms of pathogenesis used by different pathotypes of *E. coli* are described in the following table.

Pathogenesis of various *E. coli* pathotypes

E. coli pathotype	Pathogenesis
EPEC (Enteropathogenic *E. coli*)	Adheres to small bowel enterocytes with the help of 'intimin', encoded by locus for enterocyte effacement (pathogenicity island) (Clements et al., 2012) Destroys normal microvillar architecture, leading to the attaching and effacing lesion Induces cytoskeletal derangements, causing an inflammatory response and diarrhoea The process involves initial adhesion, protein translocation by type III secretion system (T3SS), and pedestal formation
EHEC (Enterohaemorrhagic *E. coli*)	Induces the attaching and effacing lesion in the colon after adhesion with the help of 'intimin', encoded by locus for enterocyte effacement (pathogenicity island) Multiplication of bacteria and secretion of Shiga toxins (Stx). Systemic absorption of Stx can lead to potentially life-threatening complications
ETEC (Enterotoxigenic *E. coli*)	Adheres to small bowel enterocytes through fimbrial adhesins (F4, F5, F6, and F18) Multiplication of the bacteria and secretion of heat-labile (LT) and/or heat-stable (ST) enterotoxins which induces watery diarrhoea
EAEC (Enteroaggregative *E. coli*)	Adheres to small and large bowel epithelia in a thick biofilm Elaborates secretory enterotoxins and cytotoxins
EIEC (Enteroinvasive *E. coli*)	Invades the colonic epithelial cell Lyses the phagosome Moves through the cell by nucleating actin microfilaments Bacteria can move laterally through the epithelium by direct cell-to-cell spread or exit and re-enter the basolateral plasma membrane
DAEC (Diffusely Adherent *E. coli*)	Elicits a characteristic signal transduction effect in small bowel enterocytes This effect results in the growth of long finger-like cellular projections that wrap around the bacteria

Clinical Symptoms

Human

Enterotoxigenic *E. coli* (ETEC)

The hallmark symptom of ETEC infection is profuse, self-limiting cholera like watery diarrhoea. The heat-labile (LT) and heat-stable (ST) enterotoxins toxins cause increased secretion of fluid and electrolytes into the intestinal lumen, leading to watery, non-bloody diarrhoea and dehydration with associated symptoms like increased thirst, dry mouth, decreased urine output, lethargy, and, in severe cases, sunken eyes and reduced skin turgor (Qadri et al., 2005). ETEC infection often leads to abdominal cramps or discomfort. These cramps may be caused by the irritation of the intestinal lining due to the enterotoxins and the increased fluid in the intestines. Some individuals with ETEC infection may experience nausea and vomiting in addition to diarrhoea. While vomiting is not as common as diarrhoea, it can contribute to further fluid and electrolyte loss. A mild fever may be present in some cases, but it is not a prominent feature of ETEC infection. The fever is usually low-grade and may not always be present. In addition to the gastrointestinal symptoms, ETEC infection can cause general feelings of malaise and weakness.

Enteropathogenic *E. coli* (EPEC)

EPEC infection is characterized by its ability to cause 'attaching and effacing' lesion in the microvilli on the intestinal epithelium. These lesions disrupt the normal structure of the intestinal lining, leading to gastrointestinal symptoms. EPEC is a common cause of watery diarrhoea in infants in developing countries with poor sanitation and limited access to clean water. Individuals with EPEC infection often experience weakness and fatigue due to fluid loss and the overall impact of the illness on the body. Dehydration is a major concern in EPEC infection. Malaise is a general feeling of discomfort or unease, which is common in EPEC-infected individuals. Nausea and vomiting can accompany EPEC infection, further exacerbating fluid and electrolyte loss. While not always present, some cases of EPEC infection may be associated with low-grade fever. EPEC infections typically last for about a week, with symptoms resolving within this timeframe in most cases. However, in some instances, particularly in young children or in cases of severe infection, the illness can persist and become chronic, leading to ongoing health issues.

Enteroaggregative *E. coli* (EAEC)

EAEC strains are known for its distinct adherence pattern to the epithelial cells in the human intestine, particularly the HEp-2 cells. This pathotype is associated with specific clinical manifestations, primarily causing persistent sporadic diarrhoea outbreaks, often in people in developing countries and individuals with compromised immune systems. EAEC bacteria adhere to the epithelial cells in the intestinal lining, forming a characteristic stacked brick-like appearance (Weintraub, 2007). This adherence is a critical feature of EAEC pathogenesis. EAEC is responsible for causing persistent and sporadic diarrhoea outbreaks, particularly in developing countries. It is a significant cause of chronic diarrhoea in these regions, and it can contribute to public health challenges. People with weakened immune systems, such as those with HIV/AIDS, are more susceptible to EAEC infections. The immunocompromised state makes them more prone to various infections, including EAEC. The hallmark symptom of EAEC infection is persistent diarrhoea, which can last for an extended period, often exceeding 2 weeks. The diarrhoea is typically watery and may be accompanied by mucus but is usually not bloody. EAEC infection can lead to various enteric dysfunctions, including. Nausea and vomiting may be present in some cases, contributing to fluid and electrolyte loss. Abdominal pain and discomfort are common symptoms. A feeling of nausea is often reported by individuals with EAEC infection. It is important to note that EAEC is known for its ability to form biofilms on the intestinal mucosa, which can further complicate the infection and contribute to its chronic nature.

Enterohaemorrhagic *E. coli* (EHEC)

EHEC strains are known for its ability to cause serious illness, including outbreaks of haemolytic uraemic syndrome (HUS). EHEC gained significant attention in 2011 when a large outbreak occurred in Germany linked to the consumption of contaminated sprouts. This outbreak was associated with a high incidence of haemolytic uraemic syndrome (HUS) and had a considerable death rate. EHEC infection can lead to a range of clinical symptoms. EHEC infection can cause the destruction of red blood cells, leading to haemolytic anaemia. This can result in symptoms such as fatigue, pallor, and jaundice. EHEC infection is known for its association with acute renal failure. This can be a severe and life-threatening complication of the infection. Thrombocytopenia is a condition characterized by a reduced number of platelets in the blood. It can lead to bleeding tendencies and bruising. While not always bloody, EHEC infection can cause diarrhoeal symptoms.

Enteroinvasive *E. coli* (EIEC)

This is phenotypically associated with Shigella infection causing inflammatory colitis preceding with diarrhoea with tinge of mucus and blood in some cases.

Diffusely Adherent *E. coli* (DAEC)

It is mainly associated with watery diarrhoea that elevates with ageing in children. It can be differentiated from other pathotypes by characteristic adhesive phenotypic property on HEp-2 cells (Servin, 2005).

Animals

Escherichia coli (*E. coli*) infections can affect a wide range of animals, including livestock, pets, and wildlife. These infections can lead to various diseases and health issues in animals.

1. **Livestock:**
   - Cattle: *E. coli* infections can cause diseases such as bovine mastitis, metritis, and respiratory infections. Mastitis, in particular, can lead to significant economic losses in dairy farming.
   - Pigs: *E. coli* is associated with diseases like oedema disease, which primarily affects piglets, and post-weaning diarrhoea, a common problem in swine production. Neonatal and postweaning diarrhoea due to EPEC strain. Oedema disease caused by STEC strain. Neonatal coli septicaemic disease due to ExPEC.
   - Calves and Lambs: Calves and lambs can suffer from neonatal diarrhoea caused by *E. coli*, leading to dehydration and high mortality rates in young animals. Neonatal diarrhoea is associated with ETEC pathotype in less than 1 week of animal's age. Dysentery is due to STEC in less than 3 weeks of age. Septicaemia is caused by ExPEC after invasion of intestinal epithelial barrier in the first few weeks of life in calves.
   - Poultry: While *E. coli* is a common part of the intestinal flora in poultry, certain strains can cause colibacillosis, leading to a range of problems such as swollen heads, cellulitis, and respiratory infections.
2. **Pets:**
   - Dogs and Cats: *E. coli* infections can cause urinary tract infections (UTIs), particularly in female dogs, as well as gastrointestinal issues. UTIs in dogs and cats can result in frequent urination, discomfort, and sometimes blood in the urine.
3. **Wildlife:**
   - *E. coli* can infect various wildlife species, and the effects can vary. For example, in marine mammals, *E. coli* infections can be linked to diseases such as sea otter stranding syndrome, while in terrestrial wildlife, it may contribute to enteric infections and environmental contamination.

Disease associated with different *E. coli pathotypes* with virulence factors associated for pathogenesis

E. coli pathotype	Virulence factors	Diseases in humans	Diseases in animals
Intestinal *E. coli*			
ETEC	LT, ST, CFAs	Traveller's diarrhoea	Neonatal diarrhoea small ruminants, post weaning diarrhoea in pigs
EPEC	Intimin, Bfp	Diarrhoea in children (Kaper et al., 2004)	Diarrhoea in pigs, dogs, and rabbits
EHEC	Intimin, Bfp, Shiga toxins	Haemolytic uraemic syndrome	Dysentery in calves and lambs, oedema disease in pigs
EAEC	Cytotoxins, AAFs	Diarrhoea in children (Nataro et al., 1998)	–

E. coli pathotype	Virulence factors	Diseases in humans	Diseases in animals
EIEC	Invasion, Ipa, haemolysin, Shiga toxin	Like shigellosis (Negroni et al., 2012)	–
DAEC	AIDA, Daa	Acute diarrhoea condition in children	–
AIEC	Cellular invasion, type 1 fimbriae	Crohn's disease (Darfeuille-Michaud, 2002)	–
Extraintestinal *E. coli* (ExPEC)			
UPEC	P & type 1 fimbriae, haemolysin, AAFs	Systemic infections and UTI (Johnson & Stell, 2000)	UTI in dogs and cats
NMEC	K1 capsule, S fimbriae	Neonatal meningitis (Gaschignard et al., 2011)	–
APEC	K1 capsule, T1 and P fimbriae	Food borne disease (Johnson et al., 2007)	Coli-septicaemia in poultry

AAF aggregative adherence fimbrae, *AIDA* adhesion in diffuse adherence, *Bfp-m* bundle forming units, *CFA* colony factor antigen, *Daa* diffuse adhesin, *LEE* locus for enterocyte effacement, *Inv* invasion plasmid antigen, *LT* heat labile, *ST* heat stable toxin, *UPEC* Uropathogenic *E. coli, NMEC* Neonatal meningitis *E. coli, APEC* Avian pathogenic *E. coli*

Diagnosis

In routine microbiology laboratories, the identification and characterization of different strains of Escherichia coli (*E. coli*) are essential for various purposes, including clinical diagnosis and epidemiological surveillance.

1. **Primary Isolation**: The clinical specimens i.e., for intestinal infections (food content in small intestine, faecal samples); for septicaemia (heart blood and spleen); for mastitis (milk samples from affected quarter); and for genital infections, (urine and uterine discharge) samples are aseptically collected from suspected animals or birds. In the laboratory, isolation of *E. coli* is performed by standard microbiological method using selective media. The isolates are confirmed by standard biochemical tests and PCR -based assay.
2. **Slide Agglutination Assays**: To differentiate between various *E. coli* strains and to identify their characteristics, slide agglutination assays are conducted. In these assays, specific antisera are used. Antisera contain antibodies that react with specific antigens found on the surface of *E. coli* strains. These antigens can include O antigens (lipopolysaccharide, associated with the cell wall) and H antigens (flagellar antigens).
3. **Molecular Diagnosis**: *E. coli* pathotypes can be identified using DNA probes or PCR assays that target the virulence factors specific to each pathotype. These molecular techniques allow for the accurate and specific detection of different *E. coli* pathotypes by analysing their genetic ch53aracteristics.

Treatment

As most of the *E. coli infections* are mainly associated with intestinal infections like diarrhoea; thus, rehabilitation of animal condition should be the prime strategy for therapeutics by oral or injectable rehydration, multivitamins, etc. Antibiotics such as polymyxin oral preparations and fluroquinolones have shown good results and remain the treatment of choice till date. With the era of excess use of antibiotics, several strains of *E. coli* are developing multi drug resistance. According to the recently published report, *E. coli* isolates were found mostly resistant to antimicrobials like streptomycin, sulphonamides, and tetracycline. Immune treatment by feeding monoclonal antibiotics containing K99 or egg yolk in powdered form in immunized hens was found effective. Phage therapy shows high efficacy against multidrug-resistant E. coli infections by using specific viruses to precisely target and kill the bacterial strains, often where antibiotics fail. While promising, its application remains largely investigational and tailored to individual cases (Pouillot et al., 2012).

Control Measures

Access to Safe, Potable Water

Accessing the safe water becomes the primary requirement for the prevention of *E. coli* infections, as contaminated water is the major source of disease transmission. According to WHO Joint Monitoring Programme for Water Supply and Sanitation in 2012, millions of people worldwide still do not have access to clean water resources.

Measures to Reduce Food Contamination Risk

Good handling procedures with relevant storage practices and temperature control during cooking play a key role in preventing disease occurrence. Nowadays, in order to reduce risk of high bacterial load, food irradiation technology is used, which is limited to various products under surveillance. Thus, this technology cannot be considered good manufacturing procedure for food industry.

Good Sanitation Practices

Implementation of strict hygienic measures to prevent cross contamination is essential to overrule the risk of multi drug-resistant pathogenic infection in hospital premises. Inappropriate use of antimicrobials, contaminated medical devices, and unhygienic environment leads to more severe situations that are difficult to treat. Thus, proper surveillance and monitoring of antibiotics are must to develop suitable strategies to combat the emerging problem of multi-drug resistance.

Mass Awareness

Risk factors should be given greater attention and awareness can be created to reduce them by experts at field level, which can benefit the community. For prophylaxis, use of probiotics belonging to the Lactobacillus comprising of safe microbes can be one approach used for competitive inhibition of pathogen to colonize gastrointestinal tract, thus preventing several *E. coli* infections to develop. It has been reported that use of these *Lactobacillus* organisms reduces the risk of vaginal and urinary tract infections, especially in humans.

Vaccination

Humans

Vaccination can be the primary prevention strategy for human against the most harmful strains, but till date, no such effective vaccine is commercially available. In some countries, for short-term prevention of ETEC strain, cholera vaccine was tried that stimulated anti-heat-labile toxin immunity to develop. UPEC strains are is responsible for more than 80% uncomplicated UTI. Several mucosal immunotherapies, including oral and vaginal bacterial preparations (e.g., Uromune, Uro-Vaxom), show long-term efficacy in preventing recurrent UTIs, while next-generation sublingual nanofiber vaccines are in development to enhance targeted urinary tract immunity. Researchers adopted genome-based vaccine development by non-conventional research vaccinology for NMEC strains and combined it with ExPEC, to develop cross-reaction.

Animals

Anti-fimbriae and anti-capsular antibodies are used to prevent the attachment of ETEC stains to enterocyte. Some of them are as listed as follows:

For Calves

- Vaccines containing K99 and F41 fimbriae.
- As ETEC infection in calves is mainly associated with rotavirus infection, combined vaccines with rotavirus antigen are also used.
- For mastitis: Vaccine containing Rc mutant of *E. coli* J5 (O111:B4) licensed in USA for use. But protective efficacy in order to prevent infection is variable.

For Pigs

- To prevent neonatal diarrhoea, prevalent adhesins like K88 and 987P and heat labile toxin are used.
- Passive immunization in piglets: Pregnant sows should be administered with killed *E. coli* K88 antigen, subunit vaccine, or oral vaccine.
- Active immunization: Done in piglets with live *E. coli* K 88 antigen that leads to the production of intestinal sIgA antibody.
- For oedema disease: Toxoid of VTe (Stx) has been found effective.

References

Al-Abri, S. S., Beeching, N. J., & Nye, F. J. (2005). Traveller's diarrhoea. *Lancet Infectious Diseases, 5*, 349–360.

Bilinski, P., Kapka-Skrzypczak, L., Posobkiewicz, M., Bondaryk, M., Holownia, P., & Wojtyla, A. (2012). Public health hazards in Poland posed by foodstuffs contaminated with *E. coli* O104:H4 bacterium from the recent European outbreak. *Annals Agricultural Environmental Medicine, 19*, 3–10.

Clements, A., Young, J. C., Constantinou, N., & Frankel, G. (2012). Infection strategies of enteric pathogenic *Escherichia coli. Gut Microbes, 3*(2), 71–87.

Darfeuille-Michaud, A. (2002). Adherent-invasive *Escherichia coli*: A putative new *E. coli* pathotype associated with Crohn's disease. *International Journal of Medical Microbiology, 292*, 185–193.

Gaschignard, J., Levy, C., Romain, O., Cohen, R., Bingen, E., Aujard, Y., & Boileau, P. (2011). Neonatal bacterial meningitis: 444 cases in 7 years. *Pediatric Infectious Disease Journal, 30*, 212–217.

Johnson, J. R., & Stell, A. L. (2000). Extended virulence genotypes of *Escherichia coli* strains from patients with urosepsis in relation to phylogeny and host compromise. *Journal of Infectious Disease, 181*, 261–272.

Johnson, T. J., Kariyawasam, S., Wannemuehler, Y., Mangiamele, P., Johnson, S. J., Doetkott, C., Skyberg, J. A., Lynne, A. M., Johnson, J. R., & Nolan, L. K. (2007). The genome sequence of avian pathogenic *Escherichia coli* strain O1:K1:H7 shares strong similarities with human extraintestinal pathogenic *E. Coli* genomes. *Journal of Bacteriology, 189*, 3228–3236.

Kaper, J. B., Nataro, J. P., & Mobley, H. L. (2004). Pathogenic *Escherichia coli. Nature Reviews Microbiology, 2*, 123–140.

Nataro, J. P., Steiner, T., & Guerrant, R. L. (1998). Enteroaggregative *Escherichia coli. Emerging Infectious Diseases, 4*, 251–261.

Negroni, A., Costanzo, M., Vitali, R., Superti, F., Bertuccini, L., Tinari, A., Minelli, F., di Nardo, G., Nuti, F., Pierdomenico, M., et al. (2012). Characterization of adherent-invasive *Escherichia coli* isolated from pediatric patients with inflammatory bowel disease. *Inflammatory Bowel Diseases, 18*, 913–924.

Pouillot, F., Chomton, M., Blois, H., Courroux, C., Noelig, J., Bidet, P., Bingen, E., & Bonacorsi, S. (2012). Efficacy of bacteriophage therapy in experimental sepsis and meningitis caused by a clone O25b:H4-ST131 *Escherichia coli* strain producing CTX-M-15. *Antimicrobial. Agents Chemotherapy, 56*, 3568–3575.

Qadri, F., Svennerholm, A. M., Faruque, A. S., & Sack, R. B. (2005). Enterotoxigenic *Escherichia coli* in developing countries: Epidemiology, microbiology, clinical features, treatment, and prevention. *Clinical Microbiology Review, 18*, 465–483.

Servin, A. L. (2005). Pathogenesis of Afa/Dr diffusely adhering *Escherichia coli. Clinical Microbiology Reviews, 18*, 264–292.

Weintraub, A. (2007). Enteroaggregative *Escherichia coli*: Epidemiology, virulence and detection. *Journal of Medical Microbiology, 56*, 4–8.

Further Reading

Drasar, B. S. (2010). Topley and Wilson's Microbiology and Microbial Infections, 8 Volume Set.

Rawat, M. (2020). *Veterinary bacteriology* (CAAST: ACLH, National Agricultural Higher Education project). ICAR- IVRI.

Tang, Y. W., & Sails, A. (Eds.). (2014). *Molecular medical microbiology*. Academic.

Erysipeloid

19

Ajay Pratap Singh

Abstract

Erysipeloid is an uncommon yet significant cutaneous infection initiated by the traumatic introduction of *Erysipelothrix rhusiopathiae* into the skin. It was not until the late nineteenth century that this microorganism was confirmed to be pathogenic in humans. This infection is attributed to the Gram-positive bacillus *Erysipelothrix rhusiopathiae* (*insidiosa*), known for its capacity to persist for extended periods in soil or decomposed organic matter. Typically, individuals with occupational exposure to raw fish, poultry, and meat products, or those in contact with contaminated animals, particularly swine, are susceptible to this condition. While Erysipeloid primarily manifests as a cutaneous infection, ranging from mild to severe forms, it can occasionally manifest systemically, particularly in rare instances among immunocompromised individuals.

Keywords

Erysipeloid · Whale finger and seal finger · Swine erysipelas · Erysipelothricosis · Rosenbach's disease · Diamond skin disease

History

Erysipelothrix, was initially described in 1873 by Tilbury Fox as "**erythema serpens**." The genus Erysipelothrix was first characterized in 1876 when Robert Koch isolated *E. muriseptica* from the blood of mice exhibiting septicemic infection. This

A. P. Singh (✉)
Department of Veterinary Microbiology, College of Veterinary Science & Animal Husbandry, U.P. Pandit Deen Dayal Upadhyaya Pashu Chikitsa Vigyan Vishwavidyalaya Evam Go-Anusandhan Sansthan, Mathura, Uttar Pradesh, India

R. Biswas et al. (eds.), *Concept of One Health and Zoonoses*,
https://doi.org/10.1007/978-981-95-6627-3_19

discovery was followed by Friedrich Loeffler's 1882 identification of a morphologically similar pathogen in the pigs afflicted with erysipelas. Subsequent investigations revealed Erysipelothrix as an etiological agent across multiple animal species, prompting early taxonomic proposals for three host-associated variants: *E. muriseptica* (murine origin), *E. porci* (porcine origin), and *E. erysipeloid* (human origin). Advanced microbiological studies later demonstrated these to be phenotypically and genotypically congruent, representing a single species that Trevisan had formally designated *E. insidiosa* in 1885. The zoonotic potential of this organism was definitively established through the work of Rosenbach, who confirmed its pathogenicity in humans. Rosenbach demonstrated the causal relationship between erysipeloid and swine erysipelas by inoculating himself with purified material from a patient's lesion. The term "**erysipeloid**" was coined by Rosenbach in 1887, encompassing various clinical morphologies such as "**erythema migrans**", "**erysipelas chronicum**", and "**diamond skin disease**" (Stackebrandt et al., 2006). In 1904, Gilchrist documented cases of erysipeloid secondary to crab bites. Klauder extensively wrote about erysipeloid and its occupational associations in the 1920s and 1940s. The first case of endocarditis associated with *Erysipelothrix* was confirmed by necropsy in 1940 by Russell and Lamb. Prausnitz purportedly documented endocarditis without autopsy in 1921. In 1944, Heilman and Herrell demonstrated the effectiveness of penicillin in *Erysipelothrix* infections through animal studies. More recent developments include associations with septicemia, subsequent endocarditis, cerebral abscesses, arthritis, and empyemas. In 1966, the final name *E. rhusiopathiae* was assigned setting aside all other proposed nomenclature.

Etiology

Erysipelothrix species are Gram-positive bacilli exhibiting several distinctive morphological features: they lack capsules, flagella, and endospores, appearing as straight or gently curved rods with blunt termini. While typically observed as individual cells, paired forms in characteristic V-shaped arrangements, or short chains, some strains display notable pleomorphism with elongated filamentous forms. Ultrastructural analysis reveals random cellular orientation in microbial aggregates, contrasting with the orderly palisades seen in related genera. Although they are culturally and morphologically similar to corynebacterium, they do not display acid-fast characteristics. Erysipelothrix species exhibit a relatively broad thermal tolerance, proliferating between 15 and 44 °C, with peak metabolic activity occurring at mesophilic temperatures (30–37 °C). The pH range conducive for growth is 7.2–7.6, spanning from 6.7 to 9.2. Erysipelothrix species exhibit facultative anaerobic metabolism, demonstrating optimal growth under microaerophilic conditions. Supplementation with 5–10% CO_2 significantly augments their proliferative capacity, a characteristic frequently utilized for laboratory cultivation.

The genus *Erysipelothrix* contained only *E. rhusiopathiae*, subdivided into 28 serovars (designated 1a, 1b, 2-26, and N) based on cell wall antigens. Subsequent recognition of significant inter-strain variability in serological and biochemical traits prompted molecular reevaluation. Molecular taxonomic studies employing DNA–DNA hybridization techniques revealed that serovars 3, 7, 10, 14, 20, 22, and 23 represented a genetically distinct cluster, leading to their reclassification as *E. tonsillarum*. Phylogenetic studies employing next-generation sequencing have expanded the Erysipelothrix genus to include four additional taxa: three provisional designations (*Erysipelothrix sp.* strains 1-3) and the formally described *E. inopinata* (Wang & Riley, 2015).

Characteristics	*E. rhusiopathiae*	*E. tonsillarum*
Fermentation of sucrose	Do not ferment	Ferment
Pathogenicity	Majority of strains (66%) are virulent	Majority of strains (96%) are avirulent
Serotypes involved	la, lb., 2, 4, 5, 6, 8, 9, 11, 12, 15, 16, 17, 19, 21, N	3, 7, 10, 14, 20, 22, 23

Epidemiology

Erysipelothrix species demonstrate extraordinary ecological versatility, maintaining viability in diverse habitats ranging from soil to marine ecosystems. Their capacity for prolonged environmental survival is complemented by broad host tropism, acting as:

- Primary pathogen in domestic animals (swine, poultry)
- Opportunistic pathogen in wild fauna (rodents, avian species)
- Commensal colonization in marine vertebrates/invertebrates

While domestic swine (Sus scrofa domesticus) represent the principal epidemiological reservoir for *E. rhusiopathiae*, surveillance data confirm significant infection rates among synanthropic rodents (particularly Rattus spp.) and avian populations. This tripartite host spectrum underscores the pathogen's remarkable cross-species adaptability. *E. rhusiopathiae* exhibits extensive shedding patterns in porcine hosts, with viable organisms excreted through intermittent fecal excretion in apparently healthy animals, urinary excretion and oronasal secretions (saliva and nasal discharge). Notably, subclinical carriage is common, with studies demonstrating persistent tonsillar colonization in asymptomatic animals, lymphoid tissue reservoirs (particularly ileal Peyer's patches). Carrier pigs can contaminate the soil with the bacterium through their feces. There is a rapid transmission of the bacterium from carrier sows to young pigs, leading to the young pigs becoming carriers themselves.

Environmental persistence of *E. rhusiopathiae* primarily stems from continuous contamination through excretions and secretions of infected animal reservoirs. The bacterium demonstrates remarkable adaptability to both organic matter and abiotic surfaces, facilitating extended survival outside host organisms. Environmental surveillance studies confirm *E. rhusiopathiae* presence in abattoir waste streams, fecal-contaminated soils—with highest recovery rates near livestock holding areas and manure-fertilized agricultural fields, and wildlife migratory pathways. Erysipeloid is characterized as an occupational disease primarily affecting individuals in specific professions. Individuals with frequent animal or animal product contact face elevated infection risk, particularly the following:

1. **Livestock Workers**
   - Swine breeders/pork producers
   - Poultry farmers
   - Veterinarians (large animal practice)
2. **Meat Processing Industries**
   - Abattoir workers
   - Butchers
   - Quality control inspectors
3. **Aquatic Industries**
   - Commercial fishermen
   - Fish processors
   - Seafood market workers
4. **Domestic Handlers**
   - Food service professionals handling raw meat/seafood
   - Household cooks preparing contaminated products

This risk is particularly high among those who handle unprocessed meats and animal products as part of their daily activities.

Erysipelothrix infection typically occurs through direct percutaneous inoculation when handling infected animals or contaminated animal products. The primary transmission route involves bacterial entry through skin abrasions, puncture wounds, or mucous membranes during occupational exposures such as livestock handling, meat processing, or fishing activities. Workers in swine and poultry operations are particularly vulnerable during routine animal care, while slaughterhouse employees and butchers face exposure when processing infected carcasses. Similarly, fishermen and seafood handlers risk infection through minor hand injuries sustained while cleaning fish or shellfish. The organism can also persist on environmental surfaces and tools, creating additional transmission risks when these contaminated fomites contact broken skin. While occupational exposure accounts for the overwhelming majority of *Erysipelothrix* infections, sporadic non-occupational cases represent exceptional epidemiological occurrences. Non-occupational *Erysipelothrix* infections occur through distinct exposure routes as follows:

1. **Immunocompromised Hosts:** Atypical presentations may arise from impaired immune defenses, allowing opportunistic invasion even with minimal bacterial exposure.

2. **Environmental Contact:** Accidental inoculation via contaminated soil, water (e.g., ponds shared by livestock and pets), or fomites, often through minor skin breaches.
3. **Zoonotic Spillover:** Close contact with infected companion animals (dogs, cats) that acquired the bacterium from livestock-contaminated environments.
4. **Foodborne Transmission:** Undercooked meat or seafood products from infected animals, though this route remains poorly documented.

Epidemiological studies demonstrate distinct *Erysipelothrix* serotype distributions across animal hosts, with clear disease associations. In swine, serotypes 1a, 1b, and 2 predominate, each linked to specific clinical manifestations: serotypes 1a and 1b are strongly associated with acute septicemic cases exhibiting high mortality, while serotype 2 typically causes chronic swine erysipelas, characterized by arthritis, vegetative endocarditis, and cutaneous diamond-skin lesions. Poultry infections follow a different serotype pattern, with outbreaks primarily involving serotypes 1, 5, 6, and 9. Among these, serotype 1 shows the highest virulence in avian hosts, often causing systemic infection, whereas serotypes 5, 6, and 9 tend to produce milder, localized cutaneous forms.

Pathogenesis

Erysipelothrix rhusiopathiae employs multiple virulence strategies to establish infection and cause erysipelas in susceptible hosts. The pathogen's disease-producing capability stems from an orchestrated interplay of the following:

Immune Evasion Tactics *E. rhusiopathiae* evades host immunity through multiple strategies: its capsular polysaccharide inhibits phagocytosis, surface proteins confer complement resistance, and biofilm formation enables persistent infection. These mechanisms collectively allow the bacterium to circumvent immune clearance and establish chronic infections.

Surface Adhesion Factors The pathogenicity of *E. rhusiopathiae* involves neuraminidase (sialic acid cleavage promoting spread and platelet dysfunction) and hyaluronidase (connective tissue degradation enabling joint invasion). These enzymes synergistically facilitate both systemic dissemination and tissue-specific damage, manifesting as thrombocytopenia and arthritis.

Tissue-Damaging Enzymes *E. rhusiopathiae* produces several tissue-damaging enzymes that enhance its pathogenicity, including phospholipases that disrupt host cell membranes, proteolytic enzymes that degrade structural proteins, and reactive oxygen species that induce oxidative stress in infected tissues. These enzymatic activities collectively facilitate bacterial spread and tissue damage during infection.

Surface Proteins Surface proteins of *E. rhusiopathiae* also play a role in its virulence. These proteins likely interact with host cells and tissues, influencing the bacterium's ability to establish infection. The specific surface proteins involved in virulence may include adhesins or factors that contribute to evasion of the host immune response. Recently, a novel virulence factor (RspA/B surface proteins) has been identified that enhance endothelial adhesion, explaining the bacterium's tropism for cardiac valves.

Host Tropism *E. rhusiopathiae* exhibits host tropism, meaning it has a preference for certain host species. This is evident in its ability to cause disease in various animals, including swine, birds, and humans.

Clinical Symptoms

In Humans

Human infection with *Erysipelothrix rhusiopathiae* manifests in three principal clinical forms. The most common presentation is **localized cutaneous erysipeloid**, characterized by well-demarcated, erythematous-to-violaceous plaques at the site of inoculation, typically on the hands or fingers, which expand centrifugally without suppuration. **A generalized cutaneous form** may develop, featuring disseminated skin lesions accompanied by systemic symptoms like low-grade fever and arthralgia, though without bloodstream invasion. The most severe manifestation is the **septicemic form**, marked by high-grade bacteremia that frequently leads to endocarditis (particularly in patients with valvular abnormalities), with potential complications including septic arthritis and, rarely, central nervous system involvement.

Localized Cutaneous Form (Erysipeloid of Rosenbach)

The localized cutaneous form, often referred to as erysipeloid of Rosenbach, usually remains confined to the site of inoculation, commonly the hands or fingers of individuals with occupational exposure (e.g., butchers, fish handlers, and veterinarians). It is characteristically self-limiting, resolving spontaneously within 2–3 weeks in immunocompetent hosts, even without antimicrobial therapy. The characteristic lesion develops at the inoculation site, most commonly affecting the fingers, hands, or other exposed areas in occupational exposures. The incubation period usually spans from 1 to 7 days post-inoculation. These skin lesions present a distinctive appearance, characterized by their bright red color and the formation of well-defined "**diamonds**" with smooth, glistening surfaces. This classic pattern often affects the hand, especially the finger webs, while excluding the terminal phalanges. The classic erysipeloid lesion exhibits distinctive centrifugal expansion, typically advancing 1–2 cm per day while showing central clearing. This progression pattern often results in annular (ring-like) or gyrate configurations, particularly in untreated cases. The localized cutaneous form of *E. rhusiopathiae* infection frequently induces profound digit inflammation, classically manifesting as "**whale finger**" (Marked circumferential edema with tense, shiny skin resembling marine mammal flippers, often affecting multiple phalanges) or "**seal finger**" (Fusiform swelling concentrated at interphalangeal joints, mimicking pinniped limb morphology). However, the infection has the potential to manifest on any exposed body part, such as the sole of the foot, face, or neck. The localized cutaneous form of *E. rhusiopathiae* infection frequently presents with distinctive neurovascular symptoms and systemic involvement burning (most pathognomonic), throbbing, or pruritic pain accompanied by low-grade fever (38–38.5 °C) in 20–30% of cases, flu-like myalgias and malaise, night sweats (suggesting cytokine release), polyarthralgia without true synovitis, firm enlargement of regional nodes (typically epitrochlear/axillary), and lymphadenopathy. Arthritis may occasionally occur in a joint adjacent to the primary lesion, typically as a reactive process rather than as a direct microbial invasion. Periarticular inflammation in erysipeloid manifests as non-suppurative arthritis that contrasts sharply with pyogenic infections caused by Staphylococcus aureus or Streptococcus pyogenes. Erysipeloid symptoms are characterized by several key diagnostic hallmarks, including the absence of purulent exudate, minimal or absent pitting edema, and notably, intense, disproportionate throbbing pain relative to the visible extent of cutaneous involvement.

Generalized Cutaneous Form

The generalized form of erysipeloid, a rare but clinically distinct manifestation of *Erysipelothrix rhusiopathiae* infection, occurs when the bacterium disseminates beyond the initial inoculation site. Unlike the self-limiting localized variant, this presentation features multiple erythematous-to-violaceous plaques that spread centrifugally, often forming annular or polycyclic patterns. Systemic involvement is hallmark, with patients exhibiting low-grade fever (38–39 °C), chills, migratory polyarthralgias, and profound fatigue. The pathogenesis likely involves either lymphatic extension or transient bacteremic seeding of distant skin sites, particularly in untreated or immunocompromised individuals. While bullous lesions may develop in severe cases, the absence of suppuration remains a key diagnostic feature distinguishing it from pyogenic infections.

Septicemic Form

The systemic form of *E. rhusiopathiae* infection manifests as a spectrum of severe complications, with infective endocarditis representing the most prevalent life-threatening presentation. Hematogenous dissemination can lead to septic arthritis (predominantly affecting large joints), osteomyelitis (notably phalangeal necrosis), and central nervous system involvement, including cerebral abscesses and septic emboli causing ischemic infarcts. Recent case reports highlight emerging associations with pulmonary effusions, particularly in immunocompromised hosts. Septicemia, while rare, carries high mortality (~40%) and typically arises secondary to endocarditis rather than localized infection. Contemporary literature documents expanding clinical manifestations, including chronic antibiotic-resistant arthritis linked to biofilm formation and intracranial abscesses. Contemporary clinical reports reveal *E. rhusiopathiae*'s emerging role in opportunistic and atypical systemic infections. Immunocompromised hosts, particularly HIV-positive patients, demonstrate prolonged bacteremia due to impaired macrophage clearance. Endocarditis complications now include acute renal failure from immune complex-mediated glomerulonephritis. *E. rhusiopathiae* may trigger lupus nephritis in SLE patients, suggesting molecular mimicry between bacterial surface proteins and glomerular antigens. Occupational exposures continue to yield unusual presentations, such as **'crayfish poisoning'** in lobster fishermen—a syndrome featuring both erysipeloid-like lesions and systemic symptoms attributed to marine zoonotic transmission. *E. rhusiopathiae* can cause rare but clinically significant musculoskeletal infections, including sacroiliitis and prosthetic joint infections (PJIs), primarily affecting individuals with occupational animal exposure or immunocompromising conditions.

In Animals

In animals, naturally occurring *E. rhusiopathiae* infections have been observed in pigs, cattle, sheep, lambs, and birds, with severe outcomes in turkeys.

In Pigs Among swine, the disease is recognized internationally by different vernacular names, including 'swine erysipelas' (English), 'rouget du porc' (French), and 'mal rossino' (Italian). Swine erysipelas, caused by *Erysipelothrix rhusiopathiae*, presents in three distinct clinical forms with varying severity and economic consequences. Notably, the disease shows remarkable parallels to human *E. rhusiopathiae* infections, with shared serotypes (1a/1b) demonstrating 98% genomic homology between porcine and human isolates The **acute septicemic form** is the most virulent, characterized by sudden onset of high fever (41–42 °C), systemic inflammatory response, and cutaneous necrosis, with mortality rates reaching 80% within 3–4 days. The **subacute cutaneous form**, classically presenting with diamond-shaped urticarial plaques, typically resolves spontaneously within a week due to effective host immune responses. The **chronic form** represents the most economically significant manifestation, primarily causing fibrinosuppurative arthritis (particularly in stifle and hock joints) and vegetative endocarditis (predominantly affecting mitral valves). These chronic lesions lead to substantial carcass condemnations at slaughter, accounting for 60–70% of total losses.

In Sheep and Cattle *E. rhusiopathiae* also causes polyarthritis in sheep and lambs. Polyarthritis begins in lambs at the age of 2–3 months. The lambs develop a stiff gait, experiencing challenges in rising and may even fall. The affected joints typically exhibit swelling, and the capsule thickens. Instances of infection in cattle leading to arthritis in calves have been observed, though such occurrences are rare.

In Birds While *E. rhusiopathiae* demonstrates broad avian host susceptibility across domestic poultry (chickens, turkeys, and ducks), game birds (pheasants, quail), and wild avian species, clinically significant outbreaks with economic consequences occur predominantly in turkey (highest incidence), layer chickens (Sporadic outbreaks in caged systems), and ducklings. While large-scale epizootics are well documented in domestic ducks, infections in chickens occur sporadically and are less commonly reported. Wild birds, especially seagulls and urban pigeons, have been identified as asymptomatic carriers of *E. rhusiopathiae*, with colonization rates reported between 22% and 33%. Sporadic cases have also been documented in more exotic species including flamingos and penguins. Adult male turkeys experience a condition termed "bluecomb" due to the development of cyanotic skin color. The turkeys exhibit lethargy, diarrhea, and ultimately succumb to the disease. Multifocal hemorrhages are commonly noted in tissues such as the pectoral and leg muscles (showing "strawberry-like" striations), pericardial and peritoneal effusions, hepatic/splenic capsular hemorrhages with "nutmeg liver" pattern.

Erysipelothrix spp., particularly *E. rhusiopathiae*, causes significant zoonotic infections in cetaceans (dolphins, porpoises, whales) with clinical and pathological parallels (diamond-shaped cutaneous lesions-pathognomonic rhomboid plaques) to swine erysipelas.

Diagnosis

Sample Collection In cases of cutaneous erysipeloid, the gold standard diagnostic specimen is a 4-mm punch biopsy taken from the lesion's advancing edge, which can be processed for direct microscopical examination using Gram stain (revealing scant Gram-positive rods), culture on Columbia CNA agar (selective for Gram-positive organisms) and histopathology (characterized by dermal neutrophilic infiltrates without suppuration). For less invasive sampling, the saline injection-aspiration technique offers a practical alternative. Inject 0.5 mL sterile saline into the lesion periphery. Gently massage the area for 30 seconds to dislodge bacteria. Aspirate fluid for direct immunofluorescence (85% sensitivity) or real-time PCR.

Isolation For the reliable diagnosis of systemic *Erysipelothrix rhusiopathiae* infections, such as sepsis and endocarditis, modern automated blood culture systems (BACTEC™, BacT/ALERT®) demonstrate excellent sensitivity (89%) when proper protocols are followed. Optimal specimen collection requires obtaining three sets of blood cultures (20 mL per set) through separate venipuncture sites to ensure adequate volume and maximize detection rates. Given the organism's fastidious nature and slow growth characteristics, cultures must be maintained for a minimum of 7 days in a 5% CO_2-enriched environment to allow sufficient time for bacterial proliferation. To enhance isolation from positive blood cultures, subculture onto blood agar plates supplemented with kanamycin (15–30 μg/mL) is recommended, as this antibiotic formulation effectively suppresses competing microbiota while permitting *E. rhusiopathiae* growth. Environmental samples (soil, manure) and heavily contaminated clinical specimens require selective media to inhibit competing flora while promoting *Erysipelothrix* growth. *Erysipelothrix* selective broth (ESB) is particularly effective, containing kanamycin (400 μg/mL) to suppress Gram-negatives, neomycin (50 μg/mL) for broad-spectrum coverage, and vancomycin (25 μg/mL) to inhibit other Gram-positives, along with 10% horse serum to enhance bacterial recovery. For solid media, modified tryptose blood agar incorporates crystal violet (1:100,000) to inhibit staphylococci and streptococci, along with sodium azide (0.02%) to eliminate catalase-positive organisms, improving isolation specificity. Packer's Agar, optimized for fecal and soil samples, includes potassium tellurite (0.04%) to suppress Enterobacteriaceae while allowing *Erysipelothrix* colonies to appear black due to tellurite reduction. Emerging chromogenic media like ERYSICHROM™ further facilitate differentiation, with *E. rhusiopathiae* forming blue colonies and *E. tonsillarum* appearing white, streamlining species identification in mixed cultures.

Mouse Protection Test The standard mouse protection test remains a definitive method for confirming *E. rhusiopathiae* virulence. In this assay, two groups of mice receive a subcutaneous injection of 0.1 mL from an 18–24-hour broth culture (containing approximately 10^7 CFU/mL of the suspected isolate). One group is concurrently administered 0.5 mL of commercial equine hyperimmune antiserum at a separate site, while the other serves as an unprotected control. Virulent strains typically cause 100% mortality in control mice within 96–120 hours, with clinical signs including ruffled fur, lethargy, and marked splenomegaly (3–5× normal weight) upon necropsy. In contrast, mice receiving neutralizing antiserum survive beyond 7 days. Blood cultures from moribund mice yield *E. rhusiopathiae* in 100% of cases, with characteristic necrotic lesions at injection sites.

Serological Tests Strains of *Erysipelothrix rhusiopathiae* demonstrate conserved antigenic profiles across human and animal isolates, exhibiting strong serological cross-reactivity that precludes the use of routine antibody-based tests for clinical identification. This antigenic similarity renders conventional serological assays unsuitable for either bacterial typing or antibody detection in patient sera due to inadequate specificity. However, fluorescent antibody techniques (both direct and indirect formats) remain valuable for visualizing the organism in tissue specimens, offering 85–92% sensitivity in formalin-fixed samples when paired with monoclonal antibodies targeting species-specific surface epitopes. For specialized applications, the surface protective antigen (SpaA) serves as a key serological marker. Growth agglutination tests and latex agglutination assays employing purified SpaA protein are routinely utilized in veterinary settings to monitor maternal antibody transfer in swine populations, with titers ≥1:160 indicating protective immunity in piglets.

Molecular Diagnosis Recent advances in molecular diagnostics have revolutionized the detection and differentiation of *Erysipelothrix* species. While conventional 16S rRNA PCR remains widely used, newer technologies offer enhanced speed, sensitivity, and specificity. Multiplex qPCR targeting the spaA (surface protective antigen) and cpsB (capsular biosynthesis) genes enables rapid species discrimination, with 100% specificity for *E. rhusiopathiae* versus *E. tonsillarum*. For point-of-care applications, the CRISPR-Cas12a-based lateral flow assay provides visual results in under 30 minutes, demonstrating 98% sensitivity in skin biopsies. Additionally, loop-mediated isothermal amplification (LAMP) offers a culture-free, field-deployable solution, particularly useful in resource-limited settings, with detection limits as low as 10 CFU/mL. These innovations—combined with whole-genome sequencing for outbreak tracing and metagenomic nanopore sequencing for direct clinical specimen analysis—are transforming *Erysipelothrix* diagnostics, enabling faster, more accurate detection and improved patient and veterinary outcomes.

Zoonotic Nexus

Erysipelothrix rhusiopathiae, historically recognized as a significant veterinary pathogen responsible for arthritis in swine and poultry, has increasingly been implicated in severe human infections, particularly among individuals with occupational or environmental exposure to animals and animal products. Emerging reports document its association with invasive conditions such as prosthetic joint infections (PJIs) and infective endocarditis, especially in high-risk groups including fishermen, butchers, veterinarians, and individuals handling wild game. Notably, a documented case involved a prosthetic knee joint infection in a patient with a history of processing wild game in the Canadian Arctic (Zautner et al., 2022). Human *E. rhusiopathiae* infections often manifest in individuals with prosthetic heart valves, with endocarditis being a life-threatening complication. Although less commonly reported, spinal involvement has also been described, including spondylitis, paravertebral abscesses, and even epidural empyema, highlighting the organism's potential for deep-seated infections beyond the skin and soft tissues.

Treatment

For uncomplicated cutaneous infections in humans, intramuscular penicillin (300,000 to one million units daily for 5–7 days) remains the first-line therapy, with alternatives including oral amoxicillin, cephalosporins (ceftriaxone, cefotaxime), macrolides (erythromycin), fluoroquinolones (ciprofloxacin), or doxycycline in penicillin-allergic patients. Severe systemic infections, such as endocarditis, require high-dose intravenous penicillin (12–24 million units daily for 4–6 weeks), sometimes combined with gentamicin, while ceftriaxone serves as an effective alternative. Notably, *E. rhusiopathiae* is resistant to vancomycin, which must be considered in empiric treatment of endocarditis. In pigs, where the infection is common, effective treatments include intramuscular ceftiofur, doxycycline, tiamulin, and penicillin.

Control and Prevention

The recommended measures include herd management, sanitation, and immunization procedures.

Biosecurity Measures Strict biosecurity protocols should be implemented on farms, particularly in swine production facilities. This includes controlling access, minimizing contact between animals and contaminated materials, and ensuring hygiene practices. Control of reservoir populations of *E. rhusiopathiae* is challenging due to its widespread distribution, diverse animal hosts, and environmental persistence.

Husbandry Practices The strategy for controlling swine erysipelas involves depopulating the entire stock, followed by the implementation of an All-In/All-Out production system. The depopulation eliminates existing infection sources, and the All-In/All-Out system minimizes the risk of disease introduction and transmission by ensuring that the population is cohesive and undergoes the production cycle together.

Hygiene Practices Promote good hygiene practices among workers, especially those with direct contact with animals. This includes regular handwashing, the use of personal protective equipment, and proper cleaning and disinfection of work areas.

Education and Training For individuals at risk of infection, preventive measures include wearing protective gear, maintaining good hygiene, and prompt treatment of injuries.

Monitoring and Surveillance Regularly monitor animals for signs of infection and promptly isolate and treat infected individuals. Conduct surveillance in human populations at risk, such as those working with animals, to detect and manage cases early.

Vaccination In animals, especially swine, vaccination is a key preventive measure. Vaccinating animals can reduce the incidence and severity of *Erysipelothrix* infections, including joint infections. In certain cases, vaccines may also be available for humans at risk of exposure, such as individuals working with animals.

References

Stackebrandt, E., Reboli, A. C., & Farrar, W. E. (2006). The genus Erysipelothrix. In M. Dworkin, S. Falkow, E. Rosenberg, K. H. Schleifer, & E. Stackebrandt (Eds.), *The Prokaryotes* (pp. 492–510). Springer. https://doi.org/10.1007/0-387-30744-3_13

Wang, Q., & Riley, T. V. (2015). *Erysipelothrix rhusiopathiae*. In *Molecular medical microbiology* (pp. 859–872). Academic. https://doi.org/10.1016/b978-0-12-397169-2.00047-0

Zautner, A. E., Tersteegen, A., Schiffner, C. J., Đilas, M., Marquardt, P., Riediger, M., Delker, A. M., Mäde, D., & Kaasch, A. J. (2022). Human *Erysipelothrix rhusiopathiae* infection via bath water–case report and genome announcement. *Frontiers in Cellular and Infection Microbiology, 12*, 981477.

Q-Fever

20

Maninder Singh

Abstract

Q-fever is considered among top 13 global priority zoonotic diseases. Despite this, it is a neglected disease and is not notified in majority of countries. The exact burden of disease is not known while the disease is of zoonotic significance for animal owners/ handlers, veterinarians and other animal occupational groups.

Keywords

Query fever (Q-fever) · *Coxiella* · Nine-mile fever

History

- Q-fever was first described in 1935 by Derrick during an outbreak of fever among slaughterhouse workers in a slaughterhouse in Brisbane, Australia. As the causative agent of outbreak was not identified, the disease was named as Query fever (Q-fever).
- Later on, in 1937, Burnet and Freeman isolated the causative agent from blood and urine of infected patients and identified it to be a *Rickettsia.*
- Simultaneously, David and Cox isolated a filterable agent from a tick from Nine Mile Creek in Montana and named it as *Rickettsia diaporica.* Coincidentally, a laboratory worker got infection by the isolated agent and showed clinical signs similar to those described by Derrick. Further studies indicated that the agent isolated by Burnet/ Freeman and David/ Cox were the same. The organism was named as *Coxiella burnetii* in the honour of discoveries of Cox and Burnet.

M. Singh (✉)
Department of Veterinary Public Health and Epidemiology, College of Veterinary Science, Guru Angad Dev Veterinary and Animal Sciences University, Bathinda, India

R. Biswas et al. (eds.), *Concept of One Health and Zoonoses*,
https://doi.org/10.1007/978-981-95-6627-3_20

Aetiology

Coxiella burnetii is a Gram-negative bacterium having obligate intracellular life cycle and grows in the phagolysosomes within the host cells. Earlier, the agent was considered as *Rickettsia*; however, the characteristics such as filterability and capability to be transmitted without vectors warranted its inclusion in *Rickettsia. Coxiella burnetii* belongs to gamma-subdivision of Proteobacteria, Family Coxeillaceae and Order Legionalles. Order Legionalles has *Coxiella, Legionella, Francisella* and *Rickettsiella* genera. *Coxiella* genus has single species, i.e. *Coxiella burnetii.*

Coxiella burnetii is a highly pleomorphic bacterium with size ranging from 0.4 to 1 μ long and 0.2 to 0.4 μ wide. Being an obligate intracellular, the growth studies in the laboratory requires embryonated eggs, laboratory animals and cell culture like the study of viruses. The cell wall of bacteria is like other Gram-negative bacteria having two membranes—outer membrane and inner membrane with an electrodense layer on the inner side of outer membrane. The outer membrane has phospholipids, lipopolysaccharide (LPS) and proteins; the LPS and proteins of outer membrane are involved in 'phase variation' phenomenon known in the *Coxiella burnetii.* (Sahu et al., 2021).

Morphologically, the bacterium has three distinct stages, which are distinguishable by electron microscopy viz., large-cell variant (LCV), small-cell variant (SCV), and small dense variant (SDC)/spore. LCV stage is a metabolically active stage found intracellularly while the other two stages are found extracellularly in the environment providing resistance to the adverse environmental conditions and are infectious in nature for the host. This strategy helps the bacteria to survive extracellularly and multiply intracellularly. SCV and LCV differ in terms of peptidoglycan content, osmotic lysis and metabolism of substrates.

The pathogen is able to survive for 30–60 days in dried milk, 150 days in soil, 7–10 months in wool, 2 months in dust and 8 months in ripened cheese having pH 4.96–5.41 and water activity 0.90–0.95 (Eldin et al., 2017).

Coxiella burnetii has been classified as bioterrorism agent of category B.

Epidemiology

Coxiella burnetii is a pathogen prevalent worldwide with exception of certain geographical areas such as New Zealand and Antarctica. As the infection in humans is primarily a zoonosis, the burden of disease in humans depends upon the disease status in animals, especially in ruminants of the region.

Coxiella burnetii is capable of infecting multiple host species and has a large number of reservoirs including ruminants, marine mammals, birds, reptiles and ticks. However, the principal reservoirs are ruminants and the disease is transmitted among animals through horizontal transmission. The disease is maintained in wild settings, especially among wild rodents and small mammals and transmission is being maintained through ticks. Among domestic animals, ruminants are the most significant reservoirs. The bacterium has been detected in bats from Brazilian

Atlantic Forest, mosquitoes (*Anopheles dirus*) and fleas (*Xenopsylla cheopis*, *Ctenocephalides felis* and *Ctenocephalides canis*); the significance of these species in disease maintenance and transmission has yet to be explored. The disease is notifiable as per World Organization for Animal Health.

In humans, the occurrence of disease is either as sporadic or as outbreak. Generally, the infection is not reported due to indistinct clinical signs, poor suspicion and inadequate diagnostic facilities in countries. In Netherlands, during 2007–2010 Q-fever outbreak, around 4000 humans were infected with mass culling of about 52,000 infected goats. Consequently, Netherlands notified coxiellosis in goats as a notifiable disease with mandatory vaccination of goats.

Transmission

Transmission of coxiellosis involves different patterns in wild settings and domestic animals. As ticks are responsible for maintaining of bacteria in wild areas, transmission among wild animals occurs mainly through tick bites. The transmission from wild animals to domestic animals such as cattle and goats through ticks is reported. Transmission among domestic animals involves direct or indirect contact with the excretions and secretions of *Coxiella*-infected animals, especially the abortion and parturient secretions. The shedding of *Coxiella burnetii* in vaginal mucus of sheep and goats for about 3 months is reported. The excretion in faeces and milk of cattle, sheep and goats is reported. The consumption of raw milk may be a contributory source for young ones and herd animals.

In humans, transmission is mainly associated with direct and indirect contact with infected animals and their excretions and secretions. Inhalation of infected aerosols during parturition of infected animals constitutes the most important hazard. Therefore, the disease is of significance in certain occupational groups such as animal keepers, veterinarians and slaughter house workers. Another important source is consumption of unpasteurized milk and milk products as the pathogen is shed for about 13 months in the milk of cattle. The pathogen is a reference used to set temperature (63 °C) for pasteurization of milk. The transmission from animals to humans by ticks is not significant. In humans, transmission by blood transfusion and sexual transmission is reported but rare.

Pathogenesis

The infectious dose of coxiellosis is very low (1–10 bacteria). The infection in the host is afflicted by small-cell variant (SCV) stage of *Coxiella burnetii*. On entry in the body, SCVs are phagocytosed by macrophages followed by fusion of phagosomes and lysosomes, leading to phagolysosomes. Phagolysosomes further fuse to form single cytoplasmic vacuole having pH between 4.7 and 5.2. The acidic pH is favourable for the growth of *Coxiella burnetii* because of passage of nutrients required for pathogens and inhibition of cell apoptosis. The acidic pH also activates the metabolic enzymes of pathogen, which protect it from host defence system; for example, superoxide dismutase of *Coxiella* protects it from host-generated free radicals. Inside the phagolysosomes, SCVs undergo binary fission and differentiate into large-cell variant (LCV), which is the metabolic active stage. In LCVs, further differentiation leads to small dense compact (SDC) structure and are released into the outer environment. SDC are further differentiated to SCVs, which can infect new cells to restart the cycle.

Depending upon the virulence, *Coxiella burnetii* have two phases—phase I and phase II, differing in the LPS complexity, and is called 'antigenic phase variation'. However, morphologically, both phases are the same. LPS is the main component of cell wall of *Coxiella burnetii* and inherits virulence for the bacteria. From phase I to phase II transition, there is partial loss of LPS leading to exposure of proteins, which were earlier hidden under LPS; this leads to proteins being the major antigenic determinants in phase II, unlike LPS in phase I. Consequently, phase I is the smooth virulent form of bacteria isolated from clinical cases while phase II is the rough laboratory-cultivated avirulent strain. This phase variation is due to chromosomal mutation and helps the bacteria to enhance replication attributes in the favourable environment. It is similar to smooth and rough transition reported in other Gram-negative bacteria. As the phases differ antigenically, the serological responses against phase I and phase II differ and are used to differentiate acute and chronic infections. In acute infection, antibodies are directed against phase II antigens, while in chronic infection, antibodies are directed against phase I antigens mainly. In both phases of *Coxiella burnetii*, plasmids (QpH1, QpRS and a 51 kb plasmid) are reported, although their significance in pathogenesis is not known.

Clinical Signs

In cattle, sheep and goats, the infections of *Coxiella burnetii* are usually inapparent and manifestations are usually of reproductive failures such as abortion and metritis. The pathogen is concentrated in uterus and placenta and is shed in the excretions and secretions of abortion and normal parturition. Generally, the abortion occurs once and the pathogen is shed for two successive parturitions in the infected animal. The persistence in the uterus may lead to fibrosis of perivascular and periglandular tissues leading to infertility and repeat breeding. Being the resistance to adverse environmental conditions, the pathogen survives for a long time in the environment.

In humans, majority of *Coxiella burnetii* infections are asymptomatic and are unnoticed and undiagnosed. The disease spectrums are acute and chronic. The average incubation period is 20 days with a range of 14–39 days. In acute cases, the symptoms are non-specific with persistent fever and flu-like signs, while some cases may progress to severe cases including pneumonia and hepatitis. About 1–5% of acute cases may progress to a chronic form with persistent foci of infection in the body. Immunosuppression, pregnancy and heart valve lesions are the predisposing factors. Endocarditis is the most significant outcome among such cases with chronic hepatitis, interstitial pneumonia and repeated abortions as other outcomes (España et al., 2020).

Diagnosis

In Animals

The diagnosis of coxiellosis in animals is carried out either to diagnose a clinical case of abortion or to determine the burden of disease as prevalence in an area. These two aims involve different approaches. In the case of abortion in cattle, sheep and goat, to determine the presence of *Coxiella burnetii*, the abortion materials such as placental cotyledons (at least 3), liver, spleen, lung and stomach contents of aborted foetus are the samples of choice (WOAH, 2022) and following diagnostic methods are employed:

Direct Demonstration of Pathogen *Coxiella burnetii* is shed in great number during abortion or parturition, and the pathogen can be demonstrated directly by staining in the smears of placental cotyledons. Smears of liver, spleen, lung and stomach contents of aborted foetus can also be used in the same pattern. The smears can be stained with any of the Stamp, Gimenez and Macchiavello staining. These methods involve staining by basic fuchsin. Stamp method is the preferred method in veterinary diagnostic laboratories. In the Stamp method, staining is performed by 0.4% basic fuchsin followed by rapid decolourization with 0.5% acetic acid and counter staining by 1% methylene blue. For cell culture method, Gimenez staining is preferred, as it is the fastest method. *Coxiella burnetii* appears as pink thin coccobacilli and are large in number. In staining, *Coxiella burnetii* can be confused with *Chlamydia abortus* and *Brucella* species. Positive staining interpretation, along with positive serological tests, is adequate for the diagnosis of *Coxiella burnetii* in such a case. If staining is negative, testing by PCR is required for a specific diagnosis.

Isolation of the Pathogen *Coxiella burnetii,* being obligate intracellular in nature, requires living systems to grow. The samples collected during abortion are processed to reduce contamination followed by inoculation in embryonated eggs or cell lines. To achieve isolation, the concentration of 10^5 bacteria per ml is recommended. For embryonated egg inoculation, the samples are inoculated in 6–7-day-old embryonated egg via yolk sac route. Yolk sacs are harvested after 10–15 days post-inoculation and are stained to determine the growth of *Coxiella burnetii*. For cell culture, human embryonic lung fibroblasts cell line is used. Inoculation in laboratory animals is preferred for highly contaminated samples. Mice and guinea pigs are the most suitable laboratory animals for *Coxiella* growth. The samples are inoculated intraperitoneally. Isolation of *Coxiella burnetii* requires biosafety level 3 laboratories.

Polymerase Chain Reaction The samples are frequently subjected to PCR for the confirmatory diagnosis of *Coxiella burnetii* in abortion cases. The method is also used in the determination of shedding pattern in various excretions and secretions, although qPCR provides more information than conventional PCR. For biosafety purposes, samples are inactivated by heating at 90 °C for 30–60 min prior to PCR. In conventional PCR, the primers targeted to multicopy insertion sequence IS1111 have been most commonly used. Due to multicopy number, the sensitivity of IS1111 PCR is high. The other targeted genes have been superoxide dismutase gene, 27 kDa outer membrane protein gene (*com1*), heat shock proteins and isocitrate dehydrogenase. For qPCR also, some of these targets have been evaluated.

Serological Tests

Enzyme-Linked Immunosorbent Assay ELISAs have been developed to detect the antibodies in serum and milk samples. Bulk tank milk samples can also be screened by this test. The test has high sensitivity and good specificity and is preferred over indirect immunofluorescence assay and complement fixation test in veterinary sector because of applicability of ELISA at large-scale population and its adaptation on different types of samples. ELISA is widely used in surveillance studies. The kits are available in the market that can detect the mixtures of anti-phase I and anti-phase II antibodies.

Indirect Immunofluorescence Assay (IFA) Indirect immunofluorescence test has found the major applications in human medicine as the test can help in the differentiation of acute and chronic Q-fever infection.

In Humans

As *Coxiella burnetii* does not grow with routine laboratory culture techniques, the diagnosis of Q-fever in humans is frequently done through indirect tests. Among these tests, IFA is the most commonly used test than ELISA being having higher sensitivity of the former. The test can differentiate between acute and chronic Q fever. In acute cases, IgG or IgM antibodies against phase I are elevated. A four-fold rise in titre of IgG or IgM against phase I indicates acute infection, while in persistent localization/chronic disease, IgG antibodies against phase II are elevated. IFA can identify these differences. Isolation and PCR assays can be performed on cerebrospinal fluid, bone marrow, bone biopsies and excised heart valve as discussed earlier.

Control and Prevention

In animals, control and prevention revolve around maintaining hygienic conditions to prevent transmission from the infected animal to other animals. Vaginal shedding in goats is longer than that in the cattle, while in cattle, shedding in milk is more important. Therefore, hygienic conditions during parturitions and post-partum should be followed. Vaccines are available for usage in animals. Vaccination reduces the shedding of the bacterium in vaginal discharges and thus reduces contamination of environment and zoonotic risk. Two vaccines for use in animals are commercially available viz., Chlamyvax (inactivated bivalent vaccine having *Chlamydia abortus* and *Coxiella burnetii* phase II) for sheep and goat, and Coxevac (inactivated nine-mile phase I strain) for cattle and goat. Coxevac is administered in non-pregnant animals and requires annual booster. In animals, treatment of infected animals is not advised (Plummer et al., 2018).

In humans, the disease is treatable with doxycycline as the drug of choice in acute Q-fever, while in chronic Q-fever, supplementation of other drugs such as hydroxychloroquine or fluoroquinolones with doxycycline is prescribed. For humans, the only licensed vaccine is Q-vax, which is a formalin-inactivated vaccine having phase I Henzerling strain of *Coxiella burnetii*; it provides protection for 5 years. The occupationally exposed persons should take proper precautions when handling abortion or normal parturition materials to prevent disease transmission. Milk and milk products should have the proper heat treatment during processing to prevent ingestion route of infection.

References

Eldin, C., Mélenotte, C., Mediannikov, O., Ghigo, E., Million, M., Edouard, S., Mege, J. L., Maurin, M., & Raoult, D. (2017). From Q fever to Coxiella burnetii infection: A paradigm change. *Clinical Microbiology Reviews, 30*(1), 115–190. https://doi.org/10.1128/CMR.00045-16

España, P. P., Uranga, A., Cillóniz, C., & Torres, A. (2020). Q Fever (Coxiella Burnetii). *Seminars in Respiratory and Critical Care Medicine, 41*(4), 509–521. https://doi.org/10.1055/s-0040-1710594

Plummer, P. J., McClure, J. T., Menzies, P., Morley, P. S., Van den Brom, R., & Van Metre, D. C. (2018). Management of Coxiella burnetii infection in livestock populations and the associated zoonotic risk: A consensus statement. *Journal of Veterinary Internal Medicine, 32*(5), 1481–1494. https://doi.org/10.1111/jvim.15229

Sahu, R., Rawool, D. B., Dhaka, P., Yadav, J. P., Mishra, S. P., Kumar, M., Vergis, J., Malik, S. S., & Barbuddhe, S. B. (2021). Current perspectives on the occurrence of Q fever: Highlighting the need for systematic surveillance for a neglected zoonotic disease in Indian subcontinent. *Environmental Microbiology Reports, 13*(2), 138–158. https://doi.org/10.1111/1758-2229.12918

WOAH (World Organization for Animal Health) (2022). Q Fever. Terrestrial manual. Assessed on 1 Aug 2022.

Zoonotic Clostridial Infection

21

Shiv Varan Singh

Abstract

Clostridium species are a group of anaerobic bacteria that can cause various diseases in both animals and humans. Some notable zoonotic Clostridium pathogens include *Clostridium perfringens (*food born zoonosis), *Clostridium botulinum (*food born zoonosis), *Clostridium tetani* (Occupational or environmental zoonosis), *Clostridium septicum* (Neglected tropical zoonosis) *Clostridium novyi* (Neglected tropical zoonosis), and *Clostridium difficile* (Direct zoonosis). These bacteria are known for their ability to produce potent toxins, leading to severe health issues such as food poisoning, gastrointestinal diseases, and other infections. Among all above zoonotic species of *Clostridium*, the *Clostridium difficile* has emerged as potent zoonotic pathogen, which involves primarily with healthcare settings, particularly among hospitalized patients who had received antibiotics.

Keywords

Clostridium botulinum · Sausage poisoning · Perfringens

History

Clostridium perfringens was earlier classified as *Bacillus welchii,* isolated from gas gangrene in the year 1891 by Dr. Theobald Smith Welch. After the identification of bacterium, it was reclassified into the genus *Clostridium* due to its spore-forming and anaerobic properties. The marked tissue-necrotizing pathogenicity of this

S. V. Singh (✉)
Department of Veterinary Microbiology, College of Veterinary and Animal Sciences, Bihar Animal Sciences University, Kishanganj, Bihar, India

R. Biswas et al. (eds.), *Concept of One Health and Zoonoses*,
https://doi.org/10.1007/978-981-95-6627-3_21

organism is reflected in the name "perfringens," which is derived from the Latin term meaning "to break through." Several decades ago, the zoonotic potential of C. perfringens was also documented. The role of *C. perfringens,* especially in animals (Livestock's diseases), has been recognized initially; later on, the importance of foodborne illness in humans was identified. Various outbreaks happened due to improperly cooked or contaminated meat products.

Understanding botulism, a severe neuroparalytic disease brought on by the toxin produced by the bacteria *Clostridium botulinum*, is closely linked to the history of this bacterium. The late eighteenth century and the early nineteenth century saw the start of the first organized investigation into botulism. Interestingly, between 1817 and 1822, German physician and poet Justinus Kerner wrote in-depth accounts of "sausage poisoning." Although he was unable to pinpoint the cause, he noted symptoms like muscle paralysis and dilated pupils, calling it "sausage poison." A notable outbreak happened in Ellezelles, Belgium, in 1895, following a funeral supper that included smoked ham.

This event was a turning point in microbiology as Professor Emile Pierre van Ermengem looked into and eventually isolated *C. botulinum.* Instead of focusing on the bacterium's physical traits, he named it for its association with sausages (from the Latin "botulus," meaning sausage). In 1793, the first foodborne botulism outbreak was reported in Belgium and was attributed to inadequately canned food. Because of this incidence, people became aware of the risks that *C. botulinum* posed.

In 1935, Hall and O'Toole isolated *C. difficile* for the first time from a healthy infant's feces. The term "*difficile*" refers to the difficulties that scientists encountered at the time when cultivating and isolating this organism. Only in 1978, after research showing *C. difficile*'s function in pseudomembranous colitis was published, did George et al. clearly relate the bacteria to human illness and identify it as the main cause of antibiotic-associated diarrhea.

Its significance as the causal agent of tetanus was confirmed in 1889 when Kitasato Shibasaburo isolated *C. tetani* from a human patient. He proved that when this bacterium was given to animals, it created a strong toxin that might cause illness. In the case of tetanus, the human–animal interaction may be traced back to veterinary records that document epidemics in horse populations. These outbreaks frequently operate as warning signs for possible human cases, especially in agricultural settings where people could come in touch with contaminated environments or diseased animals.

In 1877, Louis Pasteur and Jean-Joseph Joubert identified *Clostridium septicum*, also referred to as the "vibrion septique," in cow's blood. Robert Koch demonstrated in 1881 that *C. septicum* was caused by gas gangrene, formerly termed as malignant edema. During World Wars I and II, *C. septicum*'s zoonotic potential was identified as a primary cause of gas gangrene in soldiers.

Frederick Novy, who coincidentally discovered the anaerobic bacteria that causes septicemia in rabbits, is honored by the name *Clostridium novyi.*

Taxonomy and Etiology

The phylum Firmicutes includes the genus *Clostridium*, which is distinguished by its wide collection of Gram-positive, obligatory anaerobic bacteria. There are 164 species in this genus that have been legitimately published as of October 2022. *Clostridium* species live in soils and in the digestive tracts of animals, including humans. Some of the more significant species have the ability to spread zoonotic disease.

Clostridium perfringens, a Gram-positive, rod-shaped, anaerobic, spore-forming pathogenic bacterium belonging to the genus *Clostridium*, was originally known as *C. welchii* or *Bacillus welchii*. The technical term, perfringens, refers to the tissue rupture that takes place during gas gangrene and is derived from the Latin per (meaning "through") and frango ("burst"). Tissue necrosis, bacteremia, emphysematous cholecystitis, and gas gangrene, commonly referred to as clostridial myonecrosis, are all signs of *C. perfringens* infections.

The Gram-positive, rod-shaped, anaerobic, spore-forming, motile bacteria *Clostridium botulinum* is capable of producing the neurotoxin known as botulinum toxin, which is divided into four groups, *C. botulinum* groups I–IV. In humans and other animals, botulism is a severe flaccid paralytic sickness that can be brought on by botulinum toxin. *C. botulinum* causes wound botulism (infection of a wound by *C. botulinum*), baby botulism (intestinal infection with toxin-forming *C. botulinum*), and foodborne botulism (ingesting a preformed toxin).

The rod-shaped, Gram-positive bacterium *Clostridium tetani* is normally up to 0.5 μm broad and 2.5 μm long. The different flagella that envelop its body allow it to move. Tetanus is caused by *C. tetani*, (Oladerian et al., 2002) which is unable to grow in the presence of oxygen.

Gram-positive, endospore-forming, obligate anaerobic bacteria belong to the *Clostridium novyi* (oedematiens) class. It is present in both soil and excrement, making it ubiquitous. It is harmful, resulting in a wide range of illnesses in both people and animals. Although less frequent than *C. perfringens*, *C. novyi* has long been linked to gas gangrene associated with wounds.

Gram-positive *Clostridium septicum* is an obligatory anaerobic bacteria that forms spores. By releasing exotoxins, *Clostridium septicum* can result in gas gangrene and myonecrosis.

In addition to producing severe diarrheal infections, the bacterium *Clostridium difficile*, often known as *Clostridium difficile*, has the potential to cause colon cancer. A Gram-positive spore-forming bacterium is *C. difficile*. Anaerobic, motile bacteria, *Clostridium* species (now classified as *Clostridium*) are found throughout nature, but they are particularly common in soil. Its pleomorphic, rod-shaped vegetative cells are found in short chains or pairs. It has drawn a lot of attention since it is a prominent cause of pseudomembranous colitis and diarrhea linked to antibiotics.

Epidemiology

It was discovered that the prevalence of *C. perfringens* was 25% (70/280) in people with a history of diarrhea and 23.2% (36/155) in people in good health. Toxin type A was the most common genotype identified by genotyping isolates of *C. perfringens*. Foodborne outbreaks of *C. perfringens* were identified during the children's Panhellenic Handball Championship.

In India, the prevalence of *Clostridium difficile* (CDI) ranged from 1.2% to 29%.

Since 1973, botulism cases in the United States have been tracked by the Centers for Disease Control and Prevention (CDC). Between 2011 and 2015, an average of 162 cases per year were reported, with newborn botulism accounting for 71–88% of these cases. Foodborne botulism, wound botulism, and botulism of unknown origin were the next most common types. Types A through D were identified in the positive samples, and fresh retail fish had an overall incidence of 19% (13/67). The incidence of *C. botulinum* in demersal fish was 21% (5/24) of the samples, while it was 18% (7/39) in pelagic fish; 25% (1/4) of shrimp had *C. botulinum*, and the positive sample contained type D.

According to estimates, there are one million cases of tetanus worldwide each year. Depending on the availability of well-equipped intensive care centers, the case fatality rate might vary from 6% to 72%.

Even though *C. septicum* infections are less common than those caused by other clostridial species, such as *Clostridium perfringens*, epidemiological data show that they can nevertheless be quite harmful in some situations, particularly when paired with other infectious agents like STEC (Talukdar et al., 2017).

Transmission and Zoonotic Aspect

There are usually multiple ways that zoonotic Clostridium species are spread:

Direct Contact Direct contact with infectious animals or their surroundings can expose humans to diseases.

Foodborne Transmission A common cause of *Clostridium* infections is eating undercooked or tainted animal items.

Environmental Exposure Both humans and animals may be exposed to these bacteria if they are found in contaminated soil or water sources.

Neglected Tropical Zoonosis Certain Clostridial species infections arise as a result of early preventive and treatment neglect.

The overuse of antibiotics in animals is a common agricultural practice associated with the *C. difficile* strain. In addition to causing environmental persistence and the possibility of human transmission through food or direct contact with infected surfaces, this practice aids in the evolution of resistant strains that can contaminate meat products and agricultural runoff.

The epidemiology of C. difficile infections (CDI) has been shown to involve both direct zoonotic transmission (from animals to people) and indirect transmission (by environmental pollution). The idea that CDI should be viewed within a One Health framework—a concept that emphasizes the interdependence of environmental variables, animal health, and human health—is supported by the finding of certain sequence types linked to both human infections and animal carriers.

Pathogenesis and Virulence Factor (Toxins)

Clostridium Perfringens

Certain husbandry and environmental conditions, such as sudden feeding changes, typically to a richer diet, overeating, and voracity on high-protein and energy-rich feeds, frequently cause the toxins. As a result, peristalsis is slowed down, germs are retained in the colon, toxins are absorbed, carbohydrates are not properly digested, and a rich medium is created for the growth of *C. perfringens*. In healthy animals, the bacterium lives in the large intestine, but if it grows too much, it can spread to the small intestine and produce a lot of toxins and enteric toxicity.

Alpha Toxin (CPA) CPA, a zinc-containing 370 amino acid phospholipase C enzyme, is produced by almost all *C. perfringens* isolates. It is composed of a central loop domain with a ganglioside GM1a binding site, a catalytic N-domain made of α-helices, and a membrane-binding C-domain made of a β-sheet.

Beta Toxin (CPB) Isolates of *C. perfringens* types B and C generate CPB, a 336 amino acid pro-toxin with a 27-amino acid signal sequence that is broken down during secretion. A mature toxin of about 35 kDa is the outcome. The amino acid sequence of CPB and the β-pore-forming toxins (PFTs) of *Staphylococcus aureus* are comparable. CPB is a member of the α-hemolysin family.

***C. perfringens enterotoxin* (CPE)** All type F strains and certain type C, D, and E strains produce CPE, a 35 kDa single polypeptide. Although this toxin does not share a fundamental amino acid sequence with other toxins, it is structurally related to the aerolysin β-PFT family. The CPE protein mediates oligomerization and membrane insertion during pore formation through its N-terminal cytotoxicity domain and C-terminal receptor-binding domain.

Epsilon Toxin (ETX) ETX is categorized as a β-PFT of the aerolysin family since it shares structural similarities with the aerolysin toxin produced by *Aeromonas* sp. and is only produced by type B and type D isolates. The ETX receptor is still somewhat unknown; however, the myelin and lymphocyte (MAL) protein is becoming a more promising receptor candidate. Following binding, ETX creates a heptameric prepore, and then a β-barrel is inserted to facilitate effective pore formation.

Iota Toxin (ITX) Only strains of *C. perfringens* type E produce ITX, a binary toxin made up of an enzyme component (Ia) and a binding component (Ib). Ib is first generated as an inactive toxin of about 100 kDa, and in order to make an active Ib, a ~ 20 kDa N-terminal fragment must be removed by trypsin or chymotrypsin. Additionally, these proteases create an active form of the Ia precursor by cleaving short peptides (9–13 amino acid residues) from its N-terminus. Furthermore, some strains of *C. perfringens* type E may also activate ITX through the production of λ-protease.

Necrotic Enteritis B-like Toxin (NetB) A single-chain protein weighing 33 kDa, NetB belongs to the α-hemolysin family of β-PFTs. Among the pore-forming toxins with which NetB shares homology are *C. perfringens* delta toxin, with 40% identity, and CPB and *S. aureus* alpha hemolysin, with 38% and 31% similarity, respectively. The β-sandwich, latch, rim, and pre-stem domains are the four domains that the NetB monomer shares with other members of the α-hemolysin PFT family (Uzal et al., 2018).

Classification of current toxinotyping of *C. perfringens*

Toxinotype	Toxin produced					
	CPA	CPB	ETX	ITX	CPE	NetB
A	+	–	–	–	–	–
B	+	+	+	–	–	–
C	+	+	–	–	±	–
D	+	–	+	–	±	–
E	+	–	–	+	±	–
F	+	–	–	–	+	–
G	+	–	–	–	–	+

Note: Positive = +, Negative = –, Variable = ±

Clostridium Difficile

Two major toxins, toxin A (TcdA) and toxin B (TcdB), have been identified as the main virulence agents in *C. difficile*. These toxins cause intestinal cell damage and inflammation by interfering with cellular functions in the host tissues. Compared to TcdA, TcdB has been shown to be more effective in damaging epithelium.

Two key mechanisms have been recognized for action of TcdA and TcdB:

- These are glucosyltransferases that cause tiny Rho GTPases to become irreversibly inactive, which disrupts tight junctions and the cytoskeleton and causes cell rounding, detachment, and death.

- By disrupting the intestinal epithelial barrier, inducing proinflammatory mediators and cytokines, causing cell apoptosis or necrosis in immune and epithelial cells, and contributing to mucosal damage, the toxins themselves or their synergistic interactions with other mediators cause intestinal injury and inflammation.

C. septicum

Gas gangrene is a severe form of myonecrosis that is known to be caused by *C. septicum*. Infections generated by *C. septicum*, in contrast to other *Clostridium* species like *C. perfringens*, are frequently developed by hematogenous dissemination from the gastrointestinal tract rather than requiring damage at the infection site. Alpha toxin, one of the exotoxins the bacterium generates, is the main cause of hemolysis and tissue necrosis.

Alpha toxin (AT) is the only fatal factor produced by *Clostridium septicum*. It is a cytolytic protein with a molecular mass of about 48 kDa. It was discovered that the 48 kDa toxin was an inert protoxin (ATpro), which could be made active by trypsin cleaving its carboxy-terminal. The distance between the carboxy-terminus and the cleavage site was roughly 4 kDa. The particular activity of ATpro that was proteolytically activated was roughly 1.5 × 10 (6) hemolytic units mg^{-1}.

Trypsin-activated toxin (ATact) was hemolytic; it caused erythrocytes to discharge potassium ions prelytically, followed by hemoglobin; it caused planar membranes to form channels; and it aggregated on erythrocyte membranes to form a complex of M(r) > 210,000. These characteristics were not present in ATpro. ATact created holes that were at least 1.3–1.6 nm in diameter. We propose that alpha toxin's cytolytic action is caused by pore development on target cell membranes.

C. novyi

A rare pathogen that affects both people and animals is *C. novyi*. *C. novyi* is extremely difficult to grow due to its great oxygen sensitivity, which may help explain why it is rarely linked to infections. Similar to *P. sordellii*, *C. novyi* has been linked to a few instances of gas gangrene connected with wounds that have caused fatal infections of deep soft tissue. Because of the dark staining of subcutaneous tissue brought on by extreme blood vessel congestion, *C. novyi* has also been linked to hepatic damage and infectious necrotic hepatitis in animals, commonly referred to as black illness. Typically, subcutaneous edema, bleeding, and liver necrosis accompany infectious necrotic hepatitis.

More recently, it was shown that *C. novyi* is the cause of serious soft tissue infections among injectable drug users. Leukocytosis, edema, and refractory hypotension—all of which can result in necrotizing infection—are clinical characteristics of

C. novyi infection in injectable drug users that are comparable to those of *P. sordellii* infection. *C. novyi* types A and B generate TcnA. Infections in injection drug users and gas gangrene in people and animals are linked to *C. novyi* type A. Animal necrotic enteritis is linked to *C. novyi* type B.

Because of its deadly and edematizing effects in both in vitro and in vivo animal models, TcnA is thought to be the primary virulence factor in *C. novyi* type A and B infections. The tcnA gene has been utilized to confirm *C. novyi* type B infection in livestock with necrotic hepatitis. It is interesting to note that individuals with *P. sordellii*-induced toxic shock syndrome have extremely similar clinical manifestations to those of *C. novyi* infection in injectable drug users, indicating that toxins with comparable biochemical and biological activity may underlie both conditions.

Clostridium Botulinum

Foodborne botulism is caused by *C. botulinum,* which is an anaerobic bacterium, meaning it can only grow in the absence of oxygen. When *C. botulinum* develops and generates toxins in food before it is consumed, foodborne botulism results. Spores produced by *C. botulinum* are found in soil, rivers, and seawater, among other places.

Infant botulism occurs mostly in infants under 6 months of age. Infants contract botulism when they consume *C. botulinum* spores, which develop into bacteria that colonize the stomach and release toxins. This is distinct from foodborne botulism, which is brought on by consuming pre-formed toxins in food.

Wound botulism is rare and occurs when the spores get into an open wound and are able to reproduce in an anaerobic environment. Though they could take up to 2 weeks to manifest, the symptoms are comparable to those of foodborne botulism.

Inhalation botulism is rare and does not occur naturally; for example, it is associated with accidental or intentional events (such as bioterrorism), which result in the release of toxins in aerosols. The clinical manifestation of inhalation botulism is comparable to that of foodborne botulism. According to estimates, the median lethal dose of botulinum toxin for humans is 2 nanograms per kilogram of body weight, which is around three times higher than in foodborne instances.

Waterborne botulism could theoretically result from the ingestion of the preformed toxin. However, as standard water treatment techniques (such as boiling, disinfection with 0.1% hypochlorite bleach solution) remove the toxin, the danger is considered low.

The two categories of *C. botulinum* strains are the proteolytic group, which includes those that break down proteins, and the nonproteolytic group, which includes those that do not. *C. botulinum* type A and a few types B and F are members of the proteolytic group. Some of *C. botulinum* types B and F, as well as type E, are members of the non-proteolytic group.

BoNTs are a diverse family of proteins that are generated by *C. botulinum* and are known as toxins. BoNTs are synthesized as a single-chain polypeptide (about 150 kDa) that is broken down by a protease into dichain proteins connected by a disulfide bond. They have a similar structure. The 50 kDa light chain (LC) of the dichain protein has zinc-dependent protease activity, whereas the 100 kDa heavy chain (HC) has a C-terminal cell-binding domain (Hc) and an N-terminal translocation domain (Hn). Using the mouse biological test and certain neutralizing antisera, BoNTs are separated into nine toxinotypes (A through H and X) according to the neutralization of toxicity by those antisera. The 25 kD-Synaptosomal-associated protein (SNAP-25) is cleaved by the BoNT/A. The synaptobrevin, also known as the vesicle-associated membrane protein, is cleaved by the BoNT/B (Tiwari & Nagalli, 2021; Meurens et al., 2023)

The mode of action is the same for all forms of BoNT. Cell binding, internalization, membrane translocation of the light chain into the cytosol following disulfide bond reduction, and enzymatic cleavage of their target soluble N-ethylmaleimide-sensitive fusion protein attachment protein receptor (SNARE) are the four phases in this process.

Clostridium Tetani

The lower motor neurons, which are typically in charge of triggering voluntary muscle contractions, allow the tetanus toxin to enter the nerve terminals. The axons subsequently carry it to the brainstem and spinal cord. The poison then trans-synaptically enters the nerve terminals that produce inhibitory neurotransmitters after passing through the nervous system. Muscle rigidity and spasm may arise from a disruption in the inhibition of lower motor neurons caused by a blockage in this vesicular mechanism. By acting as a zinc-dependent metalloproteinase that targets VAMP, a protein that controls the release of neurotransmitters from nerve endings, tetanus toxin prevents normal neurological activity (Hanif et al., 2015; Lindström, & Korkeala, 2006; Marulappa et al., 2012).

Clinical Symptoms (Human and Animal) (Table 21.1)

Table 21.1 Important zoonotic diseases, their etiological agents, hosts, major symptoms, and mode of transmission

Sl. no	Etiological agent	Main toxin/causative factor	Human	Animal	Type of transmission
1.	*Clostridium perfringens* type C	CPB	Enteritis necroticans in children and adults with trypsin deficiency	Necrohemorrhagic enteritis of neonatal pigs, sheep, goats, horses, others	Foodborne zoonosis
2.	*Clostridium perfringens* type A	CPA	Very less role in human enteric disease[a]	Suggested to be involved in enterocolitis in pigs and horses, enterotoxemia in sheep and cattle and abomasitis of cattle, sheep and goats Definitive proof lacking	Foodborne zoonosis
3.	*Clostridium perfringens* type F	CPE	Food poisoning, antibiotic-associated diarrhea, sporadic diarrhea	Suggested to be involved in canine hemorrhagic gastroenteritis and colitis of horses	Foodborne zoonosis
4.	*Clostridium difficile*	TcdA, TcdB, CDT	Pseudomembranous and hemorrhagic colitis	Enterocolitis in horses, pigs, rabbits, hamsters, others	Direct zoonosis
5.	*Clostridium botulinum*	Toxins A and B (humans), BoNT1 & BoNT 2 toxin C (animals)	Infant botulism (toxicoinfection)	Possible toxico-infection in horse, cattle, and poultry	Foodborne zoonosis
6.	*Clostridium tetani*	TeNT(Tetanolysin & Tetanospasmin)	Tetanus	Tetanus	Occupational or environmental zoonosis
7.	*Clostridium septicum* (Orrell & Melnyk, 2021	Alpha-toxin and aerolysin	Gas gangrene and malignancy (especially in colon)	Cattle-malignant edema Sheep-Braxy chicken-gangrenous dermatitis	Neglected tropical zoonosis
8.	*Clostridium novyi* A	Alpha, gamma, delta, epsilon	Gas gangrene	Sheep-big head of rams Cattle & sheep- gas gangrene	Neglected tropical zoonosis
9.	*Clostridium novyi* B	Alpha, beta, zeta	Myocarditis, necrotizing hepatitis and gas gangrene	Sheep, cattle-black disease (necrotic hepatitis)	Neglected tropical zoonosis

[a]*C. novyi* type D is also known as *Clostridium haemolyticum*. However, *C. haemolyticum* and *C. novyi* are currently considered to be distinct species

Diagnosis

Isolation and Identification Depending on the suspected site of infection, get the proper clinical samples (tissue, exudate, wound, and stool).

Pre-Enrichment To enable the initial growth of *Clostridium* species, inoculate the sample into a non-selective anaerobic broth, such as fluid thioglycolate broth. Anaerobic incubation at 37 °C for a predetermined amount of time..

Selective Isolation Apply the enriched broth to selective agar plates, like Cycloserine cefoxitin fructose agar (CCFA), which is frequently used to isolate Clostridium difficile, in order to prevent the growth of other bacteria while permitting *Clostridium* to thrive. *Clostridium perfringens* can be identified using lecithinase activity on **reinforced clostridial agar** with egg yolk. Clostridial species can be grown in laboratories using Robertson's cooked beef media as a selective anaerobic medium.

Cultural and Staining Characteristics Check the plates for distinctive *Clostridium* colonies on blood agar, which are frequently big, spherical, and encircled by a hemolysis zone. Gram staining is used to verify the existence of rod-shaped, Gram-positive bacteria, which may have a single endospore. There are species that might be Gram variable.

Species Identification The different species of *clostridium* can be identified on the basis of endospore location.

Clostridium Difficile

1. Stool culture: Although the test has the highest sensitivity (90–100%; specificity: 84–100%), its results are delayed and, if used alone, could cause a delay in the diagnosis.
2. Glutamate dehydrogenase enzyme immunoassay (EIA): This test is extremely sensitive (sensitivity: 85–100%; specificity: 87–98%); it finds glutamate dehydrogenase, which is generated by *C difficile.*
3. Real-time polymerase chain reaction (PCR) assay: With sensitivity of 86% and specificity of 97%, this test is a gold standard substitute for stool culture and can be used to identify the *C difficile* gene toxin.
4. Stool cytotoxin test: When a cytopathic effect is shown to be negated by a particular antiserum (sensitivity: 70–100%; specificity: 90–100%), the test is considered positive.
5. EIA for detecting toxins A and B: Most labs utilize this test because of its great specificity (98%) and moderate sensitivity (79–80%).

Table 21.2 Diagnosis summary

Sr. No.	Genus/ Species	Cultural characteristic/ Morphology	Media used	Endospore/ Spore	Animal Inoculation	Important diagnostic test (Biochemical/Enzymatic)
1.	*Clostridium tetani*	Strict anaerobe	(**a**) Horse blood agar (alpha hemolysis to beta hemolysin /tetanolysin) (**b**) Thioglycolate medium or cooked meat broth (enrichment media)	Drumstick spores	Mice (S/C or i/m)	Neutralization/Protection test
2.	*Clostridium botulinum*	Strict anaerobe/ semitransparent with fimbriate border	(**a**) Horse blood agar (narrow zone of hemolysis)/ sheep blood agar (beta hemolysis) (**b**) Thioglycolate medium or cooked meat broth (enrichment media)	Spores present	Mice (Wasp waist appearance if toxin present) (i/v-0.3 ml or i/p-0.5 ml of serum exudates from animals)	Toxin demonstration in mice
3.	*Clostridium perfringens*	Anaerobes but microaerophilic, capsulated and non-motile	(**a**) Thioglycolate medium or cooked meat broth (enrichment media) (**b**) Rabbit/sheep blood agar (*double-zone hemolysis*)	Spores present	Mice (0.4 ml of ileal contents of dead animal inoculate i/v) dies in between 1.5 hr. and 10 hr. minute	Reverse CAMP reaction/ Nagler's reaction/toxin demonstration in mice and stormy clot
4.	*Clostridium septicum*	Cigar-shaped rods and citron forms are more common	Strict anaerobe	Spores are oval, central, or subterminal	Guinea pigs (1.2 ml culture fluid +0.3 ml f specific antitoxin for 30 min after i/p inoculation)	Toxin neutralization test in animal and stormy clot/ stormy fermentation reaction
5.	*Clostridium* novyi	Anaerobes	(**a**) Thioglycolate medium or cooked meat broth (enrichment media) (**b**) Sheep blood agar	Subterminal spore	Guinea pig (i/m homogenates) die in 1–2 days due to subcutaneous edema	Toxin demonstration in mice

6. Latex agglutination technique: Another method for identifying glutamate dehydrogenase, however, with a low sensitivity (48–59%) and a high specificity (95–96%) (Vonberg et al., 2008; Dieterle et al., 2019) (Table 21.2).

Treatment

Antibiotic Resistant and Current Scenario

Many antibiotics, such as *gentamicin* (100%), *tetracycline* tetracycline (56.2%), *lincomycin* (100%), and *erythromycin* (100%), can cause resistance in *Clostridium perfringens*. Additionally, aquatic sources of clindamycin and ceftriaxone have been documented to cause resistance (Song, et al., 2024).

Penicillin, Piperacillin, Meropenem, Ampicillin, Chloramphenicol, and Levofloxacin are among the antibiotics that *C. perfringens* is susceptible to (Tansuphasiri et al., 2005).

Numerous antibiotics, including aminoglycosides, lincomycin, tetracyclines, erythromycin, clindamycin, penicillins, cephalosporins, and fluoroquinolones, are ineffective against *Clostridium difficile* (CD).

The Infectious Diseases Society of America (IDSA) advises using penicillin and clindamycin because *C. septicum* exhibits modest resistance to vancomycin in vitro.

There is not much information about antibiotic resistance in *Clostridium novyi.*

C. botulinum is susceptible to third-generation cephalosporins, including Ampicillin, Penicillin, Amoxicillin-clavulanic acid, Cefuroxime, Cefotaxime, Ceftriaxone, Cefepime, Moxifloxacin, and Gentamicin, but resistant to aminoglycosides, sulfamethoxazole, and trimethoprim (Mazuet et al., 2016; Cordioli et al., 2024).

Metronidazole, Tetracycline, Macrolides, Clindamycin, Cephalosporin, Chloramphenicol, Erythromycin, and Penicillin G are antibiotics that can cure tetanus. In addition to benzodiazepines, muscle relaxants, tetanus immune globulin, and wound debridement, metronidazole is typically regarded as the first-line treatment for tetanus.

Vaccine Availability

The vaccines are available in different types, including whole formalin-inactivated vaccines, bacterin-toxoid vaccines, and toxoid vaccines.

Whole Formalin-Inactivated Vaccines Vaccine is achieved by culture of *C. chauvoei* strains and it is blackleg vaccine.

Bacterin-Toxoid Vaccines
Bacterin-toxoid vaccine is produced by culture of *C. perfringens* types D, B, and C strains.

For black disease vaccine, bacterin-toxoid vaccine has been produced by culture of *C. novyi* strains in glass bottles using complex media.

Vaccine against *C. septicum*, the stains are cultured in anaerobic conditions in a fermenter using complex media.

Bacterin-toxoid vaccines have good efficiency due to the bacterial cell compartments leading to a high immune response against all the infection-associated antigens.

Toxoid Vaccines Lovland and his colleagues generated a crude toxoid, or formaldehyde-inactivated *C. perfringens* (Abdolmohammadi Khiav, L., & Zahmatkesh 2021) supernatant, and shown that *C. perfringens* types A, C, or a combination of type A and C crude toxoids are highly effective in preventing infection.

Control Measures

A comprehensive approach is essential to prevent the spread of *Clostridium* spores, which can survive on surfaces for extended periods of time. To control *Clostridium* infections, important measures include strict antibiotic stewardship to minimize unnecessary antibiotic use, thorough environmental cleaning with bleach-based disinfectants, appropriate isolation precautions for infected patients, and meticulous hand hygiene with soap and water.

Control Measures at Food Production System

Clostridium can be managed in food production using physical, chemical, and biological methods. Additionally, you can properly refrigerate food and maintain acceptable temperatures.

Physical Treatments

Heating: Cook food to a safe temperature, especially meat and poultry
Irradiation: Use gamma rays to treat food products
High hydrostatic pressure (HHP): Use HHP to pasteurize food
Refrigeration: Keep food at or below 40 °F (4 °C) (Munir et al., 2023)

Chemical Treatments

Nitrite: Use nitrite to cure meat products. However, nitrite can have carcinogenic effects.
Organic acid: Use organic acids to treat food.
Natural antimicrobial compounds: Use natural antimicrobial compounds to treat food.

Biological Treatments

Bacteriophages: Use bacteriophages to treat food

Enzymes: Use enzymes to treat food

Control Measures at Food Processing System

Proper cooking, time-temperature management, and cleanliness procedures are all ways to prevent Clostridium from growing in food processing systems.

Proper Cooking

Meat and poultry should be cooked to their recommended safe internal temperatures, and leftovers should be reheated to a minimum of 165 °F (74 °C), to effectively eliminate pathogenic microorganisms and reduce the risk of foodborne illness.

Time-Temperature Control

Keep hot food above 140 °F (60 °C) and cold food below 40 °F (4.4 °C)

Besides the above precaution and preventive measures, safe food handling, washing of hands before, during, and after preparing food with use of separate cutting boards and utensils for raw and cooked foods are necessary. Use a food thermometer to check the temperature of food for prevention of spoiling of food due to temperature variation.

References

Abdolmohammadi Khiav, L., & Zahmatkesh, A. (2021). Vaccination against pathogenic clostridia in animals: A review. *Tropical Animal Health and Production, 53*(2), 284.

Aldape, M. J., Bayer, C. R., Rice, S. N., Bryant, A. E., & Stevens, D. L. (2018). Comparative efficacy of antibiotics in treating experimental Clostridium septicum infection. *International Journal of Antimicrobial Agents, 52*(4), 469–473.

Cordioli, B., Rizzardi, A., Guolo, A., Ferro, T., Bacchin, C., Garbuio, M., et al. (2024). Antimicrobial susceptibility of Clostridium botulinum group III field strains isolated in Europe from animal outbreaks. *Veterinary Microbiology, 298*, 110281.

Diab, M. (1999). *Lexicon orthopaedic etymology* (p. 128). CRC Press. ISBN 9789057025976.

Dieterle, M. G., Rao, K., & Young, V. B. (2019). Novel therapies and preventative strategies for primary and recurrent Clostridium difficile infections. *Annals of the New York Academy of Sciences, 1435*(1), 110–138.

Hajra Hanif, H. H., Anjum, A., Naeem Ali, N. A., Asif Jamal, A. J., Muhammad Imran, M. I., Bashir Ahmad, B. A., & Ali, M. I. (2015). Isolation and antibiogram of Clostridium tetani from clinically diagnosed tetanus patients. *American Journal of Tropical Medicine and Hygiene, 93*(4), 752–756. https://doi.org/10.4269/ajtmh.15-0040

Lindström, M., & Korkeala, H. (2006). Laboratory diagnostics of botulism. *Clinical Microbiology Reviews, 19*(2), 298–314. https://doi.org/10.1128/cmr.19.2.298-314.2006. PMC 1471988. PMID 16614251.

Maczulak, A. (2011). Clostridium. In *Encyclopedia of microbiology* (pp. 168–173). Facts on File. ISBN 978-0-8160-7364-1.

Marulappa, V. G., Manjunath, R., Babu, N. M., & Maligegowda, L. (2012). A ten year retrospective study on adult tetanus at the epidemic disease (ED) hospital, Mysore in southern India: A review of 512 cases. *Journal of Clinical and Diagnostic Research: JCDR, 6*(8), 1377.

Mazuet, C., Yoon, E. J., Boyer, S., Pignier, S., Blanc, T., Doehring, I., et al. (2016). A penicillin- and metronidazole-resistant Clostridium botulinum strain responsible for an infant botulism case. *Clinical Microbiology and Infection, 22*(7), 644–6e7.

Meurens, F., Carlin, F., Federighi, M., Filippitzi, M. E., Fournier, M., Fravalo, P., et al. (2023). Clostridium botulinum type C, D, C/D, and D/C: An update. *Frontiers in Microbiology, 13*, 1099184.

Munir, M. T., Mtimet, N., Guillier, L., Meurens, F., Fravalo, P., Federighi, M., & Kooh, P. (2023). Physical treatments to control Clostridium botulinum hazards in food. *Food, 12*(8), 1580.

Oladerian, I., Meir, D. E., Ojelade, A. A., Olaolorn, D. A., Adeniran, A., & Tarpley, J. L. (2002). Tetanus- a continuing problem in the developing world. *World Journal of Surgery, 26*(10), 1282–1285. https://doi.org/10.1007/s00268-002-6497-z

Orrell, K. E., & Melnyk, R. A. (2021). Large clostridial toxins: Mechanisms and roles in disease. *Microbiology and Molecular Biology Reviews, 85*(3), 10–1128.

Peck, M. W. (2009). Biology and genomic analysis of Clostridium botulinum. *Advances in Microbial Physiology, 55*, 183–265, 320. https://doi.org/10.1016/S0065-2911(09)05503-9. ISBN 978-0-12-374790-7. PMID 19573697.

Song, X., Zhong, Z., Bai, J., Pu, T., Wang, X., He, H., et al. (2024). Emergence of genetic diversity and multi-drug resistant Clostridium perfringens from wild birds. *BMC Veterinary Research, 20*(1), 300.

Talukdar, P. K., Udompijitkul, P., Hossain, A., & Sarker, M. R. (2017). Inactivation strategies for Clostridium perfringens spores and vegetative cells. *Applied and Environmental Microbiology, 83*(1), e02731–e02716.

Tansuphasiri, U., Matra, W., & Sangsuk, L. (2005). Antimicrobial resistance among Clostridium perfringens isolated from various sources in Thailand. *Southeast Asian Journal of Tropical Medicine and Public Health, 36*(4), 954–961.

Tiwari, A., & Nagalli, S. (2021). Clostridium Botulinum. In *StatPearls*. StatPearls Publishing. PMID 31971722. Retrieved 2021-09-23.

Uzal, F. A., Navarro, M. A., Li, J., Freedman, J. C., Shrestha, A., & McClane, B. A. (2018). Comparative pathogenesis of enteric clostridial infections in humans and animals. *Anaerobe, 53*, 11–20.

Vonberg, R. P., Kuijper, E. J., Wilcox, M. H., Barbut, F., Tüll, P., Gastmeier, P., et al. (2008). Infection control measures to limit the spread of Clostridium difficile. *Clinical Microbiology and Infection, 14*, 2–20.

Part III

Viral Zoonoses

Rabies

22

Ripan Biswas and Indranil Samanta

Abstract

Rabies continues to pose a major global public health issue, especially in developing countries, as it is the deadliest viral zoonotic disease. The rabies virus predominantly targets mammals, with dogs being the main carriers and sources of infections in humans. Rabies is most often transmitted to humans through the bite of an animal that is infected with the virus. However, other modes of transmission, such as scratches or exposure to contaminated saliva, are also possible. The virus targets the central nervous system, leading to a range of neurological symptoms, including increased anxiety, restlessness, sensory disturbances, and difficulties with movement. Once clinical manifestations emerge, the rabies virus invariably culminates in a fatal outcome. Administering post-exposure prophylaxis promptly, which includes cleaning the wound and providing rabies vaccine and immunoglobulin, has been shown to be a highly effective method for preventing the disease from advancing. Interventions to control the spread of rabies emphasize vaccinating domestic dog populations, promoting public education on rabies prevention strategies, and guaranteeing availability of post-exposure prophylaxis for individuals who have been exposed to the virus. The elevated expenses associated with rabies treatment in certain areas highlight the necessity for readily available and economical post-exposure prophylaxis to avert avoidable mortalities from this preventable malady.

R. Biswas (✉)
Department of Veterinary Public Health and Epidemiology, West Bengal University of Animal and Fishery Sciences, Kolkata, West Bengal, India

I. Samanta
Department of Veterinary Microbiology, West Bengal University of Animal and Fishery Sciences, Kolkata, West Bengal, India

R. Biswas et al. (eds.), *Concept of One Health and Zoonoses*,
https://doi.org/10.1007/978-981-95-6627-3_22

Keywords

Rabies · Hydrophobia · Deadliest zoonosis · Lyssa

History

Rabies virus is a zoonotic pathogen with both wild and domestic animal reservoirs. Rabies is among the oldest and most fatal infectious diseases known to humanity, with its origins tracing back to the dawn of civilization. The disease acquired the name "Rabies" due to its spectacular and unusual clinical manifestations of violence or rage. In Latin, the word "rabere" is associated with rage or raving, whereas in Sanskrit, "rabhas" conveys the idea of violence, and in Greek, the terms "lyssa" or "lytta" are linked to madness. Animal rabies was first documented in the pre-mosaic Eshmuna code of Babylon, which dates back to the twenty-third century BC. Before 2300 BC, Aristotle was the pioneer in recognizing rabies in the regions of Egypt and Greece. In the sixth century BC, canine rabies was documented in the Avesta, and by the fourth century BC, it was also noted in Greece. Our understanding of canine rabies can be traced back to the first century BC, as documented in the Susrutasamhita. In 1804, Georg Gottfried Zinke discovered that rabies is an infectious disease transmitted through the saliva of dogs that are infected. In 1881, Louis Pasteur demonstrated that the rabies virus has a specific affinity for nervous tissue. In 1885, he once again made a breakthrough by creating the first attenuated rabies vaccine. This was achieved by using rabies-infected rabbit spinal cords, which were air-dried to inactivate the virus. On July 6 of the same year, he administered this vaccine to Joseph Meister, a 9-year-old boy who had been bitten by a rabid dog. Each year, on the 6th of July, World Zoonoses Day is recognized. In 1889, Victor Babes suggested the simultaneous use of serum containing rabies antibodies along with the vaccine as a method for post-exposure prophylaxis. In 1903, the virus's structure was jointly described by Remlinger and Riffat Bay. The first oral vaccination program was initiated in Switzerland in 1978 and later expanded to other European nations. In 2022, the Government of India also sanctioned the use of oral vaccines for stray dogs.

Etiological Agent and Ecology

Rabies is characterized as a neurotropic virus with a unique "bullet" shape, featuring an enveloped surface and an unsegmented, negative-sense single-stranded RNA configuration. With dimensions of 180 nm in length and 75 nm in width, it features a genome that measures 12 Kb. The virus is classified under the *Lyssavirus* genus, which is part of the Rhabdoviridae family and the Mononegavirales order. The genome is composed of five distinct proteins: nucleoprotein (N), phosphoprotein (P), matrix protein (M), glycoprotein (G), and polymerase (L). Genomic sequencing and phylogenetic research have identified seven unique genotypes within the *Lyssavirus* genus. These include the classical rabies virus (genotype I), the Lagos bat/rabies-related virus (genotype 2), the Mokola virus (genotype 3), the Duvenhage virus (genotype 4), European bat Lyssavirus 1 (genotype 5), European bat Lyssavirus 2 (genotype 6), and the Australian bat virus (genotype 7). Outside the host, the virus does not survive and it can be inactivated by U-V ray, sunlight, heat, desiccation, phenol, and formalin. Rabies virus becomes quickly inactivated at 60 °C and remains viable for some days at 4 °C. The virus remains viable for a long period at 4 °C in infected brain tissue suspended in 50% glycerol. It can be preserved at or below -70 °C and in freeze-dried state, that is, lyophilization. The virus exhibits sensitivity to lipid solvents such as chloroform, ether, and acetone, in addition to iodine preparations, ethanol, β-propiolactone, detergents, soap, quaternary ammonium compounds, and proteolytic enzymes. The glycoprotein G, which forms the surface spikes, plays a crucial role in virulence, pathogenesis, and immunity. Virus binds with the acetyl choline receptors of nerve cells at neuromuscular junction, which persuades hemagglutination inhibiting (HI), antibodies, and activated cytotoxic T-cell immunity. The virus has the property of hemagglutination at 0–4 °C and pH 6.2 with goose RBC. The hemagglutination antigen is distinct for each species and is not the same as the antigens present in viruses associated with rabies.

Street Virus

The rabies virus that is extracted from a human or animal with a natural infection is referred to as the street virus. Negri bodies, which are intracytoplasmic inclusion bodies, can be observed in the brains of animals that succumb to street virus infection.

Fixed Virus

After several serial intracerebral passages in rabbits, the virus experiences certain changes. This type of rabies virus is called the fixed virus and it does not multiply in extraneural tissues. Negri bodies are typically not observable in the brains of animals that succumb to fixed virus infections. This fixed virus is utilized in the production of vaccines.

Chicken Embryos

The rabies virus can be cultivated in chick embryos, with the yolk sac being the typical site for inoculation. Through repeated passage of the virus in chick embryos, attenuated vaccine strains such as Flury and Kelev are developed.

Tissue Culture

The rabies virus is capable of propagation in various primary and continuous cell cultures, including chicken embryo fibroblasts and hamster kidney cell lines. Vaccines are produced using fixed virus strains that have been adapted to grow in human diploid cells, chicken embryos, and Vero cell lines.

Host range and susceptibility: While all mammals are highly prone to contracting rabies and can pass the disease to other vulnerable hosts, there is notable diversity in susceptibility among different mammal species. Carnivorous mammals act as the primary reservoir of rabies. Domestic animals like dog, cat, cattle, sheep, goat, and camel are highly susceptible while raccoons, skunks, foxes, and bats are also susceptible of rabies. In the natural world, rabies manifests through two related cycles: the sylvatic cycle, which predominantly involves dogs and cats, and the urban cycle, which features animals like foxes, raccoons, jackals, wolves, badgers, mongooses, and bats serving as vectors or reservoirs. A 10-year-old Australian boy in India was diagnosed with rabies after being bitten by a rhesus macaque.

Epidemiology

Rabies is a common disease worldwide; however, several islands and countries have managed to remain rabies-free, thanks to rigorous quarantine protocols or their natural isolation by water. Several locations within the Asian subcontinent, including Bahrain, Cyprus, Hong Kong, Maldives, Japan, Singapore, Malaysia, Qatar, Timor-Leste, and certain Indian islands are considered rabies-free. In American subcontinent, Uruguay, Trinidad and Tobago, Saintkitts and Nevis, Jamaica, Falkland, Belize, Barbados, Bahamas and Antigua and Barmuda are free from rabies. In the European subcontinent, the United Kingdom, Spain (excluding Melilla and Ceuta), Norway (excluding Svalbard), Portugal, Malta, the Isle of Man, Iceland, Greece, Gibraltar, Finland, Albania, and Macedonia are all regions that are free of rabies. Seychelles, Reunion, Libya Mauritius, Verde, Congo, and Cape are presenting rabies-free status among the African countries. The Oceania islands, including Fiji, Vanuatu, the Cook Islands, French Polynesia, Guam, New Zealand, the Solomon Islands, New Caledonia, and Papua New Guinea, have all been recognized as rabies-free. Rabies is prevalent in numerous countries, with a few exceptions such as Australia and Antarctica. According to WHO, a nation can declare itself rabies-free if it has not reported any locally acquired cases of rabies in humans or animals over the past two years, thanks to effective surveillance and import controls. Geographical distribution of rabies is described in the following table.

Name of the virus	Species	Reservoir	Name of the countries from which virus is isolated
Classical rabies virus (RABV)	Rabies lyssavirus	All mammals	Global
Lagos bat virus (LBV)	Lagos bat lyssavirus	Bats	Central African Republic, Ghana, Senegal, Ethiopia, France, Nigeria, South Zimbabwe, and Africa.
Mokola virus (MOKV)	Mokola lyssavirus	Rodents and domestic animals	Central African Republic, Cameroon, South Africa Ethiopia, Nigeria, and Zimbabwe
Duvenhage virus (DUVV)	Duvenhage lyssavirus	Bats	Kenya and South Africa
European bat lyssavirus 1 (EBLV-1)	European bat 1 lyssavirus	Bats	Denmark, the Netherlands, Belgium, France, Germany, Slovakia, Spain, Poland, Russia, and Ukraine
European bat lyssavirus 2 (EBLV-2)	European bat 2 lyssavirus	Bats	Denmark, the Netherlands, Finland, France, Norway, Germany, Switzerland, and the United Kingdom
Australian bat lyssavirus (ABLV)	Australian bat lyssavirus	Bats	Australia

Seven main genotypic linages have been recognized globally and, under every linage, several variants have appeared with a specific mammalian host in a particular geographic region. Across continents such as Europe, the Americas, Africa, and Asia, the most extensive lineage is sustained in the wild by domestic dogs and various wild animals, including foxes, raccoon dogs, skunks, and jackals. In the Americas, various rabies virus strains are transmitted by bats, with the vampire bat being a significant source of rabies infections for humans and domestic animals in Latin America.

As per the reports of WHO, more than 3 billion people are affected by the dogs, although many human cases as well as dog rabies remain unreported due to the inadequate surveillance and monitoring system. Globally, an estimated 60,000 people die each year, with over 30,000 of these deaths occurring in Asia. This data indicates that every 15 min, an individual in Asia passes away. More interestingly, India alone loses more than 20,000 lives annually due to Rabies. It is also predominant in Bangladesh, Nepal, Bhutan, Myanmar, Indonesia, and Thailand. The primary factor contributing to the spread of rabies in Africa is the dog population and the growth of urban areas. While rabies remains present in Europe, the incidence of rabies in humans has been reduced to zero in several countries, thanks to the enforcement of animal vaccination policies, especially for dogs. Due to their smaller size and interactions with both pet and stray dogs, children face a greater likelihood of being bitten. For this reason, rabies should be considered an overlooked pediatric illness in countries where it is endemic. Inadequate surveillance system due to limited healthcare facilities, medical and veterinary infrastructures, cultural, religious, and social taboos leads to become a burden of rabies. More than 99% of human rabies cases globally are caused by the classical rabies virus (RABV) found in dogs.

Rabies is mainly transmitted by saliva of infected animals. Virus can enter into the host via wounds or cut but it cannot penetrate intact skin. The likelihood of death is influenced by where the bite occurs, the concentration of the virus in the saliva, and how severe the infection is. There have been instances of rabies in humans that do not involve bites, such as when the virus is inhaled in aerosol form at high concentrations, through organ and corneal transplants, or when rabies-infected saliva or materials like brain tissue from an infected animal come into contact with open wounds, abrasions, or mucous membranes. Only a single human-to-human case has been documented throughout the world by human bite. Few air-borne cases have been recorded in laboratory facilities during vaccine production and in caves, roosted by infected bats. The Center for Disease Control and Prevention (CDC) reported that there were few cases of rabies that occurred due to consumption of unpasteurized milk from rabid cow. Transplacental transmission of rabies has been documented in animal, but yet to be reported in humans. Dog meat is a delicious item in some countries where dog is slaughtered so slaughtering of infected dog is also a mode of transmission.

Pathogenesis

The incubation period for rabies is quite inconsistent, ranging from weeks to several years, with an average duration of 20–90 days. This variation is influenced by several factors, including the location of virus entry, the amount of virus present, the specific strain or species of the virus, and the host's immune system effectiveness.

After the biting or scratching by the rabid animals, the virus gets inoculated into subcutaneous or muscle tissues. The rabies virus often lingers in muscle tissue for a time, which might allow for post-exposure treatment and enable the host's immune system to eliminate the virus.

The virus attaches to receptors, primarily the nicotinic acetylcholine receptor (nAChR), as well as the neuronal cell adhesion molecule (NCAM) and the tumor necrosis factor receptor (TNFRSF16), located on myocytes and sensory and motor neurons at the neuromuscular junction, through its viral glycoprotein. After that, the virus multiplies in the myocytes and macrophages. Once the virus attaches to the receptors, it begins its journey to the central nervous system by traveling through the muscle spindles of sensory nerves or the neuromuscular junctions of motor nerves. The virus travels inward along the nerves at a rate of 3 mm per hour or between 12 and 100 mm per day, eventually reaching the brain. The specific region of the brain that will be affected is determined by the motor neurons that supply the entry point. After the virus infiltrates the central nervous system, it disperses outward, predominantly through the parasympathetic nervous system. This pathway is essential for the virus to infect the salivary glands and skin, and it also affects the dorsal root ganglia that align with the nerves at the point of entry. In this dorsal root ganglia, the virus starts its replication, which causes inflammation. Clinical manifestations appear with pain, paresthesia, and pruritus. Subsequently, the virus impacts peripheral nerves, leading to weakness associated with paralytic rabies.

Various factors or mechanisms allow the virus to evade the adaptive immune system's detection within the nervous system. The activation of the Fas ligand leads to the inactivation or reduced efficiency of T cells by triggering apoptosis in antigen-activated T cells, thereby aiding in the conclusion of the immune response. The rabies virus takes advantage of various natural substances released by uninfected nerve cells, such as vasoactive intestinal peptide, calcitonin-gene-related peptide, and norepinephrine, to impair the activity of migratory T cells. The movement of migratory T cells into the central nervous system via the blood–brain barrier is being limited or blocked due to the presence of histocompatibility antigen, α chain G (HLA-G), which is a non-classical, immunosuppressive histocompatibility antigen. Additionally, the increased expression of programmed cell death ligand 1 (also known as B7H1) further inhibits T-cell proliferation and induces apoptosis. The transport of migratory B cells is disrupted, leading to an immunoreactive response. When rabies leads to limited neuroinflammation, it prevents T cells from crossing the blood–brain barrier into the central nervous system (CNS). Despite the CNS being an immune-privileged area, the rabies virus can still replicate within a fully immunocompetent cellular environment.

In summary, the surface spikes are made up of glycoprotein G, which is vital for pathogenesis, virulence, and immune response. This glycoprotein enables the virus to attach to acetylcholine receptors in neural tissues, initiates the formation of hemagglutination inhibiting (HI) and neutralizing (protective) antibodies, and promotes cytotoxic T-cell immunity.

Clinical Manifestations in Animals

In animals, the incubation period typically lasts from 20 days to 3 months, although it has been observed to vary from just a few days to as long as 6 months, and in some exceptional instances, even longer. Once clinical symptoms appear, it is nearly always deadly. Rabies can manifest a wide array of clinical symptoms and should be considered in any animal exhibiting sudden behavioral changes or lower motor neuron paralysis. The initial clinical signs are often non-specific, which includes anxiety, restlessness, anorexia or an improved appetite, nausea, diarrhea, low-grade fever, dilation of the pupils, hyperactive, and extreme salivation.

Dog

Incubation period of dog is usually with an average of 15–60 days, but some exceptions have been also recorded with 6–18 months. Dogs can exhibit two forms of rabies: the aggressive type and the silent type.

Furious Rabies Furious form of rabies can be divided into prodromal stage or stage of melancholy and stage of excitation.

Prodromal stage/ stage of melancholy: In this phage, dogs mainly change its behavior and it lasts 1–3 days. Dogs may show the biting tendency to the both animate and inanimate objects like sticks, straw, mud, doors, cage, feeding utensils, etc., without any provocation. The dog may show imaginary fly catching stance. Rabid dog can roam for a long distance in a circular way and can spread the disease over the area. Nervousness, anxiety, and fever may be noted during this time. Due to the paralysis of pharyngeal and laryngeal muscles, the animal fails to drink water but may lick water and loss of appetite develops. Appearance of long stranded salivation is seen from the mouth.

Acute neurological phase/ stage of excitement: This stage may last 1–7 days. Classical "mad dog syndrome" appears in this stage with excessive excitability and irritability. At the onset of the disease, the dog may put itself in a dark place due to photophobia. At this stage, the paralysis of the vocal cords causes the dog to emit a raspy bark or growl, and sometimes it may not be able to bark at all. Therefore, changes in the dog's vocal tone should not be the only criterion for deciding to isolate the animal. The animal exhibits significant drooling, with saliva and foam forming at the mouth's corners. Additionally, muscle tremors and redness may be observed in the animal. Sexual excitement may be observed. At the end of this stage, the dog loses its ability to bark and shows hanging of lower jaw, protrusion of tongue and downing of head. The dog will gradually exhibit increased drooling, raised fur, occasional pupil enlargement, fear of water and air, difficulty swallowing, inhalation spasms, rapid breathing, blood vomiting, and will eventually fall into a coma and die. In cases of furious rabies, the typical duration from the appearance of symptoms to death is 7.8 days.

Dumb rabies/ Paralytic rabies: The dumb form includes prodromal stage and stage of paralysis. The paralytic stage starts after completion of the prodromal stage. In this form, absence of excitement or irritation is observed. Paralysis of tongue, lower jaw, larynx, and hind quarters may be seen. In its most common form, the disease leads to paralysis, leaving the animal with its mouth agape and unable to swallow. The dog withdraws itself from being seen or disturbed. The dog may not bite but effort to feed it is dangerous. The dumb form is as infectious as the furious form. Once an animal begins to show signs of rabies, it typically does not live for more than 7 days. In this form, the clinical symptoms emerge slowly, including quadriplegia, difficulty speaking, trouble swallowing, excessive saliva production, inspiratory spasms, respiratory failure, fear of water or air, coma, and ultimately, death. In cases of dumb or paralytic rabies, the typical duration from the appearance of symptoms to death is 11 days.

Cattle

Prodromal stage: At this stage, diagnosing rabies clinically is quite challenging, and this phase can last from a few hours up to 2 days. Body temperature may increase 1-2 °C with some other non-specific signs like malaise, anorexia, and sudden cessation of lactation. Animals exhibit clear signs of depression, which owners might mistakenly identify as other conditions, such as rumen or rectal impaction, or the initial phase of FMD. Animal changes the bellowing sound. It is advisable not to handle animal's mouth at this point and hypodermic medications in keeping isolation is recommended as the virus can be present in saliva as early as 4–5 days before symptoms appear.

Furious Form

There is a marked change in behavior. The animal becomes nervous, often shows persistent tenesmus, anorexia, pruritis, ataxia, lameness, excessive drooling salivation, and aggression, although the animal may attempt to drink water but incapable of drinking water without exhibiting choking symptoms and quickly becoming dehydrated. Animal owners and veterinarians frequently notice excessive salivation, which may indicate that a foreign object is lodged between the teeth or in the pharynx. It is best to refrain from attempting to remove it yourself. There may be trembling or twitching of ears and animal bellow relentlessly in low pitched voice due to paralysis of vocal cord. If a bird or any other animal is tossed in front of rabid cattle, it jumps up and chases it. The animal may butt a drum, box, or other loose object. Increase of sexual desire may be seen in both male and female cattle. Urine is expelled in small quantities at short intervals. The excitement phase seldom extends beyond 2 days. The animal rapidly loses strength and passes away within a few hours due to failure of the heart or lungs.

Clinical Manifestations in Humans

In humans, the incubation period typically ranges from 1 to 2 months, but it can be as brief as 7 days or extend to several years, even reaching up to 19 years. Due to their smaller body size, children typically have a shorter duration than adults. The duration of the incubation period is influenced by factors such as the location of the bite, the severity of the bite, the number of wounds, the volume of virus introduced, the species of the animal that inflicted the bite, the protection offered by clothing, and the treatment administered. The incubation period tends to be shorter when there is severe exposure, bites occur on the face, neck, head, and upper limbs, or when bites are inflicted by wild animals.

The progression of the disease can be categorized into three stages: the prodromal stage, the acute neurological stage, and the coma stage.

Prodromal stage: This stage lasts 2–10 days with some nonspecific like anorexia, malaise, nausea, vomiting, headache, sore throat, fever, photophobia, sexual excitement, a prolonged erection of the penis and spontaneous ejaculation of semen occur rarely.

Acute Neurological Phase

This phase lasts between 2 and 7 days, during which patients may exhibit symptoms such as nervousness, anxiety, hallucinations, involuntary actions, tearing, dilated pupils, and increased salivation and sweating.

Hydrophobia

One distinctive symptom is the difficulty in swallowing water despite experiencing intense thirst. While patients might manage to eat dry foods, they struggle with liquids. When they try to drink, they experience such painful spasms in the throat and voice box that it leads to choking or gagging, causing them to fear even the sight or sound of water (hydrophobia), which can then trigger convulsions.

Coma

When patients survive the acute neurological phase, they typically enter a comatose state that can last from several hours to a few days. However, death usually follows within 1–6 days, primarily due to respiratory failure.

Milwaukee Protocol

In 2004, a group of medical experts, under the leadership of Dr. Rodney Willoughby, developed the MP after a 15-year-old girl named Jeanna Giese was hospitalized in Milwaukee following a rabies diagnosis. Following discussions with the researchers at the Centers for Disease Control and Prevention in Atlanta, an innovative procedure was developed and put into practice. The patient was put into a medically induced state and administered an antiviral combination of ketamine, ribavirin, and amantadine. The doctors proposed that if rabies pathology is linked to neurotransmitter imbalances in the central nervous system, then reducing brain activity could lessen harm while the patient's immune system mounts an effective response.

Diagnosis of Rabies in Animals

Various laboratory methods can be employed, as they differ in terms of efficiency, specificity, and reliability. Rabies virus is primarily concentrated in the thalamus, pons, and medulla, which is why it is recommended to gather brain tissues from areas such as the brain stem, Ammon's horn, thalamus, cerebral cortex, cerebellum, and medulla oblongata. It is essential to always wear personal protective equipment (PPE) like gloves, a face shield, and a mask, and to take measures to prevent the spread of aerosols. The occipital foramen and the retro-orbital path are chosen for obtaining brain tissues.

Direct Fluorescent Antibody (DFA) Test

This examination is regarded as the benchmark test (gold standard test), endorsed by both the WHO and the World Organization for Animal Health (WOAH). This test boasts a high sensitivity and specificity, ranging from 96% to 99%, and its results can be interpreted within 2 h.

Direct Rapid Immunohistochemistry Test (dRIT)

This test can be employed as an alternative to the direct fluorescent antibody test for routine rabies diagnosis, given its equivalent sensitivity and specificity. This test can be carried out in labs that lack a fluorescence microscope.

Demonstration of Inclusion Bodies (Negri Bodies)

Negri Adelchi (1876–1912), an Italian physician and pathologist, discovered structures that are round or oval and purplish-pink in color, featuring distinctive basophilic inner granules. These structures, now known as Negri bodies, were identified in brain tissue impression smears through the use of Seller's stain, which is composed of basic fuchsin and methylene blue dissolved in methanol. Due to the lack of access to immunofluorescence and biological facilities, this test remains frequently utilized. Negri bodies and rabies antigens are typically detected in animals or humans suffering from rabies, but they are seldom observed in bats. A negative result does not provide conclusive evidence.

Mice Inoculation Test

Supernatant of homogenized brain tissue in buffered solution containing antibiotics is inoculated intracerebrally in a state of anesthesia, involving 5–10 mice aged 3–4 weeks or a group of newborn mice that are 2 days old. When mice are infected, they develop encephalitis and eventually die. Over a period of 28 days, the mice are monitored daily, and each mouse carcass is analyzed using DFA or dRIT to detect the presence of rabies antigen.

Cell Culture

Neuroblastoma cell lines, like CCL-131 from the American Type Culture Collection, are utilized for routine rabies diagnosis. These cells are cultured in Dulbecco's modified Eagle's medium (DMEM) with an addition of 5% fetal calf serum (FCS) and are incubated at 37 °C in an atmosphere containing either 0.5% or 5% CO2. This approach matches the sensitivity of the mouse inoculation test.

Fluorescent Antibody Virus Neutralization Test (FAVN)

In vitro, a fixed quantity of the rabies virus, specifically, the challenge virus standard (CVS-11) strain adapted for cell culture, is neutralized prior to being introduced to BHK21 cells, which are vulnerable to the virus.

Enzyme-Linked Immunosorbent Assay (ELISA)

For international transport of dogs or cats, the ELISA is an acceptable test, provided that a kit approved by WOAH is utilized. Additionally, ELISA techniques are advantageous for tracking vaccination efforts in wildlife, as long as the kit has been validated for the specific wildlife species being examined.

Polymerase Chain Reaction (PCR)

PCR assays serve as precise diagnostic tools for detecting ribonucleic acid (RNA) from Lyssavirus in suspect samples, with the advantage that they do not necessitate the presence of a live virus for accurate identification. Both conventional RT-PCR and Real-time RT-PCR methods can be utilized to identify rabies genetic material.

Diagnosis of Rabies in Human

Identifying the classical furious form of human rabies, which accounts for approximately 80% of all rabies cases, relies on its unique clinical signs and symptoms and seldom poses challenges in diagnosis. The paralytic form of rabies, accounting for approximately 20% of human rabies cases, poses a diagnostic challenge because it is often indistinguishable from Guillain–Barre syndrome (GBS) and other neuroparalytic issues. This confusion is exacerbated by the continued use of the Semple-type anti-rabies vaccine in countries like Mongolia, Myanmar, and Pakistan. Biological samples such as serum, cerebrospinal fluid, saliva, and tissues from highly innervated areas like the skin are appropriate for ante-mortem diagnosis, and all these samples should be treated as potentially infectious.

Magnetic Resonance Imaging (MRI)

On brain MRI scans, there have been reports of heightened signal intensities in gray matter regions, alongside observations of normal characteristics. As early as the third day after onset, MRI scans can detect indistinct hyperintensities on T2-weighted images in regions such as the brain stem, hypothalamus, hippocampi, subcortical and deep white matter, and cortical gray matter, even when the patient remains conscious.

Direct Fluorescent Antibody (DFA) Test (Corneal and Skin Test)

This examination is regarded as the gold standard test, endorsed by both WHO and WOAH. During the advanced stages of rabies, the presence of the virus antigen can be identified in the corneal epithelium using a simple fluorescent antibody test. In corneal epithelial cells, the antigen is usually observed as round to oval inclusions located within the cytoplasm. Although the sensitivity of this test is relatively low at 30%, a similar method is applied in a skin biopsy called a Nuchal biopsy, which has a sensitivity ranging from 50% to 70%. This procedure involves extracting a 5–6 mm diameter piece of skin from the back of the neck along the hairline. The sample must include at least 10 hair follicles and be deep enough to reach the subcutaneous layer to capture the cutaneous nerves located at the base of the hair follicles.

Direct Rapid Immunohistochemistry Test (dRIT)

In routine rabies diagnosis, this test can be utilized as an alternative to the direct fluorescent antibody test, as it demonstrates equivalent sensitivity and specificity. It is suitable for laboratories that do not have access to a fluorescence microscope.

Enzyme-Linked Immunosorbent Assay (ELISA)

The detection of the rabies N antigen can be achieved through immunohistochemical methods and enzyme immunoassays, also referred to as rapid rabies enzyme immunodiagnosis (RREID).

Polymerase chain reaction (saliva and CSF): As described earlier.

Treatment

Treatment in Animal

Rabies is always fatal once the clinical signs develop, despite any attempts at treatment. A quick response is essential for minimizing the exposure to a rabies case. No treatment is endeavored and rabid animals are euthanized. Veterinarians who meet or suspect rabies cases should follow their national and/or local guidelines for disease reporting.

Treatment in Human

Once a patient develops symptoms, no treatment is known to get outcome. Patients may be offered either a treatment in a critical care facility that may comprise antiviral agents, immunotherapy, and neuroprotective therapy or with sedatives, tranquilizers, and analgesics as needed.

Management Control and Prevention

Primary prevention serves as the most effective defense against rabies. This approach encompasses vaccinating animals to eliminate rabies, fostering public awareness, and encouraging individuals to practice responsible dog ownership.

Secondary prevention aims to avert the development of clinical disease by utilizing post-exposure prophylaxis (PEP) or pre-exposure prophylaxis (PrEP), with PEP being capable of nearly completely preventing infection in almost all instances. In areas with elevated risk, incorporating PrEP into the recommended vaccination schedule for children should be considered. The current protocol involves administering 4 or 5 vaccine doses on days 0, 7, 14, and either 21 or 28. Individuals at high risk of rabies exposure, such as veterinarians, researchers, and bat handlers, should receive PrEP. Anti-rabies treatment, a form of specific prophylaxis, is typically administered following exposure to an infection and has the potential to save lives. This consists of the following:

- Wound care: To decrease the viral load in wounds, it is recommended to wash them with soap, ideally under running water, for a minimum of 5 minutes. Once the wound has been cleaned, any remaining virus can be neutralized by using a virucidal agent such as alcohol (400–700 ml/liter), tincture, or a 0.01 percent aqueous solution of iodine or povidone iodine. To prevent further trauma and the potential spread of the virus into deeper and surrounding tissues, biting wounds should not be immediately closed with sutures. To prevent bacterial infections, it is advised to use antibiotics and tetanus toxoid for managing wounds.
- Anti-rabies vaccines: In 1885, Louis Pasteur pioneered the creation of a weakened rabies vaccine by utilizing the spinal cords of rabbits that had been infected with the virus. Anti-rabies vaccines are used by using principle such as (i) nerve tissue vaccines and (ii) non-nerve tissue vaccines.

  Nerve tissue vaccines: These vaccines are developed using the nervous tissues of animals that have been infected with a stabilized virus. Though Pasteur vaccine was also neural vaccine, neural vaccines have some disadvantages, as these consist of crude products, so it is capable of producing severe fatal reaction. These are poor antigens due to presence of nucleocapsid part of the virus where the quantity of glycoprotein G remains low. These vaccines are encephalitogenic, which can cause post-vaccination allergy. According to WHO, nerve tissue vaccines should be phased out because they are prone to causing significant adverse reactions and are less effective in generating immunity compared to newer vaccines.

  Semple vaccine: In 1911, Sir David Semple developed this vaccine by using sheep brain, which was also known as nerve tissue vaccine (NTV) for human use in India. The solution consists of a 5% mixture of sheep brain that has been infected with a fixed virus and subsequently deactivated using phenol at a temperature of 37 °C.

Beta propiolactone (BPL) vaccine: This vaccine is just the modified form of Semple vaccine as beta propiolactone is used to inactivate the virus instead of phenol.

Suckling mouse brain vaccine: The brain of less than 9 days suckling mice has been used to produce anti-rabies vaccine, as it is believed to lack a neuroparalytic effect due to the minimal or nonexistent presence of myelin in neonatal animals.

Non-nerve tissue vaccines.

Egg passage vaccine

Duck embryo vaccine (inactivated vaccine): The rabies virus is cultivated in embryonated duck eggs, and then beta propiolactone is used to inactivate the virus. However, this vaccine was unable to maintain its effectiveness due to inadequate immunogenicity.

Chick embryo vaccine (live attenuated): Low egg passage (LEP) and high egg passage (HEP) vaccine are the two vaccines, which were used in animals. Flury strain was used for these vaccines.

Low egg passage (LEP) vaccine: The attenuation was done by 40–50 passages for immunizing the dog population.

High egg passage (HEP) vaccine: The attenuation was done by 180 at passage level for immunization of cattle and cat.

Nowadays, all these vaccines are not used because of availability of different cell culture vaccines.

Cell culture vaccines: These vaccines are safer and more efficient than the previously discussed vaccines. WHO recommends cell culture because of potency and cost effectiveness. These vaccines are known as rabies vaccines based on cell culture and embryonated eggs (CCEEV). Various cell lines, such as human diploid cells, chick embryo cells, Vero cells, baby hamster kidney cells, and duck embryo cells, are utilized for virus cultivation. The naming of vaccines is based on the cell line used, such as the baby hamster kidney cell vaccine (BHKV), human diploid cell vaccine (HDCV), purified chick embryo cell vaccine (PCECV), purified duck embryo cell vaccine (PDEV), purified primary hamster kidney cell vaccine (PPHKCV), and purified Vero cell vaccine (PVRV). Following cultivation in cell culture or embryonic eggs, the virus is collected, concentrated, purified, inactivated, and then freeze-dried. In certain CCEEVs, stabilizers such as human albumin or processed gelatin are utilized. Rabies vaccines for humans are not available in multidose vials for either intramuscular or intradermal administration. These vaccines have a shelf life of at least 3 years and should be stored at temperatures between 2 and 8 °C, ensuring that they are shielded from sunlight.

Hyperimmune serum: The use of rabies immunoglobulin (RIG) is advised, and its application is contingent upon the wound category as defined by WHO, specifically being necessary for category III wounds (Table 29.2). Administering RIG directly into the wound is designed to neutralize the virus present there and halt its spread, given that it may take up to 14 days for the body to build sufficient immunity after receiving a post-bite vaccination. Human rabies immunoglobulin

(HRIG) and equine rabies immunoglobulin (ERIG) are supplied as hyperimmune serums, with dosages set at 20 IU/Kg and 40 IU/Kg of body weight, respectively. Before administering the RIG to the patient, ensure that it reaches a temperature between 25 °C and 30 °C. It is important to avoid using multiple needle injections on the wound(s).

It is intriguing that new tools might lead to the use of monoclonal antibodies (mAbs) as an alternative to anti-rabies virus immunoglobulin (RIG) for post-exposure prophylaxis. The WHO has suggested employing monoclonal antibody "cocktails" that include at least two antibodies targeting the rabies virus.

Antibodies in the humoral immune response are crucial for defending against rabies. An anti-rabies neutralizing antibody level of 0.5 IU/ml or higher is regarded as providing adequate protection. By the 14th day of a post-exposure treatment, healthy people reach this stage, regardless of whether rabies immunoglobulin is administered at the same time.

Types of contact/rabies exposure

Category of exposure	Description	Post-exposure prophylaxis (PEP)
Category I	Touching or feeding of animals, licking on intact skin, contact of intact skin with secretions or excretions of rabid animals or persons	Not considered as exposure so no treatment or PEP is required
Category II	Nibbling of uncovered skin, Minor scratches or abrasions without bleeding	Administering wound care and the rabies vaccine without delay is essential
Category III	One or more transdermal bites or scratches, as well as licks on broken skin, can lead to contamination of the mucous membrane with saliva from licks and exposure to bats	Wound management The anti-rabies vaccine and rabies immunoglobulin should be administered at separate locations promptly. Rabies immunoglobulin remains effective if administered within 7 days following the initial vaccine dose

Intramuscular schedules: Administer a single intramuscular dose of 2.5 IU following the outlined schedules.

Essen schedule (1-1-1-1-1): Administer a single dose of the vaccine on the following days: 0, 3, 7, 14, and 28. A delay of one day for succeeding vaccinations due on days 3, 7 and 14 and of 2–3 days on day 28 is acceptable under unavoidable circumstances.

Zagreb schedule (2-1-1): On day 0, one dose is administered in the right arm and another in the left arm, followed by a dose in the deltoid muscle on days 7 and 21.

Updated Thai Red Cross (TRS) schedule (2-2-2-0-2):

Administering 0.1 ml of the reconstituted vaccine intradermally is required at two sites per visit, with each site located on the deltoid area, approximately an inch above where the deltoid muscle inserts, on days 0, 3, 7, and 28. Day 0 is defined as the date on which the initial dose of the anti-rabies vaccine is administered, and it may differ from the date when the rabies exposure or animal bite actually occurred.

Pre-exposure vaccination regimens:

- Over the course of one-week, intradermal injections are given at two separate sites on days 0 and 7.
- Over the same period, intramuscular injections are administered at a single site on days 0 and 7.

Modified post-exposure vaccination regimen (for people immunized previously):

- Over 3 days, administer intradermally at two sites: days 0 and 3
- Over 3 days, administer intramuscularly at one site: days 0 and 3
- Administer intradermally at four sites on a single day: day 0

Post-exposure vaccination regimen (for people not previously immunized):

- A one-week intradermal regimen at two sites: Scheduled for days 0, 3, and 7
- A two-week intramuscular regimen at a single site: Scheduled for days 0, 3, 7, and at some point, between days 14 and 28
- A three-week intramuscular regimen: Scheduled for day 0 at two sites, day 7 at one site, and day 21 at one site

Individuals who have undergone a minimum of two sessions of either intramuscular or intra-dermal rabies vaccination, whether as a preventive measure before exposure or as a treatment after exposure, are regarded as having prior immunization against rabies. Pre-exposure vaccines are recommended for the people like animal handler, kennel personnel, veterinarians, veterinary nurses, wildlife workers, and cave workers.

Animal Vaccination

Vaccination is primarily used to manage rabies in both domestic and wild animals, thereby minimizing the threat to humans. The goal of eradicating rabies transmitted by dogs is a key focus for the Food and Agriculture Organization (FAO), World Organization for Animal Health (WOAH/OIE), World Health Organization (WHO), and the Global Alliance for Rabies Control (GARC). These entities are working together to achieve the vision of ending human rabies caused by dogs by the year 2030, utilizing a One Health strategy that considers the interdependence of animals, humans, and their environments. It is essential for all dogs and cats to receive a rabies vaccination at the age of 3 or 4 months (depending on state legislation). A booster shot should follow 1 year later, and thereafter, vaccinations must be administered every 3 years with approved inactivated vaccines or every year with recombinant vaccines.

Oral Vaccine

Europe and the USA have been using oral rabies vaccines (ORVs) to control rabies in wildlife since 1978. ORV in dogs serves as an additional strategy to enhance vaccination rates during large-scale parenteral dog vaccination efforts, particularly in situations where achieving 70% vaccination coverage is hindered by the existence of free-roaming dog populations. There are two types of oral rabies vaccines (modified live vaccines and vector-based vaccines) currently being used under commercial license for the vaccination of various wildlife species.

Frequently Asked Questions (FAQ)

Is it possible to transmit the disease from vaccinated dog?

While proper vaccination is effective against rabies, the lack of laboratory-confirmed protection levels means that post-exposure prophylaxis is recommended regardless of the vaccination status of the dog that bit someone.

Is vaccination required after biting by a rat?

Under normal circumstances, rabies is not transmitted by domestic rats but some rat bite rabies cases have been recorded from Asian countries. So, post exposure vaccination is recommended.

Is rabies transmitted through consuming meat from an infected animal?

Vaccination is suggested after consumption of raw meat, but, for cooked meat, PEP is not required.

What will be the decision after wild animal exposure?

Exposure to any wild animal will be classified as category III, necessitating the use of RIG and vaccination for its management.

Is human-to-human transmission possible?

There is no good documentation regarding this matter, though few cases have been documented resulting from infected organ transplantation. For individuals who have had close contact with the secretions of a rabies patient, PEP might be administered as a preventive measure.

Do vaccines fail to develop the protection against rabies?

Vaccine failure can result from the vaccine's poor quality, the animals' health issues, insufficient doses, or improper administration.

Is re-vaccination started after deviation of vaccination schedule?

To ensure effective immunity against rabies, the initial three doses of post-exposure vaccines, administered on days 0, 3, and 7, should be completed within a maximum of 10 days. So, 1–2 days deviation does not require restart of vaccination. If one or more doses are failed on the scheduled date of first three doses, then inoculation of additional doses should be considered.

Is milk consumable from an infected cow?

The Centers for Disease Control and Prevention (CDC) indicated that there have been a few instances of rabies linked to the consumption of unpasteurized milk from a cow with rabies. Nonetheless, consuming unpasteurized milk from a rabid cow or goat is regarded as a form of exposure.

Can an infected cow shed the virus in their saliva?

Indeed, a cow infected with rabies can transmit the virus through its saliva, posing a risk of infection to humans and other animals. Therefore, veterinarians should use personal protective equipment, such as gloves, goggles, and masks, when dealing with animals suspected of having the disease.

When cattle are considered to be exposed to rabies?

If a calf is nursed by a cow and vice versa, then it should be considered as exposed to rabies. PEP should be given to the exposed animal.

Bibliography

Barrat, J., Picard-Meyer, E., & Cliquet, F. (2006). Rabies diagnosis. *Developmental Biology (Basel), 125*, 71–77.

Fooks, A. R., Cliquet, F., Finke, S., et al. (2017). Rabies. *Nature Reviews. Disease Primers, 3*, 17091. Published 2017 Nov 30. https://doi.org/10.1038/nrdp.2017.91

National guideline for rabies prophylaxis. (2019). National Centre for Control Disease.

OIE terrestrial manual 2018. World Organization for Animal Health

Rupprecht, C. E., Fooks, A. R., & Abela-Ridder, B. (2019). *Laboratory techniques in rabies* (5th ed.). World Health Organization.

Singh, R., Singh, K. P., Cherian, S., et al. (2017). Rabies - epidemiology, pathogenesis, public health concerns and advances in diagnosis and control: a comprehensive review. *The Veterinary Quarterly, 37*(1), 212–251. https://doi.org/10.1080/01652176.2017.1343516

Japanese Encephalitis

23

Ripan Biswas and Indranil Samanta

Abstract

Japanese encephalitis is a viral illness transmitted by mosquitoes. It is caused by the Japanese encephalitis virus, which is part of the Flaviviridae family. This disease ranks as the most prevalent vaccine-preventable encephalitis in Asia, notably in rural and agricultural zones where rice fields and pig farms offer ideal conditions for the Culex mosquito, the vector responsible for transmission. Japanese encephalitis predominantly impacts children, potentially causing significant neurological complications such as seizures, paralysis, and difficulties with cognitive functions. Severe instances have a high rate of mortality. While the majority of infections are asymptomatic, approximately 1 in 250 individuals experiences acute encephalitis. Efforts in vaccination, controlling vectors, and public health initiatives contribute to the prevention of Japanese encephalitis. Despite the presence of effective vaccines, the disease continues to be a significant public health concern in the regions it affects.

Keywords

JE · Japanese B encephalitis · Arbovirus B · Mosquito-borne encephalitis virus

R. Biswas (✉)
Department of Veterinary Public Health and Epidemiology, West Bengal University of Animal and Fishery Sciences, Kolkata, West Bengal, India

I. Samanta
Department of Veterinary Microbiology, West Bengal University of Animal and Fishery Sciences, Kolkata, West Bengal, India

R. Biswas et al. (eds.), *Concept of One Health and Zoonoses*,
https://doi.org/10.1007/978-981-95-6627-3_23

History

The Japanese encephalitis virus (JEV) is a flavivirus transmitted by mosquitoes and is widely found across numerous countries worldwide. Records of encephalitis outbreaks in Japan date back to the 1870s. About every ten-year intervals, major outbreaks were reported and in 1924 and more than 6000 encephalitis cases were presented in Japan. Initially, this form of encephalitis was labeled as "type B encephalitis" to distinguish it from type A encephalitis, known as von Economo's encephalitis lethargica, or sleepy sickness. In 1935, a patient with encephalitis was found to have a prototype of the Nakayama strain isolated from them. In 1938, Mitamura reported transmission of virus by mosquito (*Culex tritaeniorynchus*). Initially, the virus was grouped with other flaviviruses as a type B arbovirus within the Togaviridae family. However, in 1985, the classification was revised, and the term "type B" was replaced. The virus was then identified as the Japanese encephalitis virus (JEV) and placed in a different family, Flaviviridae, as part of the Flavivirus genus.

Etiological Agent and Ecology

The Japanese encephalitis virus (JEV) is an arbovirus that is part of the Flavivirus genus within the Flaviviridae family. This virus is characterized by a single-stranded RNA genome with a positive sense, enveloped structure, and is approximately 11 kb in size. The virus is composed of a lipoprotein envelope that encases the nucleocapsid and core. The genome is enclosed within the capsid, which is referred to as the capsid protein (C). The virus's outer layer features an envelope protein (E) that facilitates the virus's entry into the host cell. The E protein is referred to as a protective antigen. Furthermore, the genome also codes for a membrane protein (M) and seven non-structural proteins (NS1, NS2A, NS2B, NS3, NS4A, NS4B, NS5), among which NS3 functions as a helicase and NS5 acts as a polymerase. The virus shares a close relationship with the West Nile virus (WNV), the dengue fever virus, the St. Louis encephalitis virus (SLEV), and the Murray Valley encephalitis virus (MVEV). JEV is characterized by a single serotype and is divided into five genotypes (GI–GV). The virus can be rendered inactive at a temperature of 40 °C, and it can be completely eradicated by heating it to 56 °C for a duration of 30 min. The virus thrives in an alkaline environment with a pH range of 7–9. It is vulnerable to ultraviolet and gamma rays. Disinfectants such as common lipid solvents, detergents, iodine, iodophors, phenol, 70% ethanol, 2% glutaraldehyde, 1% sodium hypochlorite, or 3–8% formaldehyde can be effectively used against it.

Host Range and Vectors

The persistence of the Japanese encephalitis virus (JEV) in the natural environment is supported by a complex cycle involving pigs as amplifying hosts, ardeid water birds like herons and egrets serving as reservoir hosts, and mosquitoes acting as vectors. Humans and equine are classified as final hosts since the viremia present in their peripheral blood is both minimal and temporary. Studies in laboratory setup and natural condition reported that bats might have the potential to transmit the virus. Local snake of Korea, *Elapherufodorsata* and *E. schrenkii*; Cobra (*Najanaja*); freshwater turtles (*Trionyxsinensis*) also may act as hosts for the virus. Seropositivity has been recorded in cattle from southern and eastern India.

Numerous *Culex* mosquito species are capable of transmitting JEV, with *Culex tritaeniorhynchus* serving as the main vector in Southern, Eastern, and Southeastern Asia. In Northern Australia, the key vector is *Culex annulirostris.* In a larger range, *Culexvishnui* subgroup (*C. tritaeniorhynchus, C. vishnui, C. pseudovishnui*) has been recognized as the major vectors of JEV. In India, researchers have identified the virus in 16 mosquito species, including *C. whitmorei, C. gelidus, C. epidesmus, Anopheles subpictus, A. peditaeniatus, Mansoni aindiana,* and *M. uniform*, which are widespread across the country.

Epidemiology

There are three distinguished epidemiological regions for JEV. The first endemic region includes Southern India, Sri Lanka, Southern Thailand, Malaysia, Southern Vietnam, Philippines, and Indonesia. The second endemic region includes Nepal, Northern India, Bangladesh, Southern China, Northern Thailand, Myanmar, and Northern Vietnam. The third endemic region includes Korea, Northern China, Taiwan, Japan, and the southern edges of Russia. The disease has also been recorded from Pakistan, Laos, Singapore, and maritime Siberia. Large outbreaks take place mainly during the summer season in northern India, Nepal, China, Japan, and the Korean peninsula. On the other hand, sporadic cases attain its peak during the rainy season in the tropical regions, mainly the southern part of Vietnam, Thailand, Indonesia, and Sri Lanka. JEV principally affects children though all age groups of people may be affected.

Reports indicate that children aged 3–15 have infection rates that are five to 10 times higher than those of older individuals, primarily due to their lower immunity levels. In India, this disease poses a significant health concern for children, with outbreaks having been documented in numerous states since 1955. In 1973, a significant outbreak occurred in the Bankura district of West Bengal state. This disease primarily affects rural farming regions where mosquitoes that carry the virus thrive near pigs, wading birds, and ducks, although cases have also been reported in urban and suburban areas. Precipitation and temperature are key elements that significantly impact vector populations, thereby influencing the overall disease burden. The prevalence of JEV has risen among humans due to conditions such as rice cultivation and pig farming. Ardeid water birds are capable of spreading the virus over long distances. Males are more susceptible because of having more exposure with agriculture. Mosquitoes breed prolifically after the monsoon in paddy field. It is activated at twilight and the infection rate of pigs also becomes high followed by humans.

Transmission

Japanese encephalitis is a mosquito-borne zoonotic illness, primarily spread by *C. tritaeniorhynchus*. In the natural transmission cycle, several mosquito species from the *Culex* genus are involved, with pigs, birds, and bats acting as susceptible reservoir hosts. Humans and horses, on the other hand, are considered dead-end hosts. The virus is initially transmitted by mosquitoes from birds to animals like pigs, which serve as amplifier hosts. Subsequently, mosquitoes feed on these animals, mainly pigs, and pass the virus to humans due to the close proximity of these animals to human living areas. Ardeid water birds play a crucial role in maintaining JEV within their ecosystems and contribute to its spread into new geographical locations. Pigs serve as the primary natural carriers for transmitting the virus to humans, as they experience extended periods of viremia with a high concentration of the virus. While the virus generally does not lead to encephalitis in these natural hosts, it can cause abortions in pregnant sows.

Pathogenesis

The virus replicates in the dermis layer, the primary site of infection after the mosquito bite and then it reaches the local lymph nodes. Although the virus is capable of replicating in monocytes and dendritic cell lineages, its replication is significantly more proficient in neuronal cell lines. It initially replicates in peripheral areas, leading to a temporary presence in the bloodstream before it penetrates the central nervous system (CNS). In this early stage of viremia, the virus also spreads to a variety of non-neural tissues, such as connective tissues, skeletal muscles, cardiac muscle, smooth muscles, lymphoreticular tissues, and both exocrine and endocrine glands. The virus penetrates the blood–brain barrier to infect neurons and astrocytes from the bloodstream, although the exact mechanism remains unexplored. Clinical manifestations only appear after infecting the CNS cells; otherwise, the signs may be mild to asymptomatic. The propagation of virus in the neural cells causes death of the cells, and intensive inflammatory responses damage the neurons.

Clinical Manifestations in Animals

Pigs are regarded as the primary host, with humans following, in the transmission to mosquitoes, as they serve as an amplifier host. Generally, pigs remain asymptomatic but anorexia and pyrexia may be seen in a few cases. In large outbreak, decrease in reproductive performance, stillbirth and congenital deformity may be noticed. The disease in pigs causes high mortality in newborn piglets and zero in adults.

In horses, the incubation period lasts between 8 and 10 days. The disease in horses is rare and most of the infections are subclinical. Horses in poor health may exhibit symptoms such as fever, loss of appetite, difficulty moving, and grinding of teeth. Although blindness, coma, and death can occur, the fatality rate is generally low.

Clinical Manifestations in Humans

In humans, the incubation period ranges from 5 to 15 days, and the disease has a mortality rate of about 25–30%. Humans normally remain asymptomatic and less than 1% infected person can show clinical manifestation. Malnourished children from rural area are mainly affected by the infection and primary symptoms are fever, chills, muscle aches, headache, and vomiting. In severe cases, the disease pattern may include high fever, headache, weakness, acute encephalitis, stiff neck, changes in mental status, flaccid paralysis, seizures, convulsions, coma, and succumbing to death; 50–70% patient may show neurologic squeal such as seizures, paresis, movement disorders, a Parkinsonian syndrome with tremors, and rigidity and/or mental retardation after surviving from the disease. Abortion can occur in pregnant women.

Diagnosis

In clinical practice, distinguishing this disease from other types of acute encephalitis syndrome (AES) is challenging, which is why laboratory confirmation is crucial.

Detection of Antibody

The World Health Organization recommends the use of the specific IgM antibody capture ELISA (MAC-ELISA) as a primary diagnostic method. It is widely used for the quick detection of the disease in serum or cerebrospinal fluid samples.

Detection of Antigen

Reverse passive hemagglutination, immunofluorescence or staphylococcal coagglutination test is done to identify JEV in CSF peripheral blood of patient by using polyclonal or monoclonal antibodies. The plaque reduction neutralization test (PRNT) is regarded as the definitive method for diagnosing flavivirus infections.

Isolation of virus: Viruses can be isolated by introducing the clinical sample into the brains of suckling mice. Commonly used for virus isolation are chick or duck embryo cells, as well as LLCMK2, C6/36, PK, and AP61 cells. However, due to the temporary and low levels of viremia, virus isolation is not the preferred method for virus identification.

Molecular test: Viral RNA can be identified using methods such as reverse transcriptase PCR, Real Time PCR, and reverse transcription loop-mediated isothermal amplification (RT-LAMP) combined with a lateral flow dipstick (LFD).

Prevention and Control Strategies

Vector control and vaccination are the main keys for the prevention of disease.

Vector control can be done by including the following measures:

- Elimination of potential mosquito breeding areas.
- Relocation of pigs and practice of environmental sanitation.
- Reduction of larvae breeding source by intermittent irrigation system.
- Use of mosquito net for humans as well as pigs may be a classical control measure.
- Use of larvivorous fishes.
- To prevent mosquito bites, apply insect repellent and dress in long-sleeved shirts and pants.
- Make sure to get vaccinated before traveling to an area where the disease is common.
- Location of pig farm should be away from the human habitat.
- In areas where the disease is common, it is important to routinely vaccinate pigs.
- Strong surveillance and monitoring system should be set along with sustainable campaigning program in community for making awareness.

Vaccination

In various Asian nations, vaccines are available for humans, pigs, and horses. There are two vaccine types: modified live vaccines, which are developed using hamster or swine kidney tissue culture or a hamster lung (HmLu) cell line, and inactivated vaccines, which are produced in mouse brain, chick embryo, or cell lines such as Vero cells.

Type of vaccine	Strain	Efficacy
Inactivated mouse brain vaccine	NakayamaBeijing-1	91%
Inactivated primary hamster kidney cells vaccine	P-3	85%
Live attenuated primary hamster kidney cells vaccine	SA 14–14-2	>95%

Treatment

There is no antiviral medication available to prevent the disease, and treatment primarily involves supportive care. It is essential to apply the standard management strategies for viral encephalitis. The documented therapeutic benefits of antioxidants, including minocycline, arctigenin, fenofibrate, and curcumin, are noteworthy.

Bibliography

Filgueira, L., & Lannes, N. (2019). Review of emerging Japanese encephalitis virus: New aspects and concepts about entry into the brain and inter-cellular spreading. *Pathogens, 8*(3), 111. Published 2019 Jul 26. https://doi.org/10.3390/pathogens8030111

Kulkarni, R., Sapkal, G. N., Kaushal, H., & Mourya, D. T. (2018). Japanese encephalitis: A brief review on Indian perspectives. *Open Virology Journal, 12*, 121–130. Published 2018 Aug 31. https://doi.org/10.2174/1874357901812010121

Sehgal, A., & Dutta, A. K. (2003). Changing perspectives in Japanese encephalitis in India. *Tropical Doctor, 33*(3), 131–134. https://doi.org/10.1177/004947550303300304

Solomon, T., Dung, N. M., Kneen, R., Gainsborough, M., Vaughn, D. W., & Khanh, V. T. (2000). Japanese encephalitis. *Journal of Neurology, Neurosurgery, and Psychiatry, 68*(4), 405–415. https://doi.org/10.1136/jnnp.68.4.405

Tiwari, S., Singh, R. K., Tiwari, R., & Dhole, T. N. (2012). Japanese encephalitis: A review of the Indian perspective. *The Brazilian Journal of Infectious Diseases, 16*(6), 564–573. https://doi.org/10.1016/j.bjid.2012.10.004

Turtle, L., & Solomon, T. (2018). Japanese encephalitis—The prospects for new treatments. *Nature Reviews. Neurology, 14*(5), 298–313. https://doi.org/10.1038/nrneurol.2018.30

Influenza

24

Ripan Biswas and Indranil Samanta

Abstract

Influenza is a highly infectious respiratory condition caused by influenza viruses, specifically types A, B, C, and D. Of these, influenza A viruses are particularly concerning due to their ability to cause pandemics and infect a diverse array of hosts, including humans, birds, and pigs. Bird flu, scientifically referred to as avian influenza, is a contagious respiratory illness triggered by influenza A viruses. These viruses are typically present in wild bird species but can also spread to domestic poultry and, on rare occasions, pose a threat to human health. Highly virulent strains, such as H5N1 and H7N9, present substantial public health threats due to their capacity to cause severe illness and high fatality rates. Swine influenza, which is an infection caused by influenza A viruses in pigs, can also spread to humans, as demonstrated by the H1N1 pandemic in 2009. Influenza viruses from birds and pigs have the ability to merge their genetic material, resulting in new strains that could lead to widespread outbreaks. Effective prevention and control strategies for influenza rely on robust surveillance, vaccination programs, and stringent biosecurity measures.

Keywords

Flu · Bird flu · Swine flu · Influenza

R. Biswas (✉)
Department of Veterinary Public Health and Epidemiology, West Bengal University of Animal and Fishery Sciences, Kolkata, West Bengal, India

I. Samanta
Department of Veterinary Microbiology, West Bengal University of Animal and Fishery Sciences, Kolkata, West Bengal, India

R. Biswas et al. (eds.), *Concept of One Health and Zoonoses*,
https://doi.org/10.1007/978-981-95-6627-3_24

History

Influenza is an acute respiratory illness, which can spread very fast and produce outbreaks.

- In the sixth book of "Epidemics," Hippocrates detailed a highly infectious illness that exhibited symptoms similar to flu.
- In 1357, an outbreak of flu was recorded in Florence, Italy, termed "influenza di freddo."
- In 1580, the first flu pandemic was recorded which began with Asia and later spread to Africa, Europe, and Americas.
- In 1729, influenza pandemic was documented in Russia and spread across the world within three years.
- In 1781, an outbreak began in China and then extended to Russia, eventually making its way to Europe and North America over the course of the next year.
- Between 1830 and 1833, another pandemic emerged, originating in China. It spread to the Philippines, Indonesia, and India. After impacting Russia and Europe, it eventually made its way to North America.
- The first modern flu pandemic, known as the Russian Flu, occurred in 1889, affecting 40% of the global population and resulting in approximately 1 million deaths.
- The Spanish Flu (H1N1) of 1918 was the most lethal pandemic of the twentieth century, affecting one-third of the global population and resulting in the deaths of 20–50 million people. This outbreak was triggered by a novel strain of the influenza A virus that originated in birds (H1N1).
- In 1933, scientists at the National Institute for Medical Research in London successfully extracted the influenza A virus from the nasal secretions of humans who were infected.
- The influenza vaccine was first authorized for use by the military in the United States in 1945, followed by approval for civilian use in 1946.
- In 1952, the World Health Organization established the inaugural system for tracking the spread of influenza virus strains.
- The Asian Flu pandemic, caused by the H2N2 virus, began in Singapore in 1957. It is estimated that the outbreak resulted in 1.1 million deaths globally, with 116,000 fatalities occurring in the United States.
- The Hong Kong flu (H3N2) emerged in China in 1968 and subsequently spread to Southeast Asia and the United States.
- In 1978, the first trivalent flu vaccine was developed.
- In 2009, Swine influenza, a new influenza H1N1 virus, was detected in the United States.

Etiological Agent and Ecology

Influenza viruses are classified within the Orthomyxoviridae family and are divided into four unique types: *Alphainfluenza virus* (influenza A), *Betainfluenza virus* (influenza B), *Gammainfluenza virus* (influenza C), and *Deltainfluenza virus* (influenza D). The viruses range in size from 80 to 120 nm. Their classification relies on the unique characteristics of their internal nucleoprotein and matrix proteins. These viruses possess segmented genomes and are composed of negative-strand RNA. The RNA genome, which is segmented, codes for 11 proteins. These include nine structural proteins: polymerase proteins such as polymerase basic protein 1 (PB1), PB1-F2, PB2, and polymerase acidic protein (PA); surface glycoproteins like haemagglutinin (HA) and neuraminidase (NA); nucleocapsid protein, specifically nucleoproteins (NP); matrix proteins (M1); and ion pore protein (M2). Additionally, there are two nonstructural proteins, NS1 and NS2. The function of these various proteins is discussed here:

HA	Facilitates the virus's entry into the host cell
NA	Discharges and disseminates viral particles from target cells
PA	Facilitates the virus's entry into the host cell. Involves endonuclease activity and assists in the cap-snatching process
PB1	Viral mRNA transcription by 5′ cap snatching mechanism
PB2	Allows for unprimed replication of viral mRNA
PB1-F2	Host cell apoptosis
M1	Functions in connecting viral ribonucleoprotein to the cell membrane and ensures stability
M2	Facilitates the transfer of viral ribonucleoprotein from the endosome to the cytoplasm
NP	Facilitates the movement of viral ribonucleoprotein into and out of the nucleus, as well as the virus's replication process
NS1	Inhibits the production of interferon beta and host protein production
NS2	Facilitates the nuclear import of viral ribonucleoprotein

Although these viruses share a common genetic origin, they have diverged genetically, which has led to the reassortment or exchange of viral RNA within the same genus. Reassortment does not occur across the genus or type. Viruses are classified based on their matrix proteins and internal nucleoproteins, which remain constant for each viral type. Influenza viruses A, B, C, and D are classified based on their unique internal nucleoprotein and matrix proteins, which are distinct for each type.

Representation	Influenza A	Influenza B	Influenza C	Influenza D
RNA segments	8	8	7	7
Viral proteins	10	11	9	9
Subtype	Various combinations of 16 HA & 9NA	No subtype	No subtype	No subtype
Host	Humans, swine, equine, avian, canine, marine mammals	Humans only	Humans, swine, and canine	Cattle and pigs
Occurrence	Frequent	Low	Rare	Rare
Outbreaks	Pandemic/epidemic	Epidemic	Sporadic	Sporadic
Mutation/ Reassortment	High (antigenic shift/ antigenic drift)	Low (only antigenic drift)	No reassortment	No reassortment
Reservoir	Wild aquatic birds	No	No	No

Influenza A is characterized by glycoprotein spikes known as H or HA (Hemagglutinin) and N or NA (Neuraminidase), featuring distinct globular "head" and "stalk" formations. HA assists in the virus's entry into the host cell, while NA aids in the virus's release. These three integral membrane proteins—HA, NA, and M2 (an ion channel protein)—are situated on a matrix composed of M1 (matrix protein), which surrounds the genetic material, nucleoprotein (NP), and two non-structural proteins, NS1 and NS2, also known as nuclear export protein (NEP). There are 18 different HA and 11 different NA and more than 130 combinations, which have been identified in nature. Currently, the most frequently circulating types in humans are H1, H2, and H3, in combination with N1 and N2. Through electron microscopy, influenza A and influenza B viruses appear almost identical. Influenza A and influenza B viruses possess two glycoproteins, namely HA and NA, whereas influenza C and influenza D viruses contain a single glycoprotein called hemagglutinin-esterase fusion (HEF). In 2016, influenza D was assigned to a new genus, whereas previously, it was regarded as a subtype of the influenza C virus. The organization of the viral genome and the functions of its proteins play a crucial role in enabling both antigenic drift and antigenic shift. In influenza A and influenza B, antigenic drift is driven by the gradual build-up of point mutations in the HA and NA genes. As a result, it is essential to continually update human influenza vaccines to align with the changing strains.

Antibodies targeting HA neutralize the virus's ability to infect. As a result, virus strains undergo repeated amino acid alterations at the antigenic sites. These relatively minor modifications accumulate through a process known as antigenic drift. Antigenic drift refers to a gradual genetic change caused by mutations, allowing the virus to adapt effectively within the human population.

A reassortment of genetic material of viruses from different species that occurs resulting in variability of HA spikes is called antigenic shift. However, the basic structure of the virus remains the same. This process only takes place in influenza A virus. Antigenic shift can be described as the complete renewal of surface antigen of the virus. Influenza pandemic takes place when antigenic shift occurs and a the novel viral strain that spreads rapidly within an immunologically naive human population.

Epidemiology

Influenza virus infection is a highly contagious airborne disease that can manifest as a pandemic, epidemic, outbreak, or occasionally as sporadic cases. Influenza A virus is associated with pandemics involving both animals and humans, whereas influenza B virus typically leads to epidemics, causing illness and death in humans. Seasonal epidemics are primarily caused by influenza A and influenza B. Currently, the most prevalent seasonal flu strains in humans are influenza A H1N1 and influenza H3N2. Influenza C is less common and usually results in mild symptoms or none at all. The influenza virus is most prevalent during winter. Globally, there are two flu seasons because the southern and northern hemispheres experience winter at different times. Consequently, the WHO recommends two different vaccine strains each year for the two hemispheres.

Bird species, particularly those found along shores and in aquatic environments, serve as the primary carriers of influenza A. The infection remains asymptomatic in reservoir host, which dispenses the virus into the water bodies where virus can survive for few weeks. Every year, 30% of new juvenile flock is infected and becomes reservoirs. Aged persons, pregnant women, and infant are highly susceptible to influenza infection. Low relative humidity and cold are the idle conditions for the spread of influenza virus. Chronic lung disease, chronic obstructive pulmonary disease, asthma, heart disease, chronic kidney disease, diabetes, chronic metabolic disorder, obesity, anemia, liver diseases, AIDS, steroids treatment, long-period aspirin therapy, and chemotherapy may act as the cofactor for influenza virus infection. Since 1918, four pandemics (1918, influenza AH1N1; 1957, influenza AH2N2; 1968, influenza AH3N2 and 2009, influenza AH1N1) have occurred when the influenza A virus came from either avian or swine population though those viruses successively became seasonal epidemic strains in subsequent years.

Avian Influenza/Bird Flu

Avian influenza (AI), commonly referred to as bird flu or highly pathogenic avian influenza (HPAI), is a zoonotic viral disease that affects poultry, swine, and humans. Viruses that cause severe illness and high death rates in poultry are classified as highly pathogenic avian influenza (HPAI), whereas those that result in milder symptoms in poultry are known as low pathogenic avian influenza (LPAI). The H5N1 strain of the influenza A virus is responsible for the disease. In 1996, the H5N1 strain was first isolated from a goose in Guangdong Province, China (A/goose/Guangdong/1/96). Since then, the H5N1 subtype has been detected in domestic poultry or wild birds across 61 countries. The first human cases of Asian H5N1 were identified in 1997 during a poultry outbreak in Hong Kong. Highly pathogenic avian influenza (H5N1) impacts both commercial and household poultry, while wild migratory birds facilitate the spread of the virus across international borders. The avian influenza virus subtypes A(H5N1), A(H7N9), and A(H9N2) have the potential to infect humans, although most human cases have been linked to the H5N1 and

H7N9 strains. Transmission is typically associated with direct or indirect contact with infected live or deceased poultry. The persistence of zoonotic avian influenza is attributed to its extensive, undetected presence in waterfowl. The initial step toward a pandemic is the transmission from animals to humans, while the eventual widespread of outbreak occurs when the virus spreads between humans. For a virus to spread from one person to another, it must adapt to a new host by undergoing mutation or reassortment.

Transmission

Wild birds, especially migratory ducks and geese, are the main reservoir host for avian influenza. Domestic ducks act as the consistent intermediaries between the reservoir and domestic poultry. The transmission of virus occurs through contaminated water body. While most influenza A virus subtypes do not cause symptoms in birds, the H5Nx and H7Nx strains can evolve into highly pathogenic avian influenza (HPAI) forms. These strains can lead to death in chickens, turkeys, and certain domestic duck breeds by developing a new cleavage site in the hemagglutinin (HA) protein. Backyard farming takes a very crucial role in transmission of avian influenza virus in multiple domestic poultry species because in backyard farming system, multiple species are reared together. Beside this, from live poultry markets, the transmission happens in multi species. The spread of the virus to commercial poultry farms can also happen due to inadequate biosecurity measures. Humans primarily contract the avian influenza virus through three main routes: direct contact with infected birds or poultry products, exposure to contaminated surfaces, and inhalation of virus-laden aerosols.

Mixing Vessel Host

In mixing vessel host, co-infection occurs simultaneously by two or more influenza A subtypes and reassortment takes place, resulting in a new genotype/phenotype. The most common receptor for influenza A virus is α-sialic acid. In the human respiratory system, sialic acid α-2, 6 galactose receptors (SAα 2,3 Gal) are found, which are typically referred to as human receptors. In contrast, sialic acid α-2, 3 galactose receptors (SAα 2,3 Gal) are mainly located in the avian intestinal tract and are known as avian receptors. To put it simply, humans exclusively have α-2, 6 linked sialic acid receptors, while birds only possess α-2, 3 linked sialic acid receptors. However, pigs contain both types of receptors, making them a mixing vessel for generating various avian and human reassortments.

Pathogenesis

Due to previous infection or vaccination, humans carry partial immunity against influenza A virus but if this pre-exposure is not ensued, then the transmission and replication of virus occur because of novel HA or NA. Because of this novel antigen, a "cytokine storm" is triggered and successively severe symptoms appear with fever, chills, headache, muscle pain, sore throat, malaise, anorexia, and pneumonia.

Clinical Manifestations in Poultry

The OIE states that a 14-day incubation period is required at the flock level for high-pathogenicity avian influenza. All virulent avian influenza strains are designated as highly pathogenic notifiable avian influenza (HPNAI) viruses. Strains with H5 or H7 and an HA0 cleavage site are classified as HPNAI viruses, while those with H5 or H7 but without an HA0 cleavage site are identified as low-pathogenicity notifiable avian influenza (LPNAI) viruses. Avian influenza strains that do not have H5 or H7 and are not highly pathogenic to poultry are considered low-pathogenicity avian influenza (LPAI) viruses.

When the condition is mild, symptoms manifest as ruffled feathers, a drop in egg production, and slight respiratory issues. In contrast, the severe form affects the respiratory system more intensely and involves other organs, resulting in swelling and congestion in the comb and wattle, puffiness beneath the eyes, profound lethargy, coughing, sneezing, neurological symptoms, diarrhea, hemorrhages on the hock, and a sharp decrease in egg production. Initially, mortality rates may be inconsistent for a few days, but this is followed by total mortality within 48 h.

Clinical Manifestations in Humans

The avian influenza virus typically incubates in humans for a period of 3–5 days. Younger individuals are particularly at risk and often experience more severe symptoms. Unlike seasonal influenza, zoonotic avian influenza is characterized by differences in viral load behavior, the virus's ability to spread beyond the respiratory tract, and the onset of hypercytokinemia. Symptoms may include a high fever (exceeding 38 °C), general malaise, sore throat, cough, and muscle aches. Early symptoms can also involve abdominal pain, chest pain, and diarrhea. The infection can rapidly progress to serious respiratory issues, such as difficulty breathing, pneumonia, acute respiratory distress syndrome (ARDS), and changes in mental status or seizures.

Diagnosis in Poultry

- Identification of viral RNA can be achieved through reverse transcription polymerase chain reaction (RT-PCR) or real-time reverse transcription polymerase chain reaction (qRT-PCR).
- Viruses are isolated using specific pathogen-free embryonated eggs, and HA sub-typing is subsequently performed through hemagglutination and hemagglutination inhibition tests, as well as RT-PCR and real-time RT-PCR.
- The presence or absence of antibodies to influenza A virus infection is identified using agar gel immunodiffusion tests. The hemagglutination inhibition test is conducted to ascertain whether antibodies indicating influenza A virus infection belong to different subtypes, such as H5 and H7.

Diagnosis in Man

Diagnosis of avian influenza virus infections in man is similar to that of poultry though RT-PCR is commonly done.

Prevention and Control Strategies for Poultry

- Good biosecurity and hygiene including prevention of contact with domestic or wild birds, contaminated fomites and water sources should be followed.
- All-in/all-out flock management is very useful to prevent the infection in poultry farm.
- Insect and rodent control are recommended.
- Depopulation of infected flock is suggested along with cleaning and disinfection.
- Movement control, quarantine, and sometimes vaccination are to be taken as control measure.

Prevention and Control Strategies for Humans

- Avoid direct contact with birds that are infected or at risk, as well as their secretions or droppings, without protection.
- Select PPE that is appropriate for the worker and use PPE properly.
- Use respiratory protection.
- Proper cleaning and disinfection should be done in contaminated area.
- Good hand hygiene and decontamination practices are recommended.
- Receive your flu vaccination annually. Each year, the trivalent inactivated vaccines and the live attenuated influenza vaccine are formulated to target the strains most likely to cause illness in the upcoming winter. The trivalent vaccines include two strains of influenza A and one strain of influenza B.

- Implementing biosecurity measures and ensuring that workers adhere to these protocols are crucial to preventing the entry of avian influenza and other diseases into poultry populations.
- Participate in health surveillance and monitoring programs.

Treatment

Poultry: Infected poultry flocks with HPAI virus are normally depopulated, but in case of LPAI infection, birds are segregated and supportive treatment is done.

Neuraminidase inhibitors, such as oseltamivir, zanamivir, and peramivir, are the most frequently used treatments for influenza. Antiviral medications are not given to animals for human influenza treatment to prevent the development of drug resistance.

Swine Influenza/Swine Flu

Swine flu or H1N1 subtype of influenza A virus has appeared as one the most infectious diseases in the recent years as pandemic or epidemic throughout the world. In 2009, the disease was declared as pandemic by WHO and this originated from Mexico. The virus was first isolated from pig in the USA in the year 1930. There are five subtypes of influenza A that have been identified as causing disease: H1N1, H1N2, H2N3, H3N1, and H3N2. Out of these subtypes, only H1N1 influenza A strain was isolated from humans while rest of the four subtypes were found in pigs exclusively. It is intriguing that pigs can contract a range of avian influenza viruses, yet only the H1 and H3 hemagglutinins are capable of becoming enzootic swine viruses. The present Eurasian H5N1 avian virus can infect the pigs sporadically but replicate poorly.

Waterfowl serve as the natural reservoir hosts. We have previously discussed the distinctive role of pigs as a "mixing vessel" for influenza viruses due to their vulnerability to infections from both avian and human influenza viruses.

Individuals who handle pigs, such as pig farmers and pork processors, as well as those who work closely with pigs, face a heightened risk of contracting infections from pigs via airborne particles. Human-to-human transmission occurs when people inhale droplets expelled during coughing and sneezing, through direct contact, or by touching contaminated surfaces. The 2009 H1N1 swine flu in humans was not considered a zoonotic disease because it spread only from person to person. It is also now understood that consuming cooked pork does not transmit the virus.

Oseltamivir and Zanamivir are the two drugs that are approved and drugs of choice.

Prevention and Control

Prevention in pigs is done by applying proper facility management like disinfection, movement restriction of herd, and vaccination.

Prevention of transmission to humans is done by following the very basic principle as if the virus does not reach an individual mucus membrane cells, which include hand hygiene practice with soap and sanitizer, avoiding crowds, maintaining physical distance (approximately 6 feet) and use of N95 mask.

According to the World Health Organization, it is recommended that healthcare professionals such as doctors, nurses, paramedics, and ambulance staff who might come into contact with suspected or confirmed swine flu cases be vaccinated.

Bibliography

Bouvier, N. M., & Palese, P. (2008). The biology of influenza viruses. *Vaccine, 26*(Suppl 4(Suppl 4)), D49–D53. https://doi.org/10.1016/j.vaccine.2008.07.039

Cohen, J., Enserink, M., & Swine flu. (2009). After delays, WHO agrees: The 2009 pandemic has begun. *Science, 324*(5934), 1496–1497. https://doi.org/10.1126/science.324_1496

Dangi, T., & Jain, A. (2012). Influenza virus: A brief overview. *Proceedings of the National Academy of Sciences, India Section B: Biological Sciences, 82*(1), 111–121. https://doi.org/10.1007/s40011-011-0009-6

Mahapatra, A. (2009). Swine flu pandemic: Mission accomplished? *ACS Chemical Biology, 4*(7), 485–486. https://doi.org/10.1021/cb9001552

Peiris, J. S., de Jong, M. D., & Guan, Y. (2007). Avian influenza virus (H5N1): A threat to human health. *Clinical Microbiology Reviews, 20*(2), 243–267. https://doi.org/10.1128/CMR.00037-06

Peteranderl, C., Herold, S., & Schmoldt, C. (2016). Human influenza virus infections. *Seminars in Respiratory and Critical Care Medicine, 37*(4), 487–500. https://doi.org/10.1055/s-0036-1584801

Shriner, S. A., & Root, J. J. (2020). A review of avian influenza a virus associations in synanthropic birds. *Viruses, 12*(11), 1209. Published 2020 Oct 23. https://doi.org/10.3390/v12111209

Kyasanur Forest Disease (KFD) 25

Maninder Singh

Abstract

Kyasanur forest disease (KFD) is an emerging zoonotic disease in Southern India, especially in Western Ghats. The disease is an occupational hazard for the forest dwellers and hunters in these areas, as the transmission occurs through an endemic cycle where the hard ticks play an important role. The disease is expanding its geographical distribution into the surrounding states of India due to deforestation and other climatic factors.

Keywords

KFD · *Haemaphysalis* · Monkey fever · Monkey disease

History

- Kyasanur forest disease virus (KFDV) was first identified in 1957 from an outbreak in monkeys (black faced langur and red-faced bonnet monkey) in Kyasanur forest area of Shimoga (Shivamogga) district, Karnataka, India. Concurrent cases of febrile illness and deaths in humans were reported.
- From 1957 to 2012, KFDV was reported from more districts of Karnataka, but restricted to Karnataka state.
- From 2012, KFDV from Tamil Nadu, Kerala, Goa and Maharashtra along with Karnataka is getting reported.

M. Singh (✉)
Department of Veterinary Public Health and Epidemiology, College of Veterinary Science, Guru Angad Dev Veterinary and Animal Sciences University, Rampura Phul, Bathinda, India

R. Biswas et al. (eds.), *Concept of One Health and Zoonoses*,
https://doi.org/10.1007/978-981-95-6627-3_25

Aetiology

KFD is caused by KFDV of family Flaviviridae and is an arbovirus. The virus belongs to a serocomplex that includes tick-borne encephalitis, Omsk haemorrhagic fever and louping-ill viruses. KFDV is a single-stranded RNA virus; ssRNA is of positive polarity encoding a single polypeptide which after post-translational cleavage can generate three structural proteins and seven non-structural proteins.

Epidemiology

KFDV is an arthropod borne virus, which is being transmitted and maintained by ticks of *Haemaphysalis* species in an area. Small mammals, especially rodents (squirrels, rats) and shrews, are the natural reservoirs of KFDV, while monkeys are amplifier hosts in the life cycle of KFDV, as the viremia is much pronounced in monkeys than in rodents. The infection may occur in domestic animals including cattle. The role of ground-dwelling birds in the life cycle is also reported. Antibodies against KFD have been demonstrated in bats also, although their role in the transmission cycle is not established. Humans are the accidental dead-end hosts of KFDV. The infection among the animals and humans is transmitted through ticks. Various *Haemaphysalis* species have been documented to maintain infection with the most important as *H. spinigera*. Other *Haemaphysalis* species of significance is *H. turtuis*. Trans-stadial and trans-ovarian transmissions maintain the infection among endemic areas.

KFDV is endemic in the Western Ghats of India. The area has dense forest areas providing the ecological requirements for the propagation of virus. The virus was first time identified in 1957 from an outbreak in monkeys (black-faced langur and red-faced bonnet monkey) in Kyasanur forest area of Shimoga district, Karnataka, India. There were concurrent cases of febrile illness and deaths in humans. Since then 400–500 human KFD cases were reported annually in Karnataka. From 1957 to 2012, KFD cases were reported from Shimoga, Chikmagalur, Uttara Kannada, Dakshina Kannada and Udupi districts of Karnataka. However, from 2012, KFDV has been reported from some other neighbouring states also. In 2012, KFDV activity was reported from monkeys of Nilgiris, Tamil Nadu. However, there have not been further reports of KFDV from Tamil Nadu state. During 2013–2015, there were multiple detections from Kerala involving humans, monkeys and ticks. In 2016, Maharashtra state of the country had an outbreak in Dodamarg taluk, Sindhudurg district with case fatality ratio of 2.3% in humans. This was the first outbreak in the state and the virus activity was detected in humans, monkeys and ticks. In 2017, Banda taluk, Sindhudurg district had an outbreak with a CFR of 12.3%. Goa reported the first human outbreak in 2016 in Dharbandora taluk with another outbreak in 2017 involving bigger areas.

The viruses similar to KFDV are reported in Saudi Arabia and China. In Saudi Arabia, the strain is responsible for Alkhurma haemorrhagic fever.

Transmission

Tick, *Haemaphysalis* species, is responsible for the transmission of KFD. The tick has four-stages in its life cycle viz., egg, larva, nymph and adult. The adults tick generally feeds on adult animals, especially cattle; after taking its blood meal on cattle, adult tick falls down and lays eggs on the ground which hatch to liberate larvae. The larvae feed on small mammals and monkeys and drop on the ground to change into nymphs. Nymphs further feed on small mammals, birds and accidentally humans leading to human infections. After feeding, nymphs fall from the host on the ground, mature to adult and repeat the cycle. The nymphal stages in the life cycle of tick are during the dry season (November–May) and thus the human cases are mostly reported in November–May duration. The nymphs are extremely anthropophilic in nature. Frequently, human outbreaks are preceded with the outbreaks in monkeys in the endemic areas. In humans, transmission also occurs through direct contact with secretions of the dead monkey. Infected tick can directly transmit the infection to humans and large animals. In cattle, viraemia is less pronounced while in monkey, viremia is much pronounced. Humans are dead-end hosts.

Deforestation is one of the most significant factors associated with the expansion of disease in humans. The transmission is vector borne, i.e. tick, which thrives on small mammals of forest areas, and the cases are concentrated in the forest areas. Deforestation led to exposure to the infected tick or propagate migration of small mammals towards the human dwellings leading to human infections. The occupational exposure in hunters, animal keepers and forest dwellers is frequently reported.

Pathogenesis

The pathogenesis of KFDV is not well known despite being detected since 1957 due to the lack of suitable in vivo and in vitro models of disease. The proposed pathogenesis is that on entry in the body through tick bite or contact with infected animals, the virus infects antigen-presenting cells (APCs) such as macrophage and dendritic cells; the entry in the cells is through receptor-mediated endocytosis via protein E as indicated in other flaviviruses. APCs present virus to T cells, which leads to host immune response by Type I interferon production.

Clinical Sings

The incubation period of KFD in humans is 3–8 days and clinical presentation is generally biphasic. The onset of disease is sudden with fever, chills and headache, body ache especially on the neck, upper and lower back and extremities. The rise in temperature is continuous and lasts 5–12 days. Conjunctival inflammation of both sclera and palpebra is frequently reported. Gastrointestinal symptoms may appear. Haemorrhagic manifestations may occur in certain cases with the haemorrhages from the nose, gums, sputum and stool. Enlargement of cervical lymph node is reported. Majority of cases recover within 1–2 weeks without any complications. In about 20% of patients, after 1–2 weeks, the fever reappears with certain neurological symptoms such as headache, neck stiffness, tremors and mental disturbance. Case fatality rate is 2–10%, especially in newly infected areas where herd immunity is not prevailing.

The disease among animals is much pronounced in monkeys with the clinical course similar to that of humans. The monkeys have high viraemia after infection, leading to haemorrhagic manifestations and mortality. In cattle, viraemia is lower.

Diagnosis

The diagnosis involves the virus isolation, molecular detection of virus and/or demonstration of antibody in serum. For the virus isolation, traditionally, the serum from infected humans are injected into suckling mice. Other tests widely used during the early years of disease have been hemagglutination inhibition, virus neutralization and complement fixation test. The preferred samples for these tests are blood of patients (in acute phase), blood and tissues from dead monkeys and tick pools. As KFD virus belongs to biosafety level 4, the development of diagnostic tests was not kick-started till 2006 when BSL-3 facilities at the ICMR-National Institute of Virology, Pune, were established. After the establishment of the facility, molecular antigen detection assays such as real-time PCR were developed and are frequently used in confirmation of disease cases in the endemic areas. IgM and IgG ELISA to detect seroconversion have been developed for the monitoring and surveillance purposes.

Control and Prevention

The prevention of KFD revolves around the prevention of exposure to the tick bite. To control tick population in affected areas, forest floors can be treated with gamma-hexachlorocyclohexane. The usage of tick repellents such as N, N-dimethyl-meta-toluamide (DEET) and dimethyl phthalate helps in the prevention of tick bite. Further, wearing of light-coloured clothing, long-sleeved shirt tucked into pants, long-legged pants tucked into socks and gumboots diminish the chances of tick bite in endemic areas. Tick repellents for the domestic cattle will help in the reduction of ticks on the cattle. Tourists should avoid lying down in the forest areas.

To control and prevent disease in humans in the endemic areas, formalin-inactivated tissue culture vaccine is given to the individuals. Mass vaccination programme during August–November in the endemic areas of Karnataka is employed. Two-dose regimens with the gap of one-month time interval in 7–65 years individuals are administered. However, the immunity is short lived and annual booster dose is recommended for 5 years after the last confirmed case in the area. Vaccination is usually administered within a range of 5 km of the affected area.

Bibliography

Mourya, D. T., Yadav, P. D., Patil, D. Y., Sahay, R. R., & Rahi, M. (2021). Experiences of Indian Council of Medical Research with tick-borne zoonotic infections: Kyasanur Forest disease & Crimean-Congo haemorrhagic fever in India with One Health focus. *The Indian Journal of Medical Research, 153*(3), 339.

Munivenkatappa, A., Sahay, R. R., Yadav, P. D., Viswanathan, R., & Mourya, D. T. (2018). Clinical & epidemiological significance of Kyasanur forest disease. *The Indian Journal of Medical Research, 148*(2), 145.

Shah, S. Z., Jabbar, B., Ahmed, N., Rehman, A., Nasir, H., Nadeem, S., Jabbar, I., Rahman, Z. U., & Azam, S. (2018). Epidemiology, pathogenesis, and control of a tick-borne disease-Kyasanur forest disease: Current status and future directions. *Frontiers in Cellular and Infection Microbiology, 8*, 149.

Crimean–Congo Hemorrhagic Fever 26

Maninder Singh

Abstract

Crimean–Congo hemorrhagic fever is a severe viral illness caused by the CCHF virus. It is transmitted mainly through tick bites or contact with infected animal blood. CCHF virus is endemic in Africa, the Balkans, the Middle East, and Asia. The disease is characterized by fever, myalgia, and hemorrhagic manifestations in severe cases. The disease has a case-fatality rate ranging from 10% to 40%. Preventive measures emphasize tick control, safe animal-handling practices, and implementation of infection control protocols. Prompt diagnosis and supportive management are essential, despite the lack of widely accessible specific treatment or vaccine. Crimean–Congo hemorrhagic fever persists as a substantial public health concern owing to its elevated fatality rate and capacity for epidemics.

Keywords

CCHF · Congo fever · Hemorrhagic fever

History

- Crimean–Congo hemorrhagic fever (CCHF) was first identified in the Crimea region of the former Soviet Union in 1944, and in the Belgian Congo (present-day Democratic Republic of the Congo, DRC) in 1956. These independent reports were later linked in 1969, resulting in current virus naming.

M. Singh (✉)
Department of Veterinary Public Health and Epidemiology, College of Veterinary Science, Guru Angad Dev Veterinary and Animal Sciences University, Rampura Phul, Bathinda, India

R. Biswas et al. (eds.), *Concept of One Health and Zoonoses*,
https://doi.org/10.1007/978-981-95-6627-3_26

Etiology

CCHF is a disease that can be transmitted from animals to humans, primarily through tick bites. The causative agent is CCHF virus (CCHFV), which belongs to the orthonairovirus genus within the Nairoviridae family, part of the order Bunyavirales. This virus has a spherical structure measuring 80–100 nm in diameter, with a 5–7 nm-thick lipid envelope and glycoprotein projections extending 8–10 nm outward. The viral genome is composed of three segments of single-stranded RNA with negative polarity, categorized as small, medium, and large. These RNA segments interact with nucleocapsid proteins to form a ribonucleoprotein complex.

CCHF virus is categorized into several genomic clades based on its geographical distribution. West Africa is associated with Clade I, whereas Central Africa harbors Clade II. Clade III is prevalent in South and West Africa. The Middle East and Central Asia are characterized by Clade IV, and Europe and Greece are linked to Clades V and VI, respectively. Additionally, the Asian Clade is further differentiated into two distinct subgroups: Asia 1 and Asia 2. This phylogenetic categorization of the CCHF virus is determined through analysis of its small segments.

CCHF virus is a BSL-4 pathogen. Thus, the isolation of virus is restricted to few laboratories.

Epidemiology and Transmission

CCHF is found in over 30 countries across Africa, Asia, the Middle East, and certain parts of Europe. Each year, an estimated 10,000–15,000 human CCHF cases occur globally, with many going unreported or presenting as subclinical infections. The virus maintains its presence in nature through an endemic cycle involving ticks, vertebrate hosts, and back to ticks. *Hyalomma* ticks play a crucial role in this cycle, functioning as both reservoirs and vectors for CCHFV due to transstadial and transovarian transmission. Various wild animals, including herbivores and certain rodents, as well as domesticated animals like cattle, sheep, and goats, act as vertebrate hosts. These animals become infected when bitten by infected ticks, and the virus remains in their bloodstream for about a week after infection. This allows the tick-animal-tick cycle to continue when another tick feeds on the infected animal. Although infections in animals are subclinical, they can still harbor the virus and transmit it to ticks. Since animals show no clinical signs, CCHFV infections do not affect the economic aspects of livestock production. Humans, similar to animals, acquire the infection through tick bites. Cases are reported during months of increased livestock handling or heightened tick activity. Males employed in agriculture, animal husbandry, or healthcare are often disproportionately affected. Apart from *Hyalomma* species ticks, ticks of *Ixodes*, *Haemaphysalis,* and *Rhipicephalus* genera have been reported to be potential reservoirs of CCHF virus.

Animal blood or tissue contact, particularly during butchering, can also transmit the virus to humans. Furthermore, the infection can spread between individuals, notably in healthcare facilities where insufficient infection control protocols are in place, leading to hospital-associated transmission through blood and other bodily fluids.

Most avian species are typically resistant to infection, while ostriches exhibit a notable vulnerability and can display high infection rates in areas where the disease is endemic. These large flightless birds have been identified as the source of human infections in some cases. A notable incident occurred at an ostrich slaughterhouse in South Africa, where an outbreak was reported. Interestingly, despite their susceptibility, ostriches do not appear to manifest any visible signs of illness.

In the past, CCHF was primarily found in African and Asian nations. However, its geographic distribution has widened in recent times due to climate change and tick movement. New occurrences have been documented in Africa, Southern and Eastern Europe, the Middle East, India, and Asia. The first case in Spain was reported in 2013. Both Bulgaria and Spain have recorded locally acquired cases through tick bites. Since the year 2000, nine African nations (Kenya, Mali, Mozambique, Nigeria, Senegal, Sierra Leone, South Sudan, Sudan, and Tunisia) have documented their initial CCHF cases, signaling a considerable spread across the continent.

In India, the first case of CCHF was identified in 2010 in the Gujarat state. Subsequently, between 2010 and 2019, a total of 34 outbreaks were documented across the states of Gujarat, Rajasthan, and Uttar Pradesh.

In countries where CCHF is already endemic, a surge in human cases has been observed. Iraq has experienced a substantial increase, with 986 confirmed CCHF cases reported between 2021 and 2023, representing a significant jump from previous years. Concurrently, Pakistan has seen a continuous upward trend in CCHF incidents. From 2014 to 2020, the country has documented over 350 confirmed cases, with notable increases occurring during the specific months linked to the management of livestock.

Pathogenesis

The exact mechanisms of disease progression in humans remain elusive. The infection impacts various organ systems, with the liver and central nervous system (CNS) being primary targets. In the CNS, the disease induces gliosis, meningitis, and meningoencephalitis. Astrocytes, a specific type of glial cell, have been identified as the main cellular targets for infection within the CNS. This illness is also marked by abnormalities in blood coagulation and the development of petechial rash. Studies have revealed that monocyte-derived dendritic cells highly vulnerable to CCHF infection. Upon infection, these cells secrete cytokines capable of activating human endothelial cells, which may contribute to the occurrence of vascular leakage.

Clinical Signs

In animals: Infections in animals are only subclinical and therefore are not responsible for any economic losses.

In humans: The incubation period in humans varies based on the mode of transmission. Tick bites typically result in 1–3 days of incubation, with a maximum of 9 days. Exposure to contaminated blood or tissues usually leads to 5–6 days of incubation period, not exceeding 13 days. Symptoms emerge abruptly, characterized by fever, muscle pain, vertigo, neck discomfort and rigidity, back pain, headache, eye soreness, and light sensitivity. Initial symptoms may include nausea, vomiting, diarrhea, stomach pain, and sore throat, followed by drastic mood changes and disorientation. After 2–4 days, agitation may give way to drowsiness, melancholy, and fatigue, while abdominal pain might concentrate in the upper right quadrant, accompanied by noticeable liver enlargement. Additional clinical indicators include rapid heart rate, swollen lymph nodes, and a rash caused by skin bleeding, visible on internal mucous membranes like the mouth and throat, as well as on the skin. The small spots may progress to larger bruise-like patches and other bleeding manifestations. Liver inflammation is common, and severely affected individuals may experience rapid kidney decline, sudden liver failure, or lung failure after the fifth day of illness.

The disease is associated with high case-fatality ratio (CFR). CFR ranges from 5% to 80%. In Afghanistan, a recent study reported a CFR of 25.4% among hospitalized patients while the outbreaks in United Arab Emirates and China have a CFR of more than 70%. Turkey has reported a CFR of 10%.

Diagnosis

Although it is feasible to isolate viruses through cell culture, this method is seldom employed for diagnostic purposes due to its time-consuming nature and the need for stringent biosafety level 4 (BSL-4) protocols. As a result, the diagnosis of Crimean–Congo hemorrhagic fever (CCHF) primarily relies on a combination of molecular and serological techniques. Molecular diagnostics: Reverse transcriptase polymerase chain reaction (RT-PCR) is the most effective method for identifying active CCHF infections, especially within the initial 10–12 days after symptom onset. This technique targets conserved regions of the CCHFV genome, enabling high detection rates of viral RNA. Quantitative RT-PCR can measure viral load, which helps to predict the infection's severity. Serological diagnostics: Enzyme-linked immunosorbent assay is utilized to detect antibodies (IgM and IgG) against CCHFV. IgM antibodies typically become detectable around 4–6 days post-infection and can persist for several months. In contrast, IgG antibodies may remain detectable for years.

Control and Prevention

At present, there is no vaccine available for humans or animals, and ribavirin remains the main treatment for human infections. Successful management of CCHF in humans depends largely on symptom monitoring and implementing tick control strategies in animals. Symptom monitoring is crucial for early detection and prompt treatment of infected individuals. For tick control in animals, application of acaricides, maintaining hygienic conditions on farms to prevent tick breeding, and educating farmers on tick control are crucial. Further, sentinel surveillance of infection in animals can give a warning sign for the chances of a human outbreak in the area. The importance of this approach is underscored by a study conducted in Cameroon, which revealed a 56.0% seroprevalence among cattle in pastoral environments. This indicates a greater chance of the existence of CCHF in the study areas and thus the risk of transmission to humans.

References

Hawman, D. W., & Feldmann, H. (2018). Recent advances in understanding Crimean-Congo hemorrhagic fever virus. *F1000Research, 7*(4), 1715. https://doi.org/10.12688/f1000research.16189.1

Temur, A. I., Kuhn, J. H., Keshtkar-Jahromi, M., Pecor, D. B., & Apanaskevich, D. A. (2021). Epidemiology of Crimean-Congo hemorrhagic fever (CCHF) in Africa-underestimated for decades. *The American Journal of Tropical Medicine and Hygiene, 104*(6), 1978–1990. https://doi.org/10.4269/ajtmh.20-1413

Nipah Virus Infection

27

Shriya Rawat and Ripan Biswas

Abstract

Nipah virus infection is a zoonotic disease that was first recognized in Malaysia in 1999. Humans can contract Nipah virus through direct contact with infected bats, pigs, or contaminated food, as well as via person-to-person transmission. The disease presents as a severe respiratory illness and encephalitis, with a high fatality rate ranging from 40% to 75%. Outbreaks of Nipah virus infection have been reported primarily in South Asia and Southeast Asia, with notable occurrences in Bangladesh and India. The clinical presentation of Nipah virus infection encompasses a range of symptoms, including fever, headache, and dizziness. In severe cases, the disease can progress to coma. Currently, there is no specific treatment or vaccine available, rendering prevention through avoidance of exposure, enhanced surveillance, and public education critical. Nipah virus infection remains a substantial public health concern owing to its high mortality rate and the risk of outbreaks.

Keywords

Nipah · Bat saliva · NiVD · Nipah Virus Disease (NiVD)

S. Rawat (✉)
Department of Veterinary Public Health and Epidemiology, College of Veterinary and Animal Sciences, Sardar Vallabhbhai Patel University of Agriculture and Technology, Meerut, Uttar Pradesh, India

R. Biswas
Department of Veterinary Public Health and Epidemiology, Faculty of Veterinary and Animal Sciences, West Bengal University of Animal and Fishery Sciences, Kolkata, West Bengal, India

R. Biswas et al. (eds.), *Concept of One Health and Zoonoses*,
https://doi.org/10.1007/978-981-95-6627-3_27

Introduction

The Nipah virus is an emerging viral disease in Southeast Asia, classified as a Biosafety Level 4 pathogen. It is caused by a negative-sense, single-stranded RNA virus with a genome of approximately 18,000 nucleotides and a pleomorphic shape. The Nipah virus, like other Paramyxoviruses, possesses a genome with six genes responsible for encoding the fusion protein (F), glycoprotein (G), matrix protein (M), nucleocapsid (N), phosphoprotein (P), and polymerase protein (L). Additionally, the phosphoprotein gene gives rise to several important accessory proteins, including C, V, and W. The Nipah virus belongs to the genus Henipavirus, which also includes Hendra virus and the recently described Cedar virus. The initial Nipah virus cases emerged in late September 1998 in villages near the city of Ipoh, located in the state of Perak, West Malaysia, which had a substantial pig farming industry. Subsequent second cluster of Nipah virus infections emerged near Sikamat, a small town situated in the different state of Negri Sembilan, during December 1998 and January 1999. The third and largest Nipah virus outbreak occurred near Bukit Pelandok in December 1998. Virologists at the University of Malaya identified the virus in the patient's brain fluid during the onset of brain inflammation in March 1999. The Nipah virus was named after the lethal human case identified in the Sungai Nipah (Nipah river village) neighborhood of Negeri Sembilan, from which the initial isolate was obtained. The cases were initially thought to be Japanese B encephalitis because 4 out of 28 patient serum samples tested positive for JE-specific IgM, and JE genetic material was detected in some patient sera. Frugivorous bats of the Pteropus genus, commonly referred to as "flying foxes," serve as the natural reservoir for the Nipah virus. The major outbreaks of Nipah virus infection have primarily occurred within the geographical regions inhabited by these carrier bats, including Malaysia, Singapore, Bangladesh, and India, though cases have also been reported in the Philippines.

Epidemiology

Fruit bats serve as the natural reservoir for Nipah viruses. The Nipah virus-carrying bats feed on date palm sap. Nipah virus can persist in sugar-rich solutions, such as fruit pulp. Humans can contract the Nipah virus by consuming contaminated date palm sap. Pteropus fruit bats, the natural reservoirs of Nipah virus, visit these fruit trees and are able to release virus-containing droplets onto the farm's soil and fruits. Pigs and other animals then consume the contaminated fruits. Pigs act as both intermediate and amplifying hosts for the virus. The combination of proximity to fruiting trees, date palm fruits, fruit bats, pigs, and humans altogether creates the conditions for the emergence and spread of this new deadly zoonotic virus infection, Nipah. Contaminated pork may be exported, leading to human infection. Close contact with Nipah virus-affected individuals can also result in the transmission of the virus to other people.

Malaysia and Singapore

In September 1998, several cases with fever, headache, and reduced consciousness were reported from Perak, Malaysia. Initially, four of the cases tested positive for Japanese Encephalitis-specific IgM antibodies, leading to the declaration of a Japanese Encephalitis outbreak. Furthermore, additional clusters emerged in the Port Dickson District, located approximately 300 km to the south, by the end of the year. In March 1999, a new virus was isolated from the cerebrospinal fluid of a patient from Sungai Nipah village. The Malaysia outbreak, which continued until May 1999, comprised 283 reported cases and resulted in 109 deaths, yielding a case-fatality rate of 38.5%. In March 1999, an outbreak was reported from Singapore involving slaughterhouse workers, with close contact with pigs or their excreta identified as a risk factor. The mass culling of over a million pigs, combined with other control measures, successfully contained the outbreak. Animals infected with the Nipah virus showed mild respiratory symptoms. The virus spread among pigs raised together in Malaysian farms and slaughterhouses, where the outbreak began. In addition, dogs were also found to be commonly infected, and their deaths on farms were another identified risk factor. The outbreaks did not involve human-to-human transmission. Investigations revealed that Pteropus bats served as the reservoir for the Nipah virus in Malaysia, which then infected pigs as amplifying and intermediate hosts through their consumption of fruit contaminated by the bats.

In March 1999, an outbreak (11 cases, one death) was reported from Singapore among slaughterhouse workers, with close contact with pigs or their excreta identified as a risk factor. While Nipah virus has been detected in *Pteropus* bat populations in Cambodia and Thailand, there have been no documented cases of human infection in these countries.

Philippines

In 2014, a Nipah virus outbreak occurred in the southern Philippines, affecting both humans and horses. The mortality rate was high. Transmission was observed between horses and humans, as well as among humans. Flying foxes were likely the source of NiV infections in horses and humans. This outbreak resulted in 17 confirmed cases and 9 deaths.

Bangladesh

The epidemiology of Nipah virus infection in Bangladesh differs significantly from that observed in Malaysia. Since 2001, Nipah virus infections in Bangladesh have exhibited a seasonal pattern, with outbreaks typically occurring during the winter months and primarily affecting 20 districts, declared as "Nipah belt" located in central and northwestern regions of the country, where the majority of spillover events have been documented. Nipah virus spreads from bats to humans in

Bangladesh mainly through drinking raw date palm sap. Pteropus fruit bats, the natural hosts of the virus, contaminate the sap with their saliva, urine, and droppings when they visit the date palm trees during the sap harvesting season (December to May). The Nipah virus outbreaks in Bangladesh are closely tied to this sap harvesting period, as these bats have been observed visiting the date palm trees and licking the sap that is being collected. Pteropus bats have been identified as the reservoir for the Nipah virus in Bangladesh. Importantly, person-to-person transmission has been identified as a key mode of Nipah virus spread in Bangladesh, observed across all outbreaks in the country. In Bangladesh, the main routes of Nipah virus transmission have been through the consumption of contaminated date palm sap and person-to-person spread. Domestic animals can also facilitate the transmission of the Nipah virus from bats to humans.

The Nipah virus demonstrated a nosocomial transmission pattern, spreading to healthcare workers and resulting in cluster outbreaks within hospital settings. This occurred through coughing and sneezing in poorly ventilated areas, as well as high-risk contact with bodily fluids or shared beds.

India

In early 2001, an outbreak of infectious febrile illnesses occurred in and around Siliguri, a city in northern West Bengal. The outbreak was identified as caused by the Nipah virus, which was first reported in India. The Nipah virus outbreak was sudden and severe, with 66 confirmed cases of encephalitis and at least 43 deaths. The cases were adults with no known exposure to pigs or animals. Evidence showed the outbreak spread through hospital-acquired transmission. Pigs did not play a role, and the virus transmitted primarily between people, especially in hospital settings. In 2007, a second outbreak of Nipah virus was reported in Belechuapara village, near the Bangladesh border in Nadia district of West Bengal with five deaths out of five. The outbreaks occurred near the Nipah belt of Bangladesh. The 2007 outbreak had one person who got the disease from drinking date palm alcohol, and the others, including a healthcare worker, got it from that first case. Bats from the *Pteropus* genus were found to be the source of the Nipah virus outbreak in Kozhikode, India. Tests on 52 bat samples collected from the Kozhikode district at the National High Security Animal Diseases Laboratory in Bhopal revealed that 10 samples tested positive for the Nipah virus by RT-PCR. The 2018 outbreak resulted in 18 confirmed cases, of which 17 people died. The recent outbreaks from India was in 2019, 2021, 2023 and 2024 from only Kerala. India has experienced a total of 7 Nipah virus outbreaks up to the end of 2024.

Nipah virus transmission in Bangladesh and India primarily occurs through bat-to-human and subsequently human-to-human routes. Human-to-human transmission of the Nipah virus typically occurs through the inhalation of respiratory droplets or direct contact with bodily fluids from an infected individual. Phylogenetic studies have identified distinct NiV clades, including Malaysia, Bangladesh, and India clades. These clades reflect geographic origins and exhibit some genetic variations.

Factors contributing to the transmission dynamics and spread of Nipah virus infection include the following:

- Reservoir host: The Nipah virus's natural reservoirs are the fruit bats of the Pteropodidae family, particularly those within the Pteropus genus.
- Intermediate hosts: Domestic animals, including pigs and dogs, can serve as intermediate hosts, facilitating the transmission of the Nipah virus.
- Environmental factors: Habitat loss due to deforestation, agricultural expansion, and encroachment into the natural environments of bat populations has increased the frequency and proximity of human-animal interactions, thereby facilitating the spillover of the Nipah virus from its natural reservoirs.
- Sap contamination: In Bangladesh, the consumption of unprocessed date palm sap contaminated by bat saliva, urine, or feces serves as a primary transmission pathway for the Nipah virus from bats to humans.
- Nosocomial transmission: The Nipah virus can spread rapidly within healthcare settings through coughing, sneezing, and contact with bodily fluids of infected patients.
- Person-to-person transmission: Close personal interactions with infected persons, such as through exposure to respiratory secretions and contact with bodily fluids, can facilitate human-to-human transmission of the Nipah virus.
- Seasonality: In Bangladesh, Nipah virus outbreaks tend to coincide with the winter months, when the date palm sap harvesting season occurs.
- Lack of treatment and vaccines: There is currently no specific treatment or approved vaccine for Nipah virus infection, leading to high mortality rates during outbreaks.
- Geographical expansion: The potential for the Nipah virus to spread to new regions, particularly through international travel and trade, is a concern.
- Challenges in surveillance and detection: Poor surveillance and diagnostic capabilities in some affected areas can delay the identification and control of Nipah virus outbreaks.
- Socioeconomic factors: Poverty, limited healthcare access, and cultural traditions can affect how the Nipah virus spreads and impacts affected communities.
- Human–wildlife interactions: Human activities that encroach on bat habitats, such as agricultural expansion and development, as well as the consumption of food products contaminated by bat secretions, can heighten the risk of Nipah virus spillover from bats to human populations.
- Adaptation and evolution: The Nipah virus possesses the capacity to evolve and adapt, which raises concerns about the potential for increased transmissibility or virulence, an issue that warrants ongoing attention and monitoring.
- Climate change: Variations in climatic and environmental parameters may impact the geographic distribution and population levels of Nipah virus reservoirs, potentially altering the risk of zoonotic transmission to human populations.
- Lack of public awareness: Inadequate public knowledge and comprehension of the Nipah virus can hamper preventive and control measures.

- Weak health systems: Inadequate healthcare infrastructure, limited disease surveillance, and suboptimal infection prevention and control measures can contribute to the spread and impact of Nipah virus outbreaks.
- Sociopolitical instability: Sociopolitical instability in affected regions can hinder public health responses and impede effective management of outbreaks.

Pathogenesis

Elucidating the pathogenic mechanisms of Nipah virus infection is essential for designing effective antiviral treatments and preventive measures.

- Viral Entry: The Nipah virus enters the host primarily through inhalation of respiratory droplets, contact with the oropharynx, or direct exposure to infected bodily fluids. The viral G glycoprotein binds to host cell receptors, ephrin-B2 and ephrin-B3, facilitating the process of viral entry and subsequent infection.
- Replication and Dissemination: The Nipah virus replicates in the respiratory tract and then spreads throughout the body via the bloodstream, infecting various organs such as the brain, lungs, kidneys, and spleen.
- Vascular Endothelial Cell Infection: The Nipah virus has a tropism for vascular endothelial cells, leading to the development of vasculitis and endothelial damage. This, in turn, results in thrombosis, ischemia, and inflammation in multiple organs.
- Central Nervous System Involvement: The Nipah virus infiltrates the central nervous system, leading to the development of encephalitis, a characteristic feature of severe Nipah virus infection. The virus can gain access to the brain either through the olfactory nerves or by crossing the blood–brain barrier via infected leukocytes.
- Respiratory System Infection: NiV infection can also lead to severe respiratory illness, characterized by pneumonia and acute respiratory distress syndrome.
- Immune Response: The host's immunological response is instrumental in both mitigating the viral infection and shaping the disease pathogenesis. While a robust and appropriate immune reaction can effectively clear the virus, an exaggerated inflammatory response can exacerbate tissue damage and culminate in severe clinical outcomes.
- Persistent Infection: In some cases, the Nipah virus can establish persistent or latent infections, particularly within the central nervous system. This has the potential to result in recurrent episodes of encephalitis or delayed neurological sequelae.
- Pathological Findings: The key pathological features of Nipah virus infection encompass vasculitis, the formation of syncytia, damage to vascular endothelial cells, thrombosis, micro-infarctions, and infection of parenchymal cells in multiple organ systems.

Clinical Signs and Symptoms

Humans

The incubation period typically ranges from 4 to 18 days, but can be as long as 45 days. In humans, it exhibits a wide range of clinical manifestations, spanning from asymptomatic infection to severe encephalitis and even death. Common clinical features include fever, headache, drowsiness, malaise, cough, sore throat, and respiratory distress. Neurological complications, such as confusion, disorientation, seizures, and coma, may develop, indicative of encephalitis. Gastrointestinal symptoms, including nausea, vomiting, and abdominal pain, have also been reported. In severe cases, the infection can lead to acute respiratory distress syndrome. The diverse clinical presentation can resemble other infectious diseases, making diagnosis challenging and necessitating laboratory confirmation.

Pigs

Nipah virus infection in pigs, commonly referred to as porcine respiratory and encephalitis syndrome or barking pig syndrome, presents with a combination of respiratory and neurological symptoms. These include fever, coughing, labored breathing, nasal discharge, and occasionally neurological manifestations such as tremors, muscle twitching, and weakness. In dogs, Nipah virus infection can lead to respiratory distress, kidney dysfunction, fever, vomiting, depression, and dehydration.

Dogs

Nipah virus in dog may present with a constellation of clinical signs, including respiratory distress, kidney dysfunction, fever, vomiting, depression, and dehydration.

Diagnosis

Modern diagnostic techniques of Nipah virus infection primarily rely on laboratory methods to detect the virus or its components in clinical samples. Here are the main techniques:

Real-Time Reverse Transcription Polymerase Chain Reaction: Real-time reverse transcription polymerase chain reaction is the most sensitive and specific method for detecting NiV RNA in clinical specimens like blood, throat swabs, cerebrospinal fluid, and urine. This technique can identify the virus during the acute phase of infection, making it crucial for rapid diagnosis and effective outbreak management.

Enzyme-Linked Immunosorbent Assay: ELISA can detect antibodies against NiV in serum or cerebrospinal fluid. IgM antibodies suggest a recent NiV infection, while IgG antibodies indicate past exposure or immunity. This serological test is particularly useful in the later stages of infection or for conducting seroprevalence studies.

Virus Isolation: Culturing the virus from clinical samples in specialized cell lines represents a definitive approach to confirm Nipah virus infection. However, this method is resource-intensive, as it necessitates specialized laboratory infrastructure and poses biosafety concerns.

Immunohistochemistry: Immunohistochemistry is a technique that detects Nipah virus antigens in tissue samples using specific antibodies. It is valuable for post-mortem diagnosis, as it allows for the assessment of the virus's distribution in different organs.

Metagenomic Sequencing: Metagenomic sequencing techniques can be employed to identify the Nipah virus in clinical samples by comprehensively analyzing the complete genetic content present. This innovative approach holds utility for the detection of novel or unanticipated pathogens, and thus can provide valuable insights during outbreak investigations.

Emerging point-of-care diagnostic solutions leveraging RT-PCR and ELISA methodologies are being engineered for deployment in resource-limited settings. These tests exhibit accelerated response times and streamlined operational procedures, thereby promoting timely identification and management of Nipah virus infection.

Prevention and Control

Preventing and controlling Nipah virus infection relies on a multifaceted approach targeting various stages of transmission:

Reducing the Risk of Spillover from Bats to Humans

Minimizing Bat–Human Contact: Refrain from consuming date palm sap or fruits that may have been contaminated by bat saliva or urine. Additionally, implement measures to restrict bats from accessing and contaminating food sources.

Safe Food Handling Practices: Thoroughly wash produce prior to consumption and refrain from consuming unprocessed date palm sap.

Preventing Human-to-Human Transmission

Isolation and Infection Control: Isolating infected individuals and enforcing rigorous infection control protocols in healthcare facilities, such as the utilization of personal protective equipment and meticulous hand hygiene practices, are crucial measures to prevent the transmission of Nipah virus.

Contact Tracing: Trace and monitor the contacts of infected individuals to promptly identify potential secondary cases and prevent further transmission.

Safe Burial Practices: Avoid direct contact with the bodies of deceased NiV patients and follow safe burial practices.

Other Preventive Measures

Vaccination: While no licensed human vaccine is currently available, research and development efforts are ongoing. A Hendra virus vaccine for horses (Equivac®) has shown cross-protective efficacy against NiV in animal models and could potentially be adapted for human use.

Surveillance and Early Detection: Proactive surveillance of human and animal populations is essential for early detection and prompt response to Nipah virus outbreaks.

Public Awareness and Education: Educating communities about NiV risks, transmission routes, and preventive measures is essential for effective control.

One Health Approach: A collaborative, interdisciplinary approach involving human health, animal health, and environmental sectors is crucial for comprehensive NiV prevention and control.

By implementing these preventive and control measures, we can effectively reduce the risk of NiV outbreaks and protect public health.

Bibliography

Aditi, A., & Shariff, M. (2019). Nipah virus infection: A review [Review of Nipah virus infection: A review]. *Epidemiology and Infection*, 147. Cambridge University Press. https://doi.org/10.1017/s0950268819000086.

Chattu, V. K., Kumar, R., Kumary, S., Kajal, F., & David, J. K. (2018). Nipah virus epidemic in southern India and emphasizing "One Health" approach to ensure global health security. *Journal of Family Medicine and Primary Care, 7*(2), 275. Medknow.

Garbuglia, A. R., Lapa, D., Pauciullo, S., Raoul, H., & Pannetier, D. (2023). Nipah virus: An overview of the current status of diagnostics and their role in preparedness in endemic countries [Review of Nipah virus: An overview of the current status of diagnostics and their role in preparedness in endemic countries]. *Viruses, 15*(10), 2062. Multidisciplinary Digital Publishing Institute.

Magdum, M., Chowdhury, Md. A. T., Khandaker, M., Datta, R. K., Rayhan, Md. N., Uddin, Md. S., Bhuiyan, Md. J. I., Ahaduzzaman, Md., Kabir, Md. A. A.-M., Uddin, M. Z., Sahid, Md. E. U., & Saha, S. (2024). Nipah virus unveiled: A review article [Review of Nipah virus unveiled: A review article]. *Advances in Bioscience and Biotechnology, 15*(3), 161. Scientific Research Publishing. https://doi.org/10.4236/abb.2024.153011.

Uwishema, O., Wellington, J., Berjaoui, C., Muoka, K. O., Onyeaka, C. V. P., & Onyeaka, H. (2022). A short communication of Nipah virus outbreak in India: An urgent rising concern. *Annals of Medicine and Surgery, 82*. Wolters Kluwer. https://doi.org/10.1016/j.amsu.2022.104599

Wong, K. T., Shieh, W., Kumar, S., Norain, K., Abdullah, W., Guarner, J., Goldsmith, C. S., Chua, K. B., Lam, S. K., Tan, C. T., Goh, K. J., Chong, H. T., Jusoh, R., Rollin, P. E., Ksiazek, T. G., & Zaki, S. R. (2002). Nipah virus infection. *American Journal of Pathology, 161*(6), 2153. Elsevier BV. https://doi.org/10.1016/s0002-9440(10)64493-8

Coronavirus 28

Ajay Pratap Singh and Shyama N. Prabhu

Abstract

Coronaviruses (CoVs) constitute a phylogenetically diverse viral family recognized for causing infections of the respiratory tract and gastrointestinal system in humans and various animal species. Human coronaviruses (HCoVs) have particularly captured attention due to their potential to cause severe respiratory illness outbreaks. Middle East respiratory syndrome coronavirus (MERS-CoV), severe acute respiratory syndrome coronavirus (SARS-CoV-1), and the novel coronavirus (SARS-CoV-2) represent three dangerous human pathogens within the coronavirus family, sharing structural features like a positive-sense single-stranded RNA (+ssRNA) genome, an enveloped structure derived from host membranes, and a characteristic crown-like appearance under electron microscopy due to the prominent Spike (S) glycoproteins embedded in their envelope. Coronaviruses typically exhibit varying tissue tropism. The manifestations of coronavirus-induced lesions can vary widely, ranging from damage to respiratory and enteric epithelial tissues to more severe outcomes such as disseminated disease affecting multiple organs.

Keywords

SARS · MERS · COVID-19 · Corona virus

A. P. Singh (✉)
Department of Veterinary Microbiology, College of Veterinary Science & Animal Husbandry, U.P. Pandit Deen Dayal Upadhyaya Pashu Chikitsa Vigyan Vishwavidyalaya Evam Go-Anusandhan Sansthan, Mathura, Uttar Pradesh, India

S. N. Prabhu
Department of Veterinary Pathology, College of Veterinary Science & Animal Husbandry, U.P. Pandit Deen Dayal Upadhyaya Pashu Chikitsa Vigyan Vishwavidyalaya Evam Go-Anusandhan Sansthan, Mathura, Uttar Pradesh, India

R. Biswas et al. (eds.), *Concept of One Health and Zoonoses*,
https://doi.org/10.1007/978-981-95-6627-3_28

History

Animal coronaviruses (CoVs) have been studied since the late 1930s, with numerous strains identified in a wide range of animals, including turkeys, mice, cows, pigs, cats, and dogs. The first coronavirus to be discovered was the infectious bronchitis virus (IBV), a γ-coronavirus that caused outbreaks in chickens during the 1930s. This discovery marked the beginning of coronavirus research, which expanded in the 1940s with the identification of the murine hepatitis virus (MHV) and the porcine transmissible gastroenteritis virus (TGEV). These early findings laid the groundwork for understanding the diversity and adaptability of coronaviruses across species. In 1965, a significant milestone was reached when Dorothy Hamre at the University of Chicago and Dr. David Tyrrell in England independently isolated a novel virus, designated 229E, from human patients with common cold symptoms. This virus shared structural similarities with the avian coronavirus discovered decades earlier, highlighting the potential for cross-species transmission. Two years later, in 1967, Dr. Ken McIntosh of Harvard Medical School identified another human coronavirus, OC43. The term "coronavirus" was officially coined in 1968 derived from pattern of surface projections resembling a crown under electron microscopy. Human health faced new dangers in the twenty-first century with the emergence of multiple highly pathogenic coronaviruses. Severe acute respiratory syndrome (SARS)—the century's first major coronavirus outbreak—originated in late 2002 in Foshan, a city in China's southern Guangdong Province. This was followed by the discovery of HCoV-NL63 in the Netherlands and HCoV-HKU1 in Hong Kong, both identified in 2004. In 2012, Jeddah city in Saudi Arabia reported index case of the Middle East respiratory syndrome coronavirus (MERS-CoV) responsible for severe respiratory illness with high fatality rates. Most recently, in December 2019, novel coronavirus SARS-CoV-2, which triggered the COVID-19 pandemic, was first identified in Wuhan, located in China's central Hubei Province. This virus rapidly spread globally, underscoring the ongoing threat posed by coronaviruses to public health.

Etiology

Coronaviruses (CoVs) are characterized by an outer envelope and spherical shape, and the virions possess diameters ranging from 110 to 160 nm. They possess the largest known single-stranded RNA genome of positive polarity, spanning nearly 30–32 kilobases. This extensive genome encodes a range of structural and non-structural proteins, enabling the virus to replicate efficiently and adapt to different hosts. CoVs that infect humans and animals are classified under Coronaviridae family, subfamily Coronavirinae. The Coronavirinae subfamily is further divided into four genera: Alphacoronavirus (α-coronaviruses), Betacoronavirus (β-coronaviruses), Gammacoronavirus (γ-coronaviruses), and Deltacoronavirus (δ-coronaviruses). Among coronavirus genera, Alphacoronavirus and Betacoronavirus hold significant medical importance as they encompass known human pathogens. In contrast, Gammacoronavirus and Deltacoronavirus primarily infect avian species, though some strains have been detected in marine mammals and other animals. The current taxonomic classification acknowledges seven documented human coronaviruses (HCoVs). Among these, HCoV-229E and HCoV-NL63 belong to the Alphacoronavirus genus. The Betacoronavirus genus is more diverse and is subdivided into four lineages: A, B, C, and D. Human coronaviruses OC43 and HKU1, classified under Lineage A, typically produce mild upper respiratory infections. Lineage B harbors clinically significant betacoronaviruses, including SARS-CoV-1 (triggering the 2002–2003 outbreak) and SARS-CoV-2 (responsible for the ongoing COVID-19 pandemic). Lineage C includes MERS-CoV, which emerged in 2012 and is linked to severe respiratory illness with a high mortality rate. Lineage D, on the other hand, comprises coronaviruses that primarily infect bats and other wildlife, serving as potential reservoirs for future zoonotic spillover events. Coronaviridae family members share key structural features: enveloped virions with variable morphology (typically spherical) ranging from 80 to 220 nm in diameter. Their architecture depends on four essential proteins:

(i) Spike (S): Forms crown-like projections that mediate host-cell attachment
(ii) Membrane (M): An icosahedral matrix protein organizing virion assembly and defining shape through protein interactions
(iii) Envelope (E): Critical for morphogenesis, virion release, and intracellular trafficking via ER-Golgi localization
(iv) Nucleocapsid (N): RNA-bound capsid protein encapsulating the genome; facilitates replication and counteracts host defenses

Beneath the lipid envelope, the N protein–RNA complex forms the protective nucleocapsid core. In SARS-CoV-2, the M protein binds and stabilizes nucleocapsids during viral budding, while the E protein promotes pathogenesis and facilitates viral egress. Homotrimeric spike (S) glycoproteins mediate cellular entry through ACE2 receptor recognition.

Epidemiology

Alphacoronaviruses (α-CoVs) and Betacoronaviruses (β-CoVs), which include highly pathogenic viruses such as SARS-CoV and MERS-CoV, predominantly circulate in bat reservoirs, with occasional spillover to other mammals. Birds, on the other hand, serve as the main hosts for Gammacoronaviruses (γ-CoVs) and Deltacoronaviruses (δ-CoVs). While bats harbor a wide variety of α and β coronaviruses, these viruses often do not cause disease in bats themselves. However, they can be transmitted to other species, including humans, through zoonotic spillover events. Many human coronaviruses, ranging from those causing mild illnesses like the common cold (e.g., Human coronavirus-229E and Human coronavirus -NL63) to more severe diseases (e.g., SARS-CoV and MERS-Cov). Viral phylogenies consistently position bat coronaviruses as the evolutionary source of human coronaviruses, highlighting the importance of bats in the viral ecology of coronaviruses. Coronavirus spillover to humans commonly depends on intermediate hosts for viral adaptation prior to cross-species transmission. The 2002 SARS-CoV-1 outbreak was traced back to horseshoe bats, with masked palm civets and other wildlife (such as raccoon dogs and ferret badgers) acting as intermediate hosts. Similarly, the 2012 MERS-CoV outbreak in the Arabian Peninsula was linked to dromedary camels, which served as the intermediate host. Serological evidence of MERS-CoV neutralizing antibodies in dromedary camels from regions such as the Canary Islands, Oman, Southwest Asia, and Africa, confirms extensive enzootic circulation within these populations. Another example is HCoV-OC43, a human coronavirus believed to have originated from animals, with cattle proposed as a potential source. Wet markets, where live animals are sold and slaughtered in close proximity, have been implicated in several viral spillover events. The mixing of diverse animal species in these environments creates opportunities for cross-species transmission, facilitating the emergence of novel viruses. Current evidence indicates SARS-CoV-2 (the etiological agent of COVID-19) emerged via zoonotic spillover. While bats are considered the likely natural reservoir, the specific intermediate host that enabled the virus to jump to humans remains unidentified. However, cases of hamster-to-human transmission in Hong Kong pet shops have been documented, demonstrating the potential for animal-to-human spread in non-traditional settings. Susceptibility to SARS-CoV-2 infection under experimental conditions has been demonstrated in a variety of animal species, encompassing domestic felines, ferrets, and laboratory rodents and primates. Surveillance and sero-surveillance studies indicate that while SARS-CoV-2 infections in domestic pets (e.g., cats and dogs) are less prevalent than in humans, they do occur. Notably, several cases of SARS-CoV-2 have been reported in captive animals, such as tigers, lions, and gorillas in zoos, likely transmitted from infected zookeepers or staff.

Transmission

Human to Human

The disease spreads primarily between people through cough/sneeze droplets, personal contact, or contact with contaminated surfaces. Healthcare workers are at heightened risk, particularly when isolation protocols are breached during specific medical procedures. Evidence indicates SARS-CoV-2 can infect gastrointestinal epithelial cells. Viral RNA is detectable in stool, and infectious virus has been isolated from GI tissues. Viral RNA shedding in stool often persists long after respiratory samples become negative. The presence of SARS-CoV-2 in stool samples implies a risk of fecal–oral spread through environmental contamination. Emerging evidence indicates potential vertical transmission of SARS-CoV-2, with viral RNA identified in placenta, amniotic fluid, and—in rare instances—neonatal specimens.

Animal to Human

The transmission of coronaviruses from animals to humans, known as zoonotic spillover, is a critical mechanism in the emergence of novel viral infections. Camels, particularly dromedaries, are recognized as natural reservoirs for MERS-CoV (Middle East respiratory syndrome coronavirus). These animals can shed the virus in their bodily fluids, including nasal secretions, saliva, urine, and feces, creating opportunities for transmission to humans. Close contact with infected camels or exposure to their bodily fluids is a significant route of infection. Individuals who work closely with camels, such as caretakers, herders, or those involved in the camel trade, are at heightened risk due to frequent exposure to respiratory secretions and other fluids. In addition to direct contact, consuming raw or undercooked camel products, such as camel milk or meat, has been identified as a probable route of MERS-CoV transmission. Studies have shown that the virus can remain viable in raw camel milk, posing a risk to those who consume it untreated. This highlights the importance of proper food safety practices, such as pasteurization and thorough cooking, to mitigate the risk of zoonotic transmission.

Beyond camels, other animals have also been implicated in the transmission of coronaviruses to humans. For example, the 2002 SARS-CoV-1 outbreak was linked to civets and raccoon dogs, which likely served as intermediate hosts between bats and humans. Similarly, the origins of SARS-CoV-2 are believed to involve zoonotic transmission, though the exact intermediate host remains unidentified. Wet markets, where live animals are sold in close proximity, have been identified as potential hotspots for such spillover events due to the mixing of diverse species and the stress-induced shedding of viruses.

Pathogenesis

SARS-CoV-2 pathogenesis begins with inhaled particles binding respiratory epithelial cells, unfolding in three phases: replication, immune dysregulation, and lung destruction. Unlike MERS-CoV's DPP4-dependent entry, SARS-CoV-2 relies on its spike (S) protein which consists of two subunits, S1 and S2. Central to viral entry, the spike (S) protein (S1/S2 subunits) requires proteolytic activation by host TMPRSS2 (Transmembrane Serine Protease 2). Host protease TMPRSS2 cleaves S at the S1/S2 junction, which triggers conformational changes of spike protein, essential for infectivity. This proteolytic processing remodels the spike protein, exposing the S2 fusion subunit while positioning the S1 receptor binding domain (RBD) for **"angiotensin-converting enzyme 2"** (ACE2) engagement enabling viral entry via membrane fusion or endocytosis. SARS-CoV-2 preferentially infects type II pneumocytes due to exceptionally high ACE2 expression in these lung cells, though the receptor exists throughout multiple tissues including cardiomyocytes, renal tubules, and enterocytes. This explains multi-organ tropism of SARS-CoV-2's observed in severe COVID-19 cases. The discovery of viral RNA in ACE2-negative cells suggests supplementary entry pathways, potentially involving co-receptors like neuropilin-1.

The co-expression of TMPRSS2 and ACE2 in various tissues influences susceptibility to infection. For example, in the gastrointestinal tract, co-expression in enterocytes of the colon and ileum may contribute to enteric symptoms such as vomiting and diarrhea. In contrast, MERS-CoV, another highly pathogenic coronavirus, utilizes dipeptidyl peptidase 4 (DPP4) as its primary receptor. DPP4 is involved in various cellular processes, including glucose metabolism, immune modulation, and chemotaxis, highlighting its role in both viral entry and immune response regulation. Additionally, genetic factors, such as specific HLA alleles, have been associated with increased susceptibility to severe outcomes in SARS-CoV-1 infections, likely due to impaired viral antigen presentation and weakened immune responses.

SARS-CoV-2 directly activates the NF-κB pathway via pattern recognition receptors or indirectly through ACE2–angiotensin II–STAT3 signaling. This dual activation drives an aberrant immune response characterized by hyperactivated immune cells and a cytokine storm—marked by elevated IL-6, TNF-α, and IFN-γ—which directly mediates tissue damage, acute respiratory distress syndrome (ARDS), and multi-organ failure through uncontrolled proinflammatory cascades. Growing evidence confirms SARS-CoV-2's capacity to cause neurological sequelae beyond pulmonary disease. The virus may reach the central nervous system (CNS) through the olfactory bulb, exploiting synaptic connections. This mechanism could explain early symptoms like transient anosmia (loss of smell) and ageusia (loss of taste) observed in some patients. Similar neuroinvasive pathways have been documented in SARS-CoV-1 and MERS-CoV infections. Furthermore, evidence indicates that SARS-CoV-1 infection may dysregulate hypothalamus–pituitary signaling, altering thyroid hormone homeostasis, leading to secondary hypothyroidism. While the exact mechanisms remain unclear, systemic inflammation and viral-induced endocrine disruption are believed to play a role.

Clinical Symptoms

Human

The spectrum of coronavirus infection varies widely, ranging from asymptomatic cases to severe and life-threatening illness. While up to 80% of infections may be asymptomatic, some individuals initially without symptoms can develop clinical disease later. Most human coronavirus infections, including those caused by human coronavirus-229E, human coronavirus-OC43, and SARS-CoV, are associated with mild flu-like symptoms resembling the common cold. Clinical presentation typically features febrile respiratory illness (cough, pharyngitis, rhinorrhea, and sneezing) with systemic symptoms (myalgia and fatigue); gastrointestinal manifestations—including diarrhea, abdominal pain, and emesis—occur in subsets.

The associated clinical symptom of SARS-CoV-2 is highly variable and nonspecific. Infected individuals may experience fever, sore throat, body aches, dyspnea, and, in some instances, nausea, diarrhea, or a runny nose. The incubation period for COVID-19 is typically within 2 weeks, and while many recover without complications, others develop severe respiratory illnesses. Critical cases of COVID-19 are characterized by acute respiratory failure, multiple organ dysfunction, coagulopathy (abnormal blood clotting), acute kidney injury, heart failure, encephalopathy, and shock.

The MERS-CoV clinical spectrum primarily includes pyrexia (≥38 °C), lower respiratory tract symptoms (productive cough and tachypnea), with radiographic confirmation of pneumonic infiltrates in most cases, accompanied by enteric symptoms like watery diarrhea in approximately 35% of patients. Patients with specific comorbidities—including neoplastic diseases, chronic pulmonary disorders, diabetes mellitus, cardiovascular conditions, immunosuppression, and elderly individuals (≥65 years)—demonstrate markedly increased susceptibility to severe coronavirus disease progression and associated complications. These risk factors highlight the importance of targeted preventive measures and early intervention for vulnerable populations.

Kawasaki Disease It is a self-limiting disease condition most commonly seen in children under the age of five and characterized by inflammation of the blood vessels. Molecular mimicry and superantigen activity of certain human coronaviruses have been proposed as potential mechanisms linking these viral infections to Kawasaki disease pathogenesis, though causal relationships require further validation. Kawasaki disease typically presents with a combination of symptoms, including persistent high fever, erythematous, swelling of the lymph nodes in neck region, clear-discharge ocular inflammation, erosive stomatitis, marked cervicofacial lymphatic swelling, and/or coronary artery aneurysms.

Animals

The symptoms of coronavirus infections in animals can vary depending on the species affected and the specific strain of the virus. However, common symptoms may include the following:

1. **Respiratory Symptoms:** While clinical presentations vary, both avian and mammalian hosts tend to develop characteristic respiratory symptoms ranging from mild (sneezing, nasal discharge) to severe (dyspnea, coughing spasms).
2. **Gastrointestinal Symptoms:** Canine, feline, and porcine coronavirus infections frequently manifest with enteric clinical signs including diarrhea, emesis, and abdominal discomfort.
3. **Fever:** Elevated body temperature is a common response to viral infections and may be observed in animals with coronavirus infections.

4. **Lethargy:** Infected animals may exhibit lethargy, weakness, and decreased activity levels.
5. **Dehydration:** Severe diarrhea and vomiting can lead to dehydration, which may exacerbate other symptoms and contribute to morbidity and mortality.
6. **Neurological Symptoms:** In some cases, coronavirus infections in animals may lead to neurological symptoms, such as ataxia, seizures, and paralysis. These symptoms are more commonly observed in severe cases or with certain strains of the virus.

The disease severity spectrum varies significantly across species—while cats and golden hamsters show minimal or no clinical signs, non-human primates (particularly macaques) and guinea pigs exhibit more pronounced though still moderate symptoms.

Diagnosis

Virus Culture

A diverse range of cell lines have been utilized for isolation and propagation of human corona virus viz., Vero E6, LLC-MK2 cells, Caco-2 epithelial cell line, and HuH-6/AT. The general process of virus isolation involves inoculating the selected cell line with a clinical sample, often obtained from a patient with a confirmed coronavirus infection. The sample, such as a respiratory specimen or nasal swab, is mixed with a suitable medium and introduced to the cells in culture. The cells are then monitored for signs of viral replication, including the observation of cytopathic effects (CPE), which are visible changes in the morphology of the cells due to viral infection. Once a cytopathic effect is observed, indicating successful viral replication, the cell-culture supernatants can be collected. These supernatants contain the newly produced virus particles, and their genetic material can be analyzed using techniques like real-time RT-PCR to confirm the presence of the virus.

Molecular Detection of Human-CoV RNA by Real-Time qPCR

Reverse-transcription quantitative PCR (RT-qPCR) stands as the undisputed gold standard for diagnosing human coronavirus infections due to its exceptional sensitivity and reliability. This technique works by first converting viral RNA into complementary DNA (cDNA) through reverse transcription, followed by exponential amplification and real-time fluorescence detection of target sequences. Laboratories typically target conserved genomic regions encoding vital viral proteins—the nucleocapsid (N) gene (favored for its high conservation across variants), envelope (E) gene (critical for viral assembly), spike (S) gene (important for both diagnosis and variant monitoring), and RNA-dependent RNA polymerase (RdRp) gene (essential for viral replication). The strength of RT-qPCR lies in its unparalleled analytical sensitivity, capable of detecting fewer than 10 viral copies per microliter, along with its unique ability to quantify viral load—a crucial metric for assessing disease progression and infectiousness.

CRISPR/Cas-Based Diagnostic Methods

The emergence of CRISPR-based diagnostics (SHERLOCK and DETECTR) has revolutionized molecular testing during the pandemic, as rapid, accurate alternatives to traditional PCR. Both systems leverage CRISPR's precision but employ distinct mechanisms—SHERLOCK utilizes Cas13 to target viral RNA through an RPA amplification and T7 transcription process, while DETECTR employs Cas12 to detect DNA via streamlined isothermal amplification through RT-LAMP (reverse-transcription loop-mediated isothermal amplification). SHERLOCK's RNA-focused approach offers direct viral detection capabilities, whereas DETECTR's DNA-based system provides a faster, more cost-effective workflow. Both SHERLOCK and DETECTR employs preamplifiation step followed by CRISPR-Cas system-based detection. Since neither method requires thermal cycling, they eliminate the need for expensive thermocyclers, making them ideal for rapid, point-of-care (POC) diagnostics.

Detection of Coronavirus antigen

Detection of human coronavirus (CoV) antigens using various methods, including a rapid immunochromatographic tool, a monoclonal antibody-based capture ELISA, and reactivity on enzyme-linked immunosorbent assay (ELISA) targeting various antigen like S1 for MERS-CoV.

Demonstration of a Seroconversion

A range of serological tests detect antibodies against human coronaviruses. These include core methods like enzyme-linked immunosorbent assays (ELISA), immunofluorescence assays (IFA), chemiluminescence enzyme immunoassays (CLIA), and lateral flow assays (LFA). Commercially available specialized tests exist, such as a fluorescence-based LFA antigen kit (Anti-MERS-CoV IFT, EUROIMMUN) that utilizes the N or S protein in an immunofluorescence format. Additionally, advanced magnetic chemiluminescence enzyme immunoassays (MCLIA) have been developed, combining magnetic particle technology with chemiluminescence and enzyme reactions to achieve highly sensitive and specific detection of coronavirus-specific antibodies.

Demonstration of Virus Neutralization

To assess protective immunity against coronaviruses like SARS-CoV-2, scientists deploy various neutralization tests measuring how effectively antibodies block viral infection. The gold standard remains the plaque reduction neutralization test (PRNT), where live coronavirus is mixed with serum dilutions and added to susceptible cell monolayers (like Vero E6); after days, visible plaques formed by virus-killed cells are counted, and the antibody dilution reducing plaques by 50% (PRNT50) or 90% (PRNT90) is calculated. For faster, safer high-throughput screening, pseudovirus neutralization assays (PNA) are widely used: here, harmless engineered pseudoviruses, bearing the coronavirus spike protein and carrying a luciferase or GFP reporter gene, are incubated with serum; if neutralizing antibodies block spike-receptor interaction, the pseudovirus fails to enter cells, and the lack of reporter signal (luminescence/fluorescence) quantifies neutralization. Similarly, reporter virus particle (RVP) assays use more authentic, replication-deficient viral particles. When speed is critical and absolute titers less essential, surrogate virus neutralization tests (sVNT), like ELISA-based assays, measure how serum antibodies block the viral spike's receptor binding domain (RBD) from attaching to its human receptor (e.g., ACE2), providing a rapid correlate of functional neutralization. Each method balances factors like biosafety (PNA/sVNT vs. live PRNT), throughput, speed, cost, and the need for precise titer quantification (PRNT) versus rapid screening (sVNT/PNA) in evaluating vaccines, therapies, and immunity.

Prevention and Control

Global health security remains challenged by virulent human coronaviruses, notably pandemic SARS-CoV-2 (COVID-19) alongside epidemic threats of SARS-CoV-1 and MERS-CoV. Preventing and controlling the spread of these viruses are crucial to mitigate the impact of outbreaks and protect public health.

1. **Public Health Measures**: Promotion of hand sanitization and pulmonary hygiene, social distancing, use of personal protective equipments like face masks in crowded areas, avoiding crowded places, travel restrictions
2. **Vaccination**: Development, evaluation, and distribution of safe and effective vaccines against human coronaviruses. Ensure equitable access to vaccines globally, prioritizing high-risk populations and regions with limited healthcare infrastructure
3. **Surveillance and Testing**: Establish robust surveillance systems to monitor the spread of human coronaviruses, detect outbreaks early, and inform public health interventions. Expand access to diagnostic testing for timely identification of cases, contact tracing, and isolation of infected individuals
4. **Healthcare Infrastructure**: Strengthen healthcare infrastructure and surge capacity to manage an influx of patients during outbreaks. Ensure an adequate supply of PPE for healthcare workers to protect against occupational exposure

5. **Research and Innovation**: Support basic research to enhance understanding of human coronaviruses, including their transmission dynamics, pathogenesis, and immune responses. Facilitate the development of antiviral drugs and other therapeutics for the treatment of coronavirus infections. Explore innovative interventions, such as monoclonal antibodies, convalescent plasma therapy, and novel vaccine platforms

Bibliography

Azhar, E. I., El-Kafrawy, S. A., Farraj, S. A., Hassan, A. M., Al-Saeed, M. S., Hashem, A. M., & Madani, T. A. (2014). Evidence for camel-to-human transmission of MERS coronavirus. *The New England Journal of Medicine, 370*(26), 2499–2505. https://doi.org/10.1056/NEJMoa1401505

Coronavirus Disease (COVID-2019). *Situation report—123.* World Health Organization. Published May 22, 2020. https://www.who.int/docs/default-source/coronaviruse/situation-reports/2... Accessed 25 June 2020.

FRCOphth, A. K., Shantha, J. G., Olivia Li, J. P., Faia, L. J., Hartley, C., Kuthyar, S., Albini, T. A., Wu, H., Chodosh, J., Ting, D. S. W., & Yeh, S. (2020). SARS-CoV-2 and the eye: Implications for the retina specialist from human coronavirus outbreaks and animal models. *Journal of Vitreoretinal Diseases, 4*(5), 411–419. https://doi.org/10.1177/2474126420939723

Li, Q., Guan, X., Wu, P., Wang, X., Zhou, L., Tong, Y., Ren, R., Leung, K. S. M., Lau, E. H. Y., Wong, J. Y., Xing, X., Xiang, N., Wu, Y., Li, C., Chen, Q., Li, D., Liu, T., Zhao, J., Liu, M., Tu, W., et al. (2020). Early transmission dynamics in Wuhan, China, of novel coronavirus-infected pneumonia. *The New England Journal of Medicine, 382*(13), 1199–1207. https://doi.org/10.1056/NEJMoa2001316

Riggioni, C., Comberiati, P., Giovannini, M., Agache, I., Akdis, M., Alves-Correia, M., Antó, J. M., Arcolaci, A., Azkur, A. K., Azkur, D., Beken, B., Boccabella, C., Bousquet, J., Breiteneder, H., Carvalho, D., De Las Vecillas, L., Diamant, Z., Eguiluz-Gracia, I., Eiwegger, T., Eyerich, S., et al. (2020). A compendium answering 150 questions on COVID-19 and SARS-CoV-2. *Allergy, 75*(10), 2503–2541. https://doi.org/10.1111/all.14449

Bibliography

Monkeypox

29

Ripan Biswas and Maninder Singh

Abstract

The 2022 multi-national outbreak of mpox (previously known as monkeypox) has shaken the global community, which was just beginning to recover from the aftermath of the COVID-19 pandemic. This outbreak has occurred in various non-African regions, accompanied by reports of human-to-human transmission. The present manuscript provides a comprehensive review of the causative agent, epidemiology of mpox in both animal and human populations, modes of transmission, clinical manifestations, diagnostic approaches, as well as strategies for disease control and prevention. Notably, the World Health Organization has recently recommended the use of the term "mpox" instead of "monkeypox" to avoid any potential social stigmas associated with the previous nomenclature, which is reflected in the terminology used throughout this work.

Keywords

Monkeypox · Orthopox · mpox

History

Mpox was first identified in 1958 through isolating the virus from non-fatal outbreaks among cynomolgus macaques at the University of Copenhagen in Denmark, where the monkeys had been imported from Singapore. The

R. Biswas (✉)
Department of Veterinary Public Health and Epidemiology, West Bengal University of Animal and Fishery Sciences, Kolkata, West Bengal, India

M. Singh
Department of Veterinary Public Health and Epidemiology, College of Veterinary Science, Guru Angad Dev Veterinary and Animal Sciences University, Rampura Phul, Bathinda, India

R. Biswas et al. (eds.), *Concept of One Health and Zoonoses*,
https://doi.org/10.1007/978-981-95-6627-3_29

cynomolgus macaques exhibited a vesiculopapular skin eruption across their bodies. In the ensuing years, comparable outbreaks of the disease were documented among captive monkey populations in the United States and the Netherlands. The first documented isolation of the virus from a human case occurred in 1970, when the pathogen was identified from a 9-month-old infant in the Democratic Republic of the Congo during a smallpox surveillance program. Thereafter, mpox cases were reported in other African countries. The Democratic Republic of the Congo has documented human mpox infections since 1970. Cameroon documented its initial mpox case in 1979, followed by the Central African Republic in 1984 and Gabon in 1987. Between 1970 and the end of the twentieth century, the global incidence of laboratory-confirmed human monkeypox cases was less than 1000, with the majority originating from the Democratic Republic of the Congo and the remainder from seven other Central and West African nations. Over the course of the twenty-first century, the Democratic Republic of the Congo has witnessed a significant surge in the number of monkeypox cases, accompanied by a global expansion of the disease's geographical footprint. In the subsequent years, the Democratic Republic of the Congo and South Sudan documented their initial human monkeypox cases in 2003 and 2005, respectively. Several Central and West African nations, including the Central African Republic, Sierra Leone, Liberia, Cameroon, and Nigeria, have reported human mpox cases after long intervals of several decades. In Nigeria, the disease re-emerged in 2017 following a 39-year hiatus and has continued to be documented in the country since then. The first mpox cases outside Africa occurred in the United States in 2003, linked to imported Gambian pouched rats from Ghana. The virus spread from rodents to prairie dogs and then to humans, resulting in 47 cases across six states. Notably, mpox had not been reported in Ghana prior to this. In 2017, a major outbreak occurred in Nigeria, and from 2018 to 2021, sporadic cases in non-African countries were tied to travel to Nigeria.

In May 2022, mpox cases emerged in non-endemic countries, starting in the United Kingdom. While the first case was linked to travel from Nigeria, subsequent cases had no such history, suggesting community transmission. The outbreak spread rapidly, particularly among men who have sex with men, likely due to direct contact with pox lesions. This marked the first simultaneous occurrence of cases in endemic and non-endemic regions, prompting the World Health Organization to declare mpox a Public Health Emergency of International Concern on July 23, 2022. By December 23, 2022, 83,497 confirmed cases and 72 deaths were reported across 110 countries. The top 10 affected countries—United States, Brazil, Spain, France, Colombia, United Kingdom, Germany, Peru, Mexico, and Canada—accounted for 85.7% of global cases. In Africa, Nigeria reported the most cases, followed by the Democratic Republic of the Congo and Ghana. In South Asia, India reported 23 cases, with initial cases linked to travel to the United Arab Emirates, while others had no travel history.

Etiology and Characteristics

Monkeypox is a viral illness caused by a poxvirus, which are large viruses approximately 200–250 nanometers in size with a double-stranded DNA genome and a lipid-protein envelope. The *Poxviridae* family is an ancient virus family that has been observed in both vertebrate and non-vertebrate organisms. The *Poxviridae* family is divided into two subfamilies: *Chordopoxvirinae,* which encompasses poxviruses infecting vertebrates, and *Entomopoxvirinae*, which includes poxviruses that target insects. The *Chordopoxvirinae* subfamily encompasses 10 distinct genera, in addition to numerous unclassified species. Monkeypox virus is classified within the *Orthopox* genus, which is a part of the *Chordopoxvirinae* subfamily. The Orthopox genus comprises several notable viral species, including variola major virus, variola minor or vaccinia virus, cowpox virus, camelpox virus, raccoon pox virus, and skunk pox virus, among others. It is believed that orthopoxviruses have descended from ancestors associated with rodents, and some orthopoxviruses, such as monkeypox and cowpox viruses, continue to maintain rodents as their reservoir hosts. Orthopoxviruses exhibit a high degree of genetic and antigenic similarity, enabling cross-protection against infection by different orthopoxvirus species within the genus.

Poxviruses possess a linear double-stranded genome spanning approximately 200 kilobases, encoding roughly 200 genes. Roughly half of the viral genes are conserved and are critical for viral replication. The poxvirus genome encompasses all the necessary genes for viral replication, transcription, assembly, and egress, but relies on host cell ribosomes for mRNA translation. The conserved genes reside in the central region of the genome, while the non-conserved genes are situated within the inverted terminal repeats at the genome ends, where they play pivotal roles in virus–host interactions and confer host specificity to poxviruses.

MPXV is divided into two distinct clades: West African and Congo Basin (Central African). These clades differ genetically by approximately 5% and display unique epidemiological and clinical characteristics. The Central African clade is more virulent, with a higher case fatality rate of around 10%, compared to the West African clade, which has a case fatality rate of 1–3%. The ongoing outbreak is linked to the West African clade. To address concerns of cultural and social sensitivity, the WHO has reclassified the clades as clade I (Congo Basin) and clade II (West African). Clade II is further subdivided into clade IIa and clade IIb, with the current outbreak strains identified as clade IIb. Studies show that lineage B.1 of clade IIb is the predominant strain circulating in the 2022 outbreak.

Central African mpox strains are more deadly than West African strains. This is linked to certain genes, like BR-203, BR-209, and COP-C3L. In Central African strains, the BR-203 protein is complete, which helps the virus spread by stopping lymphocyte cell death. However, the BR-203 protein is shorter in West African strains. BR-209 is a protein that binds to interleukin-1β and prevents inflammation. The impact of amino acid differences in BR-209 between strains is not well understood. Additionally, the West African strains lack the complement control protein COP-C3L, which inhibits the host's immune response.

Epidemiology

Mpox, a viral zoonotic disease, first emerged outside of Africa in 2003 in the United States, stemming from the importation of Gambian pouched rats from Ghana and the subsequent infection of prairie dogs and their owners. A 2017 outbreak in Nigeria led to sporadic cases in non-African nations, primarily linked to international travel, until May 2022, when multiple cases were reported in non-endemic countries, with the initial cases surfacing in the United Kingdom. The May 2022 outbreak, which disproportionately affected the male homosexual population, represented the first concurrent spread of the virus in both endemic and non-endemic regions, leading the WHO to declare a Public Health Emergency. By the end of 2022, a total of 83,497 cases and 72 associated deaths were reported worldwide, with the United States, Brazil, and Spain experiencing the highest burdens. In Africa, Nigeria, the Democratic Republic of the Congo, and Ghana reported the greatest number of cases. Additionally, India recorded 23 cases, some of which were linked to international travel. Among orthopoxviruses, MPXV and cowpox virus possess a broad host range, while species like variola virus and camelpox virus are restricted to a single host. The host range in orthopoxviruses is primarily determined by accessory non-conserved regions at the genome's ends, which play a crucial role in virus–host interactions. Evolutionary evidence suggests that orthopoxviruses descended from a common ancestral virus that likely possessed a full set of accessory genes, enabling a broad host range. In contrast, the smallpox virus, which has the smallest genome among orthopoxviruses, is restricted to a single host species, likely due to lineage-specific losses of accessory genes. Certain accessory genes, such as E3L, K3L, K1L, C7L, CP77, and C16L, play a key role in shaping the host range of orthopoxviruses. These genes achieve this by interacting with various host antiviral proteins. For instance, E3L and K3L are associated with protein kinase R, while K1L, C7L, and CP77 interact with Mammalian Sterile Alpha Motif Domain-containing 9 and its paralog SAMD9L. Additionally, C16L is known to inhibit human Zinc-finger antiviral proteins. Poxviruses are double-stranded DNA viruses with a low mutation rate, ranging from 2×10^{-6} to 1×10^{-5} nucleotide substitutions per site per year, significantly lower than RNA viruses like SARS-CoV. This equates to approximately 1–2 nucleotide changes in the genome per year. Their DNA polymerase also has 3′-5′ exonuclease proofreading activity, which contributes to their genetic stability. However, poxvirus genomes undergo other evolutionary changes, such as gene loss, which is a key characteristic. The loss of accessory genes often results in a narrower host range, while the presence of a full set of accessory genes supports a broader host range. For instance, West African strains of MPXV have lost 17% of their accessory genes from 2008 to 2017, coinciding with increased human-to-human transmission. Gene recombination, facilitated by the large genome size of poxviruses, can enable the acquisition of host range genes. Additionally, poxviruses can duplicate accessory genes, potentially broadening their host range. These mechanisms underscore MPXV's potential to adapt to new hosts, including animals and humans.

Transmission

Animal-to-animal monkeypox transmission is uncertain, but close contact seems to be involved. This was seen in the 2003 US outbreak, where prairie dogs living with infected animals became infected, likely through respiratory droplets. Animal-to-human transmission happens through direct or indirect contact with infected animals, their bodily fluids, lesions, or respiratory secretions, including bites and eating undercooked meat. Human-to-human transmission can occur through direct contact with infectious materials, such as respiratory secretions or body fluids, including semen. Transmission may also happen during prolonged close contact or congenitally from mother to child. Mpox has been detected in human feces, but not in a form capable of replication. Transmission has occurred in healthcare settings through needle injuries. Smallpox vaccination may provide some protection against mpox. Following the eradication of smallpox and the discontinuation of vaccination in 1980, cases of mpox increased significantly in Africa. For unvaccinated individuals, the estimated secondary infection rate is 9.3%.

Clinical Symptoms

In Humans

Symptoms of human mpox are similar to but less severe than those of smallpox. After an incubation period of 5–21 days, a prodromal phase occurs with fever, fatigue, headache, and swollen lymph nodes, especially in the neck, groin, and under the jaw. This is followed by a rash phase, where the rash progresses from macules (flat spots) to papules (bumps), vesicles (blisters), pustules (pus-filled bumps), crusts, and finally falls off. The rash usually appears 1–5 days after fever and commonly affects the face, trunk, palms, soles, and oral cavity, but may also involve the genitals and conjunctiva. Patients remain contagious after the rash develops. Secondary infections can lead to serious complications such as sepsis, gastrointestinal issues, pneumonia, brain inflammation, eye problems, and dehydration. Symptoms typically resolve within 2–5 weeks. Possible alternative diagnoses include viral infections like chickenpox, bacterial diseases like staph infections, and sexually transmitted infections like syphilis. While usually self-limiting, mpox can lead to serious complications and death, especially in immunocompromised individuals, children, or those with secondary infections. Smallpox vaccination can help reduce the severity and risk of death from mpox. Mpox is diagnosed worldwide using PCR tests that target specific genetic markers, which can also distinguish between Central and West African mpox strains. Other diagnostic tests, such as ELISA, virus neutralization, immunohistochemistry, and electron microscopy, are also used. Skin samples are often used for studying the virus's genetic material because the rash stage has high levels of the virus and is very infectious.

In Animals

Non-human primates are highly susceptible to mpox, showing symptoms similar to humans, like a rash and scabs, and can die from it. Prairie dogs, which amplified the virus during the 2003 US outbreak, are now used as lab models. However, wild rodent populations may not display any mpox symptoms.

Diagnosis

Mpox is usually diagnosed using a type of PCR test that can measure the amount of the virus. This test looks for specific genes like C3L, TNFR, and E9L. The C3L and TNFR genes can help distinguish between the Central and West African strains of the virus. In resource-limited settings, PCR testing may not always be practical due to cost and technical difficulties. Instead, serological methods like Orthopox genus-based ELISA are commonly used. However, interpreting these test results requires careful consideration of the epidemiological context. IgM antibody tests, along with patient symptoms and outbreak information, can aid diagnosis in resource-poor regions. Additional diagnostic methods include virus neutralization tests, immunohistochemistry, and electron microscopy. Viral DNA analysis is commonly performed on skin samples due to the high viral loads and infectivity associated with the development of the rash.

Prevention and Control

Monkeypox can spread from rodents to people and between people. This requires a "One Health" approach involving both human health and animal health professionals. Minimizing contact with animal hosts, especially small mammals, is crucial. This requires raising awareness in areas where the disease is endemic and among those with occupational exposure. Monitoring domestic animals is necessary to determine the range of species that can be infected. Smallpox vaccines (e.g., JYNNEOS and ACAM2000 in the USA) offer some cross-protection against mpox. Antiviral medications such as cidofovir, brincidofovir, and tecovirimat are used to treat monkeypox, in addition to vaccinia immunoglobulins. Improving diagnostic capabilities and educating the public, particularly in developing nations, are key for controlling endemic and new diseases like monkeypox. Raising public consciousness of clinical manifestations and transmission pathways enables well-informed decision making. A coordinated, cross-sectoral "One Health" strategy is crucial for successful monkeypox mitigation.

References

Cowen, E. W., Tkaczyk, E. R., Norton, S. A., & Leslie, K. (2023). Mpox—A rapidly evolving disease. *JAMA Dermatology, 159*(4), 424). American Medical Association. https://doi.org/10.1001/jamadermatol.2023.0041

da Silva, S. J. R., Kohl, A., Pena, L., & Pardee, K. (2023). Clinical and laboratory diagnosis of monkeypox (mpox): Current status and future directions [Review of Clinical and laboratory diagnosis of monkeypox (mpox): Current status and future directions]. *iScience, 26*(6), 106759. Cell Press.

Doshi, R. H., Guagliardo, S. A. J., Dzabatou-Babeaux, A., Likouayoulou, C., Ndakala, N., Moses, C., Olson, V. A., McCollum, A. M., & Petersen, B. W. (2018). Strengthening of surveillance during monkeypox outbreak, Republic of the Congo, 2017. *Emerging Infectious Diseases, 24*(6), 1158). Centers for Disease Control and Prevention. https://doi.org/10.3201/eid2406.180248

Paparini, S., Whelan, I., Mwendera, C., Hayes, R., Maatouk, I., Lewis, R., Nunez, M. P., Mozalevskis, A., Wi, T., & Orkin, C. (2024). Prevention of sexual transmission of mpox: A systematic review and qualitative evidence synthesis of approaches [Review of Prevention of sexual transmission of mpox: A systematic review and qualitative evidence synthesis of approaches]. *Infectious Diseases, 56*(8), 589. Taylor & Francis. https://doi.org/10.1080/23744235.2024.2364801

Parker, S., & Buller, R. M. L. (2013). A review of experimental and natural infections of animals with monkeypox virus between 1958 and 2012 [Review of A review of experimental and natural infections of animals with monkeypox virus between 1958 and 2012]. *Future Virology, 8*(2), 129. Future Medicine. https://doi.org/10.2217/fvl.12.130

Prevention. (2024). https://www.cdc.gov/poxvirus/mpox/prevention/index.html

Riser, A., Hanley, A., Cima, M. J., Lewis, L., Saadeh, K., Alarcón, J., Finn, L., Kim, M., Adams, J. N., Holt, D. B., Feldpausch, A., Pavlick, J., English, A., Smith, M. E., Rehman, T., Lubelchek, R., Black, S. R., Collins, M., Mounsey, L., et al. (2023). Epidemiologic and clinical features of Mpox-associated deaths—United States, May 10, 2022–March 7, 2023. *MMWR Morbidity and Mortality Weekly Report, 72*(15), 404. Centers for Disease Control and Prevention. https://doi.org/10.15585/mmwr.mm7215a5

Sadeuh-Mba, S. A., Yonga, M. G., Els, M., Batéjat, C., Eyangoh, S., Caro, V., Etoundi, A., Carniel, É., & Njouom, R. (2019). Monkeypox virus phylogenetic similarities between a human case detected in Cameroon in 2018 and the 2017–2018 outbreak in Nigeria. *Infection Genetics and Evolution, 69*, 8. Elsevier BV. https://doi.org/10.1016/j.meegid.2019.01.006

References

Yellow Fever

30

Ajay Pratap Singh

Abstract

Yellow fever virus is a mosquito-borne flavivirus that causes infections in human beings, with symptoms ranging from mild non-specific illness to severe disease with jaundice, hemorrhage, and death. YFV is endemic to tropical areas of Africa and South America, where, according to the World Health Organization (WHO), there are approximately 80,000–200,000 YFV cases worldwide with approximately 30,000 deaths annually. The case fatality rate (CFR) for yellow fever (YF) ranges from 20% to 50%.

Keywords

Yellow fever · 17D Vaccine · Jaundice · American Pestilence · Barbados Distemper · Continua Putrida Icteroides Caroliniensis · Yellow Jack

History

The earliest description of yellow fever is found in a Mayan manuscript in 1648. Yellow fever virus (YFV) evolved from other mosquito-borne viruses about 3000 years ago, Yellow fever virus is believed to have originated in Africa. Genetic studies have shown that the virus likely emerged from primates in the African rainforest, where it is maintained in a cycle between non-human primates and mosquito vectors, primarily species of the Aedes and Haemagogus genera. Historical records and genetic evidence suggest that YFV was likely introduced to the American

A. Pratap Singh (✉)
Department of Veterinary Microbiology, College of Veterinary Science & Animal Husbandry, U.P. Pandit Deen Dayal Upadhyaya Pashu Chikitsa Vigyan Vishwavidyalaya Evam Go-Anusandhan Sansthan, Mathura, Uttar Pradesh, India

R. Biswas et al. (eds.), *Concept of One Health and Zoonoses*,
https://doi.org/10.1007/978-981-95-6627-3_30

continent during the transatlantic slave trade. The first recorded outbreaks of yellow fever in the Americas occurred in the seventeenth century, shortly after the arrival of African slaves. The term "yellow fever" was first coined by Griffin Hughes in 1750 in his work titled "Natural History of Barbadoes" (1750). Yellow fever was a significant problem in colonial settlements in the Americas and West Africa during the eighteenth and nineteenth centuries. The first recorded epidemic believed to be yellow fever occurred in the Yucatan in 1648. Between 1668 and 1893, the United States experienced more than 135 yellow fever epidemics in cities such as Philadelphia, New York, Charleston, and New Orleans. Josiah C. Nott (1804–1873), a physician from Alabama, United States, strongly suspected that microorganisms transmitted by insects were responsible for yellow fever and shared his theory in a publication in 1848. Louis D. Beauperthuy (1807–1871) was a French physician who, during his work in Venezuela, suggested in 1854 that mosquitoes might be involved in the transmission of yellow fever. Carlos J. Finlay, a Cuban physician, published his groundbreaking theory on mosquito transmission of yellow fever in 1881. His work laid the foundation for the later confirmation that mosquitoes are the primary vectors of the disease, significantly advancing the understanding of yellow fever transmission. In 1869, during a yellow fever outbreak in Trinidad, Charles Kingsley, an English author and naturalist, noted a puzzling coincidence: the deaths of howler monkeys seemed to align with the occurrence of yellow fever outbreaks. Henry Rose Carter, a physician with the Marine Hospital Service, conducted important epidemiological studies on yellow fever outbreaks in the late nineteenth and early twentieth centuries. One of his significant findings was the observation of a lag time, typically 2–3 weeks, between the appearance of the first (index) case of yellow fever in a community and the onset of secondary cases. Carter termed this period the "extrinsic incubation period". In the year 1898, Spanish–American War broke out over Spain's colonial possessions in the Caribbean and the Pacific. In that war, American soldiers began to suffer from outbreaks of yellow fever. The high mortality rate among troops prompted George Miller Sternberg, Surgeon General of the United States Army to form a board, known as the Yellow Fever Commission to investigate the cause of the disease and find methods for its prevention. Yellow Fever Commission was led by Major Walter Reed, who was an army pathologist and bacteriologist along, with Major James Carroll, Major Aristides Agramonte, and Major Jesse Lazear. In June 1900, a team in Cuba tested Finlay's mosquito hypothesis on yellow fever transmission, using soldiers and immigrants as volunteers. Lazear's initial exposure did not result in illness due to timing, but subsequent experiments confirmed mosquito-borne transmission. They built Camp Lazear, disproving the fomite theory and confirming mosquito transmission through controlled experiments, validating their hypothesis. The causative agent of yellow fever, the yellow fever virus (YFV), was first isolated by Hideyo Noguchi, a Japanese bacteriologist, who isolated the virus from a Ghanaian patient named Asibi in 1927. In the 1930s, Max Theiler and his colleagues developed a live-attenuated vaccine strain called 17D. Theiler was awarded the Nobel Prize in Physiology or Medicine in 1951 for his work on the yellow fever vaccine. The second live-attenuated yellow fever vaccine was developed from a strain of yellow fever virus isolated in Senegal in 1927.

Etiology

YFV is the prototypic member of the genus Flavivirus, family Flaviviridae. Mature YFV virions are icosahedral and comprise a nucleocapsid, composed of capsid (C) protein subunits, surrounded by a lipid bilayer derived from host membranes. The viral envelope is studded with dimers of envelope (E) glycoprotein and membrane (M) protein. The diameter of the virion is approximately 40 nm, with surface projections of 5–10 nm. The E glycoprotein is the major component of the virion surface and possesses most of the biologic activity. It is a positive-sense single-stranded RNA virus with a genome of approximately 11 kb that consists of a 5′ untranslated region (UTR), followed by a single open reading frame (ORF), and a 3′UTR. The ORF is further divided into three structural proteins (C, prM/M, and E) and seven non-structural proteins (NS1, NS2A, NS2B, NS3, NS4A, NS4B, and NS5). The YFV is classified as having a single serotype. To date, 7 YFV genotypes have been described, including 2 in South America and 5 in Africa, namely West Africa genotype I (Nigeria, Cameroon, and Gabon), West Africa genotype II (Senegal, Guinea, Ivory Coast, and Ghana), East and Central African genotype (Sudan, Ethiopia, Central African Republic, and Democratic Republic of the Congo), East African genotype (Kenya), and Angola genotype (Angola).

Transmission

Yellow fever virus (YFV) affects both humans and nonhuman primates in tropical regions of Africa and the Americas. It spreads among primate hosts through the bites of infected mosquitoes, mainly through their saliva. Yellow fever is transmitted primarily by mosquito genera *Haemagogus* and *Sabethes* in Central and South America, and by Aedes mosquitoes (Subgenera *Diceromyia* and *Stegomyia*) on the African continent. The natural cycle of YFV involves transmission between forest-dwelling mosquitoes and wild primates in both continents. Human transmission occurs in three cycles: sylvatic, intermediate, and urban. The sylvatic cycle occurs in rainforests, where the virus is passed between monkeys and forest mosquitoes, occasionally infecting humans who enter these areas. Sylvatic yellow fever has been demonstrated to exhibit increased transmission risk of the YFV in regions characterized by extensive forest fragmentation, thereby facilitating virus spread. The intermediate cycle leads to small outbreaks in rural villages when infected mosquitoes feed on both monkeys and humans. The urban cycle, the most concerning, occurs when the virus enters densely populated areas, where urban mosquitoes like *Aedes aegypti* transmit it between humans, leading to large-scale epidemics.

Host Range

The mosquito, the true reservoir of YF, is infected throughout its life, and can transmit the virus transovarially through infected eggs. Man and monkeys, on the other hand, play the role of temporary amplifiers providing a source of infection for mosquitoes but not maintaining the virus long term. Marsupials and ungulates could become infected by Jungle yellow fever. In Africa, animals such as cows, horses, sheep, goats, pigs, camels, and dogs, which exhibit high levels of antibody seropositivity to the YFV, serve as incidental hosts in the transmission of yellow fever.

Pathogenesis

Yellow fever is a severe viral infection with multiple pathophysiological impacts on the body. The virus primarily replicates within hepatocytes, leading to cellular disruption, metabolic impairment, and may induce apoptosis or necrosis of infected hepatocytes. This cellular damage releases viral particles and debris into the bloodstream, causing systemic effects. One significant consequence of YF is renal injury, although its pathogenesis is not fully understood. The virus can infect renal cells directly, leading to inflammation and tissue damage. The systemic inflammatory response further exacerbates this damage through immune-mediated mechanisms, resulting in endothelial dysfunction, glomerular injury, and tubular damage. YF also induces bone marrow suppression, reducing the production of blood cells such as leukocytes (leading to leukopenia) and platelets (leading to thrombocytopenia). This suppression arises from direct viral effects on bone marrow cells and the systemic inflammatory response. In severe cases, liver damage leads to hyperkalemia (elevated potassium levels), impairing cardiac function and causing arrhythmias. These cardiac issues, along with fluid redistribution and tissue edema, decrease the heart's ability to pump blood effectively, compromising tissue perfusion. Endothelial infection by YFV triggers immune and inflammatory responses, including the release of proinflammatory cytokines like interleukin-6 (IL-6) and interleukin-8 (IL-8), and the generation of free oxygen radicals. These mediators disrupt coagulation and fibrinolysis, leading to abnormal blood clotting, microthrombi formation, and tissue damage. Increased vascular permeability causes plasma leakage, resulting in edema and extracellular fluid accumulation, which further impairs tissue perfusion and oxygen delivery. Ultimately, these processes can lead to hypotension, shock, and multi-organ dysfunction, characterized by hypoxia and inadequate oxygen supply to vital organs and tissues, exacerbating cellular injury and organ failure.

Clinical Symptoms

Symptoms in Humans

YF symptoms range from a mild, non-specific, febrile sickness to a fulminant illness with frequent lethality with pathologically indicative characteristics. Yellow fever in humans typically progresses through several stages.

Period of Infection (Classical YF)

This is most common in occurrence. The onset of yellow fever starts suddenly, typically within 3–6 days after being bitten by an infected mosquito. The initial phase of yellow fever infection is characterized by fever (39–40 °C), chills, intense headache, lower back pain, generalized muscular pains, nausea, vomiting, and photophobia and hyperemia of skin.

During the initial phase of yellow fever infection, additional symptoms may include a slow pulse (Faget's sign), dark-colored urine, and febrile convulsions in young children. The patient may serve as a source of infection for mosquitoes during this phase, which normally lasts 3–5 days, because the virus is present in high titers in the blood.

Period of Remission

During the remission period of yellow fever, which typically lasts 12–24 h, several positive changes occur in the patient's condition. During this remission period, the body clears the virus with the help of antibodies and cellular immune responses. First, there is a noticeable decrease in body temperature, the intense headache subsides. Furthermore, the general condition of the patient tends to improve, with symptoms such as muscle pains and nausea diminishing. In some cases, particularly those with milder or "abortive" infections, individuals may recover fully during this remission period, experiencing a resolution of symptoms without progressing to more severe stages of the disease.

Period of Intoxication

In approximately 15–25% of yellow fever patients, the disease progresses to a more severe stage known as the "period of intoxication." During the hepatorenal phase of yellow fever, there is a notable rise in temperature along with the reappearance of generalized symptoms such as severe muscle weakness, nausea, vomiting, and epigastric pain. Jaundice becomes apparent during this phase, accompanied by a hepatic induced "bleeding diathesis" and "coagulopathy" characterized by petechiae, ecchymoses, and bleeding from mucous membranes. This phase also involves a bleeding disorder resulting from reduced synthesis of clotting factors and consumption coagulopathy, along with platelet dysfunction. Severe bleeding manifestations may include characteristic black vomit (hematemesis), melena (black, tarry stools), or metrorrhagia (uterine bleeding). Renal dysfunction is a significant complication during the hepatorenal phase of yellow fever, characterized by a sudden increase in albuminuria and diminishing urine output. In severe cases, death can occur within 7–10 days after the onset of symptoms.

Symptoms in Animals

Yellow fever infection in non-human primates such as monkeys develops into viscerotropic infection characterized by hepatitis. In areas where yellow fever is endemic, outbreaks among monkey populations can occur simultaneously with human outbreaks. Monkeys infected with YFV can experience similar symptoms to those of humans, including fever, jaundice, and hemorrhage. Asian monkeys such as Rhesus and Cynomolgus macaques tend to exhibit symptoms resembling human yellow fever infection, while African monkeys may have a silent or asymptomatic infection. Wild-type YFV can cause neurotropic infection in rodents such as mice, hamsters, and guinea pigs. In rodent models, intracerebral (IC) or intraperitoneal (IP) inoculation of YFV can result in various neurological manifestations, including encephalitis, neuronal damage, and behavioral changes.

Diagnosis

Detection of Yellow Fever Virus (YFV)-Specific IgM Antibodies It is a crucial diagnostic tool for confirming recent infection with the virus. The detection of YFV-specific IgM antibodies is typically performed using serological assays such as enzyme-linked immunosorbent assay (ELISA) or immunofluorescence assay (IFA). Commercially available IgM capture ELISA (Tariki Fiebre Amarilla IgM YF-ELISA), sandwich ELISA (MyBiosource, Inc., San Diego, CA, USA) and immunofluorescence-based kits (EUROIMMUN AG, Lübeck, Germany) are used in diagnostic laboratories.

Plaque Reduction Neutralization Test (PRNT) It is considered the gold standard in laboratory diagnosis. It offers greater specificity compared to other assays, although it may have lower sensitivity.

Histopathological and Immunohistochemical Techniques Detection of viral antigens is a critical technique in the field of virology, offering significant benefits to reference laboratories. This method is invaluable for providing supportive diagnoses, confirming the presence of specific viruses in clinical samples. Additionally, it plays a crucial role in investigating epizootics, helping to track and manage outbreaks of viral diseases in animal populations.

Virus Isolation YFV can be isolated from blood collected during the initial febrile illness and from post-mortem tissues. The virus can be propagated in a variety of cell lines, including monkey epithelial and kidney fibroblasts (MA-104, Vero,

LLC-MK2); rabbit- (MA-111) and baby hamster-derived lines (BHK); and Ae. pseudoscutellaris (AP-61) and Ae. albopictus (C6/36) mosquito cells.

Molecular Detection of YFV RNA in Tissue Samples Detection of yellow fever virus (YFV) RNA in tissue samples is a valuable diagnostic approach, particularly for confirming active infection and studying viral dynamics. Quantitative reverse transcription polymerase chain reaction (qRT-PCR) assays that target short regions of the YFV genome are highly effective for this purpose.

Prevention and Control

Prevention and control measures for yellow fever (YF) encompass a range of strategies aimed at reducing transmission and mitigating the impact of outbreaks.

Yellow Fever Vaccine The most effective preventive measure is vaccination. The yellow fever vaccine provides long-lasting immunity against the virus and is recommended for individuals living in or traveling to areas where yellow fever is endemic. Vaccination campaigns target at-risk populations to ensure widespread coverage and herd immunity. The introduction of live-attenuated yellow fever vaccinations has helped to curb the disease's spread. The FNV vaccine strain was developed from the wild-type French viscerotropic virus via 128 serial passages in mouse brain tissue. However, its administration was linked to a high incidence of encephalitis in youngsters, emphasizing the need for safer alternatives.

In response, the 17D vaccine strain was created via 176 serial passes in mouse and chick embryo tissue cultures. This lengthy attenuation process eliminated viscerotropism and mosquito competence, making it safe for human usage. The 17D vaccination is widely considered as one of the most safe and effective live-attenuated viral vaccines ever created. The yellow fever vaccine strains YF 17DD (passage 195) and 17D-204 (passage 204) are the primary strains used in current immunization efforts against yellow fever.

Adverse Events Associated with the Yellow Fever (YF) Vaccine While rare, adverse events associated with the yellow fever (YF) vaccine can occur with varying severity, ranging from mild such as fever, headache, or muscle aches to more serious.

Yellow Fever Vaccine-Associated Viscerotropic Disease (YEL-AVD)

Yellow fever vaccine-associated viscerotropic disease (YEL-AVD) mimics the symptoms of wild-type yellow fever, usually emerging 2–8 days post-vaccination. Early symptoms are non-specific and include fever, headache, malaise, muscle aches, joint pain, nausea, vomiting, and diarrhea. As the illness advances, jaundice may appear along with thrombocytopenia, elevated liver enzymes, and increased bilirubin and creatinine levels. Severe cases can escalate to hypotension, bleeding, acute kidney and respiratory failure, ultimately leading to multiorgan system failure.

Yellow Fever Vaccine-Associated Neurotropic Disease (YEL-AND)

Yellow fever vaccine-associated neurotropic disease (YEL-AND) involves neurological complications such as post-vaccinal encephalitis and autoimmune disorders impacting the central or peripheral nervous system. These conditions, like acute disseminated encephalomyelitis or Guillain–Barré syndrome, manifest through a variety of neurological symptoms.

Vector Control Controlling mosquito populations is crucial for preventing YF transmission. This includes strategies such as eliminating mosquito breeding sites, using insecticides to kill adult mosquitoes, and implementing environmental management practices to reduce mosquito habitats.

Improved Disease Surveillance Enhanced surveillance systems are essential for early detection and response to YF outbreaks. This involves monitoring for suspected cases, conducting laboratory testing to confirm diagnoses, and tracking the spread of the virus to identify high-risk areas.

Outbreak Prevention Early intervention is key to preventing YF outbreaks. Rapid response teams can be mobilized to implement control measures such as vaccination campaigns, vector control activities, and public health education initiatives to contain outbreaks and prevent further transmission.

Community Engagement Engaging communities in YF prevention efforts is vital for ensuring compliance with vaccination programs, vector control measures, and outbreak response activities. Community education and communication campaigns can raise awareness about YF transmission, symptoms, and prevention strategies.

Bibliography

Group IV Viruses. (n.d.). https://doi.org/10.1016/b978-0-12-415895-5.00042-8

Missed opportunities for Yellow Fever Surveillance in Uganda, July 2015–May 2016. (n.d.). https://doi.org/10.1016/j.ijid.2016.11.291

Monath, T. P., & Vasconcelos, P. F. da C. (2014). Yellow fever [Review of Yellow fever]. *Journal of Clinical Virology, 64*, 160. Elsevier BV. https://doi.org/10.1016/j.jcv.2014.08.030

Nwachukwu, W., Yusuff, H. A., Nwangwu, U. C., Okon, A. E., Ogunniyi, A., Imuetinyan-Clement, J., Besong, M., Ayo-Ajayi, P., Nikau, J., Baba, A., Dogunro, F. A., Akintunde, B., Oguntoye, M., Kamaldeen, K., Fakayode, O., Oyebanji, O., Emelife, O., Oteri, J., Aruna, O., et al. (2020). The response to re-emergence of yellow fever in Nigeria, 2017. *International Journal of Infectious Diseases, 92*, 189. Elsevier BV. https://doi.org/10.1016/j.ijid.2019.12.034

Norovirus and Sapovirus

31

Ajay Pratap Singh

Abstract

Norovirus ranks as the second leading cause of childhood diarrhea globally, following rotavirus. Both noroviruses and sapoviruses, members of the Caliciviridae family, are significant pathogens responsible for gastroenteritis. Human noroviruses (HNoV) are particularly notorious for triggering gastroenteritis outbreaks worldwide and are a major contributor to severe diarrhea in children. In contrast, Sapovirus typically induces milder symptoms and is less commonly linked to large-scale outbreaks. Noroviruses are highly contagious and frequently cause self-limiting gastroenteritis, especially in communal settings like hospitals, schools, and nursing homes. Sapovirus infections, on the other hand, primarily affect young children and are generally less severe.

Keywords

Norovirus · Sapovirus · Winter vomiting disease

History

The earliest documented case of illness attributed to this virus traces back to 1929, when it was termed "**Winter vomiting disease**." The name gained wider recognition after a notable outbreak in Norwalk, Ohio, USA, during the winter season. In 1968, another incident at a Norwalk elementary school affected half of the students, prompting further investigation. In a critical experiment, rectal swabs from affected

A. P. Singh (✉)
Department of Veterinary Microbiology, College of Veterinary Science & Animal Husbandry, U.P. Pandit Deen Dayal Upadhyaya Pashu Chikitsa Vigyan Vishwavidyalaya Evam Go-Anusandhan Sansthan, Mathura, Uttar Pradesh, India

R. Biswas et al. (eds.), *Concept of One Health and Zoonoses*,
https://doi.org/10.1007/978-981-95-6627-3_31

individuals were filtered and orally administered to healthy male prisoners in Maryland, resulting in gastroenteritis symptoms. This confirmed the oral transmission of the pathogen. Dr. Albert Z. Kapikian and his team made a groundbreaking discovery in the 1970s by identifying the Norwalk virus, the first known norovirus strain and the first virus proven to cause gastroenteritis. Electron microscopy in 1972 revealed the virus's structure, measuring 27–35 nm in diameter. Subsequent research in the 1990s classified Norwalk virus and related strains under the new genus Norovirus within the Caliciviridae family. Sapovirus was first detected in Great Britain in 1975 and later named after the prototype strain, Sapporo virus, identified in Japan in 1977. The International Committee on Taxonomy of Viruses (ICTV) officially recognized the genus Sapovirus in 2000.

Etiology

Noroviruses (NoVs) and sapoviruses (SaVs) are non-enveloped, enteric viruses classified within the family Caliciviridae. The viral capsid features an icosahedral structure with 90 dimers of the VP1 protein, forming 32 cup-shaped depressions—hence the name Caliciviridae, derived from the Latin "**calix**" (cup). The genome consists of a linear, positive-sense RNA (~7.6 kb) organized into three open reading frames (ORFs). ORF1 encodes nonstructural proteins, while ORF2 and ORF3 code for structural proteins VP1 and VP2, respectively. This family includes five genera: Norovirus, Sapovirus, Lagovirus, Nebovirus, and Vesivirus. The Norovirus and Sapovirus genera are represented by their prototype strains, Norwalk virus and Sapporo virus, respectively. Noroviruses are classified into genogroups (GI to GVI) based on VP1 amino acid diversity. Human-infecting strains primarily belong to genogroups I, II, and IV, while genogroups III, V, and VI predominantly include bovine, murine, and canine noroviruses, respectively. Within this classification system, genogroup I contains the Norwalk virus as its reference strain and is divided into eight genotypes. Genogroup II encompasses 19 genotypes, three of which (11, 18, and 19) are associated with porcine strains. The classification of Sapovirus recognizes five established genogroups. Among these, genogroups I, II, IV, and V primarily affect human populations, while genogroup III consists of porcine enteric sapoviruses.

Epidemiology

Norovirus and Sapovirus, collectively referred to as enteric caliciviruses, exhibit a wide host tropism, infecting diverse mammalian species such as humans, swine, bovines, ovines, and murines. Norovirus and Sapovirus are excreted in large quantities in the feces of infected individuals. Sewage contamination of water bodies, such as coastal areas and harbors, serves as a route for introducing these viruses into the environment. Shellfish, such as oysters, clams, and mussels, are filter feeders that can accumulate virus particles from contaminated water. Enteric Calicivirus infections are not seasonal and can occur year-round, but outbreaks tend to peak during colder months. Foodborne transmission is a significantly more common route of gastroenteritis outbreaks compared to waterborne transmission, with food accounting for a higher percentage of outbreaks at 14% compared to water at 1.5%. Epidemiological surveillance has confirmed numerous shellfish-linked gastroenteritis outbreaks across multiple global regions, particularly in coastal communities where shellfish harvesting occurs in waters that may be contaminated with sewage or other sources of human waste. Food service workers have been identified as transmission sources in several norovirus outbreaks at quick-service restaurants. Globally, these pathogens account for an estimated 64,000 annual cases of childhood diarrhea and 200,000 fatalities among children under five in resource-limited regions. Particularly virulent GII.4 genotype infections demonstrate higher association with clinical severity and fatal outcomes. Infections caused by GII.4 strains of norovirus demonstrate higher association with clinical severity and fatal outcomes. Enteric caliciviruses demonstrate significant epidemiological versatility, causing outbreaks across diverse institutional environments. High-transmission settings include the following:

(i) Medical care institutions (due to immunocompromised populations)
(ii) Educational establishments (through close contact among children)
(iii) Military installations (from congregate living conditions)
(iv) Maritime passenger vessels (amplified by confined spaces)

These environments particularly favor transmission due to the viruses' environmental stability and efficient person-to-person spread in densely populated settings.

Transmission

Norovirus spreads through multiple well-characterized pathways, with recent studies revealing critical details about its persistence and transmission efficiency:

(i) **Food and Waterborne Transmission**: Contaminated produce and shellfish represent over 60% of documented foodborne outbreaks, with studies showing that the virus can survive on refrigerated berries for up to 10 days. Recreational water sources have been found to contain detectable norovirus RNA in 28% of samples during peak outbreak seasons.
(ii) **Surface Transmission (Fomites)**: The virus demonstrates remarkable environmental stability, remaining infectious for 14 days on stainless steel and 7 days on common fabrics. Contact studies reveal 30–50% transfer efficiency to human skin upon touching contaminated surfaces.
(iii) **Direct Human-to-Human Spread**: A single infected individual can transmit the virus to 14–33% of close contacts within 48 h, particularly in superspreader events. Notably, 30% of healthcare workers continue viral shedding for 3 weeks or longer after symptom resolution.
(iv) **Aerosol Transmission**: Airborne spread occurs through projectile vomiting, with hospital studies detecting viral RNA in 40% of air samples collected more than 3 m from vomiting incidents.

Pathogenesis

Norovirus and sapovirus share similarities in transmission, symptomatology, and pathogenesis. Norovirus initiates infection through specific interactions between its viral capsid proteins and fucosylated histo-blood group antigens (HBGAs) expressed on intestinal epithelial cells. This binding occurs via precise molecular interactions between the viral VP1 capsid protein's P-domain and HBGA carbohydrate moieties. The binding triggers conformational changes in the viral capsid that facilitate viral entry. Strain-specific variations in VP1 determine HBGA binding affinity, explaining differential susceptibility among populations. Intestinal bacteria expressing histo-blood group antigens (HBGA) may act as co-factors during norovirus infection by facilitating attachment and entry into B-lymphocytes. Recent reports have identified sialic acids as binding factors for porcine sapovirus. Several studies have suggested a potential association between proton pump inhibitors (PPIs) and an increased risk of norovirus infection due to reduced stomach acid production, creating a more conducive environment for viral survival and transmission. Reduction of intestinal flora through antibiotic use has been shown to correlate with decreased norovirus titers in mice, indicating that increased gut flora may create facilitating conditions for norovirus infection. Once inside the cell, norovirus replicates rapidly, leading to cell death and shedding of viral particles into the intestinal lumen. Current understanding of norovirus replication suggests that primary viral amplification occurs in the small intestine, though recent studies indicate broader cellular tropism beyond enterocytes. HBGA expression patterns in the small intestine (highest in duodenal enterocytes) correlate with norovirus tropism. Murine norovirus (MNV) demonstrates a distinct tropism for dendritic cells, macrophage and B cells. This immune cell tropism, particularly the establishment of long-term reservoirs in lymph node macrophages, may contribute to the virus's ability to cause prolonged or recurrent infections despite robust immune responses. The destruction of intestinal epithelial cells disrupts normal gut function, leading to characteristic symptoms of gastroenteritis. Inflammation and immune responses may contribute to tissue damage and symptom severity. Additionally, norovirus may affect the enteric nervous system, leading to dysregulation of gastrointestinal motility and exacerbating symptoms.

Clinical Symptoms in Humans

Asymptomatic Infection Asymptomatic norovirus shedding represents a critical yet often overlooked factor in outbreak dynamics. Surveillance data indicate that silent shedders—individuals who excrete the virus without symptoms—are surprisingly common, including 11.7–49.2% of children in community settings, 22% of healthy adults, and 30% of healthcare workers even after clinical recovery. These carriers typically shed virus for a median of 14 days, with fecal viral loads reaching 10^5–10^7 copies per gram—concentrations sufficient to transmit infection. Notably, GII.4 strains, associated with more severe outcomes, demonstrate significantly prolonged shedding compared to other genotypes, amplifying their epidemic potential.

Symptomatic Infection Noroviruses continue to represent the leading cause of acute viral gastroenteritis globally, responsible for approximately 18% of all diarrheal cases worldwide. Pediatric populations bear the highest disease burden, with children under 5 years experiencing incidence rates of 120 cases per 1000 person-years—nearly double that of older age groups. Among geriatric population (≥65 years), infection carries a threefold increased risk of hospitalization compared to younger adults, while immunocompromised patients face particularly severe outcomes, exhibiting median symptom durations of 9 days versus 2.5 days in immunocompetent hosts. While sapoviruses share genetic kinship with noroviruses, their epidemiological profile differs markedly. These pathogens account for just 4–8% of viral gastroenteritis cases and demonstrate two distinctive characteristics: first, a high rate of asymptomatic carriage (occurring in ~60% of infections) and second, pronounced winter/spring seasonality in outbreak patterns.

Clinical Symptoms in Animals

The enteric calicivirus exhibits a spectrum of clinical effects in infected animals. While many infections in animals are asymptomatic or are mild, some can result in severe clinical manifestations. Animal noroviruses demonstrate distinct host-specific pathogenicity patterns. Among livestock species, bovine (BoNoV), porcine (PoNoV), and ovine (OvNoV) noroviruses have been detected in both clinically affected animals and asymptomatic carriers. In cattle, viral shedding occurs with equal frequency in diarrheic and healthy herds (32–41% prevalence). Swine populations show particularly high asymptomatic carriage rates (up to 58% in surveillance studies). Sheep exhibit intermittent shedding regardless of clinical status. This contrasts sharply with murine norovirus-1 (MNV-1), which causes systemic disease in immunocompromised mouse models characterized by lethal multiorgan disease through: (i) encephalitis/meningitis via retrograde transport, (ii) Kupffer cell-dependent hepatitis, and (iii) type I interferon-mediated vasculitis and, (iv) pneumonia. Additionally, sapovirus has been linked to diarrhea in swine and minks. The experimental infection of newborn calves with bovine norovirus led to severe diarrhea, intestinal lesions, and the norovirus antigen localization within the apical cytoplasm of small intestinal villous epithelial cells. The virus was detected in intestinal contents and feces, without evidence of systemic infection.

Norovirus preferentially targets absorptive enterocytes in the small intestinal villi, where viral replication disrupts cytoskeletal architecture and impairs nutrient absorption, leading to characteristic watery diarrhea. Following a median incubation period of 24 h (range 15–36 h), infection typically manifests as an abrupt onset of projectile vomiting (occurring in 82% of cases, predominantly on the first day), followed by frequent watery stools (4–8 episodes/day) accompanied by abdominal cramps, nausea, and low-grade fever. Disease severity is influenced by both host and viral factors: vulnerable populations like infants, elderly individuals (>65 years), and immunocompromised patients experience more prolonged and severe symptoms, including rare cases of hypovolemic shock from fluid losses exceeding 10%

body weight. Blood group polymorphisms further modulate susceptibility, with non-secretors demonstrating partial resistance. Viral determinants also contribute, as evidenced by GII.4 strains inducing more frequent vomiting (2.3× higher than other genotypes) and polymerase mutants enhancing virulence. While most cases resolve within 48–72 h, atypical presentations occur in high-risk groups (neonates and immunocompromised hosts). Mortality remains rare (0.001% overall), but rises significantly in institutional settings (0.1%), primarily due to delayed rehydration in patients with cardiorenal comorbidities. Norovirus infections in neonates, particularly premature infants, can lead to serious complications such as necrotizing enterocolitis, which can result in mortality. Neonates with norovirus-associated necrotizing enterocolitis may present with symptoms such as diarrhea, abdominal distention, apnea, vomiting, and blood in stool, with a mean duration of illness around 5 days. While necrotizing enterocolitis typically affects the small bowel, cases of ischemia involving the colon have also been reported, even in adults with norovirus gastroenteritis. Norovirus infections can lead to various unusual manifestations, including respiratory symptoms preceding gastrointestinal illness in some cases, detection in endotracheal aspirates of preterm infants, and associations with hemophagocytic lymphohistiocytosis, renal dysfunction, hemolytic-uremic syndrome, ischemic colitis, and transient hepatocellular injury. Additionally, central nervous system manifestations such as encephalopathy and seizures have been reported, particularly in children, although outcomes are generally benign.

Sapovirus gastroenteritis typically manifests after an incubation period of 12–96 h (median 24–48 h), with clinical presentation generally milder than norovirus infection and rotavirus infections. The most common symptoms include watery diarrhea (3–5 episodes/day for 2–4 days), intermittent vomiting (more prevalent in children), mild-to-moderate abdominal cramping, low-grade fever (<38.5 °C) in approximately 30% of cases. Fever is rare, and diarrhea typically resolves within 1 week, although longer durations have been reported. Sapovirus infections rarely result in severe outcomes, with fewer than 5% of cases requiring hospitalization. Mortality is exceptionally uncommon (≤0.1%) and predominantly occurs in elderly long-term care residents with multiple comorbidities.

Diagnosis

A. Detection of Viral Antigen Norovirus antigen detection methods are routinely employed for rapid diagnostics, especially during outbreak investigations, due to their ability to quickly identify the viral capsid protein VP1.

Enzyme Immunoassays (EIAs) These screening tools utilize genotype-specific antibodies, primarily targeting the GI.1 and GII.4 strains, to capture and detect viral antigens. Commercially available enzyme immunoassay (EIA) kits such as Ridascreen Norovirus (R-Biopharm) and IDEIA Norovirus (Oxoid) are commonly used for this purpose. Recent advancements have led to the development of nanoparticle-enhanced EIAs, which demonstrate a 25% improvement in detecting the GII.6 strain. Additionally, multiplex EIAs have been introduced, enabling the simultaneous detection of norovirus alongside other common enteric pathogens like rotavirus and astrovirus, thereby streamlining the diagnostic process in clinical and public health settings.

Rapid Lateral Flow Immunoassays (LFAs) Lateral flow assays (LFAs) are widely used for rapid norovirus detection, employing immunochromatographic strips coated with colloidal gold-labeled antibodies that bind to viral VP1 antigens. These tests provide results within 15–30 min, making them ideal for point-of-care and outbreak settings. Two commonly used commercial kits include SD Bioline Norovirus (Standard Diagnostics), which shows a sensitivity of 50–70% (improving during high viral-load outbreaks), and Ridaquick Norovirus (R-Biopharm), which offers high specificity (95%) but may miss low-titer samples. A key advantage of LFAs is their simplicity—they require no specialized lab equipment, allowing deployment in field settings and resource-limited areas. However, their major limitation is poor sensitivity in asymptomatic carriers or convalescent-phase patients, where viral shedding may be intermittent or low. Recent innovations aim to overcome these challenges. Fluorescent LFAs, for instance, enhance sensitivity to ~80% by using fluorescent nanoparticle labels instead of colloidal gold, enabling more reliable detection of low-concentration samples.

B. Detection of Viral RNA Accurate detection of norovirus (NoV) and sapovirus (SaV) RNA is critical for clinical diagnosis, outbreak investigations, and surveillance.

Real-Time Reverse Transcription PCR (RT-PCR) RT-PCR remains the gold standard for norovirus detection, offering exceptional sensitivity (>95%) and specificity. Commercial RT-PCR kits such as SmartNorovirus enable differentiation between GI and GII genogroups, while Ridagene Norovirus provides viral load quantification, which is valuable for outbreak tracking and severity assessment. However, these methods require costly thermocycling equipment and trained personnel, limiting their use in resource-constrained settings.

Isothermal Amplification Methods Isothermal amplification methods provide a practical alternative to RT-PCR. The NASBA-based Swiftgene Norovirus GI/GII assay (Kainos Labs, Japan) operates at a constant 41 °C, eliminating the need for thermal cycling while achieving 90% sensitivity compared to traditional RT-PCR. A groundbreaking advancement is the LAMP-CRISPR platform, which integrates loop-mediated amplification (LAMP) with CRISPR-Cas13 for rapid, visual detection. This system identifies as few as 10 RNA copies/μL within 30 min, making it ideal for point-of-care testing. Lateral flow nucleic acid assays like HybriDetect (Milenia Biotech) can be paired with RT-LAMP. These dipstick tests convert amplified RNA into visible bands, requiring no instrumentation.

Zoonotic Potential

Despite the detection of animal noroviruses (NoVs) and sapoviruses (SaVs) in livestock (cattle, pigs) and wildlife, direct evidence of cross-species transmission to humans remains limited. A qualitative risk assessment has categorized bovine noroviruses at level 0, indicating that they are considered to pose minimal risk to humans. While animal noroviruses and sapoviruses have not been directly detected in humans, there is evidence indicating a potential risk of transmission between animals and humans. While no confirmed cases of human infection by animal noroviruses (NoVs) or sapoviruses (SaVs) have been documented, several lines of evidence suggest a plausible risk of cross-species transmission.

The evidence suggests a complex interplay between human and animal norovirus (NoV) strains, potentially influencing future spillover events. For instance, the detection of human GII.4 in 8% of farmed pigs highlights a possible reservoir role for pigs in the transmission dynamics of human NoVs. Additionally, the co-detection of human GI/GII and bovine GIII NoVs in oyster farms underscores the environmental contamination risks posed by multiple viral strains. Moreover, the presence of human-like NoV strains in urban mice further complicates the epidemiological landscape, suggesting urban wildlife as potential carriers or transmitters of human NoVs. Furthermore, the partial RdRp homology between bovine GIII NoVs and human strains hints at genetic relatedness, possibly facilitating genetic exchange and recombination events between animal and human NoV strains.

Prevention and Control

The prevention and control of norovirus infections involve several key measures, as outlined below:

1. **Personal Hygiene Practices**: Appropriate hand hygiene is a key strategy to control norovirus transmission during outbreaks, although evidence is primarily descriptive. Studies using feline calicivirus (FCV) suggest that ethanol and 1-propanol are more effective than 2-propanol in reducing viral contamination on hands. Patients presenting with vomiting and/or diarrhea should be isolated and managed with contact precautions until norovirus infection is either confirmed or ruled out. Gown and glove use, based on observational studies, is recommended for preventing norovirus transmission, while masks are advised for staff expecting vomitus exposure.
2. **Environmental Hygiene**: Healthcare facilities and food preparation sites are among the many settings where norovirus can survive. Frequent decontamination of commonly touched surfaces, including door handles, light switches, and countertops, should be carried out using chlorine-based disinfectants to minimize the risk of virus transmission. Adequate ventilation in indoor spaces helps to reduce the concentration of airborne viruses.

3. **Safe Food Handling**: Adherence to strict food hygiene protocols is critical in curbing the spread of enteric viral infections. This involves meticulous handwashing with soap and water prior to food preparation—particularly following restroom use, diaper changes, or contact with raw meat, poultry, or eggs. Additionally, uncooked ingredients must be stored and handled separately from ready-to-eat items to avoid cross-contamination. Proper cooking of foods, especially meats, poultry, and eggs, to their recommended internal temperatures kills harmful viruses. Additionally, refrigerating perishable foods promptly and keeping kitchen surfaces and utensils clean and sanitized helps prevent the spread of viruses.
4. **Vaccination:** Currently, there is no vaccine available for norovirus and sapovirus infections. Current strategies in norovirus vaccine research are largely concentrated on the use of virus-like particles (VLPs), which have exhibited significant capability in stimulating strain-specific immune protection. Various routes of administration, including intranasal and oral, have demonstrated safety and immunogenicity in clinical trials. While challenges exist, such as antigenic diversity and the need for multivalency, progress has been made toward developing effective vaccines, with some candidates showing protective efficacy against norovirus infection and symptoms in human challenge studies.

Bibliography

Epidemiology of Foodborne Norovirus Outbreaks – United States, 2009–2015. (n.d.). https://doi.org/10.14252/foodsafetyfscj.2017028

Increasing Rates of Gastroenteritis Hospital Discharges in US Adults and the Contribution of Norovirus, 1996–2007. (n.d.). https://doi.org/10.1093/cid/ciq163

Lee, L. E., Cebelinski, E., Fuller, C. C., Keene, W. E., Smith, K., Vinjé, J., & Besser, J. M. (2012). Sapovirus outbreaks in long-term care facilities, Oregon and Minnesota, USA, 2002–2009. *Emerging Infectious Diseases, 18*(5), 873. Centers for Disease Control and Prevention. https://doi.org/10.3201/eid1805.111843

Organoids as a New In Vitro Model of Human Norovirus Infection. (n.d.). https://doi.org/10.4167/jbv.2020.50.3.168

Part IV

Parasitic Zoonoses

Fasciolosis and Fasciolopsiasis

32

Ripan Biswas

Abstract

Fasciolosis and fasciolopsiasis are neglected, food-borne zoonotic trematode infections caused by *Fasciola* spp. (liver flukes) and *Fasciolopsis buski* (giant intestinal fluke), respectively, and are transmitted through the ingestion of metacercariae-contaminated aquatic vegetation. These parasitic diseases are predominantly endemic in low-income, agrarian regions, particularly in Asia, where close human-animal-environment interactions facilitate transmission. Both infections contribute to significant human morbidity and impose substantial economic losses on livestock production systems. The shared ecological niches, dependence on freshwater snail intermediate hosts, and plant-borne transmission pathways highlight the necessity for integrated surveillance, improved food safety practices, and coordinated One Health-based control and prevention strategies.

Fasciolopsiasis is a neglected foodborne, snail-transmitted parasitic zoonotic disease caused by *Fasciolopsis buski*, the largest intestinal trematode of human. The infection occurs after consumption of raw or improperly cooked aquatic plants like water chestnut, water bamboo, lotus, water caltrop, etc. While often overlooked in global health agendas, it remains endemic in Southeast Asia and the Indian subcontinent, particularly in areas where poverty, poor sanitation, and the consumption of raw aquatic plants intersect.

R. Biswas (✉)
Department of Veterinary Public Health and Epidemiology, West Bengal University of Animal and Fishery Sciences, Kolkata, West Bengal, India

R. Biswas et al. (eds.), *Concept of One Health and Zoonoses*,
https://doi.org/10.1007/978-981-95-6627-3_32

Keywords

Fasciolosis · Liver fluke · Fasciolopsiasis · Intestinal fluke · Giant intestinal fluke

Fasciolosis

Etiology

Fasciola hepatica (2–3 cm) and *Fasciola gigantica* (4–10 cm) are the two trematodes commonly predilected in bile ducts of liver or gallbladder. A single liver fluke can pass about 40,000 eggs per day.

Host

The definitive hosts of *Fasciola* spp. are primarily domestic ruminants, particularly cattle, sheep, and goats; humans may become infected accidentally. Freshwater lymnaeid snails, including *Lymnaea truncatula*, *Lymnaea luteola*, and *Lymnaea tomentosa*, act as the intermediate hosts and are essential for parasite development. A range of wild mammals, such as deer, rabbits, hares, wild boars, and rodents, serve as reservoir hosts and contribute to the maintenance of the parasite in endemic ecosystems. Although birds and ducks are not true hosts of *Fasciola* spp., they may facilitate passive dispersal of infected snails or aquatic vegetation, thereby influencing the geographical spread and transmission dynamics of fasciolosis.

Geographical Distribution and Epidemiology

Presently, Peru, Bolivia, Chile, Ecuador, Cuba, Egypt, Spain, France, Portugal, and the Caspian area are mostly affected with human fasciolosis. Around 81 countries have reported the infection from different continents. It has been estimated that *Fasciola hepatica* infects more than 300 million cattle and 250 million sheep worldwide, while the global economic burden of fasciolosis in livestock, largely attributed to *Fasciola* spp., including *F. gigantica*, is approximately US $3 billion annually. Fasciolosis causes economic losses in animal farming in the form of decreased reproductive performance, mortalities, retarded growth, weight loss, decreased milk production, increased culling of cattle, and infertility. *Fasciola hepatica* demonstrates a cosmopolitan distribution, which is largely attributable to its broad host spectrum and the marked ecological plasticity of its lymnaeid intermediate hosts, enabling survival and transmission across diverse climatic zones and environmental conditions. In contrast, *Fasciola gigantica* exhibits a comparatively restricted geographical distribution, primarily reflecting the limited ecological adaptability and narrower habitat requirements of its intermediate snail hosts, which constrain its establishment and transmission to specific tropical and subtropical regions. Streams, wetlands, pastures, climate change, rainfall, new habitation, new irrigation projects, export/import of livestock, and migration of animals can introduce the parasites into the new region. The rural communities and the population below the poverty line are affected more by these parasites.

Transmission Cycle

Unembryonated eggs of *Fasciola* spp. are passed in the feces of the definitive host from adult flukes inhabiting the bile ducts. When deposited in freshwater, the eggs undergo embryonation, and under optimal temperature and moisture conditions, they hatch within 10–15 days to release the ciliated larval stage, the miracidium. The miracidium is actively motile and survives for only a few hours unless it encounters and penetrates a suitable freshwater lymnaeid snail, which serves as the intermediate host. Within the snail, the miracidium transforms sequentially into sporocysts, rediae, and finally cercariae, undergoing extensive asexual multiplication. Mature cercariae emerge from the snail and swim freely in water for a short period before losing their tails and encysting on aquatic vegetation or other submerged substrates, forming metacercariae, the infective stage for the definitive host. The metacercariae may remain viable for several months under favorable environmental conditions. Infection of the definitive host, including humans as accidental hosts, occurs following ingestion of raw or inadequately washed aquatic plants bearing metacercariae. After excystation in the duodenum, the immature flukes penetrate the intestinal wall, migrate through the peritoneal cavity, traverse the liver parenchyma, and ultimately enter the bile ducts, where they mature into adult flukes and begin egg production, thus completing the life cycle.

Mode of Transmission

Transmission of the parasite occurs through ingestion of aquatic plants contaminated with metacercariae. Infection is commonly acquired when raw or inadequately washed aquatic plants, or their fruits or bulbs, are consumed, particularly when they are peeled or bitten with the teeth and inadvertently swallowed by the definitive host or humans as accidental hosts.

Disease Manifestations

Animals

In animals, three symptomatic patterns (acute, subacute, and chronic) are seen.

In the acute phase, the most common symptoms are fever, diarrhea, anemia, hepatomegaly, and distended abdomen with colic.

In the subacute pattern, extensive tissue damage occurs due to burrowing by the parasites resulting in severe anemia, liver failure, and followed by death in 8–10 weeks.

In the chronic pattern, submandibular edema or "bottle jaw" is commonly found in animals. In this pattern, biliary colic, colic, pale mucous membrane, hepatomegaly, jaundice, and decreased milk amount are noticed.

Human

In the case of humans, the symptoms can be categorized as acute, chronic, halzoun, and ectopic.

Acute form: In the acute form, malaise, irregular fever, mild jaundice, anemia, and pain under the right costal line are observed. This form is considered without any echogenic images in the gallbladder.

Chronic form: If the symptoms of acute form persist for more than 4 months, it is considered a chronic form. Biliary obstruction, cholelithiasis, cholecystitis, fibrosis, liver abscesses, cirrhosis, hyperplasia of duct epithelium, and subcapsular hemorrhages are observed in this form. Normocytic and normochromic anemia is very much associated with the chronic form. Intrahepatic stones, along with cholangitis and cholelithiasis, establish the landscapes of the chronic phase of fasciolosis in humans.

Halzoun form: In Arabic, "Halzoun" means snail. This syndrome was first described in Lebanon in 1905. It is an acute allergic reaction in the pharynx and upper respiratory tract in those who have consumed raw or undercooked caprine/ovine liver. It occurs due to the attachment of larvae of *Fasciola* in the pharynx.

Ectopic form: Sometimes the parasite is rarely found in ectopic foci like lungs, peritoneal cavity, intestinal wall (cecum, ascending colon), ocular cavity, spinal cord, lymph nodes, subcutaneous tissue, etc.

Diagnosis

I. Detection of *Fasciola* eggs in stool or of duodenal or biliary aspirates is widely practiced.
II. FLOTAC (Flotation in Centrifuge), a recent diagnostic assay, may be useful for human and animal fasciolosis.
III. Falcon Assay Screening Test-Enzyme Linked Immunosorbent Assay (FAST-ELISA) can be effective for diagnosing fasciolosis.
IV. CT scan, Urltasound Sonography (USG), MRI, and Magnetic Resonance Cholangiopancreatography (MRCP) can be used as imaging tools.

Treatment

WHO has recommended triclabendazole at 10 mg/kg body weight per dose for one or two doses, separated by 12–24 h for the treatment of acute and chronic human fascioliasis. In 2019, the United States Food and Drug Administration (USFDA) also approved a two-dose regimen for humans who are 6 years old and older. In animals, triclabendazole, albendazole, clorsulon, and closantel are commonly used.

Prevention and Control

- Regular treatment of infected animals with appropriate drugs and dose can control further transmission.
- Snail control can be done by molluscicide, such as metallic salts (copper sulfate).
- Rational grazing of animals helps reduce fluke infestation in animals.
- Aquatic plants should be washed and boiled properly.
- Ingestion of fresh and raw aquatic plants is to be avoided.

Fasciolopsiasis

Etiology

Fasciolopsis buski is generally oval shaped, elongated, and without a cephalic cone fluke. The length of the parasites varies from 2 to 10 cm. The fluke is dorsoventrally flattened with blunt anterior end. *Fasciolopis buski* can produce about 25,000 eggs per day. The eggs are not embryonated when passed through faces. It normally remains in the upper region of the small intestine but can be found in the stomach and lower regions of the intestine due to heavy infestations.

Host

Swine and humans are the definitive host and multiple species of planorbid snails (*Lymnaea, Indoplanorbis, Segmentina, and Achatina* spp.) and serve as intermediate host. Swine acts as the reservoir host, but in some regions, swamp buffalo, ducks, and chicken are also suspected reservoir host.

Geographical Distribution

The disease is endemic in several countries including China, Thailand, Taiwan, Vietnam, Indonesia, Malaysia, Formosa Japan, Bangladesh, and India. About 10 million people are infected by this Asiatic trematode. In India, the endemic areas are Bengal and Assam.

Life Cycle

The parasites release the immature, unembryonated eggs in the intestine of the host, and then the eggs are passed out of the host body through faces. The eggs become embryonated in water or dump soil within 15 days. After 3–7 weeks, the eggs are hatched into the first stage of larva, which is called miracidia, at an optimal temperature of 27° to 32 °C. Miracidia penetrate the planorbid snails and undergo three developmental stages, i.e., sporocysts, redia, and cercariae in the lymph space of snail. Within 3 days, redia larvae are formed inside the sporocyst. The mother redia may generate daughter redia. Redia is released in the body of snail, and in about 30 days after snail infection, it changes to cercaria. The cercariae come out from the snail, and after swimming of few hours, the cercariae are encysted as metacercariae on aquatic plants such as water chestnut, water caltrop, lotus, and bamboo. The hosts (mainly swine and man) become infected after ingestion of those aquatic plant contaminated with metacercariae, which are also called the infective stage of the life cycle. Then the metacercariae are excysted in the duodenum and jejunum. The adult stage starts here.

Mode of Transmission

Transmission of parasite takes place by ingesting the infected aquatic plant. When infected plants or its fruits or bulbs are peeled with the teeth and swallowed by the primary host, the disease may develop.

Disease Manifestations

The disease mostly remains asymptomatic, but in case of heavy load of parasites severe epigastric abdominal pain, chronic diarrhea, anemia, ascites, asthenia, and allergic responses like symptoms are shown. Localized inflammation, hemorrhage, and abscess may occur at the attachment site of the parasite to the intestinal mucosa. Eosinophilia is commonly seen. In the case of heavy infestation, an edematous condition is developed from protein losing enteropathy, and sometimes intermittent obstruction occurs due to huge number of flukes in the intestine.

Diagnosis

The disease may be identified by stool sample examination for the presence of eggs and adult fluke.

Fasciolopis buski eggs cannot be distinguished from the eggs of *Fasciola hepatica.*

Treatment

Tetrachlorethylene, thymol, or hexylresorcinol are the drugs of choice.

Prevention and Control

- Aquatic plants should be washed and boiled properly.
- Snail control strategy can be effective.
- Ingestion of fresh and raw aquatic plants is to be avoided.
- Sewage and waste management should be implemented properly in those pig farms, where edible aquatic plants are grown in the adjacent water body.
- Raw aquatic plants should not be offered to the pigs.

References

https://www.who.int/news/item/26-07-2018-human-fascioliasis-review-provides-fresh-perspectives-on-infection-and-control

https://www.who.int/news/item/26-07-2018-human-fascioliasis-review-provides-fresh-perspectives-on-infectionand-control

https://www.who.int/news-room/questions-and-answers/item/q-a-on-fascioliasis

https://www.cdc.gov/dpdx/fasciolopsiasis/index.html

Ridha MR, Indriyati L, Andiarsa D, Wardhana AH (2021) A review of Fasciolopsis buski distribution and control in Indonesia. Veterinary world, 14(10):2757–2763. https://doi.org/10.14202/vetworld.2021.2757-2763

Paragonimiasis

33

Souti Prasad Sarkhel, Sunanda Sarkhel,
and Tuhina Banerjee

Abstract

Paragonimiasis is a major parasitic disease caused by lung flukes belonging to the genus *Paragonimus*, presenting a significant public health challenge in regions such as Southeast Asia, Africa, and the Americas. This genus encompasses more than 50 species, with the *Paragonimus westermani* complex, *Paragonimus skrjabini* complex, *Paragonimus ohirai* complex, and *Paragonimus mexicanus* complex being the most widely prevalent globally. Notably, the Indian subcontinent displays remarkable species diversity, featuring several documented taxa, including *Paragonimus miyazakii manipurinus* n. subsp., *Paragonimus hueitungensis*, *P. skrjabini*, *Paragonimus heterotremus*, *Paragonimus compactus*, and *P. westermani*, which are primarily found in the north-eastern regions of India. Lung flukes have a complex multihost life cycle involving turreted freshwater snails, and crabs/crayfish as their first and second intermediate hosts, respectively. While pulmonary form of paragonimiasis is most common, infection often mimics chronic tuberculosis, and symptoms can be absent or vague in mild cases. Ectopic paragonimiasis frequently involve organs such as the brain, skin, and abdomen, and other vital organs, especially in cases caused by *P. skrjabini*. Accurate diagnosis requires a combination of parasitological, radiological, serological, and molecular techniques. Understanding

S. P. Sarkhel (✉)
Department of Veterinary Parasitology, FVAS, Institute of Agricultural Sciences, Banaras Hindu University, South Campus, Mirzapur, Uttar Pradesh, India
e-mail: souti.sarkhel@bhu.ac.in

S. Sarkhel
Department of Optometry and Vision Sciences, Amity University, Gurgaon, Haryana, India

T. Banerjee
Department of Microbiology, Institute of Medical Sciences, Banaras Hindu University, Varanasi, Uttar Pradesh, India

R. Biswas et al. (eds.), *Concept of One Health and Zoonoses*,
https://doi.org/10.1007/978-981-95-6627-3_33

the multifaceted nature of this disease is crucial for control, particularly in endemic areas where consumption of undercooked crabs/crayfish poses a major transmission risk. In this context, this chapter explores paragonimiasis, highlighting its various forms, public health significance, parasite resilience, clinical aspects, diagnostic challenges, and the need for alternative mitigation strategies.

Keywords

Paragonimus · Zoonosis · Zoonotic helminths · Phylogeny · Pathogenesis · Lung flukes · Snail-borne parasitic diseases

Introduction

Paragonimiasis, a parasitic lung disease caused by trematodes of the genus *Paragonimus*, poses a significant public health concern in endemic regions of Southeast Asia, Africa, and the Americas (Blair, 2024). Within the class Trematoda, these lung flukes belong to the order Plagiorchiida, suborder Troglotremata, and the family Paragonimidae (Blair et al., 1999; Vainutis et al., 2022). The suborder Troglotremata also includes other zoonotic genera like *Nanophyetus* and *Troglotrema*. Members of these families typically infect piscivorous mammals and birds, completing a complex life cycle involving multiple hosts. A defining feature of these flukes is the production of short-tailed, styleted cercariae (microcercous xiphidiocercariae), which emerge from infected snails and encyst as metacercariae in second intermediate hosts (Doanh et al., 2020). In this regard, *Nanophyetus* species primarily utilise fish as their second intermediate host, whereas *Paragonimus* species encyst within crustaceans such as crabs and crayfish. This unique life cycle biology contributes to a significant public health concern, particularly in endemic regions where food habits involve the consumption of undercooked or raw fish and shellfish, as well as the meat of paratenic hosts such as wild boars and deer (Blair et al., 1999; Nakano et al., 2009; Yoshida et al., 2016; Irie et al., 2017; Nakashima et al., 2021; Blair, 2024).

The consumption of improperly cooked freshwater crabs or crayfish, or the contaminated utensils containing metacercaria is the primary route of human infection (Procop, 2009). While in hunting dogs, ingesting infected crabs/ crayfish, or the meat of paratenic hosts such as wild boars and deer can lead to paragonimiasis. In some cultures (Korea and Japan), the traditional use of raw crab juice as medicine poses a unique risk factor (Yun, 1960). Similarly, certain African societies associate raw crab consumption with fertility beliefs. Paragonimiasis primarily affects marginalised populations in subtropical and tropical regions (Malvy et al. 2006). Low socio-economic status and food insecurity often drive the consumption of freshwater crabs in Africa and the Americas, leading to infection. In Asia, raw or undercooked crabs are considered delicacies, dishes such as “drunken crab” in China, “Kung Ten” and “Nam Prik Poo” in Thailand, and “Ke Jang” in Korea highlight this culinary tradition, while also reflecting the associated disease transmission risks (Garcia, 2001).

Paragonimus flukes, predominantly reside in pairs within the lung parenchyma tissues of carnivorous mammals, inducing granulomatous lesions that mimic tuberculosis and cancer (Blair, 2022). Human paragonimiasis manifests in various forms, with pulmonary and cutaneous presentations being the most common. However, ectopic/extrapulmonary infections, where the parasite lodges in tissues outside the lungs (heart, brain, spinal cord, or eyes), are particularly worrisome owing to their serious pathologies. These ectopic cases are more prevalent in areas endemic for *P. skrjabini*, and where culinary practices involve raw or undercooked shellfish (Blair et al. 1999; Blair, 2019; Blair, 2024).

Against this backdrop, this chapter comprehensively explores paragonimiasis, covering its diverse clinical forms, pathobiology, diagnosis, treatment strategies, and public health considerations. Through our in-depth discussions, it explores the intricate life cycle of *Paragonimus*, its impact on human health, and the multiple factors driving transmission and geographical distribution. A thorough understanding of these aspects is important for formulating effective control protocols and prevention strategies for this much neglected tropical parasitic disease.

Aetiology and Host

Paragonimiasis, a disease caused by lung flukes (*Paragonimus*), is a public health concern in parts of Asia, Africa, and North America. This complex parasite has a multistage life cycle involving multiple host species. Freshwater snails act as the first intermediate host, followed by crabs or crayfish. Definitive hosts, where the parasite reproduces, are typically fish-eating mammals and birds such as dogs, cats, mongooses, civet cats, mustelids, and wild carnivores (Blair, 2022). However, some *Paragonimus* species can infect humans, causing human paragonimiasis (Blair et al. 1999; Yoshida et al. 2019; Zhou et al. 2021).

Paragonimus parasites exhibit remarkable resilience because they exploit natural predator–prey relationships to ensure their survival. This is achieved, in part, through the utilisation of paratenic hosts, such as rodents, deer and pigs (Doanh et al. 2013). These animals become infected by ingesting crustaceans harbouring encysted metacercariae (dormant parasite stage) of *Paragonimus*. The parasite can remain dormant in the muscles or other tissues of the paratenic host throughout its entire lifespan, but no further development occurs. However, the life cycle can resume if a suitable definitive carnivore animal, such as a tiger, preys upon the infected paratenic host. The ingested dormant parasites can then become active again, migrate to the preferred predilection site, which is the lungs, mature into adult flukes, and complete their life cycle. While some laboratory studies suggest the potential role of frogs and hens as paratenic hosts (Xie et al. 1985; Zhou, 2021), further investigation is necessary to definitively confirm their involvement in the *Paragonimus* life cycle.

Distribution

Paragonimus is a euryxenous parasite that can complete its life cycle in multiple animal and bird species. More than 50 species of the *Paragonimus* genus have been identified as causing paragonimiasis in animals and birds across continents. Among the most commonly reported zoonotic species that infect humans are *Paragonimus westermani*, *P. heterotremus*, *P. skrjabini*, *P. miyazakii*, and *P. ohirai* in Asia and *Paragonimus kellicotti* in America and *Paragonimus africanus* and *Paragonimus uterobilateralis* in Africa (Blair et al., 1999; Chai, 2013a; Blair, 2024). In India, however, a wider range of species have been identified, including *P. westermani*, *P. heterotremus*, the *Paragonimus skrjabini* complex, *P. hueitungensis* (syn. *P. skrjabini*?), *P. compactus*, and *P. miyazakii manipurinus* n. subsp (Singh et al. 2012). Humans can contract the disease by consuming raw or partially cooked meat of crabs, crayfishes, or paratenic hosts such as pigs and rodents. Estimates by the World Health Organisation (WHO, 1995) suggest that approximately 20 million individuals in Asia, the Americas, and Africa could be at risk for contracting paragonimiasis. Therefore, elucidating the parasite's biology, life cycle, and transmission dynamics between hosts, along with a comprehensive understanding of clinical features, diagnosis, and public health implications, remains crucial for effective disease control.

Taxonomy

As per the latest taxonomic revisions, the genus *Paragonimus* belongs to the family Paragonimidae, under the superfamily Troglotrematoidea, the sub-order Troglotremata within the order Plagiorchiida, which belongs to the class Digenea and subphylum Trematoda under the phylum Platyhelminthes (Blair et al., 1999; Vainutis et al., 2022). However, as per the old literatures, these flukes were included in a separate family, Troglotrematidae, under the order Plagiorchiida. This superfamily Troglotrematoidea now encompasses two families: Paragonimidae and Troglotrematidae Odhner, 1914. The genus *Paragonimus* includes approximately 50 described species, which are further classified into 7 distinct clades on the basis of DNA sequence homology (Fig. 33.1). These trematodes are known primarily for infecting the lungs of various mammalian, avian, and even human hosts across the globe.

Morphology

Paragonimus, commonly referred to as "lung flukes", bear a striking resemblance to small coffee beans in terms of size and shape (Fig. 33.2). Freshly, these parasites are plump and ovoid and display a reddish-brown hue. Generally, they range in length from 7 to 16 mm, width from 3 to 8 mm, and thickness from 3 to 5 mm in the dorsoventral axis. However, in some regions, particularly Northeast Asian countries,

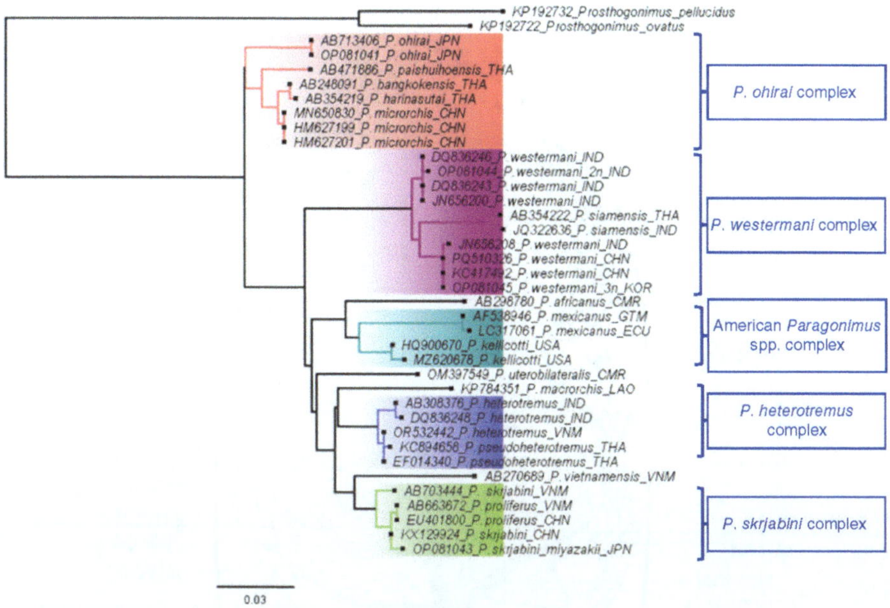

Fig. 33.1 Illustration of the maximum likelihood phylogenetic tree generated using the Kimura 2-parameter with invariant sites (K2P+I) evolutionary model, based on the ITS2 sequences of selected *Paragonimus* species highlighting the major clades

natural populations of *P. westermani* show variations in morphology due to having three types of ploidy levels (diploid, triploid, and tetraploid) within their chromosomes (Agatsuma et al. 1992; Terasaki et al. 1995; Blair et al. 1999). Compared with their diploid counterparts, the eggs, larvae, and adults of Tri- and tetraploid *P. westermani* parasites are larger (Blair, 2022). These fleshy parasites are densely covered in scale-like spines that are morphologically informative for differentiating between closely related species. For example, *Paragonimus westermani* has bifid spines placed singly on the tegument, whereas in closely related *Paragonimus siamensis* species, these bifid spines are arranged in clusters resembling comb teeth (Miyazaki, 1991). On the other hand, *P. kellicotti* is characterised by very large spines with multiple sharp points (Sugiyama et al. 2013).

The ventral and oral suckers of these parasites are almost equal in size; the oral sucker is subterminal, whereas the ventral sucker is placed slightly anterior to the middle of the fluke's body. The pharynx is small, the oesophagus is short, and the intestinal caeca is simple and slightly wavy and extends up to the posterior of the body. The genital pore opens posteriorly (postacetabular) to the ventral sucker.

Paragonimus differ from other troglotrematid parasites in that they lack cirrus and the cirrus sac (Blair, 2008). Testes have 5–6 lobes that are symmetrical or asymmetrical and located in the hind body. The ovary is six lobed, submedian, and pretesticular. Its dendritic vitelline glands are highly branched and occupy a significant area in the lateral fields from the level of the pharynx to the posterior margin of the

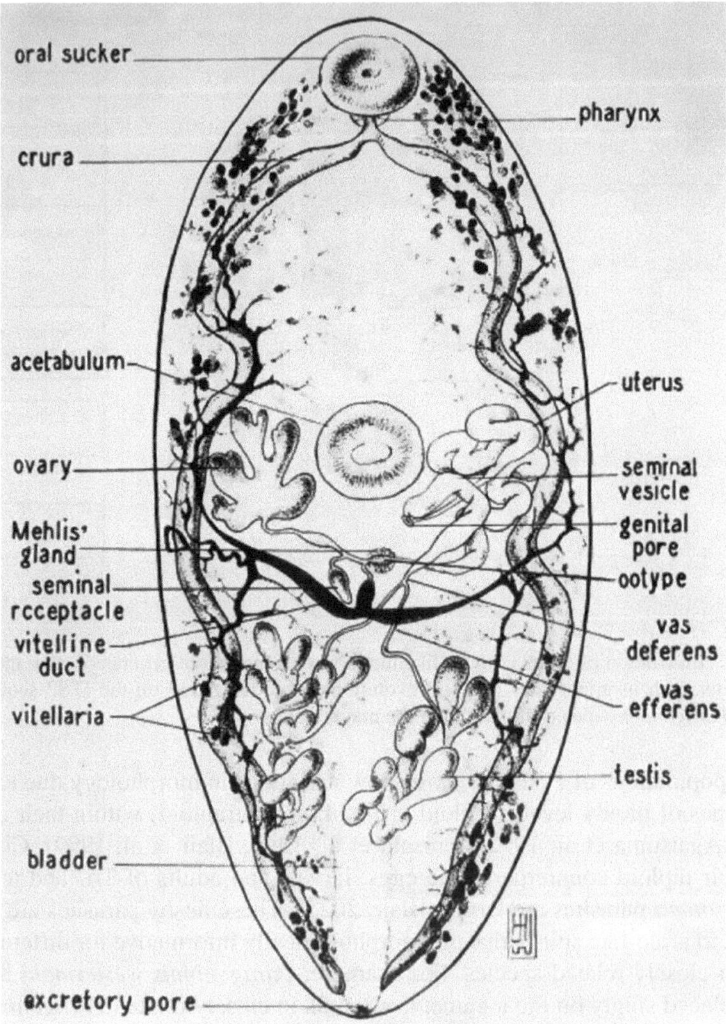

Fig. 33.2 An illustration of an adult *Paragonimus westermani* specimen, illustrating the major anatomical structures. (Image courtesy of the CDC PHIL, https://phil.cdc.gov/Details.aspx?pid=25321)

body. The uterus is strongly coiled. The excretory vesicle is tubular and long and extends up to the oesophagus. *Paragonimus* eggs are characterised by their ovoid vase shape and golden–brown colour with characteristic thickening at their abopercular end. Diploid forms produce eggs measuring approximately 80 × 44 μm (Fig. 33.3), whereas triploid forms produce larger eggs with an average size of 92 × 50 μm.

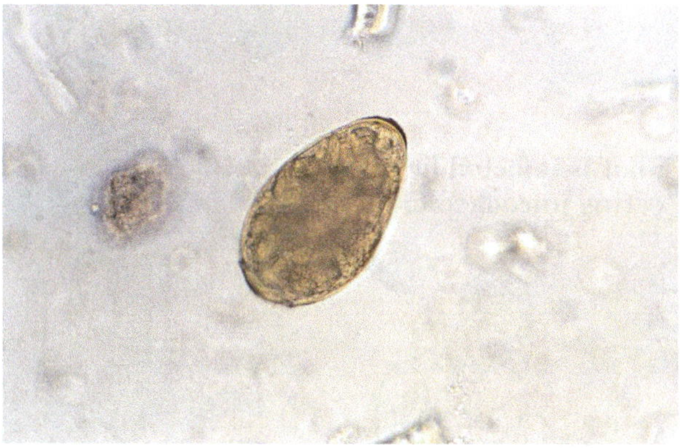

Fig. 33.3 Typical vase-shaped, golden yellow egg of *Paragonimus westermani* observed under high power (magnification at 400×) in an unstained wet mount preparation. (Image courtesy of CDC PHIL, https://phil.cdc.gov/Details.aspx?pid=1534)

Life Cycle

Adult *Paragonimus* flukes reside primarily in the lungs of carnivorous animals and birds, with some species also infecting humans. These worms typically exist in pairs within hazelnut-shaped cysts, where they reproduce sexually. However, some *Paragonimus* species exhibit polyploidy, where individuals have more than two sets of chromosomes. Notably, triploid (3n = 33) and tetraploid (4n = 44) parasites can reproduce parthenogenetically, forming cysts and laying eggs even without a mate. These eggs are fertile and contribute to parasite transmission in the environment.

The life cycle of *Paragonimus* flukes involves two different intermediate hosts (Fig. 33.4). Several susceptible species across several families of freshwater snails act as the first intermediate hosts for *Paragonimus* flukes. These families include Pleuroceridae, Thiaridae (genera *Thiara*, *Melanoides*, and *Tarebia*), Semisulcospiridae (*Semisulcospira*), Assimineidae (*Angustassiminea*), Pomatiopsidae (genera *Pomatiopsis* and *Oncomelania*), Pachichylidae (*Brotia*), Hydrobiidae (*Aroapyrgus*), and Amnicolidae (Blair et al., 1999; Kong et al. 2015; Yoshida et al., 2019). However, the parasite's ability to infect a particular snail species (competency) is highly influenced by coevolutionary relationships. Crustaceans, specifically crabs (*Eriocheir*, *Geothelphusa*, and *Potamon*) and crayfish (*Cambaroides* and *Procambarus*), are the second intermediate hosts (Blair et al., 1999; Kong et al. 2015; Zhou et al., 2022; Dubey, 2023). Definitive hosts include dogs, cats, foxes, wild carnivores, humans, and primates (cynomolgus monkeys). Additionally, pigs, rodents, frogs and also birds can harbour the parasite as paratenic hosts (Blair et al., 1999). These paratenic hosts play crucial roles in maintaining the parasite population, although parasite development is arrested within them.

Within the definitive hosts, eggs laid within the lung cysts are coughed up and expelled in faeces or sputum. Upon reaching water bodies, these eggs develop into miracidia larvae that actively seek and penetrate suitable snail hosts. Inside the snail hosts, the miracidia undergo a series of asexual reproductive stages. This process ultimately produces many short-tailed cercariae with a unique morphology. These cercariae are characterised by a prominent stylet, an accessory boring organ, and a

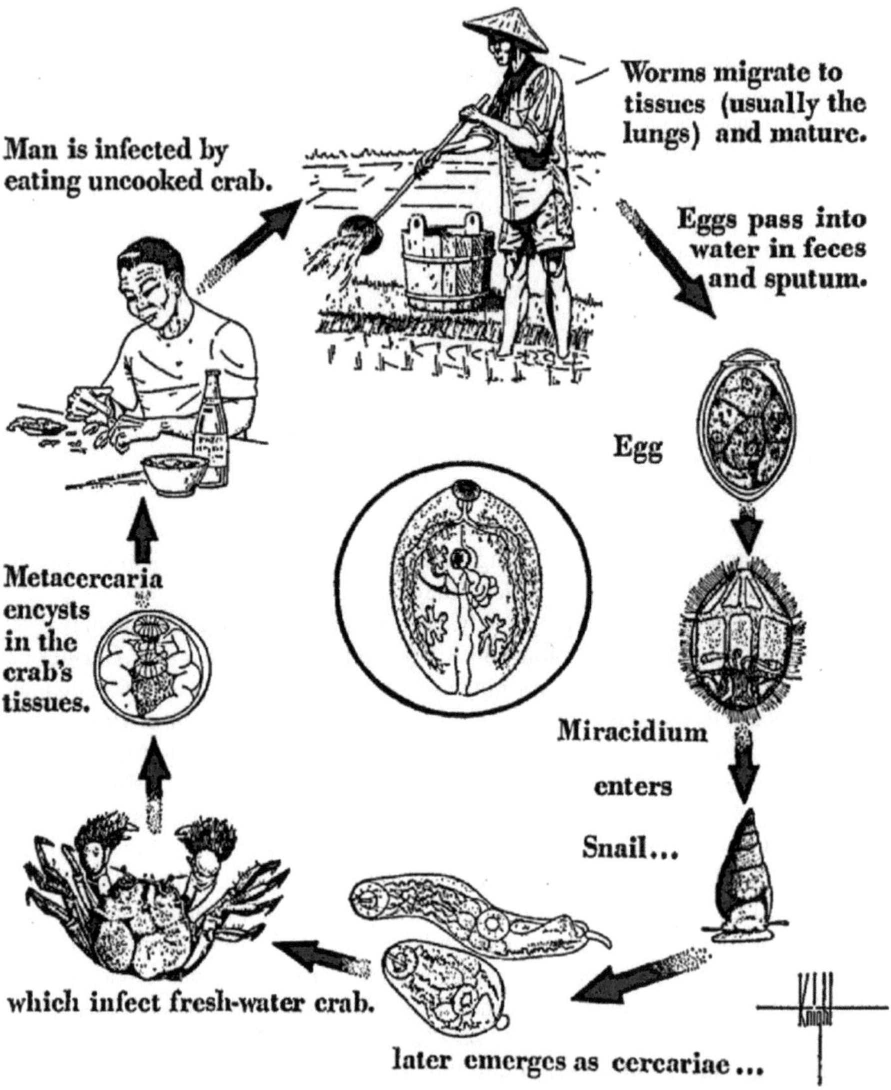

Fig. 33.4 Depiction of the life cycle of *Paragonimus westermani*, highlighting the various intermediate and definitive hosts involved, as well as the distinct developmental stages of the parasite. (Reference: Figure 10–24 from Noble et al. (1989))

short, stumpy tail. Owing to this distinctive morphology, *Paragonimus* cercariae are classified as microcercous xiphidiocercariae. When released from the snail host, the non-swimming cercariae sink to the bottom of water bodies, where their target crustaceans, such as crabs and crayfish, are usually present. Although they live in aquatic environments, these hosts often reside in burrows along water banks or paddy field edges, mainly staying near the substrate at the bottom. This overlap in microhabitat increases the likelihood of cercariae reaching suitable second intermediate hosts and facilitates the development of infective metacercarial larvae within the muscles, heart, or gills of crustaceans.

Definitive hosts acquire infection by consuming raw or undercooked crustaceans harbouring infective metacercariae. Additionally, some species can infect humans through the consumption of raw wild boar meat containing dormant metacercariae. Within the definitive host's digestive tract, gastric juices, particularly bile salts, trigger the secretion of proteases by metacercariae (Chung et al., 1995; Intapan and Maleewong, 2001). The newly released juveniles migrate through the small intestine, peritoneal cavity, diaphragm, and pleural cavity to reach the lung parenchyma tissues. This migration is facilitated by the parasite's cysteine protease enzymes, which play essential roles in its nutrition and protein catabolism, as well as in host tissue invasion and the evasion of host immune responses (Sajid and McKerrow, 2002; Chung et al. 2005; Na et al. 2006; Cancela et al., 2008; Grote et al. 2018). In the lungs, paired worms form cysts and begin egg production, reaching maturity within 3–5 weeks. The prepatent period, the time from infection to egg excretion, is typically 8–10 weeks. Adult *Paragonimus* can live in the lungs for up to 20 years, although their egg-laying capacity diminishes over time. Importantly, in paratenic hosts such as pigs and rodents, the parasites remain dormant in the muscles and other tissues and do not reach maturity.

Pathogenesis and Pathology

Paragonimiasis has a vastly different clinical picture in animals than in humans. While animal infections are often asymptomatic, heavy parasite burdens can lead to the formation of inflammatory granulomas and fibrous encapsulation of parasites within the affected tissues. These encapsulated masses, which are commonly found in the lungs, subcutaneous spaces, and inguinal regions, may contain pairs of adult flukes. In lung infections, these encapsulated areas can develop fistulas, which are abnormal connecting passages. These fistulas allow parasite eggs to reach the bronchioles and be expelled through the airways (Dubey, 2023).

In contrast, human paragonimiasis manifests with a wide range of clinical presentations depending on the parasite load and the site of infection. During the early stages, the migration of juvenile flukes through host tissues usually occurs without symptoms (Kong et al., 2015). However, heavy parasite burdens can trigger flu-like symptoms such as fever, muscle aches, fatigue, abdominal pain, and diarrhoea. Additionally, blood tests may reveal elevated levels of eosinophils and immunoglobulin E (IgE) (Kojima et al., 1972; Nakamura-Uchiyam et al., 2001; Singh et al., 2012).

The most common form of human paragonimiasis involves lung and pleural cavity infection (Choi, 1990). Patients with this form of disease often experience respiratory symptoms such as pleuritic chest pain, chronic cough, dyspnoea, and blood-tinged sputum. Pathological examination of these patients frequently reveals pleural effusion, hydropneumothorax, emphysema, and infiltration of inflammatory cells in the lung tissue (Procop, 2009; Singh et al., 2012; Ahn et al., 2021).

The early migration of juveniles is associated with localised bleeding and eosinophilic infiltration along the tracts of migrations. Young *Paragonimus* worms also release cysteine proteases as they burrow through tissues. These enzymes help them migrate and avoid host immune defences. In the late phase of infection, the presence of parasite eggs and excretory–secretory products triggers a strong immune response mediated by type 2 helper T (TH2) cells (Ikeda, 1998; Kong et al., 2015). This response is characterised by eosinophilia and the formation of granulomas, which can be either cystic or solid. These granulomas contain live or dead parasites, parasite eggs, necrotic tissue, and Charcot–Leyden crystals (Singh et al., 2012). Interestingly, the parasite secretes molecules such as cysteine proteases and several antioxidant enzymes that play complex roles in promoting its survival. These cysteine proteases suppress the degranulation of eosinophils and promote programmed cell death of eosinophils, whereas antioxidant enzymes potentially help in the detoxification of the reactive oxygen species produced by the host immune system (Shin and Lee, 2001; Shin et al., 2001, 2005; Chung et al., 2008). Following the resolution of the host inflammatory response, fibrosis typically occur, encasing the granuloma and isolating it from surrounding healthy tissue.

A distinct form of the disease, known as cerebral paragonimiasis, can occur when the parasite migrates to the brain. Early migration can cause cerebral oedema with or without a midline shift or haematoma (Cha et al. 1994). Chronic infections may lead to the formation of calcified granulomas, resulting in symptoms such as partial seizures, headaches, and vision loss in specific regions of the visual field (Kong et al., 2015).

Furthermore, *Paragonimus skrjabini* species are mostly associated with ectopic paragonimiasis in humans (Blair, 2014). In *P. skrjabini* infection, the parasite migrates to unusual locations in the body instead of maturing in the lungs. This aberrant migration leads to a distinct set of symptoms, including abdominal pain, diarrhoea, migratory lumps under the skin, and even cerebral haematoma.

Clinical Signs and Symptoms

In susceptible hosts, *Paragonimus* infection typically results in a chronic condition that resembles pulmonary tuberculosis or lung cancer (Lee et al., 2014). Nonetheless, in cases of mild infection, clinical symptoms are usually absent. Spiny adult worms, along with their eggs and excretory–secretory products, can cause significant pathologies in lung tissues (Blair et al., 1999; Procop, 2009; Dubey, 2023). In some cases, migration to aberrant sites can result in significant pathologies, particularly in vital organs, as observed in ectopic infections. Paragonimiasis can be broadly categorised into three major types in both animals and humans: pulmonary, pleuropulmonary, and ectopic paragonimiasis (Singh et al., 2012; Sugiyama et al., 2013).

Pulmonary Paragonimiasis

Pulmonary paragonimiasis is the most common type of paragonimiasis (Singh et al., 1986, 2012). It occurs when a person ingests infected metacercarial larvae by consuming raw or partially cooked crustaceans or the meat of paratenic hosts such as wild boars and rodents. These larvae migrate inside the definitive host species, with newly emerged juveniles reaching the pleural spaces of the host a few weeks post infection. They then migrate to the lungs in pairs, forming hazelnut-shaped cysts with a diameter of 1–2 cm (Choi, 1990; Blair, 2024). The cysts later establish communication with the bronchioles, and the worms' excretory–secretory products, including the eggs, are mostly released into the external environment after being coughed up by the host or through the faeces if the discharges have been swallowed by the host (Dubey, 2023; Blair, 2024). In the case of *P. westermani*, triploid and tetraploid individuals are significantly larger than those with diploid chromosomes can cause the formation of cysts in the lungs and shed eggs, which are reproduced parthenogenetically. Parasite metabolites and eggs induce chronic cough in individuals, but in symptomatic cases, occasional haemoptysis with rusty brown sputum accompanied by paroxysmal coughing may indicate infection (Liu et al., 2008; Dubey, 2023). Haemoptysis may be exacerbated during physical activity, necessitating hospitalisation and blood transfusion. The onset of cough, chest pain, and recurrent haemoptysis generally occurs after an incubation period of 8–16 weeks. Diploid forms of *P. westermani* are predominantly responsible for paragonimiasis. In endemic regions, patients are often asymptomatic, as evidenced by high seropositivity rates in these areas, indicating mild infection. Conversely, triploid forms of *P. westermani* are linked to the development of larger pulmonary cysts associated with marked respiratory pathologies (Blair, 2024).

Diagnosing pulmonary paragonimiasis can be challenging because its clinical and radiological features often overlap with those of pulmonary tuberculosis and lung cancer, making it necessary to conduct a differential diagnosis. Microscopic examination of sputum may reveal the presence of characteristic golden brown, operculated *Paragonimus* eggs. Pneumothorax and secondary bacterial infection may become common sequelae in chronic cases (Procop, 2009; Song et al., 2023).

Pleuro-pulmonary Paragonimiasis

A few weeks after infection, young flukes continue moving around the pleural cavities to find their prospective mates, after which they settle in pairs within the cysts in the lung tissue. The excretory–secretory products of these tissue-invasive juvenile worms contain different proteolytic enzymes, including different variants of cysteine proteases that serve multiple crucial functions in parasite survival. These cysteine proteases not only facilitate the migration of young flukes by dissolving host tissues but also play a significant role in preventing antibody-dependent helminthotoxicity by breaking down the host IgG (Chung et al. 1997). Additionally, they prevent inflammation in our body tissues by preventing certain cells (eosinophils) from releasing chemicals (degranulations) and by causing them to die (apoptosis) (Min et al., 2004; Shin et al., 2005).

Continuous migration of the young/unpaired worms through the pleural cavity can lead to various health issues, such as chest pain, pleural effusion, pleuritis, and pneumothorax (Procop, 2009; Blair, 2024). Pleural effusion is a common condition that primarily affects children, causing fever and fluid build-up in both lungs (Sugiyama et al., 2013). *Paragonimus* eggs are rare in pleural paragonimiasis, but the pleural aspirate typically displays high levels of eosinophilic cells, increased protein content, decreased glucose levels, low pH, and high levels of lactate dehydrogenase (Jeon et al., 2005; Yaguchi et al., 2016; Song et al., 2023). Laboratory findings may indicate peripheral eosinophilia, leukocytosis, and highly elevated serum levels of total IgE (Coogle et al., 2022). The infection may sometimes affect both the lungs and the pleural cavities, causing symptoms such as chest pain, difficulty breathing, chronic cough, and recurrent coughing of rusty-brown eggs (Singh et al., 1986; Jeon et al., 2005; Procop, 2009; Blair, 2024).

Ectopic Paragonimiasis

Paragonimiasis primarily manifests as a pleuropulmonary disease with an array of respiratory symptoms that vary depending on the phase of the infection. If juvenile worms fail to reach the lungs during tissue migration, or if they are unable to find a partner, they may reach abnormal locations outside the lungs (Blair, 2024). This phenomenon, known as atypical worm migration, can evoke pronounced tissue reactions (Wang et al., 2026). Ectopic paragonimiasis has been described in various locations, including the brain, spinal cord, liver, muscle, skin, eyes, testes, and even the scrotum (Oh, 1969; Kim et al., 2004). Although rare, the presence of adult worms in these aberrant sites can lead to complications and, in some cases, even fatalities, particularly when the parasite is lodged in the central nervous system (Amaro et al., 2016).

Cerebrospinal Paragonimiasis

Cerebrospinal paragonimiasis is a severe neurological disease caused by the migration of juvenile *Paragonimus* into the brain and spinal cord. These flukes penetrate the central nervous system through bony openings, leading to the formation of various lesions, such as arachnoiditis, intracerebral abscesses, and granulomas (Choo et al., 2003; Mathison and Pritt, 2023). The initial presentation can be acute and dramatic, mimicking meningitis with symptoms such as fever, headache, convulsions, epileptic episodes, neck stiffness, and vomiting (Procop, 2009; Singh et al., 2012; Sugiyama et al., 2013). In severe cases, fatality can occur (Choo et al., 2003), primarily from herniation associated with elevated intracranial pressure (Procop, 2009).

The chronic stage of the disease presents a more diverse clinical picture. Patients may develop organic mental syndrome, epilepsy, or syndromes resembling stroke (infarction) and tumours (Xia et al., 2014; Amaro et al., 2016; Moon and Kim,

2022; Kim et al., 2022). Seizures, headaches, visual disturbances, and motor and sensory disorders are the most frequent complaints (Kusner and King, 1993). Hemiplegia (paralysis on one side), hemihypesthesia (numbness on one side), homonymous hemianopia (loss of half the visual field in both eyes), optic atrophy (wasting of the optic nerve), and mental deterioration are the five cardinal signs of advanced disease (Oh, 1969).

Cutaneous Paragonimiasis

Cutaneous paragonimiasis cases are occasionally reported from various parts of the world, although they are less common. This disease is caused by the migration of juvenile worms through subcutaneous tissues, resulting in the formation of nodules along migratory tracts. The nodules initially appear on the chest or abdominal walls but eventually move towards the lower limbs, genitals, and inguinal region. These nodules are characterised by their mobility, firm consistency, nontenderness, and size, which varies between 2 and 5 cm in diameter (Singh et al., 2012; Sugiyama et al., 2013). However, in rare cases, nodules may remain at the same location without further migration. Palpable fibrous bands connecting these nodules may sometimes be present. The most common locations for these nodules are the chest, abdomen, and lower back, but they can also be found on the head, neck, extremities, scrotum, vulva, and other areas (Schuster et al., 2007; Procop, 2009; Sim et al., 2010; Singh et al., 2012b; Jiang et al., 2015).

Abdominal Paragonimiasis

The ingestion of raw or partially cooked crustaceans or the meat of paratenic hosts harbouring infective metacercariae initiates the parasite's life cycle within suitable definitive hosts. The infective metacercariae excyst in the intestine, releasing juvenile flukes. These juveniles then penetrate the intestinal wall and migrate through the peritoneal cavity, diaphragm, and pleural cavity to reach the lungs, their preferred predilection site. Occasionally, during this migration, juvenile flukes may wander through the peritoneal cavity, leading to the formation of multiple nodules within the abdomen. This can cause peritonitis and, occasionally, abdominal distension. Additionally, if the flukes migrate through the liver, they can sometimes cause hepatomegaly (Jeon et al., 1999; Procop, 2009; Sugiyama et al., 2013; Kim et al., 2017; Choi et al., 2022).

Ocular Paragonimiasis

Ocular paragonimiasis is a condition caused by invasion of the eye by the *Paragonimus* parasite. This invasion can lead to a range of vision problems, such as impaired vision, optic nerve degeneration, involuntary eye movements (nystagmus), abnormal pupil responses, and even loss of vision in half of the visual field (hemianopsia). In severe cases, adult flukes may lodge within the eye, causing repeated episodes of intense pain, inflammation of the uvea, and significant visual loss (Wang et al., 1984; Xia et al., 2019; Liu and Huang, 2025). Paragonimiasis can also affect vision through cerebral paragonimiasis, which affects the occipital lobes of the brain responsible for vision processing. This can manifest as visual disturbances such as double vision, blurred vision, and even blindness. Treatment often requires surgical extraction of the parasite and management of the associated inflammation to prevent further damage and alleviate symptoms. For cases where the worm has invaded the posterior chamber of the eye, vitrectomy has been shown to be effective. Additionally, medications such as praziquantel, ocular steroids, and mydriatics are administered to manage the infection and associated inflammation and improve visual comfort (El-Tantawy, 2025).

Urogenital Paragonimiasis

Urogenital paragonimiasis displays a range of clinical presentations that depend on the specific site of the infection. The ectopic migration of juvenile flukes from the lungs can reach various organs, including the adrenal glands, kidneys, urinary bladder, genitourinary system, and subcutaneous tissues surrounding the genitalia (Kang et al., 1988; Lin and Chen, 1993; Yue et al., 2014; Jiang et al., 2015; Liang et al., 2018; Dubey, 2023). This aberrant migration can mimic tuberculous epididymitis, with patients experiencing symptoms such as tenderness, pain, fever, and microhematuria due to the presence of parasite cysts in the kidney pelvis (Chen et al., 2013; Wang et al., 2015). Scrotal paragonimiasis is a rare paediatric manifestation of ectopic paragonimiasis and is categorised as a cutaneous infection. This form typically presents as migrating subcutaneous nodules ranging from 1 to 3 cm, although some may be larger. These nodules can be single or multiple, with variable shapes and poorly defined borders, whereas the overlying skin appears normal (Liang et al., 2018). Vesical paragonimiasis, an exceptionally rare form of ectopic infection limited to the bladder, is less severe than its cerebral form and may remain asymptomatic for extended periods. However, when symptoms do arise, they commonly manifest as gross haematuria and odynuria (Yue et al., 2014).

Cardiovascular Paragonimiasis

Cardiovascular paragonimiasis is a rare yet potentially life-threatening parasitic infection that primarily affects the heart. This severe condition is uncommon, but it can also affect children and be associated with pleuropulmonary paragonimiasis. Patients with cardiovascular paragonimiasis may experience a range of symptoms, including cough, chest pain, shortness of breath, abdominal swelling, fatigue, weight loss, diarrhoea, and night sweats. Physical examination findings usually reveal jugular venous distention, ascites, and low blood pressure. Imaging tests often reveal an enlarged heart, lung nodules, pericardial tamponade, pericardial effusion, and pleural effusion (Sah et al., 2019; Al Bishawi et al., 2021; Jiang et al., 2023; Bai et al., 2025). Laboratory tests may reveal elevated white blood cell counts and eosinophilia. Managing cardiovascular paragonimiasis depends on the severity of the condition. While surgical treatment may be necessary in some cases, especially for patients with large pericardial effusions, praziquantel is the mainstay of therapy (Jiang et al., 2023). Praziquantel is effective in alleviating symptoms, and after a successful course of the drug, spontaneous absorption of pleural effusion typically occurs in uncomplicated cases. However, some patients with severe complications, such as pericarditis and myocarditis, may rarely experience fatal congestive heart failure despite praziquantel therapy (Singh et al., 2012).

Diagnosis

Patent paragonimiasis, a parasitic infection caused by mature *Paragonimus* flukes, requires a multipronged approach for definitive diagnosis. This typically involves a combination of parasitological, radiological, molecular, and serological techniques. The gold standard for diagnosis remains the identification of characteristic eggs extracted from sputum or faeces (Luo et al., 2025). These eggs are readily distinguished by their ovoid shape, golden–brown colouration, and thickened shell at the ab-opercular end. However, egg detection can be challenging due to intermittent egg release (Kong et al., 2015), and low parasite burden, or in ectopic infections. Additionally, diagnosing *P. skrjabini* infection can be challenging because, unlike other *Paragonimus* species, finding eggs in sputum or adult flukes in biopsies is uncommon in humans, as they are not ideal hosts for this parasite (Wu et al., 2019; Liu and Huang, 2025). Biopsies from ectopic sites may reveal signs of tissue death (necrosis) and inflammation, which can be suggestive of *P. skrjabini* infection, but a definitive diagnosis might require additional investigations.

Advanced imaging techniques such as CT and MRI play crucial roles in definitive diagnosis, particularly in extrapulmonary paragonimiasis. These modalities excel at visualising and localising lesions, allowing differentiation between paragonimiasis and other neurological conditions that can mimic symptoms, such as tumours or complications from haemorrhagic stroke. CT and MRI scans can reveal characteristic multiple cystic lesions with ring enhancement, often described as "grape clusters" or "soap bubble appearence" (Razek et al., 2011; Singh et al., 2012; Kim et al., 2022). Radiological examinations offer valuable insights, especially in chronic cases. Plain chest X-rays may show a distinctive "bunch of grapes" pattern with ring-shaped opacities, indicative of the contiguous cavities caused by the flukes. Notably, this pattern can also be observed in cerebral lesions detected via CT or MRI. In chronic brain infections, plain radiographs may additionally reveal pathognomonic calcifications resembling soap bubbles within the tissue (Kinoshita and Koga, 1986).

Analysis of cerebrospinal fluid during the acute stage can provide valuable diagnostic clues. Elevated levels of eosinophils and proteins often signal parasitic inflammation. However, definitive diagnosis often relies on serological tests that detect the presence of *Paragonimus* antigens or *Paragonimus*-specific antibodies in the blood. Two main types of antibody tests are employed: those targeting IgM and those targeting IgG antibodies. Among these methods, specific IgG antibody tests are highly sensitive and specific, unlike tests that rely solely on IgM antibodies. IgM antibody tests are most useful during the early stages of infection because of the rapid increase in these antibodies during the initial immune response (Nakamura-Uchiyam et al. 2001). However, they may not be specific to paragonimiasis and can be positive in other parasitic infections (Procop and Neafie, 2018). In contrast, specific IgG antibodies take longer to develop (approximately 3–4 weeks after infection) but offer greater diagnostic accuracy. Similarly, tests targeting egg antigens can take even longer to become positive (approximately 10 weeks after infection) (Kong et al., 1997). Importantly, the presence of IgG antibodies does not necessarily indicate an active infection. They may remain elevated even after successful treatment due to persistent egg granulomas. Following successful treatment, antibody levels typically decrease gradually and may become undetectable within 6–28 months (Cho et al., 2000). However, in some cases, persistent egg granulomas can cause sustained IgG antibody titres. Therefore, interpreting serological results alongside other diagnostic techniques and clinical presentations remains crucial for optimal diagnosis and management. Early and accurate diagnosis through a combination of techniques, including cerebrospinal fluid analysis, serological testing, and advanced imaging modalities such as CT and MRI, facilitates the prompt initiation of appropriate treatment strategies, leading to a favourable prognosis.

Treatment

In animals, treatment options for paragonimiasis depend on the specific anthelmintic drug and treatment regimen employed. Fenbendazole, for example, can be administered orally at a dosage of 50 mg/kg body weight, divided into two daily doses for a 2–3 week period (Dubey, 2023). Albendazole and niclofolan, however, require high doses administered over extended durations (Keiser and Utzinger, 2005).

For human paragonimiasis, praziquantel is the recommended treatment with high efficacy (Chai, 2013b). In accordance with the WHO guidelines (WHO, 2011), the recommended dosage is 25 mg/kg body weight, which is administered three times daily for three consecutive days. This regimen typically leads to clinical improvement within a few weeks. However, caution is advised when coadministering praziquantel with rifampicin in patients suspected of having pulmonary tuberculosis. Rifampicin, a potent inducer of drug-metabolising enzymes, can significantly reduce the plasma concentration of praziquantel, potentially rendering it ineffective (Ridtitid, 2002). Therefore, alternative treatment regimens should be considered in such cases. Triclabendazole is another effective option (Keiser et al., 2005). A single dose of 20 mg/kg body weight, preferably divided into two administrations on the same day, effectively eliminates parasite eggs from sputum. Compared with other medications, triclabendazole offers advantages such as lower dosages, shorter treatment durations, and broad-spectrum activity against *Fasciola* parasites.

In cases of cerebral paragonimiasis, treatment strategies vary on the basis of the severity and location of the lesions. Significant oedema surrounding cerebral lesions may necessitate the use of corticosteroids and mannitol for management. Surgical removal is a feasible option for solitary, superficial lesions (Finsterer and Auer, 2013; Amaro et al., 2016). Nonetheless, timely detection is essential for successful management, even in chronic infections, whether through surgical intervention or appropriate medication.

Control and Prevention

Paragonimiasis, an important snail-borne parasitic disease (SBPD), poses a major public health concern in regions endemic to zoonotic parasitic infections (Lu et al., 2018). The consumption of undercooked or raw crabs and crayfish in these areas can increase the risk of disease transmission (Blair et al., 1999). Therefore, it is recommended that such items be discouraged from being fed to companion animals. Strict avoidance of uncooked, partially cooked, or pickled crab or crayfish products is crucial to prevent transmission.

A holistic strategy for eradicating snails in endemic areas is essential for effectively managing outbreaks and the prevalence of SBPDs (WHO, 2011; Sokolow et al., 2015). Physical control methods, such as environmental manipulation, have been employed, but they can be disruptive and require ongoing management. These methods involve eliminating natural water bodies in high-risk zones and regulating nearby human settlements. Additionally, proper drainage and strategic environmental engineering interventions have shown promise in specific locations. Mechanical disruption of snail habitats via equipment or manual tools can also reduce snail populations. However, these methods are not suitable for all ecosystems.

Biological control offers a more targeted approach by utilising natural predators and competitors of snails. Predatory fish such as black carp (*Mylopharyngodon piceus*) have been successfully used to reduce snail populations that transmit diseases such as clonorchiasis and opisthorchiasis (Hung et al., 2013; Lu et al., 2018). However, some biological control agents, such as certain fish species, can have unintended consequences by disrupting the ecological balance or becoming invasive themselves. Predatory prawns and water bugs have been identified as potential biocontrol agents because of their documented ability to prey on intermediate snail hosts of SBPDs. Sciomyzid fly larvae, Khaki Campbell, and Muscovy ducks also show potential for snail population control (Teo, 2001; Ndlela and Chimbari, 2000).

Molluscicides are chemicals or plant extracts that can control snails and are used to eradicate snail-borne parasitic diseases in endemic areas (Djirmay et al., 2024). Molluscicides may be synthetic or natural, and many plant molluscicides have been studied. These chemical or botanical agents are intended to target snails, which serve as intermediate hosts for many parasites. While previously used options such as copper sulfate have raised safety concerns, niclosamide (Bayluscide®) is currently recommended by the WHO due to its relative safety (Tchounwou et al., 1991). However, molluscicides can harm nontarget organisms and disrupt ecosystems, requiring careful application (Wan et al., 2025), especially in areas with high biodiversity.

Given the limitations of existing control methods, research into alternative approaches is crucial. The snail-eating leech *Helobdella austinensis* presents a promising option. Laboratory studies suggest that this leech species effectively targets a broad range of SBPD-carrying snails without harming other organisms in the ecosystem. Further research and field trials are necessary to fully evaluate the potential of *H. austinensis* as a biocontrol agent for SBPDs (Saglam et al., 2023).

Conclusion

Paragonimiasis poses a significant public health challenge in endemic regions, particularly where consumption of raw or undercooked crustaceans is widely prevalent. Hence, a deeper understanding of its various clinical presentations, intricate life cycle, and pathology is crucial to effectively control this disease. While it can resemble chronic tuberculosis or pulmonary malignancy in some instances, paragonimiasis can also be symptomless. The damage caused by adult worms and their secretions to lung tissue is notable, but the most severe complications arise from ectopic infections involving vital organs. The diagnosis of paragonimiasis thus necessitates a comprehensive approach that combines parasitological, radiological, molecular, and serological techniques. A thorough medical history is essential for diagnosis, especially in patients with unexplained tumour masses. Important factors to consider in suspected cases include living in endemic mountainous areas, a history of consuming raw or undercooked crustaceans, unexplained eosinophilia, and the presence of tumour masses in visceral organs or in the CNS. Existing methods for controlling paragonimiasis, a snail-borne parasitic disease, have limitations. Therefore, the exploration of new biocontrol agents presents a promising path towards a more sustainable and effective approach. Therefore, the prevailing challenges in combating paragonimiasis highlight the urgent need for continued research into such alternatives, the importance of a detailed medical history for diagnosis and management, the potential requirements for surgical intervention in specific cases, and the specific criteria for diagnosing unexplained tumour masses suspected of being atypical paragonimiasis. A comprehensive strategy that integrates these considerations is vital for mitigating this neglected zoonotic disease in endemic areas.

Bibliography

Abdel Razek, A. A. K., Watcharakorn, A., & Castillo, M. (2011). Parasitic diseases of the central nervous system. *Neuroimaging Clinics of North America, 21*(4), 815–841. https://doi.org/10.1016/j.nic.2011.07.005

Ahn, C.-S., Shin, J. W., Kim, J.-G., Lee, W.-Y., Kang, I., Im, J.-G., & Kong, Y. (2021). Spectrum of pleuropulmonary paragonimiasis: An analysis of 685 cases diagnosed over 22 years. *Journal of Infection, 82*(1), 150–158. https://doi.org/10.1016/j.jinf.2020.09.037

Al Bishawi, A., Salameh, S., Ehtesham, A., Massad, I., Arachchige, S., Hatim, A., Bozom, I., & Thapur, M. (2021). Paragonimus infection: Rare manifestation with pericardial effusion: A case report and literature review. *IDCases, 24*, e01075. https://doi.org/10.1016/j.idcr.2021.e01075

Amaro, D. E., Cowell, A., Tuohy, M. J., Procop, G. W., Morhaime, J., & Reed, S. L. (2016). Cerebral Paragonimiasis presenting with sudden death. *The American Society of Tropical Medicine and Hygiene, 95*(6), 1424–1427. https://doi.org/10.4269/ajtmh.15-0902

Bai, H., Wang, Y., Yang, X., Huang, Y., Jiao, F., & Luo, Y. (2025). Pericardial effusion in pediatric Paragonimiasis: Surgery may not be necessary in cases of moderate to large effusion—A retrospective study. *PLOS Neglected Tropical Diseases, 19*(4), e0013023. https://doi.org/10.1371/journal.pntd.0013023

Blair, D., Zhengshan, C., Minggang, C., Aili, C., Bo, W., Agatsuma, T., Iwagami, M., Corlis, D., Chengbin, F., & Ximei, Z. (2005). Paragonimus skrjabini Chen, 1959 (Digenea: Paragonimidae) and related species in eastern Asia: A combined molecular and morphological approach to identification and taxonomy. *Systematic Parasitology, 60*, 1–21.

Blair, D. (2008). Family Paragonimidae Dollfus, 1939. In R. A. Bray, D. I. Gibson, & A. Jones (Eds.), *Keys to the Trematoda* (Vol. 3). CAB International and Natural History Museum.

Blair, D., Toledo, R., & Fried, B. (2014). *Digenetic trematodes paragonimiasis* (pp. 115–152). Springer New York.

Blair, D., Nawa, Y., Mitreva, M., & Doanh, P. N. (2016). Gene diversity and genetic variation in lung flukes (genus Paragonimus). *Transactions of the Royal Society of Tropical Medicine and Hygiene, 110*(1), 6–12.

Blair, D. (2019). Paragonimiasis. In *Digenetic Trematodes* (pp. 105–138). Springer.

Blair, D. (2022). Lung flukes of the genus Paragonimus: Ancient and re-emerging pathogens. *Parasitology, 149*(10), 1286–1295.

Blair, D., Toledo, R., & Fried, B. (2024). *Digenetic trematodes paragonimiasis* (pp. 203–238). Springer International Publishing.

Cancela, M., Acosta, D., Rinaldi, G., Silva, E., Durán, R., Roche, L., Zaha, A., Carmona, C., & Tort, J. F. (2008). A distinctive repertoire of cathepsins is expressed by juvenile invasive Fasciola hepatica. *Biochimie, 90*(10), 1461–1475. https://doi.org/10.1016/j.biochi.2008.04.020

Cha, S. H., Chang, K. H., Cho, S. Y., Han, M. H., Kong, Y., Suh, D. C., Choi, C. G., Kang, H. K., & Kim, M. S. (1994). Cerebral paragonimiasis in early active stage: CT and MR features. *American Journal of Roentgenology, 162*(1), 141–145. https://doi.org/10.2214/ajr.162.1.8273653

Chai, J.-Y. (2013a). Paragonimiasis. In H. H. Garcia, H. B. Tanowitz, & O. H. Del Brutto (Eds.), *Handbook of clinical neurology, volume 114 (3rd series), neuroparasitology and tropical neurology* (Vol. 114, pp. 283–296). Elsevier.

Chai, J. Y. (2013b). Praziquantel treatment in trematode and cestode infections: An update. *Infection & chemotherapy, 45*(1), 32.

Chen, S., Meng, H., Cao, M., & Shen, B. (2013). Renal artery aneurysm mimicking renal calculus with hydronephrosis. *American Journal of Kidney Diseases, 61*(6), 1036–1040. https://doi.org/10.1053/j.ajkd.2012.10.023

Cho, S.-Y., Kong, Y., Yun, D.-H., Kan, S.-Y., Kim, L.-S., Chung, Y.-B., & Yang, H.-J. (2000). Persisting antibody reaction in Paragonimiasis after praziquantel treatment is elicited mainly by egg antigens. *TheKorean Journal of Parasitology, 38*(2), 75. https://doi.org/10.3347/kjp.2000.38.2.75

Choi, D. W. (1990). Paragonimus and paragonimiasis in Korea. *The Korean Journal of Parasitology, 28*(Suppl), 79. https://doi.org/10.3347/kjp.1990.28.Suppl.79

Choi, J. W., Lee, C. M., Kim, S. J., In Hah, S., Kwak, J. Y., Cho, H. C., Ha, C. Y., Jung, W. T., & Lee, O. J. (2022). Ectopic peritoneal Paragonimiasis mimicking tuberculous peritonitis: A care report. *World Journal of Clinical Cases, 10*(16), 5359–5364. https://doi.org/10.12998/wjcc.v10.i16.5359

Choo, J.-D., Suh, B.-S., Lee, H.-S., Lee, J.-S., Song, C.-J., Shin, D.-W., & Lee, Y.-H. (2003). Chronic cerebral Paragonimiasis combined with aneurysmal subarachnoid hemorrhage. *The American Journal of Tropical Medicine and Hygiene, 69*(5), 466–469. https://doi.org/10.4269/ajtmh.2003.69.466

Chung, Y.-B., Kong, Y., Joo, J., Cho, S.-Y., & Kang, S.-Y. (1995). Excystment of Paragonimus westermani metacercariae by endogenous cysteine protease. *The Journal of Parasitology, 81*(2), 137. https://doi.org/10.2307/3283911

Chung, Y. B., Yang, H. J., Kang, S. Y., Kong, Y., & Cho, S. Y. (1997). Activities of different cysteine proteases of paragonimus westermani in cleaving human IgG. *The Korean Journal of Parasitology, 35*(2), 139. https://doi.org/10.3347/kjp.1997.35.2.139

Chung, Y. B., Kim, T. S., & Yang, H. J. (2005). Early cysteine protease activity in excretory bladder triggers metacercaria excystment of Paragonimus westermani. *Journal of Parasitology, 91*(4), 953–954. https://doi.org/10.1645/GE-475R.1

Chung, Y.-B., Kita, H., & Shin, M. H. (2008). A 27 kDa cysteine protease secreted by newly excysted Paragonimus westermani metacercariae induces superoxide anion production and degranulation of human eosinophils. *The Korean Journal of Parasitology, 46*(2), 95. https://doi.org/10.3347/kjp.2008.46.2.95

Coogle, B., Sosland, S., & Bahr, N. C. (2022). A clinical review of human disease due to Paragonimus kellicotti in North America. *Parasitology, 149*(10), 1327–1333. https://doi.org/10.1017/S0031182021001359

Cumberlidge, N., Rollinson, D., Vercruysse, J., Tchuem Tchuenté, L.-A., Webster, B., & Clark, P. F. (2018). Paragonimus and paragonimiasis in West and Central Africa: Unresolved questions. *Parasitology, 145*(13), 1748–1757. https://doi.org/10.1017/S0031182018001439

David, B., Zhi-Biao, X., & Takeshi, A. (1999). Advances in parasitology. In *Paragonimiasis and the genus paragonimus* (Vol. 42, pp. 113–222). Elsevier.

Djirmay, A. G., Yadav, R. S., Guo, J., Rollinson, D., & Madsen, H. (2024). Chemical control of snail vectors as an integrated part of a strategy for the elimination of schistosomiasis—A review of the state of knowledge and future needs. *Tropical Medicine and Infectious Disease, 9*(9), 222. https://doi.org/10.3390/tropicalmed9090222

Doanh, P. N., Horii, Y., & Nawa, Y. (2013). Paragonimus and paragonimiasis in Vietnam: An update. *The Korean Journal of Parasitology, 51*(6), 621.

Doanh, P. N., Van Hien, H., An, P. T., & Tu, L. A. (2016). Development of lung fluke Paragonimus heterotremus in rat and mice and the role of paratenic host in its life cycle. *TAP CHI SINH HOC, 37*(3). https://doi.org/10.15625/0866-7160/v37n3.6168

Doanh, P. N., Van Hien, H., & Dung, B. T. (2020). Molecular identification of morphologically similar microcercous cercariae of two trematode families Paragonimidae and Troglotrematidae concurrently found in the same snail species of the subfamily Triculinae. *Folia Parasitologica, 67*. https://doi.org/10.14411/fp.2020.008

Dubey, J. P. (2023). Endemic Paragonimus kellicotti infections in animals and humans in USA and Canada: Review and personal perspective. *Food and Waterborne Parasitology, 30*, e00184. https://doi.org/10.1016/j.fawpar.2022.e00184

El-Tantawy, N. L. (2025). Presumed trematode-induced uveitis: A narrative review. *Taiwan Journal of Ophthalmology*. https://doi.org/10.4103/tjo.TJO-D-25-00077

Eun-A Kim, Seon-Kwan Juhng, Hye Won Kim., Gang Deuk Kim, Young Whan Lee, Hyang Jung Cho, Jong Jin Won (2004) Imaging findings of hepatic Paragonimiasis: A case report Journal of Korean Medical Science 19(5) 759 https://doi.org/10.3346/jkms.2004.19.5.759.

Finsterer, J., & Auer, H. (2013). Parasitoses of the human central nervous system. *Journal of Helminthology, 87*(3), 257–270. https://doi.org/10.1017/S0022149X12000600

Grote, A., Caffrey, C. R., Rebello, K. M., Smith, D., Dalton, J. P., & Lustigman, S. (2018). Cysteine proteases during larval migration and development of helminths in their final host. *PLOS Neglected Tropical Diseases, 12*(8), e0005919. https://doi.org/10.1371/journal.pntd.0005919

Hung, N. M., Duc, N. V., Stauffer, J. R., et al. (2013). Use of black carp (Mylopharyngodon piceus) in biological control of intermediate host snails of fish-borne zoonotic trematodes in nursery ponds in the red river delta Vietnam. *Parasites Vectors, 6*, 142. https://doi.org/10.1186/1756-3305-6-142

Ikeda, T. (1998). Antibody responses to fluke cysteine proteinases in Paragonimus – and Fasciola -infected rats. *Journal of Helminthology, 72*(3), 187–191. https://doi.org/10.1017/S0022149X00016424

Intapan, P. M., & Maleewong, W. (2001). In vitro excystation of Paragonimus heterotremus metacercariae. *Journal of Parasitology, 87*(5), 1184–1186. https://doi.org/10.1645/0022-3395(2001)087[1184:IVEOPH]2.0.CO;2

Iwagami, M., Ho, L. Y., Su, K., Lai, P. F., Fukushima, M., Nakano, M., Blair, D., Kawashima, K., & Agatsuma, T. (2000). Molecular phylogeographic studies on Paragonimus westermani in Asia. *Journal of Helminthology, 74*(4), 315–322.

Jeon, K., Koh, W.-J., Kim, H., Kwon, O. J., Kim, T. S., Lee, K. S., & Han, J. (2005). Clinical features of recently diagnosed pulmonary paragonimiasis in Korea. *Chest, 128*(3), 1423–1430. https://doi.org/10.1378/chest.128.3.1423

Jeong, M.-G., Yu, J.-S., Kim, K. W., Kim, J. K., Kim, S. J., Kim, H. J., & Choi, Y. D. (1999). Retroperitoneal paragonimiasis: A case of ectopic Paragonimiasis presenting as periureteral masses. *Journal of Computer Assissted Tomography, 23*(5), 696–698. https://doi.org/10.1097/00004728-199909000-00009

Jiang, A., Ge, L., You, Y., & Shen, Z. (2015). The diagnosis of cutaneous paragonimiasis after the external migration of the parasite from a punch biopsy site. *JAAD Case Reports, 1*(5), 239–240. https://doi.org/10.1016/j.jdcr.2015.05.009

Jiang, Y.-x., Li, G.-q., Pan, C.-j., He, Z.-q., Wang, C., Mu, Q.-r., & Cao, L.-l. (2023). Pediatric paragonimiasis: a retrospective analysis of cases from a county in south-west China. *Frontiers in Pediatrics, 11*. https://doi.org/10.3389/fped.2023.1143262

Kang, S. M., Park, Y. K., Lee, J. H., Mok, J. E., & Yang, M. H. (1988). Ovarian Paragonimiasis: A case report. *Korean Journal of Pathology*, 336–339.

Keiser, J., & Utzinger, J. (2004). Chemotherapy for major food-borne trematodes: A review. *Expert Opinion on Pharmacotherapy, 5*(8), 1711–1726. https://doi.org/10.1517/14656566.5.8.1711

Keiser, J., Engels, D., Büscher, G., & Utzinger, J. (2005). Triclabendazole for the treatment of fascioliasis and paragonimiasis. *Expert Opinion on Investigational Drugs, 14*(12), 1513–1526. https://doi.org/10.1517/13543784.14.12.1513

Kim, M. J., Kim, S. H., Lee, S. O., Choi, S. H., Kim, Y. S., Woo, J. H., Yoon, Y. S., Kim, K. W., Cho, J., Chai, J. Y., & Chong, Y. P. (2017). A case of ectopic peritoneal paragonimiasis mimicking diverticulitis or abdominal abscess. The Korean Journal of Parasitology, 55(3), 313–318. https://doi.org/10.3347/kjp.2017.55.3.313

Kim, J.-G., Ahn, C.-S., Kang, I., Shin, J.-W., Jeong, H.-B., Nawa, Y., & Kong, Y. (2022). Cerebral Paragonimiasis: clinicoradiological features and serodiagnosis using recombinant yolk ferritin. *PLOS Neglected Tropical Diseases, 16*(3), e0010240. https://doi.org/10.1371/journal.pntd.0010240

Kinoshita, K., & Koga, T. (1986). Cerebral Paragonimiasis with peculiar calcified foci: A case report. *No Shinkei geka*Neurological Surgery, *14*(5), 669–672.

Kojima, S., Yokogawa, M., & Tada, T. (1972). Raised levels of serum IgE in human helminthiases. *The American Journal of Tropical Medicine and Hygiene, 21*(6), 913–918. https://doi.org/10.4269/ajtmh.1972.21.913

Kong, Y., Chung, J. Y., Yun, D. H., Kim, L. S., Kang, S. Y., Ito, A., Ma, L., & Cho, S. Y. (1997). Variation of antigenic proteins of eggs and developmental stages of Paragonimus westermani. *TheKorean Journal of Parasitology, 35*(3), 197. https://doi.org/10.3347/kjp.1997.35.3.197

Kong, Y., Doanh, P. N., & Nawa, Y. (2015). Paragonimus. In *Biology of foodborne parasites, food microbiology* (pp. 445–462). CRC Press. https://doi.org/10.1201/b18317-27

Kusner, D., & King, C. (1993). Cerebral Paragonimiasis. *Seminars in Neurology, 13*(02), 201–208. https://doi.org/10.1055/s-2008-1041126

Lee, S., Yu, Y., An, J., Lee, J., Son, J.-S., Lee, Y. K., Song, S., Kim, H., & Kim, S. (2014). A case of delayed diagnosis of pulmonary paragonimiasis due to improvement after anti-tuberculosis therapy. *Tuberculosis and Respiratory Diseases, 77*(4), 178. https://doi.org/10.4046/trd.2014.77.4.178

Liang, T., Liang, G., Du, Y., Wang, X., Wang, Y., Chen, A., Chen, Z., Du, J., Li, H., & Yu, L. (2018). Scrotal Paragonimiasis in adults. *Medicine, 97*(14), e0328. https://doi.org/10.1097/MD.0000000000010328

Lin, C.-M., & Chen, S.-K. (1993). Paragonimus calcified ova mimicking left renal staghorn stone. *Journal of Urology, 149*(4), 819–820. https://doi.org/10.1016/S0022-5347(17)36219-5

Liu, Q., Wei, F., Liu, W., Yang, S., & Zhang, X. (2008). Paragonimiasis: An important food-borne zoonosis in China. *Trends in Parasitology, 24*(7), 318–323. https://doi.org/10.1016/j.pt.2008.03.014

Liu, J., & Huang, L. (2025). Case report: Simultaneous Paragonimus skrjabini infection in twin girls with spontaneous emergence of a juvenile worm from the eyelid of the elder sister. *Frontiers in Pediatrics, 13*. https://doi.org/10.3389/fped.2025.1708963

Lu, X.-T., Gu, Q.-Y., Limpanont, Y., Song, L.-G., Wu, Z.-D., Okanurak, K., & Lv, Z.-Y. (2018). Snailborne parasitic diseases: An update on global epidemiological distribution transmission interruption and control methods. *Infectious Diseases ofPoverty, 7*(1). https://doi.org/10.1186/s40249-018-0414-7

Luo, M., Liu, Y., Yang, X., Liu, M., & Yu, C. (2025). Paragonimiasis diagnosed through detection of parasitic ova in pleural effusion: A clinical case report ABSTRACT. *Respirology Case Reports, 13*(10) 10.1002/rcr2.v13.10. https://doi.org/10.1002/rcr2.70354

Lynne, S., & Garcia, A. L. *Truant manual of commercial methods in clinical microbiology diagnostic medical parasitology* (pp. 274–305). ASM Press.

Malvy, D., Ezzedine, K. H., Receveur, M. C., Pistone, T., Mercié, P., & Longy-Boursier, M. (2006). Extra-pulmonary paragonimiasis with unusual arthritis and cutaneous features among a tourist returning from Gabon. *Travel Medicine and Infectious Disease, 4*(6), 340–342. https://doi.org/10.1016/j.tmaid.2006.01.003

Mathison, B. A., & Pritt, B. S. (2023). The landscape of parasitic infections in the United States. *Modern Pathology, 36*(8), 100217. https://doi.org/10.1016/j.modpat.2023.100217

Min, D.-Y., Lee, Y.-A., Ryu, J.-S., Ahn, M.-H., Chung, Y.-B., Sim, S., & Shin, M. H. (2004). Caspase-3-mediated apoptosis of human eosinophils by the tissue-invading helminth paragonimus westermani. *International Archives of Allergy and Immunology, 133*(4), 357–364. https://doi.org/10.1159/000077355

Moon, S. W., & Kim, T. (2022). Cerebral Paragonimiasis presenting with dementia. *Parasites Hosts and Diseases, 60*(5), 353–355. https://doi.org/10.3347/kjp.2022.60.5.353

Na, B.-K., Kim, S.-H., Lee, E.-G., Kim, T.-S., Bae, Y.-A., Kang, I., Yu, J.-R., Sohn, W.-M., Cho, S.-Y., & Kong, Y. (2006). Critical roles for excretory-secretory cysteine proteases during tissue invasion of Paragonimus westermani newly excysted metacercariae. *Cellular Microbiology, 8*(6), 1034–1046. 10.1111/cmi.2006.8.issue-6. https://doi.org/10.1111/j.1462-5822.2006.00685.x

Nakamura-Uchiyama, F., Onah, D. N., & Nawa, Y. (2001). Clinical features of paragonimiasis cases recently found in Japan: Parasite-specific immunoglobulin M and G antibody classes. *Clinical Infectious Diseases, 32*(12), e171–e175. https://doi.org/10.1086/320750

Ndlela, B., & Chimbari, M. J. (2000). A preliminary assessment of the potential of the Muschovy duck (Cairina maschata) as a biocontrol agent of schistosomiasis intermediate host snails. *Central African Journal of Medicine, 46*(10). https://doi.org/10.4314/cajm.v46i10.8568

Noble, E. R., Noble, G. A., Schad, G. A., & MacInnes, A. J. (1989). *Parasitology the biology of animal parasites* (6th ed., p. 574). Lea & Febiger.

Oh, S. J. (1969). Cerebral and spinal paragonimiasis histopathological study. *Journal of the Neurological Sciences, 9*(2), 205–236. https://doi.org/10.1016/0022-510X(69)90072-0

Procop, G. W., & Neafie, R. C. (2018). *Pulmonary pathology human parasitic pulmonary infections* (pp. 289–314). Elsevier.

Procop, G. W. (2009). North American paragonimiasis (caused by Paragonimus kellicotti) in the context of global paragonimiasis. *Clinical Microbiology Reviews, 22*(3), 415–446. https://doi.org/10.1128/CMR.00005-08

Ridtitid, W., Wongnawa, M., Mahatthanatrakul, W., Punyo, J., & Sunbhanich, M. (2002). Rifampin markedly decreases plasma concentrations of praziquantel in healthy volunteers. *Clinical Pharmacology & Therapeutics, 72*(5), 505–513.

Saglam, N., Melissaratos, D. S., & Shain, D. H. (2023). Biocontrol of snail-borne parasites with the glossiphoniid leech Helobdella austinensis. *Biology Letters, 19*(4). https://doi.org/10.1098/rsbl.2022.0484

Sah, R., Gupta, N., Chatterji, P., Krishan, S., Aggarwal, M., Sood, N., Neupane, S., Sah, S., Sah, R., Poppert, S., Ruf, M.-T., Nickel, B., & Neumayr, A. (2019). Case report: Paragonimiasis presenting with pericardial tamponade. *The American Journal of Tropical Medicine and Hygiene, 101*(1), 62–64. https://doi.org/10.4269/ajtmh.19-0018

Sajid, M., & McKerrow, J. H. (2002). Cysteine proteases of parasitic organisms. *Molecular and Biochemical Parasitology, 120*(1), 1–21. https://doi.org/10.1016/S0166-6851(01)00438-8

Schuster, H., Agada, F. O., Anderson, A. R., Jackson, R. S., Blair, D., McGann, H., & Kelly, G. (2007). Otitis media and a neck lump—current diagnostic challenges for Paragonimus-like trematode infections. *Journal of Infection, 54*(2), e103–e106. https://doi.org/10.1016/j.jinf.2006.05.011

Shin, M. H., & Lee, S. Y. (2000). Proteolytic activity of cysteine protease in excretory–secretory product of Paragonimus westermani newly excysted metacercariae pivotally regulates

IL-8 production of human eosinophils. *Parasite Immunology, 22*(10), 529–533. https://doi.org/10.1046/j.1365-3024.2000.00337.x

Shin, M. H., Kita, H., Park, H. Y., & Seoh, J. Y. (2001). Cysteine protease secreted by Paragonimus westermani attenuates effector functions of human eosinophils stimulated with immunoglobulin G. *Infection and Immunity, 69*(3), 1599–1604. https://doi.org/10.1128/IAI.69.3.1599-1604.2001

Shin, M. H., Chung, Y.-B., & Kita, H. (2005). Degranulation of human eosinophils induced by Paragonimus westermani-secreted protease. *The Korean Journal of Parasitology, 43*(1), 33. https://doi.org/10.3347/kjp.2005.43.1.33

Sim, Y. S., Lee, J. H., Hong, S. C., Chang, J. H., Kang, S. R., Yang, H. J., & Sung, S. H. (2010). Paragonimus westermani found in the tip of a little finger. *Internal Medicine, 49*(15), 1645–1648. https://doi.org/10.2169/internalmedicine.49.3495

Singh, T. S., Mutum, S. S., & Razaque, M. A. (1986). Pulmonary paragonimiasis: Clinical features diagnosis and treatment of 39 cases in Manipur. *Transactions of the Royal Society of Tropical Medicine and Hygiene, 80*(6), 967–971. https://doi.org/10.1016/0035-9203(86)90275-0

Singh, T. S., Singh, S. R., Sugiyama, H., & Devi, K. H. R. (2012). A case of cutaneous Paragonimiasis presented with minimal pleuritis tropical. *Parasitology, 2*(2), 142. https://doi.org/10.4103/2229-5070.105184

Singh, T. S., Sugiyama, H., & Rangsiruji, A. (2012). Paragonimus & paragonimiasis in India. *Indian Journal of Medical Research, 136*(2), 192–204.

Sokolow, S. H., Huttinger, E., Jouanard, N., Hsieh, M. H., Lafferty, K. D., Kuris, A. M., Riveau, G., Senghor, S., Thiam, C., N'Diaye, A., Faye, D. S., & De Leo, G. A. (2015). Reduced transmission of human schistosomiasis after restoration of a native river prawn that preys on the snail intermediate host significance. *Proceedings of the National Academy of Sciences, 112*(31), 9650–9655. https://doi.org/10.1073/pnas.1502651112

Song, Y., Lee, J., Hahn, W., Jang, Y., Na, S., Oh, S.-M., Hwang, J.-H., Lee, C.-S., Choe, Y. H., & Hwang, J.-H. (2023). Risk factors and clinical features for pulmonary paragonimiasis-associated pneumothorax. *PLOS Neglected Tropical Diseases, 17*(12), e0011828. https://doi.org/10.1371/journal.pntd.0011828

Sugiyama, H., Singh, T. S., & Rangsiruji, A. (2013). Paragonimus (ch 39). In *Molecular detection of human parasitic pathogens* (1st ed., pp. 421–433). CRC Press.

Tchounwou, P. B., Englande, A. J., & Malek, E. A. (1991). Toxicity evaluation of bayluscide® and malathion to three developmental stages of freshwater snails. *Archives of Environmental Contamination and Toxicology, 21*(3), 351–358. https://doi.org/10.1007/BF01060355

Teo, S. S. (2001). Evaluation of different duck varieties for the control of the golden apple snail (Pomacea canaliculata) in transplanted and direct seeded rice. *Crop Protection, 20*(7), 599–604. https://doi.org/10.1016/S0261-2194(01)00029-1

Vainutis, K. S., Voronova, A. N., Duscher, G. G., Shchelkanov, E. M., & Shchelkanov, M. Y. (2022). Origins, phylogenetic relationships and host-parasite interactions of Troglotrematoidea since the cretaceous. *Infection, Genetics and Evolution, 101*, 105274. https://doi.org/10.1016/j.meegid.2022.105274

Wang, W. J., Xin, Y. J., Robinson, N. L., Ting, H. W., Ni, C., & Kuo, P. K. (1984). Intraocular Paragonimiasis. *British Journal of Ophthalmology, 68*(2), 85–88. https://doi.org/10.1136/bjo.68.2.85

Wang, Y., Chen, S., Wang, W., Liu, J., & Jin, B. (2015). Renal vein thrombosis mimicking urinary calculus: A dilemma of diagnosis BMC. *Urology, 15*(1). https://doi.org/10.1186/s12894-015-0054-1

Wang, X., Wu, J., Huang, J., Zeng, Y., Xu, Z., & Li, C. (2026). Ectopic paragonimiasis presenting as massive pericardial effusion in a child: A rare diagnostic challenge. *International Journal of Infectious Diseases, 164*, 108326. https://doi.org/10.1016/j.ijid.2025.108326

WHO Control of foodborne trematode infections. Report of a WHO study group. World Health Organization Technical Report Series 1995;849:1–157. PMID: 7740791.

World Health Organization. (2011). Report of the WHO expert consultation on foodborne trematode infections and taeniasis/cysticercosis., Vientiane, Lao People's Democratic Republic 12–16 October 2009 (No. WHO/HTM/NTD/PCT/2011.3).

Wu, Y., Zhou, Y., Jin, X., Li, Y., Li, H., Dai, J., Wang, G., & Wu, C. (2019). Diagnosis and surgical management of pericardial effusion due to Paragonimiasis. *International Journal of Infectious Diseases, 83*, 102–108. https://doi.org/10.1016/j.ijid.2019.03.032

Xia, Y., Ju, Y., Chen, J., & You, C. (2014). Hemorrhagic stroke and cerebral Paragonimiasis. *Stroke, 45*(11), 3420–3422. https://doi.org/10.1161/STROKEAHA.114.007267

Xia, Y., Chen, J., & Chen, L. Y. (2019). Intraorbital Paragonimus infection. *Indian Journal of Ophthalmology, 67*(10), 1736. https://doi.org/10.4103/ijo.IJO_295_19

Xie, B. W., Yang, Y. S., Li, Q. Z., & Tang, L. Y. (1985). Rana boulengeri as the paratenic host of Pagumogonimus skrjabini. *Chinese Journal Parasitology and Parasitic Diseases, 3*, 157. (In Chinese.).

Yaguchi, D., Ichikawa, M., Kobayashi, D., Inoue, N., Shizu, M., & Imai, N. (2016). Two cases of paragonimiasis westermani diagnosed after eosinophilic pleural effusion-induced hydropneumothorax. *Respirology Case Reports, 4*(3) 10.1002/rcr2.v4.3. https://doi.org/10.1002/rcr2.154

Yoshida, A., Doanh, P. N., & Maruyama, H. (2019). Paragonimus and paragonimiasis in Asia: An update. *Acta Tropica, 199*, 105074.

Yue, X., Wei, X., Zhu, Y., Han, P., Li, X., Wei, Q., & Liu, L. (2014). Vesical Paragonimiasis diagnosed by histopathology: A case report. *Urologia Internationalis, 93*(3), 361–363. https://doi.org/10.1159/000350651

Yun, D. J. (1960). Paragonimiasis in children in Korea. *The Journal of Pediatrics, 56*(6), 736–751. https://doi.org/10.1016/S0022-3476(60)80310-1

Zhou, X. J., Yang, Q., Tan, Q. H., Zhang, L. Y., Shi, L. B., & Zou, J. X. (2021). Paragonimus and its hosts in China: An update. *Acta Tropica, 223*, 106094. https://doi.org/10.1016/j.actatropica.2021.106094

Trichinellosis

34

Tuhina Banerjee, Shruti Kaushal, and Souti Prasad Sarkhel

Abstract

Trichinellosis is a food-borne zoonotic disease in humans caused by nematode worms belonging to the genus *Trichinella*, which can infect over 100 species of mammals, birds, and poikilotherms. While pigs and rodents are key reservoirs particularly for the zoonotic *T. spiralis*, especially occurring in peridomestic settings, the parasite circulates in diverse ecological cycles. Recent research findings indicate the existence of 13 species and genotypes, categorised into capsulated and non-capsulated clades. Natural infections in animals are often asymptomatic or present with vague clinical signs, making early diagnosis difficult and contributing to their role as reservoirs. In humans, trichinellosis can manifest as mild gastrointestinal symptoms or severe systemic disease, including muscle pain, periorbital oedema, fever, and ocular or subungual haemorrhages. While fatalities are rare due to improved diagnostic and therapeutic options, severe complications such as myocarditis and encephalitis can occur. Therefore, to prevent human and animal infections, a One Health approach is crucial, involving a combination of strategies, including proper meat handling, promoting safe meat consumption, hygienic disposal of meat waste, accurate diagnosis, and effective treatment strategies. Whereas mass treatment of animals is often impractical, reducing the risk of human transmission by preventing carrion consumption by stray animals is vital through responsible disposal of animal carcasses, implementing public awareness campaigns, and stringent food safety regulations. In this context, this chapter presents a comprehensive overview of

T. Banerjee (✉) · S. Kaushal
Department of Microbiology, Institute of Medical Sciences, Banaras Hindu University, Varanasi, Uttar Pradesh, India
e-mail: drtuhina105@bhu.ac.in

S. P. Sarkhel
Department of Veterinary Parasitology, FVAS, Institute of Agricultural Sciences, Banaras Hindu University, South Campus, Mirzapur, Uttar Pradesh, India

R. Biswas et al. (eds.), *Concept of One Health and Zoonoses*,
https://doi.org/10.1007/978-981-95-6627-3_34

trichinellosis, discussing its history, taxonomy, epidemiology, clinical manifestations, diagnosis, treatment, and prevention strategies.

Keywords

Trichinella · Taxonomy · History · Phylogeny · Zoonosis · One Health · Vaccine

Introduction

Trichinella worms, the smallest known nematodes to infect humans, are responsible for a meat-borne zoonotic disease called trichinellosis, also known as trichinosis or trichiniasis (Brooks et al., 1948; Despommier, 1990; Diaz et al., 2020). While these worms can infect a wide range of mammals, swine and rats play a crucial role as reservoir animals, harbouring the parasite in areas surrounding human habitation. The transmission of *Trichinella* in nature mainly occurs through carnivory or the consumption of contaminated meat and meat products. The infection cycles independently within different ecological systems, such as sylvatic and synanthropic cycles. The transmission between these biotypes is largely influenced by increased human activities and their environmental impacts (Gottsteinet al., 2009; Omeragić et al., 2024). In most cases, individuals become infected by consuming undercooked meat from domestic and wild pigs, but in some regions, the meat of goats, horses, bears, and crocodiles can also be a potential source of infection (Rostami et al., 2017). The establishment of an infection in the host enables the exploitation by the parasite, with the host serving as both the definitive host for the adult parasites, and the intermediate host for their larval offsprings, resulting from the auto-heteroxenous life cycle of the parasite (Robles & Fernández, 2024).

The causative agent of trichinellosis once thought to be solely *Trichinella spiralis* (formerly *Trichina spiralis*) is now understood to be a diverse group of species that naturally infect pigs and other synanthropic animals in domestic cycles, as well as the feral and wild carnivore and omnivore mammals in sylvatic cycles (Omeragić et al., 2024). Cross-breeding experiments, alloenzyme analysis, PCR-RFLP, and advanced molecular phylogeny techniques have revealed the existence of 13 recognised species and genotypes of *Trichinella* (Dubey et al., 2024). These species are broadly categorised into two clades based on their ability to induce collagen capsule formation around infected host muscle cell fibres (Pozio & Zarlenga, 2013). Interestingly, these species show no significant morphological differences, even under a scanning electron microscope, highlighting the challenges of species delineation in this genus based on morphometric analyses (Pozio et al., 1992; Pozio & Zarlenga, 2013).

Trichinellosis caused by *T. spiralis* and *Trichinella pseudospiralis* is a widespread disease in terms of distribution, and is responsible for affecting approximately 10,000 people annually worldwide, with a relatively low mortality rate of about 0.2% (Pozio, 2007; Murrell & Pozio, 2011; Ram et al., 2018). While human infections are not directly linked to the presence of *Trichinella* spp. parasites in wild or domestic animals, the feed-scavenging behaviour of these animals, particularly free-range pigs, helps sustain the infection within the peridomestic cycle (Thomas et al., 2013; Pozio et al., 2021), and which significantly influences parasite transmission dynamics through the farm-to-fork route. This highlights the critical importance of understanding the global distribution of *Trichinella* spp., and the significant role of proper food preparation practices, particularly when handling free-range pig meat, which could potentially increase human exposure (Pozio et al., 2021; Pozio & Gomez Morales, 2023).

History

The discovery of the muscle-dwelling *Trichinella* parasite was accidental. In 1835, while dissecting a human corpse with tuberculosis, James Paget noticed calcified granules in the muscles that dulled his scalpel blades. Upon microscopic examination, he discovered tiny, encysted worms within these granules. He sought the assistance of Richard Owen, a distinguished anatomist, who officially described and named the parasite as *Trichina spiralis* that same year (Owen, 1835; Campbell., 1979). Further elucidation of *Trichinella* came in quick succession. In 1859, Rudolf Virchow, a prominent German physician and pathologist, characterised the parasite's life cycle and muscle cell encapsulation (Virchow, 1859). Shortly thereafter, in 1860, F. A. Zenker described the pathogenicity of these worms in humans, coining the term "trichiniasis" (later changed to the more appropriate "trichinosis"; TRICK-a-NO-sis) (Zenker, 1860; 1866; Campbell, 1983). Finally, in 1896, Alcide Railliet renamed the parasite *Trichinella spiralis* due to a naming conflict with a dipteran fly (Pozio, 2021), the hybotid dance fly *Trichina clavipes* Meigen, 1830. Subsequently, the disease became known as trichinellosis (/trik-ih-nuh-LOW-sis/), aligning with the standardised nomenclature prescribed by the World Association for the Advancement of Veterinary Parasitology (WAAVP) (Kassai, 2006).

Aetiology and Host

Trichinella parasites exhibit a wide geographical distribution, spanning all continents except Antarctica (Malone et al., 2024), and are responsible for causing trichinellosis, a potentially fatal disease in severe human cases. The genus *Trichinella* encompasses 13 taxa, of which 10 are recognised as valid species, while the remaining 3 are designated as genotypes (Zarlenga et al., 2020). These parasitic nematodes fall into two primary clades: encapsulated and non-encapsulated (Fig. 34.1). The encapsulated clade consists of species that create a collagen capsule around the infected muscle cell-larva complex in mammalian hosts. On the other hand, the non-encapsulated clade includes species that do not form a capsule within the host tissue of mammals, birds, and reptiles (Sharma et al., 2020; Marucci et al., 2021; Borhani et al., 2023).

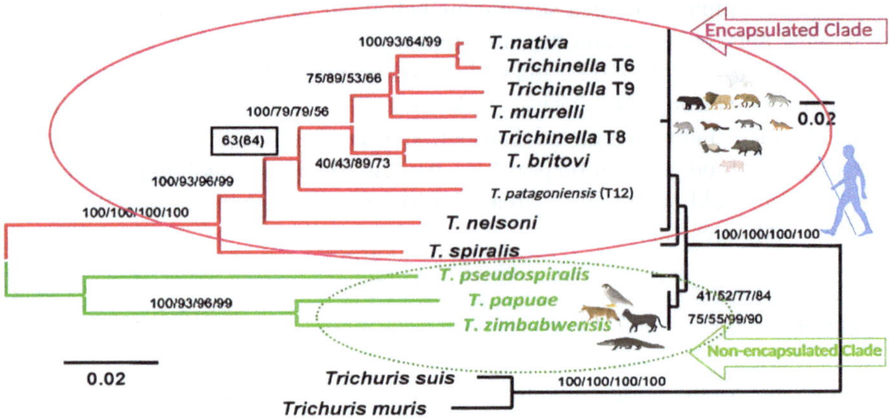

Fig. 34.1 Species of *Trichinella*. (Adapted from Pozio et al., Infect. Genet. Evol., 9: 606–616, 2009)

The encapsulated clade previously consisted of six species and three genotypes that exclusively parasitise mammals: *Trichinella spiralis* (T1), *Trichinella nativa* (T2), *Trichinella britovi* (T3), *Trichinella murrelli* (T5), *Trichinella nelsoni* (T7), *Trichinella* T6, *Trichinella* T8, *Trichinella* T9, and *Trichinella patagoniensis* (T12). A newly identified species, *Trichinella chanchalensis* (T13), was discovered in frozen wolverine tissue from Canada using a novel PCR-RFLP test targeting the worm's cytb gene. This species falls within the encapsulated clade and shares freeze-resistance properties with *T. nativa* and the *T.* T6 genotype. Encysted muscle larvae of *Trichinella nativa* and the *T.* T6 genotype have been found to remain viable in the frozen muscles of carnivores for up to 5 years, significantly longer than *T. britovi*, which can survive in frozen carnivore tissues for a maximum of 11 months and in frozen pig muscles for up to 3 weeks (Gottstein et al., 2009; Bilska-Zając et al., 2021).

The non-encapsulated clade comprises one species circulating among mammals and birds (*T. pseudospiralis*, T4) and two species infecting mammals and poikilothermic animals (reptiles): *Trichinella papuae* (T10) and *Trichinella zimbabwensis* (T11). However, all *Trichinella* spp. stages, except for the presence of capsules and potential size variations, exhibit minimal morphological differentiation across species and genotypes (Marucci et al., 2021; Bruschi & Dupouy-Camet, 2022).

Taxonomy

Nematodes of the genus *Trichinella* (/TRIK-in-el-ah/) belong to the family Trichinellidae within the order Trichocephalida (or Enoplida, depending on the classification system) under the class Enoplea (or Secernentea) of the phylum Nematoda in the kingdom Animalia (Pozio et al., 2009; ITRC, 2026).

Morphology

The adult worms (Fig. 34.2) are short-lived (Oksanen, 2022), and rarely observed in intestinal biopsies, except during patent infection, at which point they are typically embedded in the intestinal epithelium (Gardiner, 1976). Male worms measure about 1.5 mm, while females measure 3.5–4.0 mm long (Takahashi, 2021). The oesophagus occupies the anterior one-third of the adults. Male worms lack spicules and a spicular sheath, but these non-bursate nematodes have two small cloacal flaps/

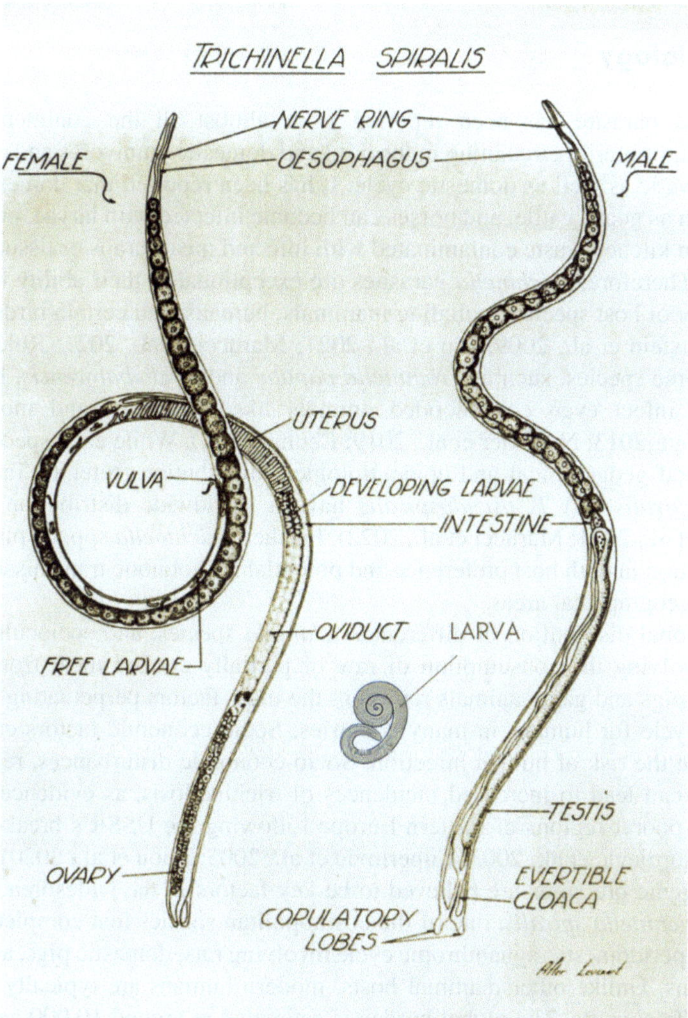

Fig. 34.2 Illustration of adult *Trichinella* spp. worms, depicting the key anatomical structures of the female (left) and male (right) specimens, along with the liberated larvae of *Trichinella* spp. (Image courtesy of CDC PHIL, https://phil.cdc.gov/Details.aspx?pid=25322)

claspers that can be mistaken as bursa. The vulva is located in the middle of the oesophageal region, with the uterus appearing fully packed with newborn larvae towards at their anterior end. In the case of muscle trichinella, the coiled larvae are about 800–1000 μm long, transforming the muscle cell into a translucent, lemon-shaped cyst measuring approximately 0.3–0.8 mm by 0.2–0.4 mm. The process of calcification of the nurse cell-larva complex (muscle cyst) typically begins approximately 6 months after the invasion and is generally completed within a year of the initial invasion (Gottstein et al., 2009; Foreyt, 2013; Taylor et al., 2015; Chaudhury, 2022).

Epidemiology

Trichinella parasite has been reported from almost all the continents except Antarctica, primarily circulating in the wild and domestic omnivores and carnivores in the sylvatic as well as domestic cycle. It has been reported that domestic herbivores such as goats, cattle, and horses can become infected with larvae, possibly by feeding on kitchen waste contaminated with infected meat scraps or tissues of food animals. Therefore, *Trichinella* parasites are exceptional in their ability to infect a wide array of host species, including mammals, humans, and certain birds and reptiles (Gottstein et al., 2009; Liu et al., 2021; Marucci et al., 2021; Roldan et al., 2023). Some species, such as *Trichinella papuae* and *T. zimbabwensis*, have been found to infect even cold-blooded animals like crocodiles and monitor lizards (Foreyt, 2013; Noeckler et al., 2019; Leung, 2024). While each species has its own limited geographical and epidemiological distribution (refer to Table 34.1), only *T. spiralis* and *T. pseudospiralis* have a worldwide distribution (Gómez-Morales et al., 2021; Marucci et al., 2022). Further, *Trichinella* spp. display significant variation in both host preference and potential for zoonotic transmission across different geographical areas.

The global distribution of different *Trichinella* species, and sociocultural food habits involving the consumption of raw or partially cooked meat from wild or domestic pigs and game animals represent the main factors perpetuating the transmission cycle for humans in many countries. Socio-economic factors can further exacerbate the risk of human infection. Socio-economic disturbances, revolutions, and wars can lead to increased incidences of trichinellosis, as evidenced by outbreaks in poorer regions of Eastern Europe following the USSR's break-up in the 1990s (Djordjevic et al., 2003; Cuperlovic et al., 2005; Zhou et al., 2020). Besides, anthropogenic pressures are believed to be key factors in the widespread distribution of *Trichinella spiralis*, one of the cosmopolitan species that completes its life cycle in a peridomestic-synanthropic cycle involving rats, domestic pigs, and human populations. Unlike other mammal hosts, modern humans are typically dead-end hosts for *Trichinella*. The global burden is estimated at around 10,000 cases annually as per the CDC estimate (CDC, 2024). Therefore, a One Health approach is crucial for effectively controlling *Trichinella outbreaks* (Pitt & Gunn, 2024). This

Table 34.1 Important *Trichinella* taxa and their geographical ranges

Clade	Species/ genotype	Geographical distribution	Major host species	Zoonotic potential	Freeze resistance
Encapsulated	*T. spiralis* (T1)	Cosmopolitan	Wild carnivores and omnivores, domestic animals	Yes	Yes, in horse muscles
	T. nelsoni (T7)	Eastern and Southern Africa	Wild carnivores and omnivores	Yes	No
	T. nativa (T2)	Holarctic: Arctic and Subarctic	Wild and marine carnivores	Yes	Yes, in carnivore muscles
	T. britovi (T3)	Palearctic: temperate areas, northern and western Africa	Wild carnivores, synanthropic mammals, humans	Yes	Yes, in carnivore and horse muscles
	T. chanchalensis (T13)	Canada	Carnivore animals	–	Yes
	T. murrelli (T5)	Nearctic (temperate areas): The USA and Southern Canada	Wild animals, horses, humans	Yes	No
	T. T6	Canada, Alaska, the USA	Wild carnivores	Yes	Yes, in carnivore muscle
	T. T8	South Africa, Namibia	Wild carnivores	–	No
	T. T9	Japan	Wild carnivores	–	No
	T. patagoniensis (T12)	Argentina	Cougars	–	Unknown
Non-encapsulated	*T. pseudospiralis* (T4)	Cosmopolitan	Mammals, birds	Yes	No
	T. papuae (T10)	Papua New Guinea, Thailand	Mammals, humans, crocodiles	Yes	No
	T. zimbabwensis (T11)	Zimbabwe, Mozambique, Ethiopia, South Africa	Nile Crocodiles, Monitor lizards	–	No

collaborative approach, involving public health, veterinary services, and the livestock sector, is essential for ensuring food safety and mitigating the risk of meat-borne zoonotic diseases like trichinellosis.

Life Cycle

Trichinella spiralis, the type species of the parasitic nematode genus *Trichinella*, passes its entire life cycle (Fig. 34.3) within a single mammalian host. The host typically serves as both the final and intermediate host in an auto-heteroxenous life cycle, during which all of its developmental stages take place within a single host. Once inside the host, active larvae are released from infected muscle cells as the result of digestion by the host's stomach juice and enzymes. These larvae then

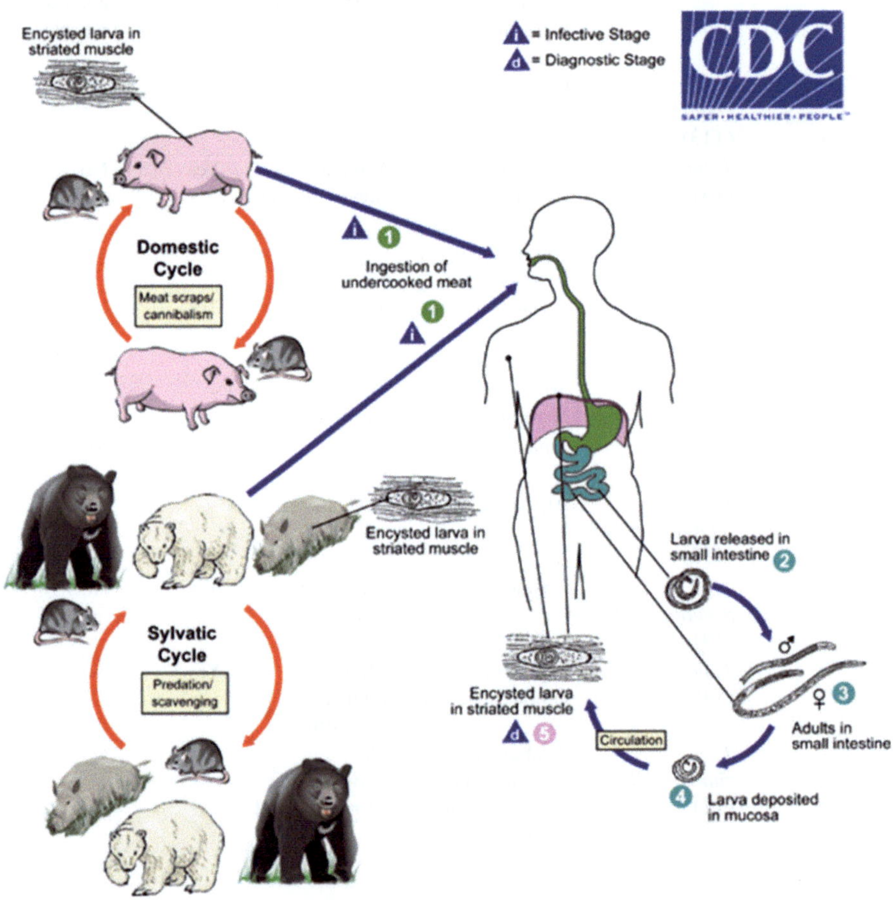

Fig. 34.3 The domestic and sylvatic life cycle of *Trichinella* spp. (Image courtesy of CDC, PHIL https://www.cdc.gov/dpdx/trichinellosis/index.html)

mature into adults within the intestinal mucosa and anchor themselves by penetrating and embedding within the intestinal columnar epithelium. The viviparous female releases newborn larvae (NBL, 110 μm in length) that then penetrate through the host's mucosal basal lamina, enter the systemic circulation, and infect the host's skeletal muscles, where they remain as intercellular muscle larvae (ML) (Dupouy-Camet & Bruschi, 2007; Gottstein et al., 2009; Ortiz et al., 2024; Lu et al., 2025).

Generally, infection occurs through the ingestion of raw or partially cooked meat of pigs and game animals containing the encysted *Trichinella* larvae **1** (muscle trichinellae). Gastric acid and pepsin in the host's stomach digest the encapsulated nurse cell-larva complex and liberate the encysted *Trichinella* larva **2**. The activated L1 larvae then penetrate into the columnar epithelium of the small intestine mucosa, where they develop into adult worms **3** within 36–40 h (Gagliardo et al., 2002; Yen & Chang, 2018; Zhou et al., 2020). These adult worms, measuring only 0.14–0.37 cm, are difficult to observe with the naked eye. They burrow deeper into the glandular crypts of the intestinal lining, specifically in the first third of the small intestine. Shortly after reaching maturity, adult worms engage in mating. Subsequently, the males die and are expelled from the host's body. Female *T. spiralis* are viviparous, and they begin to release larvae **4** as early as day 6 post-infection while remaining embedded in the intestinal mucosa. The prolific reproduction continues for 4–5 weeks, with each female releasing a staggering 1000–2000 larvae (Despommier & Hotez, 2012). Despite this high fecundity, adult females exhibit short-lived infections, and they are ultimately expelled by the host's immune defence mechanisms (Gottstein et al., 2009). The newborn L1 larvae, measuring around 0.08 mm by 7 μm, enter the bloodstream through veins and lymphatic vessels of the small intestine. They travel through the circulatory system and ultimately reach the host's striated muscles, which make up a significant portion of the host's body mass (30–40% in humans). Now, upon reaching their preferred sites, typically skeletal muscles, individual larvae penetrate muscle fibres using their stiletto apparatus to migrate into muscle cells. Furthermore, the excretory-secretory (ES) products are released from a specialised organ in the larvae called the stichosome, which secretes various tissue-lytic enzymes and tissue growth factors. These further aid in the penetration of muscle cells and initiate the process of tissue encapsulation **5**. After the invading the striated muscle cell, the L1 *Trichinella* larva establishes itself in an intracellular niche within individual myocytes, thereby becoming the largest intracellular parasite (Despommier, 1990). The excretory-secretory products of the parasite then induce a targeted transformation of the host cell, leading to microvascularisation and the formation of a dedicated circulatory rete to nourish the growing parasite (Capó et al., 1998; Gottstein et al., 2009; Sofronic-Milosavljevic et al., 2015). The transformed host cell acts as a "nurse cell", providing nutritional support for the larva's growth (Wu et al., 2013). Within this specialised environment, the larva grows to approximately 1 mm in length and adopts a coiled shape (Walden, 2021). Encapsulation of the infected cell complex begins 3–6 weeks after invasion, as a host-derived membranous structure forms a protective cyst around the nurse cell and larva. This encystment process is completed within approximately 3 months, allowing the encysted L1 larva to remain viable for extended periods within the infected host. However, *T. pseudospiralis, T. papuae*, and *T. zimbabwensis* do not encyst. In pigs, *Trichinella* muscle larvae can become infectious as early as 15 days post-infection (dpi) (Pozio, 2007b). It is important to note that these larvae are not uniformly distributed within the striated muscles. They exhibit a predilection for muscle sites with high activity and significant blood supply, such as the diaphragm, intercostal muscles, masseter muscles, and tongue. Larval concentrations in these areas are notably higher, typically 3–5 times greater than in other locations within the striated muscles (Nöckler & Kapel, 2007; Gondek et al., 2021). The ingestion of infected muscle tissues by susceptible host species maintains the cycle.

Zoonotic Importance

Trichinellosis, caused by *Trichinella* spp., retains significant zoonotic potential due to its broad host range and adaptability. While all mammals, including humans, are susceptible, species like *T. papuae* and *T. zimbabwensis* demonstrate remarkable adaptability by infecting even cold-blooded reptiles like crocodiles (Pozio et al., 2004; Roldan et al., 2023). Human infection classically occurs through the consumption of raw or partially cooked meat containing the encysted muscle larvae. Therefore, *Trichinella* infection in meat of food and game animals play important roles in precipitating the infection in the humans (Zhou et al., 2020; Malone et al., 2024).

Stomach acid triggers larval release from nurse cells within the ingested meat. Subsequently, these larvae migrate to the small intestine, invade the mucosal layer, undergo four moulting, and finally mature into adult worms. Mating pairs then produce offspring that enter the bloodstream, encyst within striated muscle tissue, and develop into infective muscle larvae, completing the life cycle (Lindquist & Cross, 2017). The clinical signs and symptoms mostly manifest with symptoms like food poisoning or typhoid (Capó & Despommier, 1996). However, it is during these times when the worms are most susceptible to the anthelmintics and can be eliminated effectively (Gottstein et al., 2009). Therefore, effective control strategies must encompass preventing livestock exposures, improving meat inspection protocols, and educating the public about the risks of consuming undercooked meat of pigs and other game animals (Zhou et al., 2020).

The dependence of the *Trichinella* worms on carnivory, coupled with the role of scavenger animals facilitating the widespread transmission, underscores the importance of ecological considerations. Even herbivores can act as accidental hosts, as evidenced by outbreaks. The recent European outbreaks linked to horse meat, wild boar meat consumption, and the historical association with pigs highlight the variable zoonotic potential across different *Trichinella* species. The Paris outbreak involving infected horse meat emphasises the need for continued vigilance, especially with evolving agricultural practices (Diaz et al., 2020; Pavel et al., 2023; Baptista et al., 2023). Transboundary tourism and homestays in the trichinella endemic regions may potentially lead to transmission in the international tourists who are unaware of risks associated with the consumption of local meat delicacies made with partially or raw meat of food or game animals (Zhou et al., 2020).

Pathogenesis

In trichinellosis, animal hosts can harbour chronic, heavy infections with minimal clinical signs. However, for humans, the minimum infective muscle larval dose of *T. spiralis* is 70–150 ingested larvae (EFSA, 2005), whereas ingestion of more than 500 encysted muscle larvae can cause serious fatalities in humans (Kassai, 1999). On the other hand, humans typically experience biphasic clinical outcomes following ingestion of infected meat (Wakelin & Goyal, 1996; Bruschi & Murrell, 2002). During the initial enteral phase, larvae are released from infected muscle tissues, where they develop into adults in the intestinal villi. Adults mate within 30–32 h of infection, and subsequently, the females thread her through a series of intestinal epithelial cells and start releasing thousands of infective new larvae from 6 days post-infection. This continues for a few weeks until the adults are expelled by the host's immune defences. The onset of symptoms following *Trichinella* infection can vary widely, ranging from as early as 12 h to as late as 2 days after consuming contaminated meat. The presence of the parasites in the gastrointestinal tract (GIT) and the excretory-secretory (ES) products released by the parasites are known to invoke a type 1 hypersensitivity reaction resulting in mastocytosis, eosinophilia, and parasite-specific IgE production (Bruschi & Murrell, 2002). The severity is often linked to the larval burden in the ingested meat. However, this initial phase frequently presents vague symptoms that mimic other illnesses like food poisoning, typhoid, etc., leading to misdiagnosis in many cases. The presence of fully developed worms embedded within the intestinal lining typically triggers inflammatory responses, potentially leading to gastroenteritis. This condition is characterised by symptoms such as nausea, vomiting, abdominal pain, secretory diarrhoea, and prominent changes in the intestinal structure, including villous atrophy, crypt hyperplasia, and goblet cell hyperplasia. Diarrhoea lasts longer than vomiting and may last up to 3 months, and when excessive, it causes dehydration due to leakage of ions, and the symptoms are similar to *Vibrio cholerae* infection. After 2 weeks or more, the symptoms resolve as the adult worms are expelled from the intestinal tract by the host's immune defence mechanisms, which is mainly attributed to the mucosal mast cell protease 1 (Knight et al., 2000; Bruschi & Murrell, 2002). Furthermore, cytokines are known to assist in the expulsion of larvae and induce hypercontractility of muscles. The newly released larvae traverse the basal lamina of the intestinal mucosa and then pass through the hepatoportal system, travel to the heart and lungs via the liver, and spread throughout the body via the arterial circulation (Capó and Despommier, 1996; Clausen et al., 1996; Dupouy-Camet and Bruschi, 2007; Gottstein et al., 2009; Sharma et al., 2014; Diaz et al., 2020; Mitic et al., 2026).

During the parenteral phase, approximately 1 week after infection, migrating newborn larvae can cause allergic and inflammatory responses as they seek to invade the skeletal muscle tissues (Wakelin and Goyal, 1996). This phase is often accompanied by fever, fever-associated chills, muscle tenderness, and myalgia. The movement of NBLs through blood vessels can cause extensive damage, leading to conditions like oedema, bilateral periorbital oedema with conjunctivitis, and subungual haemorrhages (splinter haemorrhages). Local inflammatory reactions cause the release of histamine, serotonin, bradykinin, and prostaglandins like PGE2, PGD2, and PGJ2, inducing capillary permeability, causing extravasation of fluids, electrolytes, albumins, and cellular components into surrounding tissues, leading to localised oedema. This process can also contribute to vasculitis and microthrombus formation during the acute phase of the infection. Symmetrical periorbital oedema is most peculiar in trichinellosis, ranging from 17% to 100% of patients, probably arising due to an allergic response. Skin rashes and oedemas also arise due to IgE-dependent allergic manifestations (Bruschi & Murrell, 2002; Kocęcka et al., 2003; Pozio, 2018; Pozio and Gomez Morales, 2023; Furhad & Bokhari, 2019).

The newborn larvae (NBLs) of *Trichinella* do not exhibit tissue tropism during their migratory phase. As they exit the capillaries, they have the ability to infiltrate various organs and infect different types of cells. The invasion of host cells by migrating NBLs is largely random. However, if they infiltrate cells other than skeletal muscles, they are unable to transform and induce the host cell to form a nurse cell-larva complex (Bruschi & Murrell, 2002). The parasite's inability to establish itself within other cell types triggers localised inflammatory responses, ultimately leading to the characteristic signs and symptoms associated with trichinellosis. This can result in the death of the invading larvae, or the larvae may re-enter the circulation and spread to other parts of the body. Hence, invasion of non-skeletal muscle cells in vital organs and tissues, like lungs, pleura, brain, meninges, eyes, and kidneys, can result in the manifestation of diverse pathologies like pneumonia, pleuritis, encephalitis, meningitis, brain or eye damage, subconjunctival and sublingual haemorrhages, nephritis, etc. The potential for congenital transmission of *Trichinella* through the placenta has been established in experimental studies. Evidence supporting this includes successful demonstrations in mice and guinea pigs, as well as post-mortem examinations of aborted human foetuses. In mice, vertical transmission appears to be more likely if the infection occurs during pregnancy or conception takes place within a month of infection. Several research reports indicate that larvae are capable of penetrating and developing within the foetus, but they are not typically found encapsulated within the muscle tissues during histological examination, although acid digestion methods confirmed the presence of larvae (Despommier, 1990; Dubinský et al., 2001; Bruschi et al., 2002; Boulos et al., 2005; Cui et al., 2006; Pozio, 2007b; Bruschi and Chiumiento, 2012; Roberts et al., 2013; Riva et al., 2017; ICT, 2026; Mitic et al., 2026).

Skeletal muscle tissue presents a unique niche for *Trichinella* infection. Unlike other cell types, skeletal muscle cells are permissive to NBL invasion and subsequent transformation into encapsulated muscle larvae, a hallmark of successful parasite establishment. The presence of muscle larvae from *Trichinella* infection causes various changes in the infected muscle cells. These changes include making the sarcoplasm of the cell basophilic, the disappearance of sarcomere myofibrils, and displacement of the nucleus, which increases the infected cell's size and number of nucleoli. Additionally, the infected cells become more permeable, allowing elevated levels of muscle enzymes such as creatinine phosphokinase (CPK), lactate dehydrogenase, aldolases, aminotransferases, and 1,6-diphosphofructoaldolase to circulate in the body, serving as markers for trichinella-associated myopathies. The invasion of muscle cells by larvae also triggers the infiltration of inflammatory cells, primarily eosinophils, which can cause damage to the muscles and lead to elevated serum muscle enzymes. While eosinophils exert cytotoxic effects against newborn larvae (NBLs) through antibody-dependent cell cytotoxicity (ADCC) mediated by the release of major basic protein, eosinophil cationic protein, eosinophil peroxidase, and reactive oxygen species, this process can also cause collateral damage to host muscle cells (Lee, 1991; Bruschi & Murrell, 2002; Bruschi et al., 2002; Dupouy-Camet and Bruschi, 2007; Pozio, 2007b; Sharma et al., 2014; Motran et al., 2018; Ilic et al., 2022; Lombardi et al., 2022; ICT, 2026; Mitic et al., 2026).

Immunopathology

Invading newborn larvae, armed with a specialised secretory organ called the "stichosome", secrete various factors that regulate interactions between the host and the parasite. These factors, potentially including "parakines", likely aid in penetrating muscle cells and initiating the encapsulation process, a specialised nurse cell structure critical for the parasite's long-term survival. The stichosome, located on the dorsal side of the oesophagus and occupying nearly half the mature larva's body cavity, comprises of cuboidal cells, the "stichocytes". During infection, their number increases significantly, from approximately 20 in a day-old larva to 45–55 in a mature one. These glandular cells release excretory-secretory products that not only aid in muscle cell penetration and encapsulation but also manipulate host cell machinery. This manipulation includes altering gene expression, leading to loss of contractility, induction of apoptosis initially, and later suppression of apoptosis. Additionally, infected muscle cells undergo DNA synthesis and arrest at the G_2/M phase, while normal muscle nuclei remain arrested in the G_0/G_1 phase. Furthermore, stichosome secretions like vascular endothelial growth factor (VEGF) induce micro-vascularisation and formation of a specialised circulatory network ("circulatory rete") around the infected cell, nourishing the growing parasite. This manipulation also includes upregulating genes in the insulin signalling pathway, leading to increased glucose uptake and glycogen accumulation within infected cells and causing transient hypoglycaemia. Ultimately, these multifaceted effects promote the establishment of the nurse cell, ensuring the parasite's survival within the host for a prolonged period. The larvae have the ability to persist within muscle cells for as long as 40 years until the host initiates a process that results in the gradual calcification of the nurse cell-larva complex and, eventually, the death of the arrested larvae. Calcification starts as early as 5 months after infection and is usually completed by 18 months (Gould, 1970). Typically, the calcification starts in the collagen capsule, then in the nurse cell, and later around the larva. Although the whole process may kill most of the larvae in nurse cells, some of the larvae may survive the encapsulation process and remain infective for years (Zimmermann et al., 1973; Despommier & Müller, 1976; Fröscher et al., 1988; Jasmer and Neary, 1994; Bruschi & Murrell, 2002; Wu et al., 2008; Roberts et al., 2013; Wang et al., 2022; Mitic et al., 2026).

Trichinella infections typically induce elevated IgE antibody production, peripheral eosinophilia, and mucosal mastocytosis (Wakelin and Goyal, 1996; Bruschi et al., 2008; Bruschi and Chiumiento, 2012; EL-Dardiry et al., 2021). The severity of *Trichinella* infection in humans depends on many factors, such as the number of infective larvae consumed, the specific *Trichinella* species involved, and host factors such as age, sex, ethnicity, host's immune response (Kociecka, 2000; Bruschi & Murrell, 2002). *Trichinella spiralis* is the most commonly implicated species in outbreaks worldwide due to its high pathogenicity and reproductive fecundity. Infection of *T. spiralis* can result in an increased number of newborn larvae entering the muscle tissues, triggering substantial inflammatory responses. The occurrence of fatal outcomes is often attributed to complications related to the respiratory, cardiovascular, or neurological systems (Rosca et al., 2021; Tanabe et al., 2021; Lupşe et al., 2023). Infection with the non-encapsulated species *T. pseudospiralis* produces less inflammatory reactions and pathologies in both the enteral and parenteral phases of the infection, primarily attributable to the fact that this species can elevate host plasma corticosterone (Stewart, 1989; Stewart et al., 1988; Li and Ko, 2001); therefore, they appear less virulent and pathogenic than *T. spiralis*.

Moreover, during the course of an active *Trichinella* infection, each life cycle stage can elicit a stage-specific protective immune response in the host. This specificity arises from the unique nature of cuticular and excretory-secretory antigens (ES Ags) expressed during each life cycle stage. The newborn larvae can elicit stronger and more potent immune responses than the encapsulating larvae. Despite the host's immune response, including CD4+ T-helper (Th) cells that eliminate some of the migrating larvae through antibody-dependent cellular cytotoxicity (ADCC), most of the wandering *Trichinella* larvae establish chronic infection in skeletal muscle (Bruschi & Murrell, 2002; Bruschi & Chiumiento, 2012; Bien et al., 2013; Bruschi et al., 2022).

Clinical Signs in Animals

Clinical signs in animals are variable and depend upon the host species, species of the parasite, and the level of the infection in the animals. Natural infections in pigs and other domestic/wild animals often remain asymptomatic, although severe infections can lead to mortality. Notably, *Trichinella* infections in wildlife and domestic animals frequently go undiagnosed, allowing these animals to act as reservoirs for the parasite. Young pigs are generally more susceptible than adults and may display vague clinical signs like inappetence, malaise, and diarrhoea. Across various animal hosts, infected individuals may exhibit a range of non-specific signs, like diarrhoea, fever, muscle pain, dyspnoea, and peripheral eosinophilia. In endemic regions, these clinical presentations necessitate differential diagnosis through sensitive tests like acid digestion muscle biopsy and ELISA to distinguish trichinellosis from other potential causes (Bai et al., 2022; Borhani et al., 2023; WOAH, 2023).

Clinical Importance in Humans

The clinical pictures in humans can be similar to food poisoning, typhoid, influenza, myositis in HIV infection, polyarthritis nodosa, and eosinophilic leukaemia. Patients may experience nausea, headache, fever, fever-associated chills, and occasional gastrointestinal disorders during the early stage of infection (Table 34.2). The fever may persist for 1–3 weeks depending upon the infection load and successful treatment. In humans, preferred sites of the parasites are the extraocular, masseters, tongue, larynx, diaphragm, neck, and intercostal muscle, which are mostly infected. Principal clinical features include bilateral periorbital or facial oedema, myalgia, conjunctivitis, urticarial and maculopapular exanthema, and subungual, subconjunctival, and retinal haemorrhages (Fig. 34.4). Myocarditis, pericardial pain, tachycardia, and electrocardiogram abnormalities may arise due to the transitory passage of the larvae in the cardiac muscles, although they fail to become mature or get encapsulated within the heart muscle cells (Frayha, 1981; Andrews et al., 1994;

Table 34.2 Clinical manifestations of trichinellosis caused by *T. spiralis* in humans (symptoms vary depending on the infection load, virulence of the species, and immunocompetence of the hosts)

Common clinical features	Onset of symptoms	Life cycle events of *Trichinella spiralis*
Gastrointestinal symptoms	2–24 h	Excystment of infective larvae in the gut, invasion of the intestinal mucosa
Facial oedema/fever	Days 6–7	Larviposition by gravid females, invasion of NBLs
High fever, myalgia, pain	Days 10–11	Peak muscle invasion by NBLs
Peripheral eosinophilia, circulating antibody	Day 14	Larval release decreases
Peak eosinophilia	Days 17–20	Development of NBLs
Myocarditis, neurological symptoms	Day 21	Encapsulation of nurse cell-larva complex
Respiratory symptoms	Days 23–26	Adult worms expelled from the gut
Subsiding of fever	Months 1–2	Encapsulation almost complete
Fatalities due to myocarditis/encephalitis	Month 3	
Slow convalescence, subsiding of myocarditis and neurological symptoms	Months 6–8	Calcification begins in collagen fibres around the nurse cell-larva complex
–	Year 1	Calcification completed
–	Year 6	Most larvae remain viable within the calcified cyst

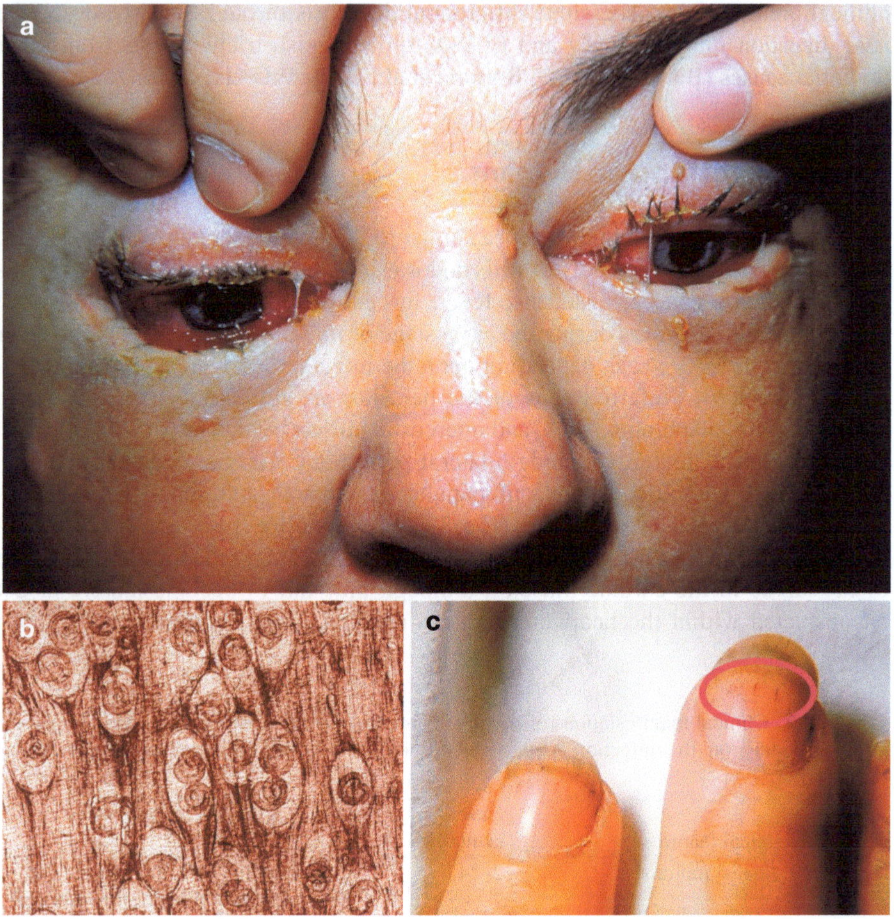

Fig. 34.4 Pathologies associated with *Trichinella* spp. (**a**) A patient exhibiting periorbital swelling and inflammation of the eyes (Image courtesy of CDC, PHIL https://phil.cdc.gov/Details.aspx?pid=342); (**b**) a muscle tissue specimen demonstrating the presence of encysted larvae of *Trichinella spiralis* (Image courtesy of CDC PHIL, https://phil.cdc.gov/Details.aspx?pid=10142); (**c**) subungual splinter-like haemorrhages in the nail beds which resolve rapidly following anthelminthic treatment. (Adapted from Diaz et al., *Wilderness Environ. Med.* 2020, 31(2):235–244; & CDC, PHIL)

Somer & Finegold, 1995; Chatterjee et al., 2000; Bruschi & Murrell, 2002; Dupouy-Camet & Bruschi, 2007; Dubey et al., 2011; Roberts et al., 2013; Nkoke et al., 2022; Tanariyakul et al., 2024).

Trichinellosis in humans often results in a moderate leucocytosis (12,000–18,000 cells/mm^3) and an increased presence of eosinophils (5–50%, 1400–8700 eosinophils/mm^3) (Capó & Despommier, 1996). When the respiratory system is affected, the parasite can invade the diaphragm and intercostal muscles, leading to breathing difficulties (dyspnea). Neuro-trichinellosis is particularly worrisome as it involves the invasion of the central nervous system (CNS). In serious cases (up to 25% of heavy infections), this can lead to meningitis and encephalitis, characterised by oedema, infarcts, and haemorrhage. Headache is very common in trichinellosis, and frequent movement of the head may cause exacerbation of the pain (Bruschi et al., 2002; Pozio et al., 2003; Bruschi et al., 2013; Gottstein et al., 2009; Diaz et al., 2020; Tanabe et al., 2021).

Trichinellosis has been reported on an irregular basis, with several small outbreaks from Uttarakhand remote regions recently documented, according to a literature search on disease prevalence (India). However, trichinellosis has now emerged as a new threat in isolated pockets of India (Sethi et al., 2010, 2013; Ram et al., 2018), and the reason that people acquire this parasitic infection may be due to social and cultural practices, lack of knowledge about hygiene in slaughterhouses, improper meat inspection, and consumption of contaminated meat (Djordjevic et al., 2003). Death is rare because of improved diagnosis and availability of treatment; however, fatalities may happen from congestive heart failure due to myocarditis, encephalitis, pneumonitis, hypokalaemia, or adrenal gland disturbances (Bruschi & Murrell, 2002; Bruschi et al., 2002; Sharma et al., 2014; Ram et al., 2018; Mohib et al., 2022; Sethi et al., 2021; Furhad & Bokhari, 2019).

Prevention and Control in Animals and Humans

Trichinellosis poses a continuing public health threat in some regions. Traditionally, thorough cooking of pork (58.5°C for 10 min) has been the cornerstone of control (Despommier et al., 1995). However, existing research findings underscore the limitations of freezing as a complementary strategy. Freezing pork chops less than 15 cm (6 in) thick at −15°C for 20 days or −23°C for 10 days and −30°C for 6 days effectively eliminates *T. spiralis*, provided the fact that the meat is less than 15 cm thick (Bruschi & Murrell, 2002). Meat having a thickness up to 50 cm in thickness should carefully be hard frozen for not less than 4 weeks. However, deep freezing may not be universally applicable as freeze-resistant species from the Arctic region, like *Trichinella* T6, *T. nativa*, *T. britovi*, and *T. chanchalensis*, may remain viable (Hill et al., 2009; Lacour et al., 2013; Pozio, 2016; Franssen et al., 2017; USDA, 2018; Gamble, 2021; Oksanen et al., 2022; Furhad & Bokhari, 2019; Bessi et al., 2024). Further, due presence of antifreeze protein molecules in wild carnivores like bears, racoons, and opossums, which prevents ice crystal formations, not only protects the animal tissues but also protects the worms and is able to protect these encapsulated worms in the carcasses also. Therefore, consumers should cook fresh ground pork or wild game to an internal temperature of 160°C (71°C) for at least 1 min which will change the colour of the meat from red to grey (Capó and Despommier, 1996; Gamble et al., 2000; McIntyre et al., 2007; Foreyt, 2013; Wilson et al., 2015; Duman & Newton, 2020; FAO, 2021; Ortiz et al., 2024).

Microwave cooking is ineffective in killing all the larvae, as cold spots may remain in large pieces of meat due to the characteristic pattern of microwave radiation. Canned meat can be rendered safe for human consumption by subjecting it to irradiation at 0.3 kGy. Additionally, proper hygiene practices for meat workers are crucial, including proper cleaning of the meat grinders after use and frequent handwashing with warm water and soap after handling raw meat. Notably, alternative processing methods like curing, drying, and smoking are ineffective against *Trichinella* larvae (Zimmermann & Beach, 1982; Capó & Despommier, 1996; Franssen et al., 2019; Noeckler et al., 2019; Ercole et al., 2020).

Mass treatment of animals is often impractical. The best strategy to avoid illness in humans and animals is to ensure that no viable *Trichinella* cysts are ingested through muscles. Public awareness campaigns that educate consumers to purchase and consume only properly inspected and controlled meat products are critical for controlling human trichinellosis. Improper disposal of domestic animal carcasses near wildlife parks or forests can result in the transmission of parasites to wild animal populations. This happens when wild animals scavenge the carcasses, enabling the parasites to enter the sylvatic cycle and potentially infect wildlife. This poses a significant risk to humans who participate in game hunting in these environments. Proper disposal of domestic animal carcasses is crucial to mitigate the spread of parasites and safeguard the well-being of both wildlife and human communities. Additionally, due to the persistence of *Trichinella* spp. in a range of animal species that can readily infect people, the prevalence of infection in wild omnivores and carnivores in the nation is extremely concerning. Therefore, controlling *Trichinella* spp. outbreaks in humans and animals require robust surveillance, biosecurity, and hygiene measures in traditional food markets, supported by a multifaceted approach encompassing rodent control, rigorous meat inspection practices, proper oversight of animal feed, and adherence to appropriate freezing or heating protocols for pork intended for human consumption (Thompson et al., 2009; Kołodziej-Sobocińska, 2019; FAO, 2021; Gamble, 2022; Sgroi et al., 2022; Dubey et al., 2024; Guzmán-Faúndez et al., 2025; Sparaciari et al., 2025; Torreggiani et al., 2025).

Vaccines

Recent advancements in *Trichinella* vaccine development involve the identification of immunogenic antigens. This process utilises a multipronged approach, including genomics, proteomics, and transcriptomics for antigen discovery (Fig. 34.5). Additionally, immunoinformatics and in silico analyses contribute to the identification and prediction of these immunogenic targets. Potential vaccination strategies for *Trichinella* spiralis infection include inactivated vaccines, recombinant proteins, subunit vaccines, epitope vaccines, viral or bacterial vectors, phage DNA vaccines, phage display vaccines, and DNA vaccines (Table 34.3). Newer vaccine delivery techniques such as nanoparticles, virus-like particles, and plant-based vaccines are showing promise in effectively delivering vaccine constructs to target cells to elicit an optimal and protective immune response against helminths (Andrade-Becerra et al., 2017; Pati et al., 2018; Perera and Ndao, 2021; Tang et al., 2022; Mohammad Hasani et al., 2023; Stachyra & Bień-Kalinowska, 2025).

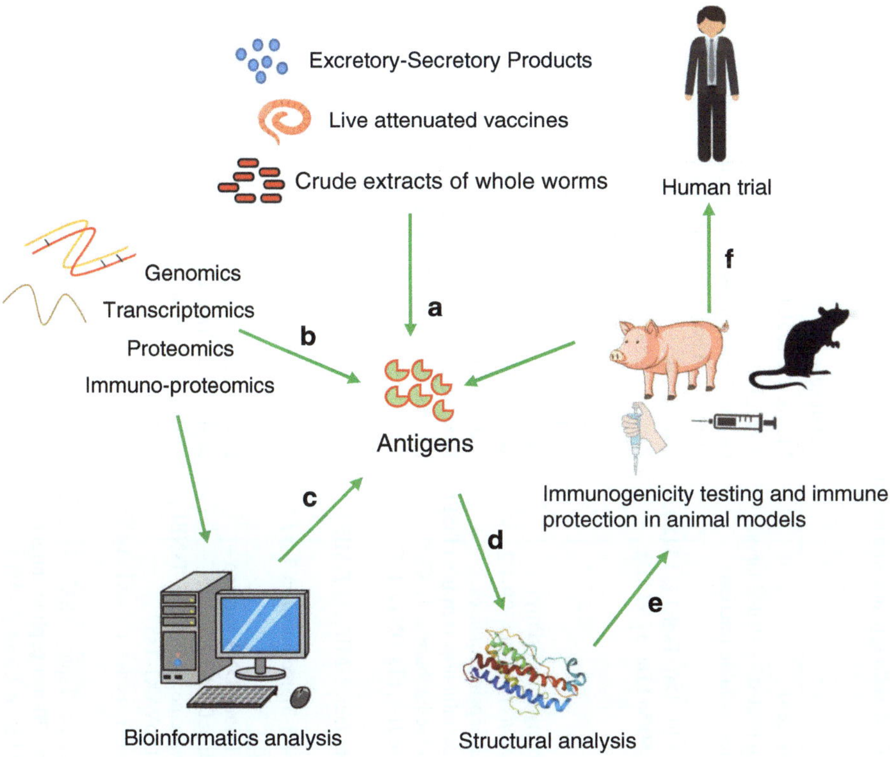

Fig. 34.5 Identification and detection of vaccine candidates in animal models through various strategies. (**a**) Vaccines against *T. spiralis* can be developed from excretory-secretory products, live attenuated strains, and crude extracts of whole worms. (**b**) Various approaches, including genomics, transcriptomics, proteomics, and immune proteomics, can be used to screen suitable vaccine candidates. (**c**) Appropriate vaccine antigens can be analysed through bioinformatics tools. (**d**) Structural modifications of antigens can be conducted to enhance the effectiveness of future vaccine candidates. (**e**) The immunogenicity and protective capabilities of vaccine antigens can be assessed both in vitro and in animal models. (**f**) Highly precise and effective vaccines can then be evaluated in human trials. (Adapted from Tang et al., Front. Microbiol. 13: 857786, 2022)

Recent studies have identified several promising antigens derived from various life stages of the parasite, as different immunogenic antigens are expressed during different life cycle stages. These antigens have demonstrated the potential to elicit strong immune responses and offer protection against infection. However, candidate vaccine immunogens against *T. spiralis* mainly focus on excretory-secretory (ES) antigens with mixed or single molecular weights, as well as other recombinant functional proteins like proteases. The excretory-secretory (ES) products of these parasites, primarily secreted by their stichocyte cells, play a crucial role in modulating the host's immune responses. Other immunogenic antigens which play roles in the *T. spiralis* intracellular processes have also been shown to elicit effective protective immune responses. The functional proteins released by parasites effectively enhance

Table 34.3 Strategies for developing potential vaccine candidates against *Trichinella* spp. utilising traditional and emerging technologies

	Vaccine approaches	Type of antigen	References
Traditional methods	1. Killed vaccine	a. Whole worm (newborn larval antigen)	Marinculic et al. (1991)
	2. Live attenuated/ inactivated/irradiated vaccine (UV/microwave)	a. Wole worm and ES antigens: attenuated/irradiated larvae, sonicated whole NBL, with soluble/insoluble fractions	Agyei-Frempong and Catty (1983), Cabrera and Gould (1964), Hafez et al. (2020), Marti et al. (1987), and Nakayama (1998)
	3. Natural antigen vaccine	a. Crude extract: HSP70, Ts-ES-1 (20 kDa), TspE1 (31 kDa), 43 kDa, 53 kDa, 48 kDa, 49 kDa, 50/55 kDa, 37.97 kDa protein (Ts87) b. ES antigens: TsES c. Surface antigens d. Others: aquaporins (AQPs), paramyosin (TsPmy)	Gamble (1985), Bi et al. (2015), Lightowlers and Rickard (1988), Nagano et al. (2008), Robinson et al. (1995), Gu et al. (2008), Wang et al. (2006), McGuire et al. (2002), Liu et al. (2014), Cui et al. (2015a, 2015b), Epstein et al. (1985), Wei et al. (2011), and Gamble et al. (1988)
Modern methods	4. Recombinant protein vaccines	a. Proteases: serine proteases (rTsSP/CTAB, rTspSP-1.2, rTspSP-1.3), cysteine proteases, cathepsin proteases b. Protease inhibitors: serine protease inhibitors (serpin): rTsSPI, rTs-serpin; cystatins (cysteine protease inhibitors): Ts-CLP c. Deoxyribonuclease-II: rTsDNase II-1, rTs-DNase II-7 d. Nudix hydrolase: Tsp10/ TsNd e. Macrophage migration inhibitory factor (MIF): (r *Ts*MIF) f. Proteasomes (rTspst) g. Others: 14-3-3 proteins (rTsp 14-3-3), recombinant Ts87 h. Heat shock proteins (rTspHSP70)	Wang et al. (2013), Feng et al. (2013), Li et al. (2013), Xu et al. (2017a), Liu et al. (2015a), Long et al. (2014), Qu et al. (2015), Yang et al. (2015), Yang et al. (2016), Fang et al. (2014), Wang et al. (2009), Song et al. (2018), Qi et al. (2018), Yue et al. (2020), and Sun et al. (2019)
	5. DNA vaccines	Parasite DNA, vectored delivery: TsPmy DNA vaccine + Montanide ISA; pcDNA3.1-TsNd; pcDNA3.1-TsE1; pcDNA3.1 (+)-Ts-NBL sp DNA vaccine; pcDNA3.1-TsNBL; T7-Tsp10; pVAX1-*Tsmif-Tsmcd*-1; pVAX1-Ts87 Phage T7-Tsp10; pVAX1-*Ub-Tsmif-Tsmcd*-1; pVAX1-Ag30 (43 kDa Ag); pcDNA3.1-TsE1	Liu et al. (2015b), Ren et al. (2011, 2013), Cui et al. (2013), Yang et al. (2008), Gu et al. (2014), Wang et al. (2017), and Xu et al. (2017b)
	6. Synthetic peptide vaccines	Peptide: multi-epitope peptide vaccine (Tspmy, TS87); 40-mer synthetic peptide (Ts glycoprotein); 30-mer peptide antigen (30-mer peptide derived from gp43 from *T. spiralis* ML)	Gu et al. (2020), Robinson et al. (1995), McGuire et al. (2002), and Gu et al. (2013)

their capacity to survive and overcome the challenges presented by the host's immune defences (Andrade-Becerra et al., 2017; Perera and Ndao, 2021; Tang et al., 2022; Stachyra and Bień-Kalinowska, 2025).

The efficacy of oral and intranasal vaccine delivery in stimulating mucosal IgA responses against mucosal pathogens, as well as in preventing *Trichinella* infections in the intestine, is well documented. Recent studies have demonstrated that DNA vaccines delivered via attenuated *Salmonella typhimurium* vectors can produce robust and long-lasting mucosal and systemic immunity. This approach efficiently eliminates intestinal *Trichinella* infections by invoking a localised immunity. For example, administering the *S. typhimurium* vectored Ts87 DNA vaccine orally substantially elevated intestinal IgA levels and conferred protective immunity against infective larvae. Furthermore, mucosal vaccines can provide comprehensive protection against trichinellosis by eliciting local immune responses at mucosal sites, preventing the parasite's establishment and maturation in the intestine, and blocking further invasion by newborn larvae through mucosal barriers (Yang et al., 2010; Wang et al., 2016; Qi et al., 2018b; Xu et al., 2025).

When animals are infected with helminths, their immune response primarily involves a type 2 (Th2) response, characterised by increased levels of specific cytokines such as IL-4, IL-5, IL-9, IL-10, and IL-13, which activates eosinophil mast cells and also elevate the levels of IgE, neutrophils, and natural killer cells. While the Th2 response can provide some protection against adult worms, Th1 and innate immunity mechanisms are crucial for protecting against the larval stages at cutaneous or mucosal surfaces. Therefore, to effectively combat helminth infections, it is important to develop immunogenic antigens that elicit a combination of Th1 and Th2 or innate immune responses. Additionally, the immunogenicity of potential vaccine candidates can be augmented through the use of adjuvants like Freund's adjuvant, cholera toxin subunit B (CTB), aluminium hydroxide, AS04, MF59, AS01B, Montanide® ISA70, and CpG oligonucleotides, among others. These adjuvants can augment or skew the immunogenicity of the antigen constructs to enhance host protective effects (Mulcahy et al., 2005; Anthony et al., 2007; Allen and Sutherland, 2014; Andrade-Becerra et al., 2017; Henry et al., 2017; Motran et al., 2018; Guerrero Manriquez and Tuero, 2021; Tang et al., 2022; Vacca and Le Gros, 2022; Amare et al., 2025; Moawad et al., 2025).

Food-borne vaccines against trichinellosis are considered essential for improving public health, controlling parasite infection, and encouraging socio-economic development. Two thorough analyses conducted by Zang et al. (2018) and Tang et al. (2022) discussed in detail the recent advancements in the development of potential vaccine candidates against *T. spiralis* in murine and swine models. They further discussed the factors influencing the effectiveness of promising immunoprotective antigens in reducing worm burdens in experimental animals. Combination vaccination in murine models has shown promise in improving the effectiveness of *T. spiralis* vaccines. However, the use of cocktail DNA vaccines is gaining attention due to their cost-effectiveness, long-lasting immunity, evoking Th1 immune response, and stability (Kozak and Hu, 2024). An ideal *Trichinella* vaccine should elicit protective immune responses capable of expelling adult worms, inhibiting the production of newborn larvae, and eliminating migrating larvae before they invade muscle cells, thereby preventing muscle pathologies and making vaccinated animals unsuitable hosts for the parasite, which effectively disrupts transmission in the synanthropic cycle.

Diagnosis

The definitive diagnosis of active trichinellosis relies on finding the nurse cell-L1 larva complex in a muscle biopsy through microscopy (Bruschi & Murrell, 2002), or confirming the presence of *Trichinella*-specific DNA using PCR. Traditional trichinoscopy method is still being used in most of the slaughterhouses and meat processing plants, during the veterinary inspection of the carcasses. In trichinoscopy, a small piece of muscle (0.2–0.5 g of muscle tissue without fat or skin, ~2–4 mm^3) is compressed between two slides and viewed under a microscope at low power (10× objective). This method is used to detect the presence of larvae within the muscle cells, indicating infection (Capó & Despommier, 1996; Dupouy-Camet & Bruschi, 2007). However, it should be noted that the method may yield false-negative results unless performed 2–3 weeks after the onset of illness.

Muscle biopsy may yield negative results, even in heavy infections, due to sampling errors, so it is important to collect biopsy materials from symptomatic muscles or preferred predilection sites. Though the method is very fast, it lacks sensitivity compared to the other available tests and fails to detect the encysted larvae in mild infection. Therefore, artificial muscle digestion using 1% HCL-pepsin followed by microscopic examination enhances the sensitivity of conventional microscopy-based methods and is therefore more preferred. Though trichinoscopy-based muscle biopsy can detect infection when there are 3–5 larvae per gram of muscles (Forbes & Gajadhar, 1999; van Knapen, 2000; Bruschi & Murrell, 2002; Bruschi et al., 2002), trichinoscopy may be useful when there are 15 or more larvae is there in per gram of meat (Vignau et al., 1997).

The sensitivity of the acid pepsin digestion is primarily attributed to the amount of sample that is almost 50–100 times larger than that used in performing the routine trichinoscopy test. Therefore, the artificial digestion method is the method of choice for routine slaughterhouse inspection as recommended by the International Commission on Trichinellosis and WAHO. However, the artificial digestion test protocol requires the digestion process to be continued for approximately 4–5 h, which causes the release of larvae from the encapsulated muscle cyst. However, this long digestion time in this protocol can cause mortality of the encysted larvae because of the lack of a complete cuticle prior to the tissue encapsulation. During the sieving process, dead larvae may become trapped, leading to non-recovery that can result in false negatives or an underestimation of larval presence in muscle tissues. Furthermore, muscle samples with a low larval burden might remain undetected altogether. Therefore, it is best to digest the finely minced part of the biopsy tissue in the solution of 1% HCl and 1% pepsin for 1 h at 37°C, which releases the larvae from the nurse cells and makes these more easily seen under the microscope.

However, younger larvae can still get destroyed by this process. Also, the acid digestion method may not be informative if done before 17–21 days of infection as the larvae are not yet resistant to digestion (Despommier, 1986; Vignau et al., 1997; Bruschi & Murrell, 2002; Bruschi & Dupouy-Camet, 2022). Further, recent studies with Baermann's method were found to be successful in facilitating higher recovery of larvae by using lukewarm (37°C) water. Here, muscle tissues are crushed and may be placed on top of the wire mesh covered with 5–6 layers of cotton gauze and kept for approximately 2.5 h. Baermann's technique is one of the most used techniques of classical parasitology for the detection of free-living larvae of *Strongyloides* and other animal parasites coming along the faeces and is also used to collect the soil nematodes or to estimate the pasture larval burdens (Jiang et al., 2012).

Histopathological examination of muscle tissues is routinely done with standard haematoxylin-eosin-stained sections that may reveal the intracellular parasite within the infected muscle cell or may reveal the basophilic changes in the muscle cells, due to invasion by the infective larva, or may reveal the dissolutions of myofilaments by the larva after 4–5 day post-invasion. The presence of an uncoiled and straight larva along with the absence of encapsulation indicates recent infection, whereas a more mature capsule with a coiled parasite indicates chronicity of the infection. In humans, muscle biopsy can be done using deltoid muscle tissue, ensuring that the biopsy sample is free of subcutaneous fat and skin. Approximately 0.2–0.5 g of muscle tissue should be collected for detection of the typical nurse cell-larva complex, or for isolation of the larvae for genotyping. Although parasites' preferred sites are not deltoid in humans, as mentioned earlier, deltoid muscle is preferred because of easy accessibility (Gould, 1970; Weatherly, 1983; Capó and Despommier, 1996; Bruschi & Murrell, 2002).

Serological tests, such as ELISA and Western blot testing, are capable of detecting antibodies associated with specific infections. These tests may yield positive results within 2 weeks of infection, with ELISA potentially detecting antibodies as early as 12 days post-infection. By 14 dpi, IFAT and ELISA targeting IgG antibodies may become positive and remain positive for years, while the sensitivity of the IgG-based ELISA may reach 100% at 50 dpi. ELISA has been the most commonly used technique due its limit of detection being as low as 1 larva per 100 g of muscle tissue and because ELISA is a lot cheaper than IFAT, or immunoblotting techniques, suggesting its wider applicability as a herd test (Bai et al., 2017; Bruschi & Dupouy-Camet, 2022; Gómez-Morales et al., 2022).

ES antigens are preferred over the crude extracts of the muscle larvae for performing ELISA as well as immunoblotting tests since they give higher specificity than the latter. ES antigens are structurally related groups of glycoproteins that can be generated by the in vitro culture of *Trichinella* larvae. A highly conserved and immunogenic antigen- TSL-1 (*T. spiralis* larvae group 1), which are a group of glycoproteins having a highly immunodominant sugar moiety, the 3, 6-dideoxyhexose sugar—d-(tyvelose), is predominantly secreted by the cuticle of the migrating larvae and the stichocytes of the muscle*t Trichinella*. Chemical synthesis of tyvelose sugars has been achieved, which now allows for the direct study of responses to this major immunogenic component. This development opens up opportunities for further research and analysis in the field of immunology. However, antigen detection tests based on circulatory antigens are not reliable (Ortega-Pierres, 1995; Bolás-Fernandez & Corral Bezara, 2006; Liu et al., 2014; Bruschi & Dupouy-Camet, 2022).

Additionally, fluctuating eosinophil levels throughout the course of a suspected infection, characterised by an initial rise, plateau, and subsequent decline, do not definitively confirm the infection. However, in conjunction with a detailed exposure history, the pattern of eosinophil fluctuation, with elevated levels of muscle enzymes like CPK and lactate dehydrogenase, may inform a clinician's decision to initiate presumptive treatment prior to getting any confirmatory diagnosis report. Therefore, characteristic immunological responses along with elevated levels of peripheral eosinophils, total IgE, and muscle enzyme levels provide clues in the diagnosis of the disease in humans (Bruschi & Murrell, 2002; Bruschi & Dupouy-Camet, 2022).

Molecular genotyping methods such as random amplified polymorphic DNA (RAPD), polymerase chain reaction-restriction fragment length polymorphism (PCR-RFLP), multiplex PCR, and reverse line blot hybridisation are needed for identification due to morphological similarities among different *Trichinella* taxa. The conventional multiplex PCR method, recommended by the International Commission on Trichinellosis, is widely used for reliable differentiation of most *Trichinella* species and genotypes. However, next-generation sequencing (NGS) offers significant advantages over conventional PCR, especially in cases of multiple-species infection. NGS provides a comprehensive genotyping solution and enhanced resolution for detecting under-represented genotypes in mixed infections. It can differentiate all known *Trichinella* taxa without requiring additional assays and is particularly useful for identifying novel species like *T. chanchalensis*, which may be missed by conventional PCR methods. NGS also enables higher throughput and enhanced multiplexing capabilities compared to traditional PCR techniques. Researchers have demonstrated that NGS significantly enhances the capability to identify rarely occurring genotypes in larval pools, including the discovery of the previously unknown species *T. chanchalensis*, which could not be identified using multiplex PCR assays. Furthermore, the availability of sequence data in GenBank for all *Trichinella* taxa suggests that the internal transcribed spacer-1 (ITS-1) primers can reliably amplify marker fragments of these taxa, enabling their rapid differentiation through the NGS method (Pozio et al., 2009; Pozio and Zarlenga, 2019; Bilska-Zając et al., 2021; Zarlenga, 2021; Bruschi & Dupouy-Camet, 2022; Thaenkham et al., 2022; Lobanov et al., 2023; Malone et al., 2025; Yu et al., 2025).

Treatment

Among the available anthelmintic agents used in animals, only the benzimidazole group of drugs has been found to be more effective against *Trichinella* worms compared to other groups such as macrocyclic lactones, levamisole, etc. The mode of action of the benzimidazole drugs is that they, upon binding with the free β-tubulin, inhibit the polymerisation of tubulin and the microtubule-dependent uptake of glucose. Further, at elevated concentrations, benzimidazoles exhibit various non-specific effects on nematodes, such as the inhibition of fumarate reductase. Although in experimental studies using the rat models, the benzimidazole group of drugs have some limited efficacies in killing the encapsulated larvae, these drugs are effective against extraintestinal developing stages, against the adults in the gut mucosal epithelium, as well as against the migrating NBLs inside the body. Particularly in pigs, mebendazole treatment has been found to be effective at the dose rate of 83–100 mg/kg/day for 3-5 days against the muscle larva stages. Mebendazole, unlike other benzimidazole drugs, has low water solubility, resulting in restricted absorption in the gastrointestinal tract and reduced bioavailability. As a result, a prolonged treatment regimen is recommended over a single-dose therapy. Studies have shown that a combined total dose of 250 mg/kg per day of mebendazole is effective in rendering the muscle trichinella in mice non-infective, but this effectiveness is achieved over a longer duration (3–5 days) compared to a single-dose therapy. In food animals, mebendazole requires withholding times of 8–14 days. Albendazole has a clear advantage over mebendazole due to its ease of achieving the required plasma concentrations. Unlike mebendazole, which has poor bioavailability and necessitates individual monitoring and dosing, albendazole offers a more straightforward approach (Pozio et al., 2001; Chai et al., 2021; Bruschi & Dupouy-Camet, 2022; Dubey et al., 2024; Fredericks et al., 2024; Moawad et al., 2025).

In humans, benzimidazole derivatives like mebendazole and albendazole are commonly used to treat parasitic infections, while thiabendazole is no longer recommended due to health concerns. Mebendazole and albendazole can eliminate adult worms from the intestine but have little effect on muscle cysts. These medications should be given orally for a prolonged duration to ensure therapeutic plasma levels are maintained and to limit invasion of muscle tissues by the NBLs and reduce the muscle damages. Both are given orally. Mebendazole is dosed at 200–400 mg thrice a day for 3 days and then 400–500 mg thrice a day for 10 days. Albendazole is dosed at 400 mg twice a day for 8–14 days. However, both drugs are contraindicated in pregnant women and children under 2 years old. A 5–7-day course of either mebendazole or albendazole has been found to be effective in reducing the clinical signs and symptoms, and is therefore the recommended course of treatment by the CDC, except in the pregnant women and children under two years of age. Whereas, for children over two years, albendazole may be administered at 5mg/kg body weight b.i.d for 10-15 days. Pyrantel, an imidazothiazole-derived tetrahydropyrimidine nicotinic agonist, exhibits a broad spectrum of action against both adult and immature gastrointestinal nematodes in humans and animals. By mimicking the action of acetylcholine with greater potency, it induces spastic paralysis in the parasites, and facilitates the expulsion of worms. Pyrantel is usually administered as a single-dose regimen ranging from 10 to 20 mg/kg for 2–3 days, and is considered safe for use in pregnant women and children. However, it is important to note that its efficacy is limited to gut dwelling stages, rendering it ineffective against tissue-dwelling *Trichinella* larvae (NBLs, muscle larvae) (Bruschi et al., 2002; Dupouy-Camet & Bruschi, 2007; Gottstein et al., 2009; Chai et al., 2021; Bruschi & Dupouy-Camet, 2022; Chaudhury, 2022; Mitic et al., 2026).

Antipyretics and analgesics are sufficient for symptomatic infections, but effective treatment is not available if the larvae invade the muscles. Severe cases can be treated with antiparasitic medications and steroids, which help resolve symptoms. Corticosteroid therapy helps in the resolution of the symptoms as in other helminth infections; as in trichinellosis, clinical signs mostly arise from the host's inflammatory responses and allergic reactions against the excretory-secretory antigens from the migrating and encapsulating larvae. Prednisolone at 40–60 mg/day for 10-14 days, relieves the inflammatory changes in the body and aids in subsiding the fever by reducing the inflammatory condition associated with the larval penetration of the host cells. However, prolonged steroid treatment is not recommended, and symptomatic treatment may be necessary for the long-term effects of the infection. Immunosuppressive drugs are useful in life-threatening situations but may promote parasite longevity in the gastrointestinal tract, and that may cause higher NBL release and their subsequent migration to the body (Capó and Despommier, 1996; Bruschi et al., 2002; Bruschi & Dupouy-Camet, 2022; Chaudhury, 2022).

The severity of *Trichinella* infections can vary significantly depending on the parasite species, larval burden, host's immune status, and initiation of therapeutic intervention. While mild cases may be managed with analgesics and antipyretics, severe infections may necessitate a combination of anthelmintic drugs and corticosteroids (Pawlowski, 1983). Notably, prompt treatment with albendazole or mebendazole can effectively eliminate developing stages in the intestine, as well as adults and migrating larvae, leading to a significant reduction in muscle pain and inflammation. However, once the larvae have formed cysts in the muscle tissue, these medications become ineffective. In cases of myopathies caused by encysted larvae, corticosteroids like prednisolone are employed to manage inflammation. At this stage, additional antipyretics and analgesics are necessary until symptoms subside. However, owing to advancements in diagnostic techniques and improved treatment protocols, fatalities due to trichinellosis have become exceedingly rare.

References

Agyei-Frempong, M., & Catty, D. (1983). The measurement of antigens released by radiation-attenuated Trichinella spiralis larvae. *Parasite Immunology, 5*(3), 289–303.

Allen, J. E., & Sutherland, T. E. (2014). Host protective roles of type 2 immunity: Parasite killing and tissue repair, flip sides of the same coin. *Seminars in Immunology, 26*(4), 329–340. https://doi.org/10.1016/j.smim.2014.06.003

Amare, G. A., Wondmagegn, Y. M., Adugna, A., Abebaw, D., Melkamu, A., Belew, H., Teffera, Z. H., Getinet, M., Jemal, M., Baylie, T., Asmare, D. S., Yenesew, A., Fenta, A., Woreket, M., Setegn, A., & Malede, B. (2025). Immunological role of IL-25 on gut-dwelling helminths. *Infection, Genetics and Evolution : Journal of Molecular Epidemiology and Evolutionary Genetics in Infectious Diseases, 136*, 105856. https://doi.org/10.1016/j.meegid.2025.105856

Andrade-Becerra, J. I., Pompa-Mera, E. N., Ribas-Aparicio, R. M., & Yépez-Mulia, L. (2017). Vaccination against Trichinella spiralis: potential, limitations and future directions. In *Natural remedies in the fight against parasites*. IntechOpen. https://doi.org/10.5772/66499

Andrews, J. R., Ainsworth, R., & Abernethy, D. (1994). Trichinella pseudospiralis in humans: Description of a case and its treatment. *Transactions of the Royal Society of Tropical Medicine and Hygiene, 88*(2), 200–203. https://doi.org/10.1016/0035-9203(94)90295-x

Anthony, R. M., Rutitzky, L. I., Urban, J. F., Jr., Stadecker, M. J., & Gause, W. C. (2007). Protective immune mechanisms in helminth infection. *Nature Reviews. Immunology, 7*(12), 975–987. https://doi.org/10.1038/nri2199

Bai, X., Hu, X., Liu, X., Tang, B., & Liu, M. (2017). Current research of Trichinellosis in China. *Frontiers in Microbiology, 8*, 1472. https://doi.org/10.3389/fmicb.2017.01472

Bai, H., Tang, B., Qiao, W., Wu, X., Liu, M., & Wang, X. (2022). The prevalence of Trichinella spiralis in domestic pigs in China: A systematic review and meta-analysis. *Animals : An Open Access Journal from MDPI, 12*(24), 3553. https://doi.org/10.3390/ani12243553

Bessi, C., Ercole, M. E., Fariña, F. A., Montalvo, F., Fassa, V., Acerbo, M., Ribicich, M. M., & Pasqualetti, M. I. (2024). Survival of Trichinella spiralis and T. Pseudospiralis in experimentally infected wild boar muscle tissue under freezing and environmental conditions. *Iranian Journal of Parasitology, 19*(4), 457–464. https://doi.org/10.18502/ijpa.v19i4.17166

Bezbaruah, R., Chavda, V. P., Nongrang, L., Alom, S., Deka, K., Kalita, T., Ali, F., Bhattacharjee, B., & Vora, L. (2022). Nanoparticle-based delivery Systems for Vaccines. *Vaccine, 10*(11), 1946. https://doi.org/10.3390/vaccines10111946

Bhattacharya, D., Das, A., Sarkhel, S. P., & Satyaprakash, K. (2024). Milk-borne parasitic zoonoses. In *The microbiology, pathogenesis and zoonosis of milk borne diseases* (pp. 295–304). Academic.

Bi, K., Yang, J., Wang, L., Gu, Y., Zhan, B., & Zhu, X. (2015). Partially protective immunity induced by a 20 kDa protein secreted by Trichinella spiralis stichocytes. *PLoS One, 10*(8), e0136189.

Bien, J., Cabaj, W., & Moskwa, B. (2013). Recognition of antigens of three different stages of the Trichinella spiralis by antibodies from pigs infected with T. Spiralis. *Experimental Parasitology, 134*(2), 129–137. https://doi.org/10.1016/j.exppara.2013.02.007

Bilska-Zając, E., Thompson, P., Rosenthal, B., Różycki, M., & Cencek, T. (2021). Infection, genetics, and evolution of Trichinella: Historical insights and applications to molecular epidemiology. *Infection, Genetics and Evolution : Journal of Molecular Epidemiology and Evolutionary Genetics in Infectious Diseases, 95*, 105080. https://doi.org/10.1016/j.meegid.2021.105080

Bolás-Fernandez, F., & Corral Bezara, L. D. (2006). TSL-1 antigens of Trichinella: An overview of their potential role in parasite invasion, survival and serodiagnosis of trichinellosis. *Research in Veterinary Science, 81*(3), 297–303. https://doi.org/10.1016/j.rvsc.2006.01.002

Borhani, M., Fathi, S., Harandi, M. F., Simsek, S., Ahmed, H., Wu, X., & Liu, M. (2023). Trichinella infections in animals and humans of Iran and Turkey. *Frontiers in Medicine, 10*, 1088507.

Boulos, L. M., Ibrahim, I. R., Said, D. E., & El-Zawawy, L. A. (2005). Congenital trichinellosis in experimentally infected mice. *Journal of the Egyptian Society of Parasitology, 35*(2), 433–445.

Brooks, T. J., Jr., Ward, J. W., & Holder, T. M. (1948). Studies on the incidence of trichiniasis in Mississippi; preliminary report. *The American Journal of Tropical Medicine and Hygiene, 28*(6), 863–866. https://doi.org/10.4269/ajtmh.1948.s1-28.863

Bruschi, F., & Chiumiento, L. (2012). Immunomodulation in trichinellosis: Does Trichinella really escape the host immune system? *Endocrine, Metabolic & Immune Disorders Drug Targets, 12*(1), 4–15. https://doi.org/10.2174/187153012799279081

Bruschi, F., & Dupouy-Camet, J. (2022). Trichinellosis. In F. Bruschi (Ed.), *Helminth infections and their impact on global public health*. Springer. https://doi.org/10.1007/978-3-031-00303-5_10

Bruschi, F., & Murrell, K. D. (2002). New aspects of human trichinellosis: the impact of new Trichinella species. *Postgraduate Medical Journal, 78*(915), 15–22. https://doi.org/10.1136/pmj.78.915.15

Bruschi, F., Dupouy-Camet, J., Kociecka, W., Pozio, E., & Bolas-Fernandez, F. (2002). Opinion on the diagnosis and treatment of human trichinellosis. *Expert Opinion on Pharmacotherapy, 3*(8), 1117–1130. https://doi.org/10.1517/14656566.3.8.1117

Bruschi, F., Korenaga, M., & Watanabe, N. (2008). Eosinophils and Trichinella infection: Toxic for the parasite and the host? *Trends in Parasitology, 24*(10), 462–467. https://doi.org/10.1016/j.pt.2008.07.001

Bruschi, F., Brunetti, E., & Pozio, E. (2013). Neurotrichinellosis. *Handbook of Clinical Neurology, 114*, 243–249. https://doi.org/10.1016/B978-0-444-53490-3.00019-4

Bruschi, F., Ashour, D. S., & Othman, A. A. (2022). Trichinella-induced immunomodulation: Another tale of helminth success. *Food and Waterborne Parasitology, 27*, e00164. https://doi.org/10.1016/j.fawpar.2022.e00164

Cabrera, P. B., & Gould, S. E. (1964). Resistance to trichinosis in swine induced by administration of irradiated larvae. *The Journal of Parasitology*, 681–684.

Campbell, W. C. (1979). History of trichinosis: Paget, Owen and the discovery of "trichinella spiralis". *Bulletin of the History of Medicine, 53*(4), 520–552. http://www.jstor.org/stable/44450948

Campbell, W. C. (1983). Historical introduction. In W. C. Campbell (Ed.), *Trichinella and Trichinosis*. Springer. https://doi.org/10.1007/978-1-4613-3578-8_1

Capó, V., & Despommier, D. D. (1996). Clinical aspects of infection with Trichinella spp. *Clinical Microbiology Reviews, 9*. https://doi.org/10.1128/cmr.9.1.47

Capó, V. A., Despommier, D. D., & Polvere, R. I. (1998). Trichinella spiralis: Vascular endothelial growth factor is up-regulated within the nurse cell during the early phase of its formation. *The Journal of Parasitology, 84*(2), 209–214.

CDC, 2024. Centers for Disease Control and Prevention. Parasites - Trichinellosis (Trichinosis). Available at: https://www.cdc.gov/trichinellosis/about/index.html

Chai, J. Y., Jung, B. K., & Hong, S. J. (2021). Albendazole and Mebendazole as anti-parasitic and anti-cancer agents: An update. *The Korean Journal of Parasitology, 59*(3), 189–225. https://doi.org/10.3347/kjp.2021.59.3.189

Chatterjee, T., Jagani, R., Sabhiki, A. K., & Chander, Y. (2000). Trichinosis. *Medical Journal, Armed Forces India, 56*(2), 161–162. https://doi.org/10.1016/S0377-1237(17)30138-7

Chaudhury, A. (2022). Trichinellosis. In S. C. Parija & A. Chaudhury (Eds.), *Textbook of parasitic zoonoses. Microbial zoonoses*. Springer. https://doi.org/10.1007/978-981-16-7204-0_40

Clausen, M. R., Meyer, C. N., Krantz, T., Moser, C., Gomme, G., Kayser, L., Albrectsen, J., Kapel, C. M., & Bygbjerg, I. C. (1996). Trichinella infection and clinical disease. *QJM : Monthly Journal of the Association of Physicians, 89*(8), 631–636. https://doi.org/10.1093/qjmed/89.8.631

Cui, J., Wang, Z. Q., & Han, H. M. (2006). Congenital transmission of Trichinella spiralis in experimentally infected mice. *Helminthologia, 43*, 7–10. https://doi.org/10.2478/s11687-006-0002-4

Cui, J., Ren, H. J., Liu, R. D., Wang, L., Zhang, Z. F., & Wang, Z. Q. (2013). Phage-displayed specific polypeptide antigens induce significant protective immunity against Trichinella spiralis infection in BALB/c mice. *Vaccine, 31*(8), 1171–1177.

Cui, J., Li, L. G., Jiang, P., Liu, R. D., Yang, X., Liu, L. N., et al. (2015a). Biochemical and functional characterization of the glutathione S-transferase from Trichinella spiralis. *Parasitology Research, 114*(5), 2007–2013.

Cui, J. M., Zhang, N. Z., Li, W. H., Yan, H. B., & Fu, B. Q. (2015b). Cloning, identification, and bioinformatics analysis of a putative aquaporin TsAQP from Trichinella spiralis. *Genetics and Molecular Research : GMR, 14*(4), 12699–12709. https://doi.org/10.4238/2015.October.19.14

Cuperlovic, K., Djordjevic, M., & Pavlovic, S. J. V. P. (2005). Re-emergence of trichinellosis in southeastern Europe due to political and economic changes. *Veterinary Parasitology, 132*(1–2), 159–166.

Despommier, D. (1986). Trichinellosis. In: Walls KW, Schantz PM (eds) Immunodiagnostics of parasitic diseases. Academic Press, New York, pp 163–181.

Despommier, D. D. (1990). Trichinella spiralis: The worm that would be virus. *Parasitology Today (Personal ed.), 6*(6), 193–196. https://doi.org/10.1016/0169-4758(90)90355-8

Despommier, D. D., & Hotez, P. J. (2012). Tissue nematodes. In *Principles and practice of pediatric infectious diseases* (pp. 1334–1341). Elsevier. https://doi.org/10.1016/B978-1-4377-2702-9.00279-8

Despommier, D. D., & Müller, M. (1976). The Stichosome and its secretion granules in the mature muscle larva of Trichinella spiralis. *The Journal of Parasitology, 62*(5), 775–785. https://doi.org/10.2307/3278960

Diaz, J. H., Warren, R. J., & Oster, M. J. (2020). The disease ecology, epidemiology, clinical manifestations, and management of trichinellosis linked to consumption of wild animal meat. *Wilderness & Environmental Medicine, 31*(2), 235–244.

Djordjevic, M., Bacic, M., Petricevic, M., Cuperlovic, K., Malakauskas, A., Kapel, C. M., & Murrell, K. D. (2003). Social, political, and economic factors responsible for the reemergence of trichinellosis in Serbia: A case study. *The Journal of Parasitology, 89*(2), 226–231. https://doi.org/10.1645/0022-3395(2003)089[0226:SPAEFR]2.0.CO;2

Dubey, M. L., Khurana, S., Singhal, L., Dogra, S., & Singh, S. (2011). Atypical trichinellosis without eosinophilia associated with osteomyelitis. *Tropical Doctor, 41*(4), 244–246. https://doi.org/10.1258/td.2011.110106

Dubey, J. P., Thompson, P. C., Fournet, V., Hill, D. E., Zarlenga, D., Gamble, H. R., & Rosenthal, B. M. (2024). Over a century of progress on Trichinella research in pigs at the United States Department of Agriculture: Challenges and solutions. *Food and Waterborne Parasitology, 36*, e00239. https://doi.org/10.1016/j.fawpar.2024.e00239

Dubinský, P., Böör, A., Kinceková, J., Tomasovicová, O., Reiterová, K., & Bielik, P. (2001). Congenital trichinellosis? Case report. *Parasite (Paris, France), 8*(2 Suppl), S180–S182. https://doi.org/10.1051/parasite/200108s2180

Duman, J. G., & Newton, S. S. (2020). Antifreeze proteins in other species. In H. Ramløv & D. Friis (Eds.), *Antifreeze Proteins Volume 1*. Springer. https://doi.org/10.1007/978-3-030-41929-5_8

Dupouy-Camet, J., & Bruschi, F. (2007). Management and diagnosis of human trichinellosis. In J. Dupouy-Camet & K. D. Murrell (Eds.), *FAO/WHO/OIE guidelines for the surveillance, management, prevention and control of trichinellosis* (pp. 37–68). World Organisation for Animal Health Press.

El-Dardiry, M. A., Abdel-Aal, A. A., Abdeltawab, M. S. A., El-Sherbini, M., Hassan, M. A., Abdel-Aal, A. A., Badawi, M., Anis, S. E., Khaled, B. A., & Al-Antably, A. S. (2021). Effect of mast cell stabilization on angiogenesis in primary and secondary experimental Trichinella spiralis infection. *Parasites & Vectors, 14*(1), 567. https://doi.org/10.1186/s13071-021-05075-9

Epstein, H. F., Miller, D. M., IIIrd, Ortiz, I., & Berliner, G. C. (1985). Myosin and paramyosin are organized about a newly identified core structure. *The Journal of Cell Biology, 100*(3), 904–915.

Ercole, M. E., Bessi, C., Pasqualetti, M. I., Ribicich, M. M., Aronowicz, T., Bonboni, A., Acerbo, M., & Fariña, F. A. (2020). Gamma radiation effect on Trichinella pseudospiralis and Trichinella spiralis infected wild boar meat. *Veterinary Parasitology, 287*, 109257. https://doi.org/10.1016/j.vetpar.2020.109257

European Food Safety Authority (EFSA). (2005). Opinion of the scientific panel on biological hazards (BIOHAZ) on the on the "risk assessment of a revised inspection of slaughter animals in areas with low prevalence of Trichinella". *EFSA Journal, 3*(6), 200. https://doi.org/10.2903/j.efsa.2005.200

Fang, L., Sun, L., Yang, J., Gu, Y., Zhan, B., Huang, J., & Zhu, X. (2014). Heat shock protein 70 from Trichinella spiralis induces protective immunity in BALB/c mice by activating dendritic cells. *Vaccine, 32*(35), 4412–4419.

FAO. (2021). Parasites in food: An invisible threat. In *Food safety technical toolkit for Asia and the Pacific no. 7.*

Feng, S., Wu, X., Wang, X., Bai, X., Shi, H., Tang, B., et al. (2013). Vaccination of mice with an antigenic serine protease-like protein elicits a protective immune response against Trichinella spiralis infection. *The Journal of Parasitology, 99*, 426–432. https://doi.org/10.1645/12-46.1

Forbes, L. B., & Gajadhar, A. A. (1999). A validated Trichinella digestion assay and an associated sampling and quality assurance system for use in testing pork and horse meat. *Journal of Food Protection, 62*(11), 1308–1313. https://doi.org/10.4315/0362-028x-62.11.1308

Foreyt, W. J. (2013). *Trichinosis: Reston, Va.* U.S. Geological Survey Circular 1388, 60 p., 2 appendixes. https://doi.org/10.3133/cir1388

Franssen, F., Swart, A., van der Giessen, J., Havelaar, A., & Takumi, K. (2017). Parasite to patient: A quantitative risk model for Trichinella spp. in pork and wild boar meat. *International Journal of Food Microbiology, 241*, 262–275. https://doi.org/10.1016/j.ijfoodmicro.2016.10.029

Franssen, F., Gerard, C., Cozma-Petruţ, A., Vieira-Pinto, M., Jambrak, A. R., Rowan, N., et al. (2019). Inactivation of parasite transmission stages: Efficacy of treatments on food of ani-

mal origin. *Trends in Food Science & Technology, 83*, 114–128. https://doi.org/10.1016/j.tifs.2018.11.009
Frayha, R. A. (1981). Trichinosis-related polyarteritis nodosa. *The American Journal of Medicine, 71*(2), 307–312. https://doi.org/10.1016/0002-9343(81)90133-9
Fredericks, J., Hill, D. E., Zarlenga, D. S., Fournet, V. M., Hawkins-Cooper, D. S., Urban, J. F., Jr., & Kramer, M. (2024). *inactivation of encysted muscle larvae of Trichinella spiralis in pigs using Mebendazole. *Veterinary Parasitology, 327*, 110140. https://doi.org/10.1016/j.vetpar.2024.110140
Fröscher, W., Gullotta, F., Saathoff, M., & Tackmann, W. (1988). Chronic trichinosis. Clinical, bioptic, serological and electromyographic observations. *European Neurology, 28*(4), 221–226. https://doi.org/10.1159/000116271
Furhad, S., & Bokhari, A. A. (2019). Trichinosis. [Updated 2023 Jul 19]. In: StatPearls [Internet]. Treasure Island (FL): StatPearls Publishing; 2025 Jan-. [Updated 2023 Jul 19]. Available from: https://www.ncbi.nlm.nih.gov/books/NBK536945/
Gagliardo, L. F., McVay, C. S., & Appleton, J. A. (2002). Molting, Ecdysis, and reproduction of Trichinella spiralis are supported in vitro by intestinal epithelial cells. *Infection and Immunity, 70*. https://doi.org/10.1128/iai.70.4.1853-1859.2002
Gamble, H. R. (1985). Trichinella spiralis: Immunization of mice using monoclonal antibody affinity-isolated antigens. *Experimental Parasitology, 59*(3), 398–404.
Gamble, H. R. (2021). Preharvest and postharvest control of Trichinella in meat. In *Trichinella and Trichinellosis* (pp. 433–453). Academic Press. https://doi.org/10.1016/B978-0-12-821209-7.00016-0
Gamble, H. R. (2022). Trichinella spp. control in modern pork production systems. *Food and Waterborne Parasitology, 28*, e00172. https://doi.org/10.1016/j.fawpar.2022.e00172
Gamble, H., Rapić, D., Marinculić, A., & Murrell, K. D. (1988). Evaluation of excretory-secretory antigens for the serodiagnosis of swine trichinellosis. *Veterinary Parasitology, 30*, 131–137. https://doi.org/10.1016/0304-4017(88)90160-4
Gardiner, C. H. (1976). Habitat and reproductive behavior of Trichinella spiralis. *The Journal of Parasitology, 62*(6), 865–870.
Gómez-Morales, M. A., Mazzarello, G., Bondi, E., Arenare, L., Bisso, M. C., Ludovisi, A., Amati, M., Viscoli, C., Castagnola, E., Orefice, G., Magnè, F., Pezzotti, P., & Pozio, E. (2021). Second outbreak of Trichinella pseudospiralis in Europe: Clinical patterns, epidemiological investigation and identification of the etiological agent based on the western blot patterns of the patients' serum. *Zoonoses and Public Health, 68*(1), 29–37. https://doi.org/10.1111/zph.12761
Gómez-Morales, M. Á., Cherchi, S., & Ludovisi, A. (2022). Serological testing for Trichinella infection in animals and man: Current status and opportunities for advancements. *Food and Waterborne Parasitology, 27*, e00165. https://doi.org/10.1016/j.fawpar.2022.e00165
Gottstein, B., Pozio, E., & Nöckler, K. (2009). Epidemiology, diagnosis, treatment, and control of trichinellosis. *Clinical Microbiology Reviews, 22*(1), 127–145. https://doi.org/10.1128/CMR.00026-08
Gould, S. E. (Ed.). (1970). *Trichinosis in man and animals*. Charles C.
Gu, Y., Li, J., Zhu, X., Yang, J., Li, Q., Liu, Z., et al. (2008). Trichinella spiralis: Characterization of phage-displayed specific epitopes and their protective immunity in BALB/c mice. *Experimental Parasitology, 118*(1), 66–74.
Gu, Y., Wei, J., Yang, J., Huang, J., Yang, X., & Zhu, X. (2013). Protective immunity against Trichinella spiralis infection induced by a multi-epitope vaccine in a murine model. *PLoS One, 8*(10), e77238.
Gu, Y., Zhan, B., Yang, Y., Yang, X., Zhao, X., Wang, L., Yang, J., Bi, K., Wang, Y., & Zhu, X. (2014). Protective effect of a prime-boost strategy with the Ts87 vaccine against Trichinella spiralis infection in mice. *BioMed Research International, 2014*, 326860. https://doi.org/10.1155/2014/326860
Gu, Y., Sun, X., Huang, J., Zhan, B., & Zhu, X. (2020). A multiple antigen peptide vaccine containing CD4+ T cell epitopes enhances humoral immunity against Trichinella spiralis infection in mice. *Journal of Immunology Research, 2020*(1), 2074803.

Guerrero Manriquez, G. G., & Tuero, I. (2021). Adjuvants: Friends in vaccine formulations against infectious diseases. *Human Vaccines & Immunotherapeutics, 17*(10), 3539–3550. https://doi.org/10.1080/21645515.2021.1934354

Guzmán-Faúndez, J., Crisóstomo-Jorquera, V., Landaeta-Aqueveque, C., & Henríquez, A. (2025). First assessment of the prevalence of Trichinella in backyard-raised pigs in central-southern Chile. *Veterinary Quarterly, 45*(1), 1–7. https://doi.org/10.1080/01652176.2025.2475986

Hafez, E. N., El-Kholy, W. A. M. S., & El-Kholy, M. A. E. M. (2020). Amelioration of Myogenin, Bcl-2 expression and DNA damages in myocytes of Trichinella spiralis-infected mice after immunization with gamma radiationattenuated larvae. *International Journal of Radiation Research, 18*(4), 699–706.

Henry, E. K., Inclan-Rico, J. M., & Siracusa, M. C. (2017). Type 2 cytokine responses: Regulating immunity to helminth parasites and allergic inflammation. *Current Pharmacology Reports, 3*(6), 346–359. https://doi.org/10.1007/s40495-017-0114-1

Hill, D. E., Forbes, L., Zarlenga, D. S., Urban, J. F., Jr., Gajadhar, A. A., & Gamble, H. R. (2009). Survival of north American genotypes of Trichinella in frozen pork. *Journal of Food Protection, 72*(12), 2565–2570. https://doi.org/10.4315/0362-028x-72.12.2565

ICT. (2026). Trichinella and Trichinellosis. In *International commission on Trichinellosis.* http://trichinellosis.org/Home_Page.html

Ilic, N., Vasilev, S., Gruden-Movsesijan, A., Gnjatovic, M., Sofronic-Milosavljevic, L., & Mitic, I. (2022). Long lasting immunity in trichinellosis—Insight from a small study group. *Journal of Helminthology, 96*, e35. https://doi.org/10.1017/S0022149X22000268

ITRC. (2026). *International Trichinella reference center, unit of foodborne and neglected parasitic diseases, Department of Infectious Diseases, Istituto Superiore di Sanità.* Rome. Available at https://trichinella.iss.it/. Accessed on 27 Jan 2026

Jasmer, D. P., & Neary, S. M. (1994). Trichinella spiralis: Inhibition of muscle larva growth and development is associated with a delay in expression of infected skeletal muscle characteristics. *Experimental Parasitology, 78*(3), 317–325. https://doi.org/10.1006/expr.1994.1033

Jiang, P., Wang, Z. Q., Cui, J., & Zhang, X. (2012). Comparison of artificial digestion and Baermann's methods for detection of Trichinella spiralis pre-encapsulated larvae in muscles with low-level infections. *Foodborne Pathogens and Disease, 9*(1), 27–31. https://doi.org/10.1089/fpd.2011.0985

Jota Baptista, C., Seixas, F., Gonzalo-Orden, J. M., et al. (2023). Wild boar (Sus scrofa) as a potential reservoir of infectious agents in Portugal: a review of two decades (2001–2021). *European Journal of Wildlife Research, 69*, 101. https://doi.org/10.1007/s10344-023-01732-9

Kassai, T. (1999). *Veterinary helminthology* (pp. 133–136). Butterworth-Heinemann.

Kassai, T. (2006). Nomenclature for parasitic diseases: Cohabitation with inconsistency for how long and why? *Veterinary Parasitology, 138*(3–4), 169–178. https://doi.org/10.1016/j.vetpar.2006.02.019

Knight, P. A., Wright, S. H., Lawrence, C. E., Paterson, Y. Y., & Miller, H. R. (2000). Delayed expulsion of the nematode Trichinella spiralis in mice lacking the mucosal mast cell-specific granule chymase, mouse mast cell protease-1. *The Journal of Experimental Medicine, 192*(12), 1849–1856. https://doi.org/10.1084/jem.192.12.1849

Kocęcka, W., Boczon, K., Pozio, E., & Von Knapen, F. (2003). Trichinella. In *International handbook of foodborne pathogens* (pp. 657–678). CRC Press.

Kociecka, W. (2000). Trichinellosis: Human disease, diagnosis and treatment. *Veterinary Parasitology, 93*(3–4), 365–383. https://doi.org/10.1016/s0304-4017(00)00352-6

Kołodziej-Sobocińska, M. (2019). Factors affecting the spread of parasites in populations of wild European terrestrial mammals. *Mamm Res, 64*, 301–318. https://doi.org/10.1007/s13364-019-00423-8

Kozak, M., & Hu, J. (2024). DNA vaccines: Their formulations, engineering and delivery. *Vaccines, 12*(1), 71. https://doi.org/10.3390/vaccines12010071

Lacour, S. A., Heckmann, A., Macé, P., Grasset-Chevillot, A., Zanella, G., Vallée, I., Kapel, C. M., & Boireau, P. (2013). Freezetolerance of Trichinella muscle larvae in experimentally

infected wild boars. *Veterinary Parasitology, 194*(2–4), 175–178. https://doi.org/10.1016/j.vetpar.2013.01.049
Lee, T. D. (1991). Helminthotoxic responses of intestinal eosinophils to Trichinella spiralis newborn larvae. *Infection and Immunity, 59*(12), 4405–4411.
Leung, T. L. F. (2024). Zoonotic parasites in reptiles, with particular emphasis on potential zoonoses in Australian reptiles. *Current Clinical Microbiology Reports, 11*, 88–98. https://doi.org/10.1007/s40588-024-00221-0
Li, C., & Ko, R. (2001). Inflammatory response during the muscle phase of Trichinella spiralis and T. Pseudospiralis infections. *Parasitology Research, 87*, 708–714. https://doi.org/10.1007/s004360100420
Li, X., Yao, J. P., Pan, A. H., Liu, W., Hu, X. C., Wu, Z. D., & Zhou, X. W. (2013). An antigenic recombinant serine protease from Trichinella spiralis induces protective immunity in BALB/c mice. *Parasitology Research, 112*(9), 3229–3238.
Lightowlers, M. W., & Rickard, M. D. (1988). Excretory–secretory products of helminth parasites: Effects on host immune responses. *Parasitology, 96*(S1), S123–S166.
Lindquist, H. A., & Cross, J. H. (2017). Helminths. In *Infectious diseases* (pp. 1763–1779). Elsevier. https://doi.org/10.1016/B978-0-7020-6285-8.00195-7
Liu, R. D., Cui, J., Wang, L., Long, S. R., Zhang, X., Liu, M. Y., & Wang, Z. Q. (2014). Identification of surface proteins of Trichinella spiralis muscle larvae using immunoproteomics. *Tropical Biomedicine, 31*(4), 579–591.
Liu, P., Cui, J., Liu, R. D., Wang, M., Jiang, P., Liu, L. N., et al. (2015a). Protective immunity against Trichinella spiralis infection induced by TsNd vaccine in mice. *Parasites & Vectors, 8*(1), 185.
Liu, P., Wang, Z. Q., Liu, R. D., Jiang, P., Long, S. R., Liu, L. N., et al. (2015b). Oral vaccination of mice with Trichinella spiralis nudix hydrolase DNA vaccine delivered by attenuated salmonella elicited protective immunity. *Experimental Parasitology, 153*, 29–38.
Liu, Y., Dong, Z., Pang, J., Liu, M., & Jin, X. (2021). Prevalence of meat-transmitted Taenia and Trichinella parasites in the Far East countries. *Parasitology Research, 120*, 4145–4151.
Lobanov, V. A., Konecsni, K. A., Scandrett, W. B., et al. (2023). Identification of Trichinella taxa by ITS-1 amplicon next-generation sequencing with an improved resolution for detecting underrepresented genotypes in mixed natural infections. *Parasites and Vectors, 16*, 466. https://doi.org/10.1186/s13071-023-06035-1
Lombardi, C., Berti, A., & Cottini, M. (2022). The emerging roles of eosinophils: Implications for the targeted treatment of eosinophilic-associated inflammatory conditions. *Current Research in Immunology, 3*, 42–53. https://doi.org/10.1016/j.crimmu.2022.03.002
Long, S. R., Wang, Z. Q., Liu, R. D., Liu, L. N., Li, L. G., Jiang, P., et al. (2014). Molecular identification of Trichinella spiralis nudix hydrolase and its induced protective immunity against trichinellosis in BALB/c mice. *Parasites & Vectors, 7*(1), 600.
Lu, Q. Q., Zheng, W. W., Zhang, Z. Y., Cong, P. K., Guo, X., Zhang, Y., Zhang, X. Z., Long, S. R., Liu, R. D., Wang, Z. Q., & Cui, J. (2025). Trichinella spiralis excretory/secretory proteins mediated larval invasion via inducing gut epithelial apoptosis and barrier disruption. *PLoS Neglected Tropical Diseases, 19*(1), e0012842. https://doi.org/10.1371/journal.pntd.0012842
Lupşe, M., Ionică, A. M., Flonta, M., Rus, M. A., & Briciu, V. (2023). Retrospective survey of human Trichinellosis in a Romanian infectious diseases hospital over a thirty-year interval-the never-ending story. *Pathogens (Basel, Switzerland), 12*(3), 369. https://doi.org/10.3390/pathogens12030369
Malone, C. J., Oksanen, A., Mukaratirwa, S., Sharma, R., & Jenkins, E. (2024). From wildlife to humans: The global distribution of Trichinella species and genotypes in wildlife and wildlife-associated human trichinellosis. *International Journal for Parasitology. Parasites and Wildlife, 24*, 100934. https://doi.org/10.1016/j.ijppaw.2024.100934
Malone, C. J., Harms, N. J., Lobanov, V. A., Scandrett, W. B., Queiroz, C. A., Voordouw, M. J., Jung, T. S., Parker, S. E., & Jenkins, E. J. (2025). Broad host specificity of Trichinella chanchalensis and minimal interspecific competition with T. Nativa and T6 in naturally co-infected

hosts. *International Journal for Parasitology, S0020-7519*(25), 00165–00161. Advance online publication. https://doi.org/10.1016/j.ijpara.2025.09.001

Marinculic, A., Gamble, H. R., Urban, J. F., Rapic, D., Zivicnjak, T., Smith, H. J., & Murrell, K. D. (1991). Immunity in swine inoculated with larvae or extracts of a pig isolate and a sylvatic isolate of Trichinella spiralis. *American Journal of Veterinary Research, 52*(5), 754–758.

Marti, H. P., Murrell, K. D., & Gamble, H. R. (1987). Trichinella spiralis: Immunization of pigs with newborn larval antigens. *Experimental Parasitology, 63*(1), 68–73.

Marucci, G., Romano, A. C., Interisano, M., et al. (2021). Trichinella pseudospiralis in a red kite (Milvus milvus) from Italy. *Parasitology Research, 120*, 2287–2290. https://doi.org/10.1007/s00436-021-07165-0

Marucci, G., Tonanzi, D., Interisano, M., Vatta, P., Galati, F., & La Rosa, G. (2022). The international Trichinella reference Centre database. Report on thirty-three years of activity and future perspectives. *Food and Waterborne Parasitology, 27*, e00156. https://doi.org/10.1016/j.fawpar.2022.e00156

McGuire, C., Chan, W. C., & Wakelin, D. (2002). Nasal immunization with homogenate and peptide antigens induces protective immunity against Trichinella spiralis. *Infection and Immunity, 70*(12), 7149–7152.

McIntyre, L., Pollock, S. L., Fyfe, M., Gajadhar, A., Isaac-Renton, J., Fung, J., & Morshed, M. (2007). Trichinellosis from consumption of wild game meat. *CMAJ : Canadian Medical Association Journal = Journal de l'Association Medicale Canadienne, 176*(4), 449–451. https://doi.org/10.1503/cmaj.061530

Mitic, I., Vasilev, S., & Gruden-Movsesijan, A. (2026). Trichinellosis: A zoonosis that still requires vigilance. *PLoS Neglected Tropical Diseases, 20*(1), e0013944. https://doi.org/10.1371/journal.pntd.0013944

Moawad, H. S., Elgendy, A. M., Mohamed, S. M. A., et al. (2025). Molecular and serological assessment of the therapeutic efficacy of Ivermectin Loaded nanoparticles on Trichinella spiralis experimentally. *Journal of Parasitic Diseases*. https://doi.org/10.1007/s12639-024-01756-1

Mohammad Hasani, S., Ghafouri, E., Kouhpayeh, S., Amerizadeh, F., Rahimmanesh, I., Amirkhani, Z., & Khanahmad, H. (2023). Phage based vaccine: A novel strategy in prevention and treatment. *Heliyon, 9*(9), e19925. https://doi.org/10.1016/j.heliyon.2023.e19925

Mohib, O., Clevenbergh, P., Truyens, C., Morissens, M., & Castro Rodriguez, J. (2022). Trichinella spiralis-associated myocarditis mimicking acute myocardial infarction. *Acta Clinica Belgica, 77*(1), 147–152. https://doi.org/10.1080/17843286.2020.1790867

Motran, C. C., Silvane, L., Chiapello, L. S., Theumer, M. G., Ambrosio, L. F., Volpini, X., Celias, D. P., & Cervi, L. (2018). Helminth infections: Recognition and modulation of the immune response by innate immune cells. *Frontiers in Immunology, 9*, 664. https://doi.org/10.3389/fimmu.2018.00664

Mulcahy, G., O'Neill, S., Fanning, J., McCarthy, E., & Sekiya, M. (2005). Tissue migration by parasitic helminths—An immunoevasive strategy? *Trends in Parasitology, 21*(6), 273–277. https://doi.org/10.1016/j.pt.2005.04.003

Murrell, K. D., & Pozio, E. (2011). Worldwide occurrence and impact of human trichinellosis, 1986–2009. *Emerging Infectious Diseases, 17*(12), 2194–2202. https://doi.org/10.3201/eid1712.110896

Nagano, I., Wu, Z., & Takahashi, Y. (2008). Species-specific antibody responses to the recombinant 53-kilodalton excretory and secretory proteins in mice infected with Trichinella spp. *Clinical and Vaccine Immunology, 15*(3), 468–473.

Nakayama, H., Inaba, T., Nargis, M., Chisty, M., Ito, M., & Kamiya, H. (1998). Immunization of laboratory animals with ultraviolet-attenuated larvae against homologous challenge infection with Trichinella britovi. *The Southeast Asian Journal of Tropical Medicine and Public Health, 29*(3), 563–566.

Nkoke, C., Djibrilla, S., Gobina, R., & Dzudie, A. (2022). A case report of Trichinella spiralis pericarditis: An unusual cause of pericardial effusion and cardiac tamponade in an immunocompetent urban black African. *European Heart Journal. Case Reports, 7*(1), ytac480. https://doi.org/10.1093/ehjcr/ytac480

Nöckler, K., & Kapel, C. M. O. (2007). Detection and surveillance for Trichinella: Meat inspection and hygiene, and legislation. In J. Dupouy-Camet & K. D. Murrell (Eds.), *FAO/WHO/OIE guidelines for the surveillance, management, prevention and control of trichinellosis* (pp. 69–97). World Organisation for Animal Health Press.

Noeckler, K., Pozio, E., van der Giessen, J., Hill, D. E., & Gamble, H. R. (2019). International commission on Trichinellosis: Recommendations on post-harvest control of Trichinella in food animals. *Food and Waterborne Parasitology, 14*, e00041. https://doi.org/10.1016/j.fawpar.2019.e00041

Oksanen, A., Kärssin, A., Berg, R. P. K. D., Koch, A., Jokelainen, P., Sharma, R., Jenkins, E., & Loginova, O. (2022). Epidemiology of Trichinella in the Arctic and subarctic: A review. *Food and Waterborne Parasitology, 28*, e00167. https://doi.org/10.1016/j.fawpar.2022.e00167

Omeragić, J., Kapo, N., Škapur, V., et al. (2024). Diversity of Trichinella species in carnivores from Bosnia and Herzegovina. *BMC Veterinary Research, 20*, 117. https://doi.org/10.1186/s12917-024-03964-6

Ortega-Pierres, M. G. (1995). Characterization of surface antigens of the nematode parasite Trichinella spiralis: Study of its role in protection mechanisms and their usefulness in the diagnosis of trichinosis. *Gaceta Médica de Mexico, 131*(1), 2–12.

Ortiz, J. B., Uys, M., Seguino, A., et al. (2024). Foodborne helminthiasis. *Current Clinical Microbiology Reports, 11*, 153–165. https://doi.org/10.1007/s40588-024-00231-y

Owen, R. (1835). Description of a Microscophc Entozoon infesting the muscles of the human body. *The Transactions of the Zoological Society of London, 1*(4), 315–324. https://doi.org/10.1111/1835.1.issue-4. https://doi.org/10.1111/j.1096-3642.1835.tb00631.x.

Pati, R., Shevtsov, M., & Sonawane, A. (2018). Nanoparticle vaccines against infectious diseases. *Frontiers in Immunology, 9*, 22–24. https://doi.org/10.3389/fimmu.2018.02224

Pavel, R., Ursoniu, S., Lupu, M. A., & Olariu, T. R. (2023). Trichinellosis in hospitalized children and adults from Western Romania: A 11-year retrospective study. *Life, 13*(4), 969.

Pawlowski, Z. S. (1983). Notes on treatment of human trichinellosis. *Wiadomości Parazytologiczne, 29*(4–6), 495–497.

Perera, D. J., & Ndao, M. (2021). Promising Technologies in the Field of helminth vaccines. *Frontiers in Immunology, 12*, 711650. https://doi.org/10.3389/fimmu.2021.711650

Pitt, S. J., & Gunn, A. (2024). The one health concept. *British Journal of Biomedical Science, 81*, 12366. https://doi.org/10.3389/bjbs.2024.12366

Pozio, E. (2007). World distribution of Trichinella spp. infections in animals and humans. *Veterinary Parasitology, 149*(1–2), 3–21. https://doi.org/10.1016/j.vetpar.2007.07.002

Pozio, E. (2007b). Taxonomy, biology and epidemiology of Trichinella parasites. In J. Dupouy-Camet & K. D. Murrell (Eds.), *FAO/WHO/OIE guidelines for the surveillance, management, prevention and control of trichinellosis* (pp. 1–35). World Organisation for Animal Health Press.

Pozio, E. (2016). Adaptation of Trichinella spp. for survival in cold climates food and waterborne. *Parasitology*, 44–12. https://doi.org/10.1016/j.fawpar.2016.07.001

Pozio, E. (2018). Trichinella and other foodborne nematodes. In Y. Ortega & C. Sterling (Eds.), *Foodborne parasites. Food microbiology and food safety*. Springer. https://doi.org/10.1007/978-3-319-67664-7_9

Pozio, E. (2021). Scientific achievements of the last 60 years: From a single to a multispecies concept of the genus Trichinella. *Veterinary Parasitology, 297*, 109042. https://doi.org/10.1016/j.vetpar.2020.109042

Pozio, E., & Zarlenga, D. S. (2013). New pieces of the Trichinella puzzle. *International Journal for Parasitology, 43*(12–13), 983–997. https://doi.org/10.1016/j.ijpara.2013.05.010

Pozio, E., & Zarlenga, D. (2019). International commission on Trichinellosis: Recommendations for genotyping Trichinella muscle stage larvae. *Food and Waterborne Parasitology, 15*, e00033. https://doi.org/10.1016/j.fawpar.2018.e00033

Pozio, E., Marucci, G., Casulli, A., Sacchi, L., Mukaratirwa, S., Foggin, C. M., & La Rosa, G. (2004). Trichinella papuae and Trichinella zimbabwensis induce infection in experimentally infected varans, caimans, pythons and turtles. *Parasitology, 128*(Pt 3), 333–342. https://doi.org/10.1017/s0031182003004542

Pozio, E., Hoberg, E., La Rosa, G., & Zarlenga, D. S. (2009). Molecular taxonomy, phylogeny and biogeography of nematodes belonging to the Trichinella genus. *Infection, Genetics and Evolution : Journal of Molecular Epidemiology and Evolutionary Genetics in Infectious Diseases, 9*(4), 606–616. https://doi.org/10.1016/j.meegid.2009.03.003

Pozio, E., & Gomez Morales, M. Á. (2023). Trichinella and Trichinellosis: From wildlife to the human beings. In *Zoonoses: Infections affecting humans and animals* (pp. 529–544). Springer International Publishing.

Pozio, E., La Rosa, G., Murrell, K. D., & Lichtenfels, J. R. (1992). Taxonomic revision of the genus Trichinella. *The Journal of Parasitology, 78*(4), 654–659.

Pozio, E., Gomez Morales, M. A., & Dupouy-Camet, J. (2003). Clinical aspects, diagnosis and treatment of trichinellosis. *Expert Review of Anti-Infective Therapy, 1*(3), 471–482.

Pozio, E., Sacchini, D., Sacchi, L., Tamburrini, A., & Alberici, F. (2001). Failure of mebendazole in the treatment of humans with Trichinella spiralis infection at the stage of encapsulating larvae. *Clinical Infectious Diseases : An Official Publication of the Infectious Diseases Society of America, 32*(4), 638–642. https://doi.org/10.1086/318707

Pozio, E., Celli, M., Ludovisi, A., et al. (2021). Animal welfare and zoonosis risk: anti-Trichinella antibodies in breeding pigs farmed under controlled housing conditions. *Parasites and Vectors, 14*, 417. https://doi.org/10.1186/s13071-021-04920-1

Qi, X., Yue, X., Han, Y., Jiang, P., Yang, F., Lei, J. J., et al. (2018). Characterization of two Trichinella spiralis adult-specific DNase II and their capacity to induce protective immunity. *Frontiers in Microbiology, 9*, 2504. https://doi.org/10.3389/fmicb.2018.02504

Qi, X., Han, Y., Jiang, P., Yue, X., Ren, H. N., Sun, G. G., Long, S. R., Yu, C., Cheng, X. C., Cui, J., & Wang, Z. Q. (2018b). Oral vaccination with Trichinella spiralis DNase II DNA vaccine delivered by attenuated salmonella induces a protective immunity in BALB/c mice. *Veterinary Research, 49*(1), 119. https://doi.org/10.1186/s13567-018-0614-y

Qu, Z. G., Ma, X. T., Li, W. H., Zhang, N. Z., Yue, L., Cui, J. M., et al. (2015). Molecular characterization of a cathepsin F-like protease in Trichinella spiralis. *Parasites & Vectors, 8*(1), 652.

Ram, H., Garg, R., Banerjee, P. S., & Singh, R. K. (2018). Present status of trichinellosis - a neglected zoonosis in India. *The Indian Journal of Animal Sciences, 88*(5), 513–521. https://doi.org/10.56093/ijans.v88i5.79904

Ren, H. J., Cui, J., Wang, Z. Q., & Liu, R. D. (2011). Normal mouse intestinal epithelial cells as a model for the in vitro invasion of Trichinella spiralis infective larvae. *PLoS One, 6*(10), e27010.

Ren, H. J., Liu, R. D., Wang, Z. Q., & Cui, J. (2013). Construction and use of a Trichinella spiralis phage display library to identify the interactions between parasite and host enterocytes. *Parasitology Research, 112*(5), 1857–1863.

Riva, E., Fiel, C., Bernat, G., Muchiut, S., & Steffan, P. (2017). Studies on vertical transmission of Trichinella spiralis in experimentally infected Guinea pigs (Cavia porcellus). *Parasitology Research, 116*(8), 2271–2276. https://doi.org/10.1007/s00436-017-5533-8

Roberts, L. S., Janovy, J., Jr., & Nadler, S. (2013). Nematodes: Trichinellida and Dioctophymatida, Enoplean parasites. In G. D. Schmidt & L. S. Roberts (Eds.), *Foundations of parasitology* (9th ed., pp. 377–390). McGraw-Hill Education.

Robinson, K., Bellaby, T., Chan, W. C., & Wakelin, D. (1995). High levels of protection induced by a 40-mer synthetic peptide vaccine against the intestinal nematode parasite Trichinella spiralis. *Immunology, 86*(4), 495–498.

Robles, M. D. R., & Fernández, R. C. (2024). Chapter 49: Trichuroidea and Trichinelloidea (Superfamilies). In S. L. Gardner & S. A. Gardner (Eds.), *Concepts in animal parasitology*. Zea Books. https://doi.org/10.32873/unl.dc.ciap0492024

Roldan, J. A. M., Corrente, M., & Otranto, D. (2023). Zoonotic pathogens of reptiles: An unregarded slithery matter. In A. Sing (Ed.), *Zoonoses: Infections affecting humans and animals*. Springer. https://doi.org/10.1007/978-3-030-85877-3_56-1

Rosca, E. C., Tudor, R., Cornea, A., & Simu, M. (2021). Central nervous system involvement in Trichinellosis: A systematic review. *Diagnostics (Basel, Switzerland), 11*(6), 945. https://doi.org/10.3390/diagnostics11060945

Rostami, A., Gamble, H. R., Dupouy-Camet, J., Khazan, H., & Bruschi, F. (2017). Meat sources of infection for outbreaks of human trichinellosis. *Food Microbiology, 64*, 65–71. https://doi.org/10.1016/j.fm.2016.12.012

Sethi, B., Butola, K. S., Arora, B., Kumar, Y., & Suri, V. (2010). Human trichinosis in remotes of Uttarakhand, India. *Indian Journal of Medical Sciences, 64*(3), 104–110.

Sethi, B., Butola, K. S., Kumar, Y., & Mishra, J. P. (2013). Multiple outbreaks of trichinellosis with high mortality rate. *Tropical Doctor, 42*(4), 243.

Sethi, P. P., Bairwa, M., Pathania, M., Meena, M., & Rajkumar, B. (2021). Trichinellosis: A case of Lifethreatening myositis and myocarditis in Garhwal Hill region. *Journal of Clinical & Diagnostic Research, 15*(12). https://doi.org/10.7860/JCDR/2021/50453.15731

Sgroi, G., D'Alessio, N., Marucci, G., Pacifico, L., Buono, F., Deak, G., Anastasio, A., Interisano, M., Fraulo, P., Pesce, A., Toscano, V., Romano, A. C., Toce, M., Palazzo, L., De Carlo, E., Fioretti, A., & Veneziano, V. (2022). Trichinella britovi in wild boar meat from Italy, 2015-2021: A citizen science approach to surveillance. *One health (Amsterdam, Netherlands), 16*, 100480. https://doi.org/10.1016/j.onehlt.2022.100480

Sharma, R. K., Raghavendra, N., Mohanty, S., Tripathi, B. K., Gupta, B., & Goel, A. (2014). Clinical & biochemical profile of trichinellosis outbreak in North India. *The Indian Journal of Medical Research, 140*(3), 414–419.

Sharma, R., Thompson, P. C., Hoberg, E. P., Brad Scandrett, W., Konecsni, K., Harms, N. J., Kukka, P. M., Jung, T. S., Elkin, B., Mulders, R., Larter, N. C., Branigan, M., Pongracz, J., Wagner, B., Kafle, P., Lobanov, V. A., Rosenthal, B. M., & Jenkins, E. J. (2020). Hiding in plain sight: Discovery and phylogeography of a cryptic species of Trichinella (Nematoda: Trichinellidae) in wolverine (Gulo gulo). *International Journal for Parasitology, 50*(4), 277–287. https://doi.org/10.1016/j.ijpara.2020.01.003

Sofronic-Milosavljevic, L., Ilic, N., Pinelli, E., & Gruden-Movsesijan, A. (2015). Secretory products of Trichinella spiralis muscle larvae and immunomodulation: Implication for autoimmune diseases, allergies, and malignancies. *Journal of Immunology Research, 2015*, 523875. https://doi.org/10.1155/2015/523875

Somer, T., & Finegold, S. M. (1995). Vasculitides associated with infections, immunization, and antimicrobial drugs. *Clinical Infectious Diseases : An Official Publication of the Infectious Diseases Society of America, 20*(4), 1010–1036. https://doi.org/10.1093/clinids/20.4.1010

Song, Y. Y., Zhang, Y., Yang, D., Ren, H. N., Sun, G. G., Jiang, P., et al. (2018). The immune protection induced by a serine protease inhibitor from the foodborne parasite Trichinella spiralis. *Frontiers in Microbiology, 9*, 1544.

Sparaciari, F. E., Firth, C., Karlsson, E. A., & Horwood, P. F. (2025). Zoonotic disease risk at traditional food markets. *Journal of Virology, 99*, e00718-25. https://doi.org/10.1128/jvi.00718-25

Stachyra, A., & Bień-Kalinowska, J. (2025). Assessing diagnostic, vaccine and therapeutic potential of selected Trichinella proteins. *Food and Waterborne Parasitology, 40*, e00283. https://doi.org/10.1016/j.fawpar.2025.e00283

Stewart, G. L. (1989). Biological and immunological characteristics of Trichinella pseudospiralis. *Parasitology Today (Personal ed.), 5*(11), 344–349. https://doi.org/10.1016/0169-4758(89)90105-1

Stewart, G. L., Mann, M. A., Ubelaker, J. E., McCarthy, J. L., & Wood, B. G. (1988). A role for elevated plasma corticosterone in modulation of host response during infection with Trichinella pseudospiralis. *Parasite Immunology, 10*(2), 139–150. https://doi.org/10.1111/j.1365-3024.1988.tb00210.x

Sun, G. G., Lei, J. J., Ren, H. N., Zhang, Y., Guo, K. X., Long, S. R., et al. (2019). Intranasal immunization with recombinant Trichinella spiralis serine protease elicits protective immunity in BALB/c mice. *Experimental Parasitology, 201*, 1–10.

Takahashi, Y. (2021). Biology of Trichinella. In F. Bruschi (Ed.), *Trichinella and Trichinellosis* (pp. 77–101). Academic Press.

Tanabe, M. B., Schilling, M., & White, A. C. (2021). Neurocysticercosis and other CNS helminthic infections. In R. Hasbun, K. C. Bloch, & A. Bhimraj (Eds.), *Neurological complications of infectious diseases. Current clinical neurology*. Humana. https://doi.org/10.1007/978-3-030-56084-3_11

Tanariyakul, M., Estaris, J., & Saowapa, S. (2024). Trichinellosis-induced eosinophilic myocarditis mimicking Hypereosinophilic syndrome. *Cureus, 16*(4), e58946. https://doi.org/10.7759/cureus.58946

Tang, B., Li, J., Li, T., Xie, Y., Guan, W., Zhao, Y., Yang, S., Liu, M., & Xu, D. (2022). Vaccines as a strategy to control trichinellosis. *Frontiers in Microbiology, 13*, 857786. https://doi.org/10.3389/fmicb.2022.857786

Taylor, M. A., Coop, R. L., & Wall, R. (2015). Veterinary helminthology. In *Veterinary parasitology* (4th ed., pp. 1–109). Wiley. https://doi.org/10.1002/9781119073680.ch1

Thaenkham, U., Chaisiri, K., & Hui En Chan, A. (2022). Molecular systematics of parasitic nematodes. In *Molecular systematics of parasitic helminths*. Springer. https://doi.org/10.1007/978-981-19-1786-8_10

Thomas, L. F., de Glanville, W. A., Cook, E. A., & Fèvre, E. M. (2013). The spatial ecology of free-ranging domestic pigs (Sus scrofa) in western Kenya. *BMC Veterinary Research, 9*, 46. https://doi.org/10.1186/1746-6148-9-46

Thompson, R. C. A., Kutz, S. J., & Smith, A. (2009). Parasite Zoonoses and wildlife: Emerging issues. *International Journal of Environmental Research and Public Health, 6*(2), 678–693. https://doi.org/10.3390/ijerph6020678

Torreggiani, C., Garbarino, C., Pupillo, G., De Lorenzi, G., Sampieri, M., Massella, E., Rugna, G., Reggiani, A., Rubini, S., Frasnelli, M., Cirasella, L., Galletti, G., Marucci, G., Celani, F., & Maioli, G. (2025). Occurrence of Trichinella spp. in Grey wolves and red foxes: Insights from wild mammal surveillance in Emilia-Romagna (Italy). *Animals, 15*(24), 3532. https://doi.org/10.3390/ani15243532

USDA. (2018). *FSIS compliance guideline for the prevention and control of Trichinella and other parasitic hazards in pork products*. https://www.fsis.usda.gov/wps/wcm/connect/2ca75475-3efd-4fa7-8f34-7393c245a1df/Trichinella-Compliance-Guide-03162016.pdf?MOD 5AJPERES.

Vacca, F., & Le Gros, G. (2022). Tissue-specific immunity in helminth infections. *Mucosal Immunology, 15*(6), 1212–1223. https://doi.org/10.1038/s41385-022-00531-w

van Knapen, F. (2000). Control of trichinellosis by inspection and farm management practices. *Veterinary Parasitology, 93*(3–4), 385–392. https://doi.org/10.1016/s0304-4017(00)00353-8

Vignau, M. L., del Valle Guardis, M., Risso, M. A., & Eiras, D. F. (1997). Comparison between two methods for diagnosis of trichinellosis: Trichinoscopy and artificial digestion. *Memorias do Instituto Oswaldo Cruz, 92*(5), 585–587. https://doi.org/10.1590/s0074-02761997000500004

Virchow, R. (1859). Recherches sur le développement du Trichina spiralis. *Comptes Rendus de l'Académie des Sciences, 49*, 660–662.

Wakelin, D., & Goyal, P. K. (1996). Trichinella isolates: Parasite variability and host responses. *International Journal for Parasitology, 26*(5), 471–481. https://doi.org/10.1016/0020-7519(96)89377-5

Walden, H. S. (2021). Trichinella. In *Foodborne infections and intoxications* (pp. 379–389). Academic Press. https://doi.org/10.1016/B978-0-12-819519-2.00017-7

Wang, Z. Q., Cui, J., Wei, H. Y., Han, H. M., Zhang, H. W., & Li, Y. L. (2006). Vaccination of mice with DNA vaccine induces the immune response and partial protection against T. Spiralis infection. *Vaccine, 24*(8), 1205–1212.

Wang, S., Zhu, X., Yang, Y., Yang, J., Gu, Y., Wei, J., et al. (2009). Molecular cloning and characterization of heat shock protein 70 from Trichinella spiralis. *Acta Tropica, 110*(1), 46–51.

Wang, B., Wang, Z. Q., Jin, J., Ren, H. J., Liu, L. N., & Cui, J. (2013). Cloning, expression and characterization of a Trichinella spiralis serine protease gene encoding a 35.5 ákDa protein. *Experimental Parasitology, 134*(2), 148–154.

Wang, L., Wang, X., Bi, K., Sun, X., Yang, J., Gu, Y., Huang, J., Zhan, B., & Zhu, X. (2016). Oral vaccination with attenuated salmonella typhimurium-delivered TsPmy DNA vaccine elicits protective immunity against Trichinella spiralis in BALB/c mice. *PLoS Neglected Tropical Diseases, 10*(9), e0004952. https://doi.org/10.1371/journal.pntd.0004952

Wang, L., Sun, X., Huang, J., Zhan, B., & Zhu, X. (2017). Heterologous prime-boost vaccination enhances Ts Pmy's protective immunity against Trichinella spiralis infection in a murine model. *Frontiers in Microbiology, 8*, 1394.

Wang, A., Liu, X., Heckmann, A., Caignard, G., Vitour, D., Hirchaud, E., Liu, M., Boireau, P., Karadjian, G., & Vallée, I. (2022). A Trichinella spiralis new born larvae-specific protein, Ts-NBL1, interacts with host's cell vimentin. *Parasitology Research, 121*(5), 1369–1378. https://doi.org/10.1007/s00436-022-07479-7

Weatherly, N. F. (1983). Anatomical pathology. In W. C. Campbell (Ed.), *Trichinella and trichinosis*. Springer. https://doi.org/10.1007/978-1-4613-3578-8_5

Wei, J., Gu, Y., Yang, J., Yang, Y., Wang, S., Cui, S., & Zhu, X. (2011). Identification and characterization of protective epitope of Trichinella spiralis paramyosin. *Vaccine, 29*(17), 3162–3168.

Wilson, N. O., Hall, R. L., Montgomery, S. P., & Jones, J. L. (2015). Trichinellosis surveillance—United States, 2008–2012. *Morbidity and Mortality Weekly Report: Surveillance Summaries, 64*(1), 1–8.

WOAH. (2023). *World organisation for animal health*. Trichinellosis.

Wu, Z., Sofronic-Milosavljevic, L., Nagano, I., et al. (2008). Trichinella spiralis: Nurse cell formation with emphasis on analogy to muscle cell repair. *Parasites and Vectors, 1*, 27. https://doi.org/10.1186/1756-3305-1-27

Wu, Z., Nagano, I., & Takahashi, Y. (2013). Trichinella: What is going on during nurse cell formation? *Veterinary Parasitology, 194*(2–4), 155–159. https://doi.org/10.1016/j.vetpar.2013.01.044

Xu, J., Bai, X., Wang, L. B., Shi, H. N., Van Der Giessen, J. W. B., Boireau, P., et al. (2017a). Influence of adjuvant formulation on inducing immune response in mice immunized with a recombinant serpin from Trichinella spiralis. *Parasite Immunology, 39*(7), e12437.

Xu, J., Bai, X., Wang, L. B., Shi, H. N., Van Der Giessen, J. W. B., Boireau, P., et al. (2017b). Immune responses in mice vaccinated with a DNA vaccine expressing serine protease-like protein from the new-born larval stage of Trichinella spiralis. *Parasitology, 144*, 712–719. https://doi.org/10.1017/S0031182016002493

Xu, N., Wang, Y., Xu, N., Xiang, Z., Wang, D., Yu, Y., Liu, M., Liu, X., Tang, B., & Bai, X. (2025). Heterologous prime-boost immunization based on a human adenovirus 5 vectored containing Trichinella spiralis cystatin-like protein elicits protective mucosal immunity in mice. *PLoS Neglected Tropical Diseases, 19*(7), e0013323. https://doi.org/10.1371/journal.pntd.0013323

Yang, J., Yang, Y., Gu, Y., Li, Q., Wei, J., Wang, S., et al. (2008). Identification and characterization of a full-length cDNA encoding paramyosin of Trichinella spiralis. *Biochemical and Biophysical Research Communications, 365*(3), 528–533.

Yang, Y., Zhang, Z., Yang, J., Chen, X., Cui, S., & Zhu, X. (2010). Oral vaccination with Ts87 DNA vaccine delivered by attenuated salmonella typhimurium elicits a protective immune response against Trichinella spiralis larval challenge. *Vaccine, 28*(15), 2735–2742. https://doi.org/10.1016/j.vaccine.2010.01.026

Yang, W., Li, L. G., Liu, R. D., Sun, G. G., Liu, C. Y., Zhang, S. B., et al. (2015). Molecular identification and characterization of Trichinella spiralis proteasome subunit beta type-7. *Parasites & Vectors, 8*(1), 18.

Yang, J., Zhu, W., Huang, J., Wang, X., Sun, X., Zhan, B., & Zhu, X. (2016). Partially protective immunity induced by the 14-3-3 protein from Trichinella spiralis. *Veterinary Parasitology, 231*, 63–68.

Yen, M. T., & Chang, E. (2018). Trichinella spiralis trichinosis, orbital infection caused by. In U. Schmidt-Erfurth & T. Kohnen (Eds.), *Encyclopedia of ophthalmology*. Springer. https://doi.org/10.1007/978-3-540-69000-9_196

Yu, S. M., Li, T. T., Fu, B. Q., & Zhang, N. Z. (2025). Molecular diagnosis of Trichinella spp.: Current status and future prospects. *Parasitology Research, 124*(11), 136. https://doi.org/10.1007/s00436-025-08591-0

Yue, X., Sun, X. Y., Liu, F., et al. (2020). Molecular characterization of a Trichinella spiralis serine proteinase. *Veterinary Research, 51*, 125. https://doi.org/10.1186/s13567-020-00847-0

Zarlenga, D. S. (2021). Molecular methods for identifying and diagnosing Trichinella; from historical perspectives to the "-omics" revolution. In *Trichinella and Trichinellosis* (pp. 395–415). Academic Press. https://doi.org/10.1016/B978-0-12-821209-7.00001-9

Zarlenga, D., Thompson, P., & Pozio, E. (2020). Trichinella species and genotypes. *Research in Veterinary Science, 133*, 289–296. https://doi.org/10.1016/j.rvsc.2020.08.012
Zenker, F. A. (1860). Ueber die Trichinen-krankheit des Menschen. *Archiv für Pathologische Anatomie und Physiologie und für Klinische Medicin, 18*(5), 561–572.
Zenker, F. A. (1866). Beiträge zur Lehre von der Trichinenkrankheit. *Deutsches Archiv für Klinische Medizin, 1*, 90–124.
Zhang, N., Li, W., & Fu, B. (2018). Vaccines against Trichinella spiralis: Progress, challenges and future prospects. *Transboundary and Emerging Diseases, 65*(6), 1447–1458.
Zhou, M., Zhang, N., Zhang, M., et al. (2020). Culture, eating behavior, and infectious disease control and prevention. *Journal of Ethnic Foods, 7*, 40. https://doi.org/10.1186/s42779-020-00076-y
Zimmermann, W. J., & Beach, P. J. (1982). Efficacy of microwave cooking for devitalizing trichinae in pork roasts and chops. *Journal of Food Protection, 45*(5), 405–409. https://doi.org/10.4315/0362-028X-45.5.405
Zimmermann, W. J., Steele, J. H., & Kagan, I. G. (1973). Trichiniasis in the U.S. population, 1966-70: Prevalence Epidemologic factors. *Health Services Reports, 88*(7), 606–623. https://doi.org/10.2307/4594863

Taeniasis and Cysticercosis

35

Tuhina Banerjee, Shruti Kaushal, Sandhya Chaurasiya, and Souti Prasad Sarkhel

Abstract

Globally, taeniasis and cysticercosis are recognized as major zoonotic diseases caused by the adult and larval stages of *Taenia* tapeworms, respectively, presenting significant public health challenges and socioeconomic burdens. Humans serve as the definitive host for three major *Taenia* species: *Taenia solium*, *Taenia saginata*, and *Taenia asiatica*. Taeniasis, generally a mild intestinal infection, is acquired through the ingestion of raw or undercooked meat containing infective metacestode larvae. In contrast, cysticercosis, a more serious disease in humans, results from the ingestion of infective *Taenia* eggs, often through faeco-oral transmission or consumption of contaminated raw vegetables. Although *T. solium* larvae are the principal cause of human cysticercosis, the larval stages of other zoonotic *Taenia* species, such as *Taenia crassiceps*, *Taenia martis*, *Hydatigera taeniaeformis*, and *Versteria* spp., can also infect humans, although less frequently. Sporadic reports of human coenurosis also highlight the zoonotic potential of *Taenia multiceps*, *Taenia serialis*, *Taenia brauni*, and *Taenia glomeratus* larvae. In humans, cysticerci of *T. solium* can affect various tissues, including muscles, eyes, and skin, but lodgement in the brain leads to the most serious form of the disease, neurocysticercosis. High prevalence regions include Latin America, sub-Saharan Africa, Eastern Europe, India, and other Asian countries. Treatment options for taeniasis include praziquantel and niclosamide, while neurocysticercosis may be treated with antiseizure medications, corticosteroids, and

T. Banerjee (✉) · S. Kaushal · S. Chaurasiya
Department of Microbiology, Institute of Medical Sciences, Banaras Hindu University, Varanasi, Uttar Pradesh, India
e-mail: drtuhina105@bhu.ac.in

S. P. Sarkhel
Department of Veterinary Parasitology, FVAS, Institute of Agricultural Sciences, Banaras Hindu University, South Campus, Mirzapur, Uttar Pradesh, India

R. Biswas et al. (eds.), *Concept of One Health and Zoonoses*,
https://doi.org/10.1007/978-981-95-6627-3_35

anthelmintics. Prevention strategies emphasise improved hand hygiene, proper food handling, and regular deworming of humans and animals. Cysvax™, the only commercially available *Taenia* vaccine, utilises the recombinant oncosphere protein TSOL18 that provides 99.5% protection against porcine cysticercosis, indicating its significant potential to reduce cases of taeniasis in humans. Against this background, this chapter provides a concise overview of taeniasis and cysticercosis from a One Health perspective, focusing on the epidemiology, pathogenesis, immunopathology, clinical manifestations, diagnosis, treatment, and prevention of zoonotic *Taenia* tapeworms.

Keywords

Human taeniasis · Zoonotic *Taenia* · Human cysticercosis · Neurocysticercosis · TSOL

Introduction

Taeniasis and cysticercosis are two major food-borne neglected tropical diseases caused by different life cycle stages of taeniid tapeworms (Hossain et al., 2023). The family Taeniidae consists of two major genera, *Taenia* and *Echinococcus*, both of which have considerable zoonotic importance (Saelens et al., 2022). Recently, the genus *Taenia* has been divided into three distinct genera: *Taenia*, *Hydatigera*, and Versteria. Nevertheless, the primary genus of concern for public health remains *Taenia*, which includes three major zoonotic parasites: *Taenia solium* (the pork tapeworm), *Taenia saginata* (the beef tapeworm), and *Taenia asiatica* (the Asian tapeworm). These species are responsible for causing taeniasis in humans, as well as cysticercosis in humans, pigs, and cattle (Hoberg, 2002; Mendlovic et al., 2021). *Taenia* tapeworms are classified within the family Taeniidae, which falls under the order Cyclophyllidea, class Cestoda (true cestodes), and phylum Platyhelminthes (Verster, 1969; Rausch, 1994).

Taenia tapeworms primarily reside in the small intestines of carnivorous animals and humans during their adult stage, while their larval forms, known as metacestodes, are typically found in various tissues of mammalian hosts. Taeniid metacestodes typically manifest in four distinct forms, cysticerci, strobilocerci, coenuri, or hydatid cysts, depending on the specific species of the parasite (Brunetti and White Jr., 2012; Deplazes et al., 2019). In contrast to *Echinococcus*, which are the smallest members of the taeniid group, adult *Taenia* parasites are significantly larger tapeworms (Nakao et al., 2010). To date, numerous *Taenia* species have been documented infecting a wide range of hosts, and differentiation among these species is generally achieved through the examination of distinctive features associated with their heads (scolex), rostellum, and the morphology of their reproductive organs in mature segments (proglottids) (González et al., 2002; Knapp et al., 2011; Eom et al., 2020).

Medically important *Taenia* species are found globally, with their life cycle predominantly sustained through close contacts among humans, cattle, and pigs. These parasites predominantly affect marginalised and impoverished populations in low- and middle-income countries (Hoberg, 2002; Wong et al., 2020), where inadequate hygiene and sanitation practices create conditions that favour their persistence and spread. It is estimated that approximately 2.5 billion individuals globally are infected with the *Taenia* parasite, resulting in around 5000 reported deaths from neurocysticercosis each year (Aung and Spelman, 2016) (Table 35.1).

History

The prevalence of taeniasis and vesicular cysticercosis in humans and animals has been historically documented, tracing back to ancient civilisations. The *Ebers Papyrus*, an Egyptian medical manuscript from around 1550 BC, documents remedies for various health issues, including information about tapeworms, and provides insight into early anatomical understanding (Stewart, 1951). The occurrence of measly pork, or muscles of pigs infected with cysticerci, was known in ancient Greece. This is evident in Aristophanes' play "The Knights", written in 424 BC. In the play, Demosthenes, a character representing a young and inexperienced politician, threatens to inspect Cleon, the leader of the demagogues, for "pimples" on his tongue, a reference to the appearance of cysticerci in infected pork (Hoeppli, 1956; Flisser, 2013a). The ancient Greeks were thus familiar with cysticerci vesicles in pigs, as Aristotle described them as "bladders that are like hailstones" (Cox, 2002). Additionally, the ancient Indian text, the *Charaka Samhita*, discussed (?) cysticercosis in humans. Therefore, the ancient literature reflects that civilisational knowledge existed about the presence of cysticerci (measly) in pork and well-known occurrences of cysticerci cysts in humans, as well as recognising intestinal taeniasis in humans.

Although the scientific understanding of parasitic infections among ancient scholars was not as advanced as it is today, but their empirical observations and morphological descriptions of tapeworms were remarkably accurate for their time. However, the relationship between taeniasis in humans and cysticercosis pigs, as manifestations of two developmental stages of the same helminth species, remained elusive for centuries. During the sixteenth and seventeenth centuries, notable scholars such as Johannes Udalricus Rumler and Thomas Wharton described structures resembling cysticerci in humans, but the parasitic nature of these entities was not recognised at that time (Cox, 2002). A significant turning point occurred in the late seventeenth century when Philip Hartmann and Marcello Malpighi provided the first detailed descriptions of these cystic vesicles during autopsies of patients with a history of seizures (Grove, 2014). However, it was Edward Tyson's seminal work in 1683 on the anatomy and physiology of tapeworms in various animals, including humans, that established the foundation for contemporary understanding of taeniid biology (Tyson, 1683). Additionally, Johann Goeze's observations in 1784 on the morphological similarities between the scolex of human tapeworms and porcine

Table 35.1 Comprehensive overview of major and minor taeniid tapeworm species with zoonotic potential

Family	Name of the parasite	Common name	Geographical distribution	Name of the metacestode (larval) stage	Predilection site in IMH	Final hosts	Intermediate host	Egg features	Zoonotic Importance
Taeniidae	*Taenia saginata*	Beef tapeworm	Cosmopolitan	*Cysticercus bovis*	Muscle	Human	Bovines	"Taeniid-type" eggs, small (30–40 um), brown colour Round, thick, and smooth radiating shell containing six-hooked oncosphere	Yes, causes only taeniasis in humans
	Taenia solium	Pork tapeworm	Cosmopolitan	*Cysticercus cellulosae*	Muscle	Human	Pigs, humans		Yes, it can cause both taeniasis and cysticercosis in humans
	Taenia asiatica (syn. *T. saginata asiatica*)	Asian *Taenia*	Southeast Asia, Africa, Eastern Europe	*Cysticercus viscerotropica*	Liver, viscera	Human	Pigs, occasionally cattle		Yes, it causes only taeniasis in humans
	Other potentially zoonotic *Taenia* species:								
	T. crassiceps	–	Northern hemisphere	*Cysticercus longicollis*	Subcutis, body cavity	Dog, cat, fox,	Small rodents	"Taeniid-type" eggs, small (30–40 um) Round, thick, and smooth radiating shell containing six-hooked oncosphere	Yes
	T. martis	Marten tapeworm	Northern hemisphere	*Cysticercus*	Body cavity, rarely CNS	Mustelids	Rodents		Yes
	Versteria spp.	–	Northern hemisphere	Cysticercus	Liver	Mustelids, mink	Rodents		Yes
	Hydatigera taeniaeformis (syn. *T. taeniaeformis*)	Feline tapeworm	Worldwide	*Strobilocercus fasciolaris*	Liver	Cat, wild felids	Small rodents		Yes
	T. multiceps	Coenurus tapeworm	Worldwide	*Coenurus cerebralis*	CNS, connective tissue	Dog, fox, hyena	Domestic, wild ruminants		Yes
	T. serialis	–	Worldwide	*Coenurus serialis*	Subcutis and intermuscular tissue	Dog, fox	Lagomorphs		Yes
	T. brauni	–	Africa	*Coenurus brauna* (African type coenurosis)	–	Dog, fox, jackal	Rodents		Yes
	T. glomeratus	–	Africa	*Coenurus glomeratus* (African type coenurosis)	–	Dog, fox, jackal	Rodents		Yes

cysts were crucial in linking cysts to the larval stages of tapeworms (Goeze, 1784). Thereafter, in 1855, Friedrich Küchenmeister's critical differentiation between *T. solium* and *T. saginata* based on scolex morphology, coupled with his controversial experimentation in criminals who were sentenced to death, further elucidated the life cycle of *T. solium* and its role in taeniasis originating from the consumption of infective pork cysticerci. Ethical limitations in human experimentation unexpectedly drove a dependence on animal studies and clinical observations, culminating in the mid-nineteenth-century recognition of the causal relationship between *T. solium* eggs and cysticercosis. This was particularly significant given the epidemiological observations by Küchenmeister suggesting an absence of the disease in religious communities such as Jews and Muslims, who abstained from pork consumption (Grove, 2014). Subsequent research in the late nineteenth century, including the work of J.H. Oliver linking *T. saginata* to "measly" beef (Oliver, 1871), and an Italian veterinarian Edoardo Perroncito's validation of this connection, further improved the understanding of these parasitic infections (Perroncito, 1877). Neurocysticercosis was first identified in Asia in 1888 during an autopsy of a porter in Madras (Armstrong, 1888). In 1912, Krishnaswamy further enhanced the clinical understanding of human cysticercosis by reporting the presence of abundant *Cysticercus cellulosae* in various tissues during an autopsy (Krishnaswamy, 1912). In 1934, McArthur documented the prevalence of neurocysticercosis among British soldiers returning from India, particularly in patients presenting with epileptic seizures (MacArthur, 1934). Additionally, British physicians of that time observed that treatments for taeniasis could, in certain instances, inadvertently worsen the symptoms of cysticercosis (Singh and Sander, 2019). This observation has been supported by more recent research in neurocysticercosis immunopathology, which has established that the dying bladder worm stages can provoke a strong immune response (Takayanagui et al., 2011), leading to significant adverse effects, including tissue damage and inflammation in vital tissues.

Aetiology and Host

Taeniasis and cysticercosis are parasitic diseases caused by different species of *Taenia*. Humans serve as the definitive host for *Taenia solium*, *T. saginata*, and *T. asiatica*, harbouring adult tapeworms. Meanwhile, cattle, domestic pigs, and other suitable animal species may serve as intermediate hosts for the larval stages of these parasites. Among the various species of *Taenia*, *T. solium* is responsible for serious illness in humans, manifesting as human cysticercosis or neurocysticercosis. While taeniasis is typically asymptomatic in cases of intestinal infection, cysticercosis often leads to fatal outcomes due to the involvement of vital organs like the eye, the heart, and, frequently, the central nervous system. Humans commonly acquire taeniasis by consuming raw or undercooked pork, beef, or pig organs that are infected with cysticerci, while in humans, cysticercosis occurs due to accidental ingestion of *T. solium* eggs in contaminated vegetables or salads. *Taenia solium* and *T. saginata* have a global presence, while *T. asiatica* is primarily found in certain

Asian countries, with higher case numbers reported in countries like Korea, Taiwan, Thailand, Indonesia, India, and China (Flisser et al., 2005; Prasad et al., 2008; Eom et al., 2009; Dixon et al., 2021; Mendlovic et al., 2021; Moudgil et al., 2025).

Several species of cestode tapeworms can inadvertently invade human tissues, including *Taenia multiceps*, *Taenia serialis*, *Taenia crassiceps*, and *Hydatigera taeniaeformis*. Nevertheless, the primary cause of cysticercosis in humans is *Taenia solium*. Coenurosis, a rare zoonotic condition that mimics the symptoms of a tumour or cerebral tuberculosis in the central nervous system, is linked to *T. multiceps* (Varcasia et al., 2022). While domestic canines serve as the definitive host for *T. multiceps*, humans can occasionally act as aberrant intermediate hosts for this parasite. Cases of coenurosis in humans have also been reported in developing countries in Africa and South America, often attributed to the uncontrolled stray dog populations. Other *Taenia* species with lesser zoonotic significance that also infect various non-human primate species include *T. martis*, *T. multiceps*, *T. serialis*, *T. brauni*, *T. glomeratus*, and *T. crassiceps*. Additionally, infections caused by *Hydatigera taeniaeformis* (syn. *T. taeniaeformis*) in humans are quite rare, with only a single historical case reported. Conversely, *T. crassiceps* and *Versteria* spp. have primarily been documented from immunocompromised patients only (Symeonidou et al., 2018; Deplazes et al., 2019; Spickler, 2020; WOAH, 2021).

Transmission

Ingestion of raw or undercooked food contaminated with live cysticercus larvae and eggs of *Taenia* spp. leads to taeniasis and cysticercosis, respectively (Zhou et al., 2020). Humans can serve as intermediate hosts for *T. solium*, wherein ingestion of taeniid eggs leads to cysticercosis. This condition usually occurs when individuals accidentally ingest *T. solium* eggs, or when adult tapeworms turn back into the host's stomach due to reverse peristalsis (Smith, 1994). Ingestion of encysted larvae in undercooked beef or pork leads to maturation into adult tapeworms in the human gut that ultimately attach to the intestinal wall and produce thousands of eggs that are excreted with the faeces. Eggs of *Taenia solium* can infect humans via the faecal-oral route due to improper hygiene maintenance/ autoinfection, or by eating or drinking contaminated milk, food, and water (Gabriël et al., 2017). Upon ingestion of taeniid eggs, the oncosphere larva is liberated within the host's intestinal lumen. The embryophore layer of the egg gets effectively degraded by gastric and intestinal secretions, coupled with the action of bile salts, facilitating the release of the oncosphere for further development and infection. The oncospheres become activated through intestinal stimulation, penetrate the intestinal wall, and migrate to various organs, where they develop into cysticerci. Larval cysts of *T. solium* infect muscle, brain, or other tissues and sometimes become the reason for seizures in humans, also known as neurocysticercosis (NCC), the most prevalent helminthic cause of adult-onset seizures (Garcia et al., 2014; Taylor et al., 2015).

Taxonomy

Tapeworms of the genus *Taenia* are classified within the family Taeniidae, which falls under the order Cyclophyllidea under the class Cestoda of the phylum Platyhelminthes, within the kingdom Animalia (Wu et al., 2022). *Taenia solium*, *T. saginata*, and *T. asiatica* are the three major *Taenia* tapeworms of humans with noted public health significance (Parija and Ponnambath, 2013; Flisser, 2013b). The Taeniidae family initially included the genera *Echinococcus* and Taenia (Lavikainen et al., 2008), both notable for their pathogenicity, complex life cycles, and zoonotic significance (Deplazes et al., 2017), making them important subjects of study in both human and veterinary parasitology. The genus *Taenia*, which includes various species of tapeworms that affect both humans and animals, has recently undergone a significant taxonomic revision. This reclassification has led to the division of *Taenia* into three distinct genera: *Taenia*, *Hydatigera*, and *Versteria* (Dumendiak et al., 2025). Each of these genera encompasses species that have their unique characteristics and life cycles, reflecting advancements in our understanding of their evolutionary relationships and biology.

General Morphology of Adult Stages, Larvae, and Ova of *Taenia* spp.

The body of the adult *Taenia* tapeworm can be anatomically divided into three primary regions: the scolex, neck, and strobila (a chain of segmented proglottids). The scolex, the anterior attachment organ of tapeworms, is essential for secure attachment to the host's intestinal wall, which prevents expulsion by peristalsis. In certain *Taenia* species, the scolex may possess a dome-shaped rostellum, which can either be unarmed or armed with rows of hooks or hooklets (Taylor et al., 2015). Rostellar hooks, when present, penetrate the intestinal mucosa, further reinforcing the attachment. Typically, *Taenia solium* has a scolex armed with hooks and a rostellum. In contrast, *Taenia saginata* features an unarmed scolex, which includes only four suckers that facilitate attachment, lacking both hooks and a rostellum. Meanwhile, *Taenia asiatica* exhibits a moderately armed scolex featuring rudimentary rostellum and hooklets (Flisser et al., 2011; 2013c) (Fig. 35.1).

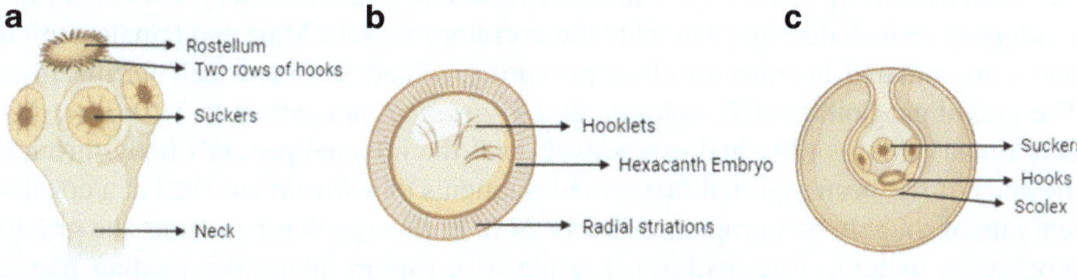

Fig. 35.1 *Taenia solium*. (**a**) Scolex magnified. (**b**) Egg. (**c**) Cysticercus. (Created with BioRender.com)

The neck of a *Taenia* is an elongated section situated between the scolex and the strobila. In *T. saginata*, the neck is longer compared to that of *T. solium*. This narrow neck region contains undifferentiated pluripotent cells that are essential for the formation of proglottids (strobilation) and enable the organism to regenerate its body segments. Proglottids, which house all the systems of the tapeworm, develop from the neck through a process called as strobilation. Constrictions appear between each segment, which facilitates the strobilation process by creating multiple segments. As the proglottids mature sexually, they move towards the posterior end of the strobila. The average length of strobila varies depending upon the species of the *Taenia*; it is largest in the case of *T. saginata*, moderate in the case of *T. asiatica*, and comparatively shorter in the case of *T.* solium (Taylor et al., 2015; Omeragić et al., 2023). The salient morphological features of various human taenia tapeworms are listed in Table 35.2.

Tapeworms lack a digestive system; therefore, they use their flattened and ribbon-like body structure, which is lined with living tissue known as the tegument. This adaptation maximises the surface area for effective nutrient absorption from the host's intestinal content (chyme). Every proglottid contains both male and female reproductive systems since it is a monoecious species. In mature segments, each proglottid contains a large number of testes, which are morphologically informative for species identification. The female reproductive system comprises a vagina, an ootype, an oviduct, a uterus, a vitelline gland, and the ovary, which is positioned in the posterior part of the proglottids. The vagina, which may bear a seminal vesicle for sperm storage, opens into the common genital atrium/sinus through the genital pore. The ootype, surrounded by unicellular Mehlis's glands, receives connections from the following: an oviduct, a blind uterus having a median stem and extensive lateral branches, and a compact vitelline gland. The oviduct connects the ovary, which is usually bilobed, to the ootype. From a morphological identification perspective, *T. solium* is characterised by the absence of a vaginal sphincter and the presence of three-lobed ovaries, distinguishing it from *T. saginata*. Genital pores facilitate the fertilisation process, which are single in each proglottid and irregularly alternate. Unlike pseudophyllidean, genital pores in *Taenia* do not allow passage of eggs to the environment. Therefore, eggs continue accumulating within the uterus, which occupies all the spaces within the gravid segments. The male reproductive system starts with the testes, which join with the vas efferens, and then the vas deferens and finally ends in the cirrus, which is generally enclosed in a pouch called a cirrus sac. The cirrus pouch in *T. solium* extends up to the excretory vessels, unlike *T. saginata*, which does not extend to the excretory vessels. Male and female genital pores are situated laterally in close proximity, which facilitates self-fertilisation. The gravid proglottids of *T. saginata* and *T. asiatica*, in contrast to *T. solium*, display remarkable motility and can actively exit the infected person's body through the anus. It has been reported that gravid segments of a *Taenia* are shed at a consistent rate of roughly 6–8 proglottids every 24 h. Following this expulsion, the gravid proglottids undergo true apolysis, leading to a rupture along the median plane, which subsequently releases a substantial number of eggs into the environment (Lumsden, 1975; Bowman et al., 2006; Ríos-Valencia et al., 2023).

Table 35.2 Major *Taenia* species responsible for human taeniasis/cysticercosis infections

Taenia species	Final hosts	Stages that infect humans	Stages present in humans	Key morphological characteristics of *Taenia* species
T. solium	Human	Metacestode larvae in undercooked pork (measly pork)	Adult tapeworms in the small intestine	Armed taeniid of humans, scolex of adult and larva: four cup-shaped suckers, rostellum armed with two circular rows of 22–32 hooks Size: 10–25 feet, width: 5–6 mm; gravid segments 10–12 mm Number of proglottids: 800–1000 Uteri: 5–12 lateral uterine branches Genital pores: irregularly alternate on the midsection of proglottids *Ovary*: two lateral lobes and one accessory central lobe Vaginal sphincter: absent
		Eggs in food or water contaminated with human faeces/faeco-oral transmission/reverse peristalsis	Cysticerci in the muscles, CNS, eyes, lungs, liver, subcutis, etc.	Type of cysticercus: musculotropic Three varieties of cysticercus.: cellulosae type (0.5–1.5 cm), intermediate type, and larger racemose type (up to 20 cm, ~ 60 ml fluids, with no scolex) Characteristics of ordinary cysticercus cyst: white ovoid up to 1.8 cm, enclosed in a thin fibrous capsule, invaginated protoscolex with armed rostellum
T. saginata	Human	Metacestode larvae in undercooked beef (measly beef)	Adult tapeworms in the intestine	Unarmed taeniid of humans, scolex of adult and larva: four sucking discs, no rostellum or hooks (unarmed scolex) Size: 10–75 feet, width: 6–7 mm, gravid segments 10–20 mm Number of proglottids: 1000–2000 Uteri: 15–35 lateral branches Ovary: two lobes Vaginal sphincter: present Genital pores: irregularly alternate on the midsection of proglottids Type of cysticercus: musculotropic, *Cysticercus bovis*
T. asiatica	Human	Metacestode larvae in raw/undercooked pork liver	Adult tapeworms in the intestine	Scolex of adult and larva: armed with two rows of rudimentary rostellum and hooklets Size: ~ 10 feet Uteri: 16–21 short lateral branches Unique posterior protuberances in the gravid proglottid Vaginal sphincter: present Genital pores: irregularly alternate on the midsection of proglottids Type of cysticercus: viscerotropic, *C. viscerotropica* Size of cysticerci: very small (1 mm), with armed scolex rudimentary rostellum and hooklets, wart-like growths on the outer surface of the cyst wall

Taenia Eggs

The eggs of human *Taenia* tapeworms have a fragile outer shell that disintegrates when the egg exits the host's body, exposing the thick radially striated embryophore coat of the oncosphere larva to the external environment. The larva (metacestode stage) of the taeniid worms within the mature eggs is often called a hexacanth larva due to its distinctive arrangement of six hooks. The embryophore layer, consisting of keratin blocks, serves as a protective covering for the hexacanth embryo, shielding it from desiccation and harsh environmental conditions (Jabbar et al., 2010; Flisser et al., 2011). Beneath the embryophore layer, a membranous lining covers the oncosphere larva; typically, the taeniid eggs containing the oncosphere larva usually have a diameter ranging from 30 to 57 μm.

Upon entering an appropriate environment in the gastrointestinal tract of a competent intermediate host and due to the action of the host's gastrointestinal juices and bile salts, the *Taenia* egg undergoes a transformative process where the embryophore layer disintegrates (Flisser et al., 2011; Mendlovic et al., 2021). This event facilitates the subsequent migration of the oncosphere into the host's skeletal muscle tissue through the circulatory system. In this stage, the hexacanth embryo continues its development and transitions from a solid structure into a vesicular form filled with an opalescent fluid, ultimately progressing into the cysticercus stage, which typically features a visible single invaginated scolex.

Epidemiology

Taeniasis and cysticercosis pose significant global health challenges, especially in regions where dietary habits involve the consumption of raw or undercooked pork and beef. Additionally, a lack of public awareness about proper hygiene practices and sanitation measures further perpetuates their high prevalence and endemic status, especially in low- and middle-income countries (LMICs) (Aung and Spelman, 2016; Dixon et al., 2021; Bhattacharya et al., 2024). Higher incidences of these diseases have been reported among individuals living in America, Latin America, sub-Saharan Africa, Eastern Europe, India, and Asia (Symeonidou et al., 2018; Ta and Blond, 2022; Wilson et al., 2025). Infection rates significantly increase in regions with poor sanitation and where pigs are raised in close proximity to human populations. In certain endemic regions, approximately 30% of epilepsy cases can be traced back to *T. solium* (WHO, 2022), particularly where people live in close contact with pigs. Amongst the three human *Taenia* spp., *T. saginata* and *T. asiatica* are usually found in remote endemic foci, while *T. solium* transmission cycles utilise frequent human-pig contact in peridomestic and peri-urban settings (Pray et al., 2019).

Humans play a vital role in the life cycle of the three major taeniid species, *Taenia saginata*, *T. asiatica*, and *T. solium*, acting as the sole definitive hosts for these medically important parasites. Among all the human *Taenia* worms, *T. solium* presents the most significant public health challenges (Garcia et al., 2020), as it can lead to cysticercosis, particularly neurocysticercosis (NCC), following the inadvertent ingestion of its eggs (Ito et al., 2003). *Taenia solium* primarily infects domestic pigs; however, incidental ingestion of viable eggs can also lead to infections in dogs and humans. Human hosts experience considerable morbidity when cysticerci migrate to aberrant sites (Ngwili et al., 2023), resulting from the inadvertent ingestion of the infective eggs of *Taenia solium*. In humans, these eggs develop into metacestodes, likely due to the similarities (?) in internal milieu between humans and porcine hosts (Pabst, 2020; Lunney et al., 2021; Deręgowska et al., 2025), which facilitates the subsequent development of the oncosphere larva within the human host. On the other hand, current evidence does not indicate that *T. saginata* causes cysticercosis in humans (McManus, 2008), although there is one report suggesting otherwise—indicating that it may potentially cause cysticercosis in some cases (Hou et al., 2023). *Taenia saginata* is also a cosmopolitan parasite, with cattle serving as its primary intermediate hosts globally (Eichenberger et al., 2020). Nonetheless, other herbivore species like animals, such as buffaloes, yaks, giraffes, zebras, llamas, and reindeer, can also act as potential intermediate hosts for this parasite (Nelson et al., 1965; Raether and Hänel, 2003).

Life Cycle

Human *Taenia* species involve humans as the definitive host and either cattle or pigs as the intermediate host, depending on the specific species (see Table 35.2). Notably, *Taenia solium* is unique in that humans can serve as both definitive and intermediate hosts. The diagrammatic life cycle of *Taenia* is described in Fig. 35.2. Humans infected with the adult larva of *Taenia* pass gravid proglottids or infective eggs with faeces **(1)**. *Taenia* eggs are notably resilient and capable of surviving for years even when subjected to harsh environmental conditions in the soil. Ingestion of the contaminated vegetation with human nightsoil or the eggs or gravid proglottids of the human *Taenia* parasites causes infection in the intermediate hosts (see Table 35.2) **(2)**. When ingested by the suitable mammalian hosts, the egg hatches and the hexacanth embryo/oncosphere comes out and penetrates **(3)** the gut wall and wanders in the body. The activated oncosphere larva goes to the striated muscles or other organs through the bloodstream, where it slowly matures into a cysticercus. A cysticercus can remain alive for many years in the suitable intermediate host species. Humans usually get the infection by eating raw or undercooked pork or beef infected with the metacestode stages (cysticerci) of the human *Taenia* species **(4)**. With the fluid-filled bladder around the scolex getting digested, leaving the scolex with little undifferentiated tissue (neck) as free, the scolex then attaches itself on the mucosa of the small intestine with the neck undergoing strobilation **(5)** and mature in the small intestine lumen **(6)**. Undifferentiated immature proglottids get mature, and fertilisation occurs. Fertilised eggs build up and turn the mature segments into gravid proglottids. Later, gravid segments of the tapeworm, filled with fertilised eggs, disintegrate (apolysis), and typical *Taenia*-type eggs are passed with the stool. The number of eggs in proglottids varies among different species of *Taenia* (generally,

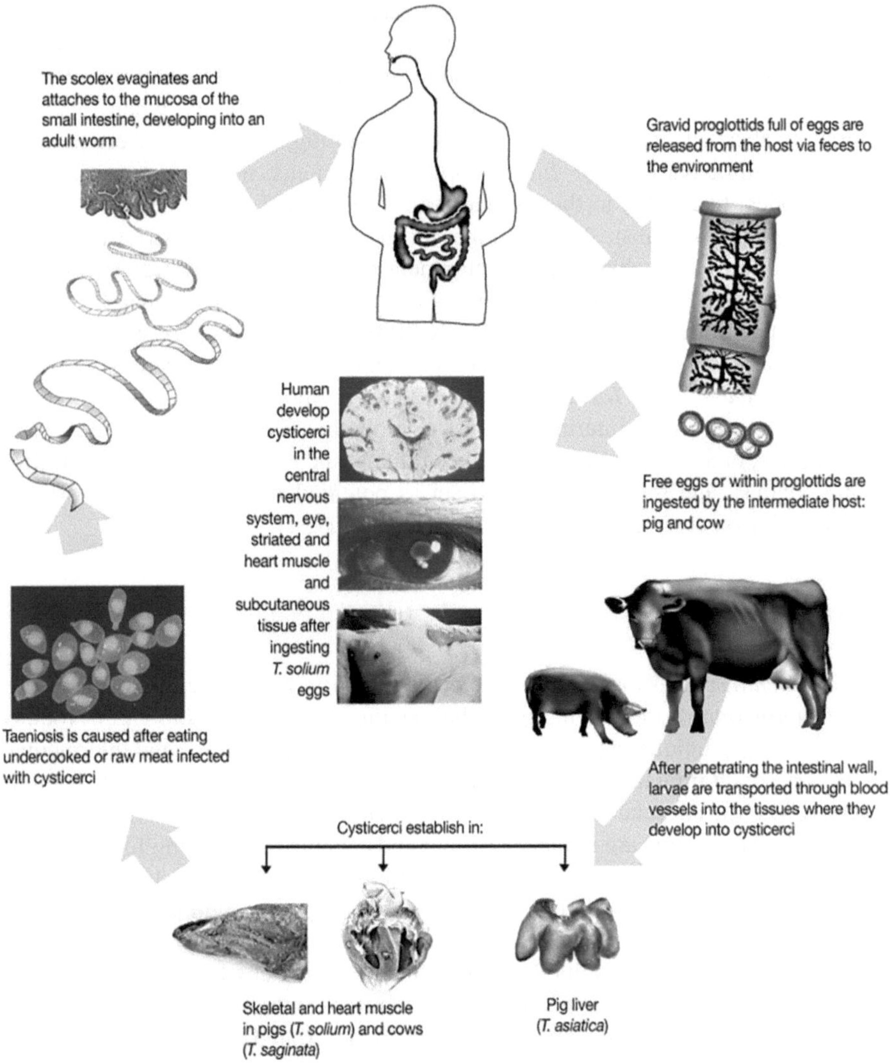

Fig. 35.2 Life cycle and host ranges of *Taenia solium* and *Taenia saginata*. (From Murrell et al., *Rev Sci Tech*, *32*(2): 559–569, 2013)

in the case of *T. saginata*, 100,000, and *T. solium*, 50,000 eggs per proglottids) (Murrell, 2013; Del Brutto and García, 2014; Taylor et al., 2015; Omeragić et al., 2023; Ortiz et al., 2024). Adult tapeworms inhabit the gastrointestinal lumen of their hosts for many years, often without causing significant harm. However, in cases of heavy worm infestation, they can cause complications such as mechanical damage to the intestinal wall, bowel obstruction, or nutritional deprivation of the host due to nutrient scavenging.

Neurocysticercosis

Neurocysticercosis (NCC) is the most prevalent parasitic disease affecting the nervous system and stands as one of the leading primary causes of epilepsy in regions such as China, Latin America, and India (Nash & Garcia, 2011). This condition often arises from the inadvertent ingestion of the eggs of the *Taenia solium*, commonly known as the pork tapeworm. This typically happens due to the contamination of vegetables or salads with human night soil (Fallah et al., 2016), or through the release of oncospheres from gravid proglottids transmitted via contaminated fingernails, particularly in cases of autoinfection. Furthermore, gravid proglottids, situated at the posterior end of *T. solium*, may get displaced into the stomach during episodes of reverse peristalsis in infected hosts (Murrell, 2005). Upon ingestion, *Taenia* eggs lose their protective capsules (embryophore) upon contact with the gastric and intestinal secretions and bile juices of the human host (Flisser et al., 2011; Mendlovic et al., 2021). The oncosphere then gets activated and penetrates the intestinal wall using its hooks, reaching its preferred sites via blood or lymphatic circulation (Verastegui et al., 2007). Upon arrival at predilection sites, the hexacanth embryos shed their hooks and develop into a single fluid-filled larval cyst, or bladder, commonly called as cysticercus cellulosae (Garcia et al., 2020), which contain an invaginated armed protoscolex (Taylor et al., 2015; Minciullo et al., 2016). This larval cyst displays preferences for certain tissues and organs in intermediate hosts. In humans, it primarily targets muscles, but it can also invade the eyes, brain, spinal cord, and other organs (Ito and Budke, 2021). Although the majority of infections are asymptomatic, the presence of cysticerci in vital organs such as the brain, spinal cord, or eyes can lead to serious complications and potentially fatal outcomes (Carpio, 2002; Gripper and Welburn, 2017; Del Brutto, 2022a; Garg et al., 2025).

In neurocysticercosis, when the developing oncosphere larva invades the central nervous system, it usually results in little or no immuno-inflammatory responses in the surrounding tissues (Carpio, 2002; Flisser et al., 2002). Hence, the larva continues to develop into a fluid-filled cyst through active- immuno-evasion and suppresion, often persisting for an extended period and frequently surviving for years within the brain tissue (Gripper and Welburn, 2017). Inside the cyst, the scolex manifests as an eccentric nodule, exhibiting morphological characteristics similar to those of the adult *Taenia solium* scolex (Garcia et al., 2003; Nash and Garcia, 2011). When the cyst undergoes degeneration, it releases fluid into the brain parenchyma, eliciting a strong immune response that can be detected through enhancement on contrast-enhanced computed tomography (CT) and magnetic resonance imaging (MRI) (Singhi, 2011; Garcia et al., 2014; Carpio et al., 2018; Prodjinotho et al., 2020). A calcified granuloma forms from the degenerating cyst, appearing on CT as non-enhancing punctate calcifications (Nash et al., 2017). Hydrocephalus can occur if the cyst enters the ventricles or cisterns of the brain (Garcia et al., 2014; Shrestha et al., 2025). Approximately 70–90% of individuals with acute signs of neurocysticercosis (NCC) typically present with seizures or headaches initially (Singhi, 2011). Those with hydrocephalus tend to have a higher mortality rate compared to individuals experiencing seizures (DeGiorgio et al., 2002).

Zoonotic Importance

Infection with *T. solium* is a major foodborne, neglected parasitic zoonosis causing significant public health challenges. Zoonotic cysticercosis/ taeniasis is having the potential to be transmitted from animals (primarily pigs, occasionally humans) to humans and vice versa. Understanding the factors facilitating the survival and dissemination of *Taenia* spp. from primary host to intermediate host through environment is necessary (Jansen et al., 2021). The mode of transmission between humans and pigs can be direct (coprophagy by pigs) or indirect (through environmental contamination with *T.* solium eggs) (Braae et al., 2016). *Taenia solium* infections in humans and other intermediate hosts are more likely to occur when food, soil, and water are contaminated; wind and invertebrates also play crucial roles in the transmission of the infective eggs (Jansen et al., 2021). Infective eggs present within gravid proglottids are passed with the stool of infected humans, contaminating the soil or water, where they remain viable for at least one year (Flisser, 2013b; Jansen et al., 2021). In certain developed countries, improperly treated sewage poses a risk of infection in cattle and other animals (Cabaret et al., 2002). Furthermore, the presence of viable Taenia spp. eggs resulting from inadequate sewage and effluent treatment, may contribute to the contamination of surface water (Jansen et al., 2021). In developing nations, inadequate hygiene practices, poor hand hygiene, and a shortage of proper latrines can lead to environmental contamination (Kungu et al., 2017). Consequently, crops, water sources, and soil may become tainted with *Taenia* eggs found in faeces. Environmental conditions, such as temperatures between 0°C and 20°C and optimum humidity, allow these eggs to survive for up to a year and frequently contaminate vegetables, water, and soil samples, thus increasing the risk of infection for both final and intermediate hosts. In resource-poor agricultural areas surrounding urban centres, the practice of using human night soil as fertiliser poses significant health risks to urban consumers (Jansen et al., 2021; Minani et al., 2025). This practice has often been linked to outbreaks of neurocysticercosis within vulnerable communities, making it a serious public health concern (Fig. 35.3).

Pathogenesis

Adult tapeworms are generally non-pathogenic or only mildly pathogenic. In contrast, the metacestode larval stages are considered pathogenic. Adult tapeworms can cause mild irritation and inflammation of the intestinal mucosa due to their scolex, especially in species where the scolex is equipped with hooks, as in the case of *T. solium*, but their overall impact is relatively minor (Heyneman, 1996). Infection in humans with the cysticerci of *T. solium* leads to a serious medical condition known as cysticercosis, which primarily affects the skeletal muscles, central nervous system (CNS), eyes, skin, lungs, heart, and other tissues. The cyst remains viable for several years, managing to evade the host's immune defences, particularly in the brain (Garcia et al., 2003; Garcia et al., 2014). The presence of the cyst

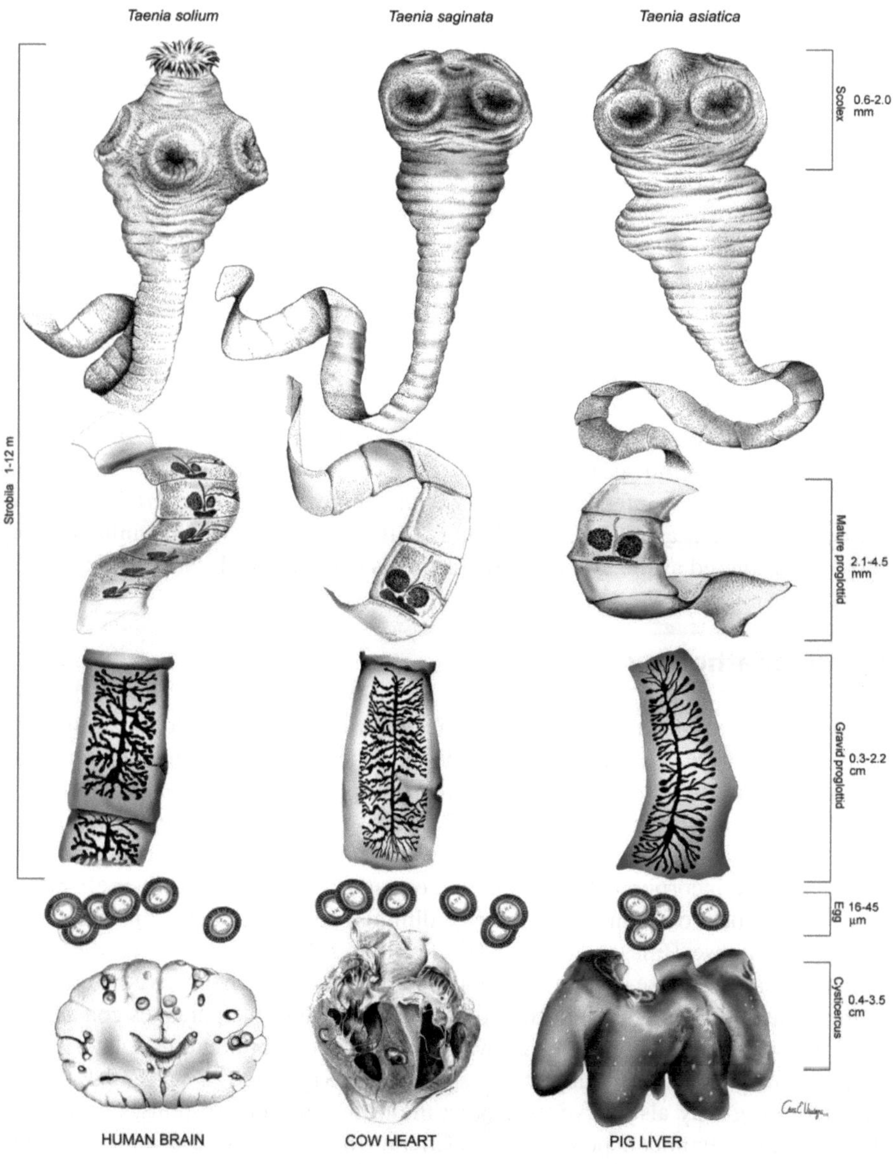

Fig. 35.3 Representation of three tapeworm species found in humans. Morphological characteristics and size variations of the different parasite stages of the life cycle are shown. (From Mendlovic et al., *Res. Vet. Sci.* 134: 69–77, 2021)

prompts the host tissue to release prostaglandins and other immunomodulatory substances, leading to the inactivation of complement proteins and cytokines (Garcia et al., 2014; Prodjinotho et al., 2020); Fan et al., 2021). In addition, an inflammatory immune response is triggered, characterised by elevated levels of IL-1β, TNF-α, and

IFN-γ (Singh et al., 2013; Prodjinotho et al., 2020). This response facilitates the migration of leukocytes across the blood-brain barrier (BBB) into the subarachnoid space via the choroid plexus. Furthermore, leukocytes can access brain tissue through the spaces surrounding blood vessels, allowing them to reach cysts and their contents (Prodjinotho et al., 2020).

Further, these cytokines and their elevated levels may stimulate astrocyte proliferation and gliosis, conditions that may potentially alter the physiological function of the hippocampus (Del Brutto et al., 2015; Prodjinotho et al., 2020). Further, when a live cyst is present, the host exhibits minimal cellular responses, characterised by a limited number of eosinophils and macrophages (Prasad et al., 2008; Garcia et al., 2014; Prodjinotho et al., 2020). In contrast, the presence of a dead cyst elicits a robust inflammatory response, surrounded by a diverse array of inflammatory cells, including leukocytes and multinucleated macrophages, which ultimately leads to the formation of a fibrotic zone (Restrepo et al., 2001; Garcia et al., 2014; Meena et al., 2016). The clinical manifestations of neurocysticercosis (NCC) may appear months to years post-infection, typically coinciding with the degenerative phase of the cysts (Garcia et al., 2018). This timing is often associated with the inflammatory response triggered as the cysts undergo necrosis (Singhi, 2011; Nash et al., 2019).

Immunopathology

In human neurocysticercosis (NCC), cysticerci primarily migrate to the central nervous system of the host, where they evade immune surveillance, and remain viable for years due to the immuno-protective properties of the blood-brain barrier (Flisser et al., 2002; Garcia et al., 2022). These cysts utilise various strategies to bypass immune detection, notably through antigenic masking of their surface coats. Especially, by incorporating host antibodies onto their tegument, they effectively camouflage their own antigens, thereby diminishing the chances of an effective immune response (Flisser et al., 2002; Prodjinotho et al., 2020). However, the inflammatory response is more pronounced in cases which involve the dead or dying parasites. Studies in murine models have revealed that parasites consistently shed their outer glycoprotein layer, which has a critical immunomodulatory function within the host's central nervous system (CNS) tissue (Alvarez et al., 2008). This shedding may also serve to redirect the host's immune responses away from the parasites' tegument, potentially enhancing the parasites' evasion of immune detection (Dell et al., 1999; Flisser et al., 2002; Alvarez et al., 2008). Most asymptomatic living cysts evoke a type 2 immune response, characterised by elevated levels of IgE and IgG1 antibodies. However, upon the initiation of treatment, this immune response shifts to a type 1 response as the cysts begin to disintegrate. This transition instigates a strong inflammatory reaction specifically targeting the decaying parasite. The involution of the parasite, whether it occurs naturally or as a result of antiparasitic treatment, leads to the development of a granulomatous immune response. This response is marked by the activation of various immune cells, including plasma cells, lymphocytes, macrophages, and eosinophils (Singh et al., 2015; Prodjinotho et al., 2020). The immune response to cysticercosis in the host is characterised by both humoral and cellular components that function synergistically against the infection (Grewal et al., 2000). In intermediate hosts, the mechanism of concomitant immunity plays a critical role, as it effectively prevents the establishment of subsequent infections by oncosphere larvae. This phenomenon was first documented by Harry M. Miller Jr. in the 1930s, when he demonstrated that rats previously infected with *Hydatigera* (*Taenia*) *taeniaeformis* metacestodes were able to prevent the establishment of new infections from subsequent experimental exposures (Miller, 1931; Lightowlers, 2010). This foundational observation laid the groundwork for the development of the initial *Taenia* vaccine, which focused on a ~45 kDa oncospheral antigen derived from *Taenia ovis*. This research culminated in the development of *T. ovis* 45W vaccine, introduced by Johnson and his team in 1989, representing the first detailed antigen vaccine against a parasitic infection (Johnson et al., 1989). This breakthrough was later recognised in the 200th edition of *Parasitology Today* as one of the 100 milestones in the field of parasitology (C).

Clinical Signs in Animals

Taeniasis in dogs and cats results from patent infections with host-specific adult *Taenia* tapeworms, which poses a minimal pathogenic risk unless heavy infestations occur. However, in cases of heavy infestations, subclinical disease may develop. Notably, dogs and cats can occasionally harbour zoonotic Taenia species, particularly metacestodes of T. solium, as documented in several reports (Fan et al., 1994; Sitali, et al. 2022). Clinical signs, when present, can vary widely depending on factors such as the severity of infection, the age and overall health of the host, and the specific breed. These signs may range from subtle manifestations such as irritability, pruritus ani, malaise, and a dull hair coat to more severe conditions like thriftiness, shaggy fur, colic, and mild diarrhoea, which may contain tapeworm segments (Taylor et al., 2015). However, in rare cases, fatal ileal obstruction can occur due to the coiling of large masses of tapeworms (Wilcox et al., 2009). On the other hand, cysticercosis is more common in herbivorous and omnivorous animals, as they are often the preferred intermediate hosts for various *Taenia* species. The larval stages of these parasites develop into cysticerci within diverse tissues, including striated muscle, heart, tongue, liver, kidney, eye, lymph nodes, subcutaneous tissue, and the central nervous system (WOAH, 2021). Notably, *Taenia solium* cysticerci, which develop in pigs, are larger than those of *Taenia saginata*, which are found in cattle. In contrast, Asian *Taenia* cysticerci are the smallest and typically remain localised to the surface and parenchyma of the pig liver (Flisser, 2013c; Galán-Puchades and Fuentes, 2000).

Clinical Importance in Humans

Taeniasis, a parasitic infection caused by adult tapeworms of the *Taenia* species, is often asymptomatic in both animals and humans. However, heavy infections in humans with *T. saginata*, *T. asiatica*, or *T. solium* can lead to mild to moderate symptoms such as dizziness, headache, stomach pain, abdominal distention, pruritus ani, nausea, and diarrhoea (Tanowitz et al., 2001). In rare cases of *T. saginata* infection, serious complications such as intestinal perforation (Hakeem et al., 2012) or biliary peritonitis may develop (Lopura et al., 2025).

A distinctive feature of *T. saginata* and *T. asiatica* infections is the daily, spontaneous expulsion of solitary proglottids from the anus, which may cause perianal itching. This phenomenon is not observed in *T. solium* infections. Additionally, diarrhoeic faeces in *T. saginata* and *T. asiatica* infections may contain gravid proglottids that exhibit unique crawling motility in freshly voided stool specimens (Mwape and Gabriël, 2014; Mendlovic et al., 2021).

Cysticercosis in humans is primarily caused by the larval stage of *Taenia solium*. In contrast to *Taenia saginata* and *Taenia asiatica*, the oncosphere larva of *T. solium* has the unique ability to invade and develop in various human tissues and organs. This can lead to a range of clinical outcomes that depend on the cyst's location, the developmental stage of the larvae (which can be vesicular, colloidal, granular, or calcified), and the overall infection load (Kraft, 2007; Garcia, 2018; Singhi and Saini, 2019; Ratcliffe et al., 2023; Rodríguez-Leyva et al., 2023). However, neurocysticercosis (NCC) represents the most severe form of manifestation, which occurs when oncosphere larvae migrate to aberrant locations like brain, crossing the blood-brain barrier and developing into cysticerci. This condition can result in significant neurological complications, particularly when the brain, spinal cord, or retina is compromised (Venkat et al., 2016). Within the central nervous system, cysts are primarily found in the cerebral parenchyma, followed by the meninges, ventricles, and spinal cord (Prasad et al., 2008). Clinical symptoms of NCC encompass a wide range of neurological disorders, notably acute seizures and paralysis. Additional manifestations include headaches, hydrocephalus, chronic meningitis, focal neurological deficits, psychological symptoms, dementia, as well as the presence of ocular or spinal cysts. Hydrocephalus is frequently observed when cysticerci localise in extra-parenchymal regions (Garcia et al., 2014; Del Brutto, 2022b).

Intraocular cysticercosis generally manifests as proptosis, diplopia, and vision loss, while extraocular cysts may resemble slow-growing tumours or nodules associated with localised inflammatory responses (Khandpur et al., 2014; Dhiman et al., 2017; Garg et al., 2025). Surgical intervention, although challenging, is often indicated in cases of ocular or orbital cysticercosis (Mehta et al., 2021). Cysticerci located in subcutaneous tissue may present as palpable nodules, whereas those in muscle tissue are typically asymptomatic (Rajput et al., 2025; Yadav and Parihar, 2025). However, heavy muscle infestations can produce symptoms such as muscular pain, weakness, or pseudohypertrophy (Prasad et al., 2008; Bandyopadhyay and Sen, 2009). The cysticercus undergo several stages of involution, each with distinct radiological characteristics that influence treatment strategies (Coyle and Tanowitz, 2009; Hurla et al., 2024). Hence, accurate diagnosis is crucial, and differential diagnosis with conditions such as tuberculosis, toxoplasmosis, or metaplastic brain tumours is vital to ensure appropriate clinical management of cysticercosis (Pretell et al., 2005). Cysts generally undergo various stages of development and involution, each with distinct radiological characteristics that inform treatment strategies. Advanced neuroimaging modalities facilitate clear visualisation of the scolex within cystic lesions, which often exhibit a "hole-with-dot" appearance on CT and MRI scans. Vital diagnostic information can also be obtained from the cyst's characteristics, including size, shape, and cyst wall morphology. Viable cysticerci usually measure under 20 mm in diameter and are well defined relative to adjacent brain parenchyma. Multiple cysticerci within the brain can produce the classic "Swiss cheese" appearance, while larger racemose cysts often resemble a "bunch of grapes" due to the fusion of vesicles. In endemic regions, cerebral cysts located in the brain parenchyma typically undergo calcification over time, resulting in a "starry night" appearance on CT imaging. Therefore, accurate diagnosis is paramount, and differential diagnosis with conditions such as tuberculosis, toxoplasmosis, or metaplastic brain tumours is essential for appropriate clinical management of the disease (Del Brutto et al., 2001; Garg, 2004; Del Brutto et al., 2017).

Prevention and Control in Animals and Humans

Neurocysticercosis (NCC) poses a substantial public health challenge, impacting an estimated 50 million individuals globally and contributing to over 50,000 deaths annually (CDC, 1993). In India, recent epidemiological investigations have indicated that epilepsy attributable to NCC results in approximately 2.1 million disability-adjusted life years (DALYs) each year (Singh et al., 2017). In India, cysticercosis remains endemic in peridomestic pig populations across the country, with select regions exhibiting particularly elevated prevalence rates (Prasad et al., 2011). Uttar Pradesh, for example, accounts for over 40% of NCC cases, indicating a localised hotspot for this infection (Manjunathachar et al., 2025). The majority of *T. solium* infections are prevalent in rural areas where pigs have access to untreated human faeces or sewage (Garcia et al., 2003). This, combined with the accessibility of contaminated pork, significantly enhances the risk of transmission.

Therefore, to mitigate the risk of intestinal *T. solium* infection in humans, it is recommended that whole pork be cooked to a minimum internal temperature of ≥63°C (≥ 145°F), verified with a food thermometer inserted into the thickest part of the meat and that the meat be allowed to rest for an additional 3 min post-cooking. Ground pork, on the other hand, should reach an internal temperature of at least ≥71°C (≥160°F) (Marie and Petri Jr, 2025). Further, it is imperative that food handlers/ chefs, particularly in endemic regions, are educated and strictly adhere to rigorous hand hygiene protocols (Carabin and Traoré, 2014). Moreover, individuals residing in endemic areas should avoid consuming raw foods or vegetables, that may be contaminated with human faecal matter, particularly in areas where night soil is used as fertilizer. Besides, one should also refrain from consuming raw or inadequately cooked pork to reduce the risk of infection.

Vaccines

Cysticercosis presents a considerable public health challenge, particularly in economically disadvantaged communities in low- and middle-income countries (LMICs). Notably, the International Task Force for Disease Eradication observed in 1993 that this disease is both preventable and potentially eradicable in endemic areas through implementing comprehensive surveillance and targeted interventions (CDC, 1993; O'Neal et al., 2011). In view of this, effective strategies for eradicating cysticercosis necessitate a multifaceted approach, incorporating systematic disease surveillance, initiatives to enhance living conditions, targeted public health education, and regulatory measures aimed at promoting the production of cysticerci-free pork. However, one of the most promising strategies for controlling cysticercosis is the development of effective vaccines for porcine cysticercosis. In this context, recent advancements in recombinant DNA technology have revolutionised vaccine development, enabling the rapid and precise identification, characterization and production of specific protective antigens. Some of the identified antigens have demonstrated efficacy in protecting pigs from infection, thereby reducing the incidence of the disease among pigs and contributing to production of cysticerci free pork. Such advancements could play a pivotal role in mitigating the risk of human cysticercosis from pork originating from endemic regions.

In the context of *Taenia solium*, several antigens have shown promise as potential vaccine candidates. Notable among these are the scolex protein antigen, the oncosphere proteins TSOL16, TSOL18, TSOL45-1A, and TSOL45-1B, as well as the muscle and tegument protein- paramyosin (Gauci et al., 2013; Butala et al., 2021; Hossain et al., 2023; Ríos-Valencia et al., 2023). Among these, TSOL18 has emerged as the most effective candidate for the development of recombinant vaccines aimed at porcine cysticercosis (PCC), achieving an impressive protection rate of 99.5% (Butala et al., 2021). This has led to a collaborative effort involving the Global Alliance for Livestock Veterinary Medicines (GALVmed), Indian Immunologicals Limited (IIL), and the University of Melbourne to scale up the TSOL18 vaccine (WHO and Thomas, 2015). Cysvax™, produced using the *Pichia pastoris* expression system and formulated with an oil adjuvant, was licensed for commercial production in India in 2016 to meet the substantial demand for this vaccine, especially in the densely populated pig farming regions of north-eastern India. Each 1-mL dose contains a minimum of 150 µg of the *Taenia solium* oncosphere antigen (TSOL 18). The primary vaccination involves a 1-mL intramuscular injection administered behind the ear to pigs at approximately 2 months of age. A booster dose is recommended 3–4 weeks later, with re-vaccination suggested every 6 months. When oxfendazole is administered concurrently with the second Cysvax vaccination, it can effectively eliminate any pre-existing viable infections, while Cysvax provides protection against future infections in vaccinated pigs (CystiTeam Group, 2019; Kabululu et al., 2020; Cysvax, 2026). This approach not only offers significant potential for establishing *Taenia*-free pig populations in affected regions but also mitigates the risk of human cysticercosis, particularly in endemic regions. Therefore, vaccinating pigs against porcine cysticercosis is a crucial strategy in combating this disease. Further, the implementation of mass vaccination programs for pigs, is essential for alleviating the global public health burden of human cysticercosis. Other potential vaccine candidates against PCC include SP3VAC, which incorporates KETc7, KETc1, and KETc12, as well as a phage-based SP3VAC vaccine that has undergone trials in Mexico (Huerta et al., 2001; Morales et al., 2008; Dixon et al., 2021; Kaur et al., 2021). However, in the case of *T. saginata*, currently, there are no vaccines available for the prevention of taeniasis in humans or cysticercosis in cattle caused by it (Butala et al., 2021). However, orthologues of genes that encode protective antigens from *T. ovis* have been identified as promising. Notably, TSA9, a homologue of *T. ovis* 45, and TSA18, a homologue of *T. ovis* 18, have demonstrated a high level of protection in cattle against challenge infections (Lightowlers et al., 1996; Okello and Thomas, 2017).

Diagnosis

The diagnosis of intestinal taeniasis is relatively straightforward through morphology-based identification of the body parts of the adult tapeworms or their eggs, which relies on various morphological characteristics such as the number, shape, and size of the scolex hooks or features of their gravid proglottids (Deplazes et al., 2019). While the scolex and rostellar hook morphology of taeniid worms generally remains consistent in larval stages, morphological characterization can sometimes become challenging due to potential variations between different developmental stages, and among different natural or accidental hosts. Therefore, laboratory diagnosis of taeniasis can be accomplished through various methods, including microscopic examination of eggs, serological assays for antibody detection, and advanced molecular diagnostic techniques (Mwape and Gabriël, 2014).

However, diagnosing based on egg morphology presents challenges, as taeniid eggs share similar/overlapping morphological characteristics, which can lead to inconclusive results when attempting to delineate species (Horak et al., 2019). To overcome this limitation, various serological tests, including complement fixation tests, immunofluorescence, hemagglutination, enzyme-linked immunosorbent assays (ELISA), and enzyme-linked immunotransfer blotting (EITB), have been developed. These tests typically employ pig serum or human cerebrospinal fluid (CSF)/ serum to detect specific antibodies against *Taenia* antigens. While serological assays, particularly Ag-ELISAs can distinguish between active and non-active NCC cases in humans, tests such as EITB or AG-ELISA should not be utilised as stand-alone diagnostic methods for NCC without accompanying neuroimaging confirmation (Jayshree and Mahadevan, 2022). Neuroimaging techniques, such as MRI and CT scans, are essential for confirming the diagnosis where neurocysticercosis (NCC) is suspected (Del Brutto, 2012; Rajshekhar, 2016; Gripper and Welburn, 2017). Additionally, serological tests are not effective in differentiating between cysts attributable to *T. solium* and *Taenia hydatigena* in pigs (Dorny et al., 2017). Therefore, traditional methods of meat and tongue inspection during post-mortem examination of pigs remain a common approach for detecting porcine cysticercosis (Lightowlers et al., 2015), particularly in resource-limited settings. However, these methods have limited sensitivity and can be influenced by the intensity of infection (Sithole et al., 2019). On the other hand, antibody-based assays generally exhibit high sensitivity; however, they may fail to differentiate between active and past infections. In contrast, antigen-based assays are valuable for estimating the burden and severity of infection. Importantly, in neurocysticercosis (NCC) patients, the negative conversion of antigen-based assays correlates closely with the complete resolution of cysticercosis infections.

In light of these challenges, the World Health Organisation and the Pan American Health Organisation recommend the enzyme immunotransfer blot (EITB) as the preferred diagnostic test for cysticercosis (PAHO, 1997). EITB assays employ a group of seven well-characterised lentil lectin-purified glycoproteins (LLGP) as the gold standard for serodiagnosis, boasting a sensitivity of 95% and a specificity of 100% (Coyle, 2014; Noh et al., 2014). Despite these impressive metrics, EITB assays have some limitations; they are time-consuming, require well-equipped laboratories, and necessitate highly skilled personnel. Furthermore, the preparation of LLGP involves the availability and handling of fresh cysts, as well as complex antigen processing and handling of pork samples. Additionally, EITB may lack the specificity to differentiate between *Taenia solium* and other related *Taenia* species, particularly *Taenia hydatigena*, which is also a zoonotic parasite and is often more prevalent than *T. solium* in free-ranging pigs (Salazar-Anton et al., 2012; Noh et al., 2014; Lightowlers et al., 2016; Gomez-Puerta et al., 2019; Trevisan et al., 2021). Consequently, recent studies have been initiated in various labs to explore suitable synthetic or recombinant antigens for the development of EITB-based or alternative, simpler techniques that maintain similar sensitivity and specificity for potential diagnostic assays.

Excretory-secretory products (ESPs) and other circulating antigens (CAgs) secreted by the parasites or peptides released through the exosomal pathways by the parasites offer promising avenues for developing sensitive antigen-based diagnostic assays. Exosomes are nanoscale extracellular vesicles that facilitate intercellular communication (Wang et al., 2020). Parasitic organisms exploit exosomes to transfer bioactive molecules to host cells, thereby modulating the host's immune response and enhancing the parasites' survival and adaptation. This cargo exchange significantly influences host-parasite dynamics, leading to alterations in the host's immunological pathways and responses. By analysing the molecular cargo within these exosomes, researchers can identify novel biomarkers for accurate and early diagnosis of parasitic infections (Menezes and Tasca, 2024; Pinheiro et al., 2024; Qadeer et al., 2024).

Recent advancements in lateral flow assay (LFA) technology have markedly increased their popularity as point-of-care diagnostic tools. When compared to conventional serology-based or PCR-based tests, LFAs present several advantages, including rapid results, user-friendliness, and cost-effectiveness, which make them ideal for point-of-care testing across various settings (Mubanga et al., 2021). These assays are versatile and can detect a wide array of analytes, including antigens, antibodies, and nucleic acids, such as those related to *T. solium*. They can be utilised in diverse environments, from healthcare facilities to remote locations, enabling accurate and timely diagnoses of neurocysticercosis (Stelzle et al., 2024; Van Damme et al., 2025; Sadr et al., 2025). Recent studies have highlighted the potential of LFAs for the rapid and precise detection of *Taenia solium*, particularly through the use of innovative technologies like quantum dots and up-converting phosphor nanoparticles, which have significantly improved the sensitivity and specificity for detecting *T. solium* antigens in neurocysticercosis (NCC) patients. By integrating these advanced technologies, LFAs can target specific *T. solium* antigens, such as rT24h, TS18 (an oncosphere-stage protein), and GP50 (a cystic-larval stage protein), thereby enhancing diagnostic accuracy (Lee et al., 2019; Zhang et al., 2021; Hossain et al., 2023). Furthermore, the incorporation of recombinant antigens, including rT24H, rES33, TsPKA-r, rAg2, and rAg1V1, holds promise for further enhancing diagnostic capabilities and expanding the applications of LFAs (Sako et al., 2000; Liu et al., 2017; Dixon et al., 2021; Manjunathachar et al., 2025).

The contemporary approach for diagnosing neurocysticercosis (NCC) involves the use of neuroimaging techniques to identify parasitic lesions within the central nervous system, along with the detection of specific antibodies or antigens associated with the parasite, *Taenia solium* (Del Brutto et al., 2017). Achieving a precise diagnosis can be complicated, as neuroimaging results may not always provide definitive indicators, and the sensitivity and specificity of immunological assays can vary significantly (Carpio et al., 2016). Nevertheless, neuroimaging remains an essential component in the accurate diagnosis of NCC. In this context, clinical manifestations, relevant epidemiological data, and results from immunodiagnostic assays play crucial roles in classifying cases as either "definitive" or "probable" NCC (Del Brutto et al., 2017).

Histopathological examination can provide confirmatory evidence for NCC by aiding in the identification of characteristic parasite structures in biopsy samples, such as the typical rostellum of *T. solium* (Del Brutto, 2022b). However, in cases involving racemose cysticerci, which are predominantly found in subarachnoid spaces, scolices are typically absent. Consequently, the identification of a distinct three-layered membrane can facilitate their accurate diagnosis (Del Brutto et al., 2017). Meanwhile older, calcified lesions may present calcareous corpuscles as additional evidence of the parasitic nature of the cysts (Nash et al., 2017). Notably, during the initiation of anthelmintic treatment, active cysts may undergo involution, subsequently transforming into calcified nodules, thus reinforcing the diagnosis of NCC (WHO, 2021). The spontaneous resolution of single enhancing lesions is considered confirmatory for NCC (Garg et al., 2000; Garg, 2002), as such phenomena are uncommon in other conditions, including brain tuberculosis, toxoplasmosis, or malignancies (Del Brutto et al., 2017). Additionally, cytochemical analysis of cerebrospinal fluid (CSF) can provide critical insights for differential diagnosis and should be integrated with clinical evaluations. In cases of retinal cysticercosis, fundoscopic examination may reveal characteristic lesions supporting the diagnosis of NCC. Lastly, diagnosing extraneural cysticercosis requires histopathological analysis or the identification of parasites within biopsy specimens (Del Brutto et al., 2017; Lobo et al., 2021). Radiological indicators such as "cigar-shaped" calcifications in the thigh or calf muscles or direct visualisation of the cysticercus in the anterior chamber may also suggest human cysticercosis (Del brutto et al., 2001).

While DNA-based diagnostic assays provide superior sensitivity and specificity, their application in diagnosing NCC is mostly limited to cases where cerebrospinal fluid (CSF) is accessible from suspected patients or individuals having extra-parenchymal neurocysticercosis (Zemmour et al., 2023). Various DNA diagnostic methods have been established to detect *T. solium* DNA from human stool samples or directly from the worm segments. Recently, loop-mediated isothermal amplification (LAMP), a DNA-based isothermal PCR technique, has emerged as a means to identify various taeniids at the species level (Nkouawa et al., 2009, 2016). Further, quantitative PCR (qPCR) has become a suitable method for identifying specific sequences of *T. solium*, allowing for effective species determination through primers targeting specific gene regions (Ng-Nguyen et al., 2017; O'Connell et al., 2020), including mitochondrial genes (such as 12S rRNA, NADH dehydrogenase 1, and cytochrome c oxidase subunit 1) and nuclear genes (like ribosomal internal transcribed spacers 1 and 2) (Deplazes et al., 2019). These molecular methods are considered the gold standard for diagnosis, as PCR-based identification tests require only a minimal volume of larval material. However, the availability of target DNA fragments is crucial for the broader application of DNA-based techniques. By integrating these advanced diagnostic tools, healthcare professionals can effectively diagnose and manage neurocysticercosis (NCC) cases, ultimately enhancing patient outcomes and strengthening public health strategies.

However, accurate species identification of *Taenia* parasites is essential for risk assessment and effective public health interventions. It is critical to rapidly and precisely identify different adult *Taenia* species infecting humans, as well as accurately identify any cysticerci from the infected individuals. While morphological examination can provide preliminary information, molecular and immunological methods deliver more precise and reliable results. Therefore, strategic application of advanced diagnostic techniques facilitates the differentiation of *T. solium* from other *Taenia* species, such as *T. saginata* and *T. asiatica*, each of which has unique epidemiological and clinical implications for preventing taeniasis/cysticercosis in risk prone communities (Table 35.3 and Figs. 35.4 and 35.5).

Table 35.3 Diagnostic techniques for human taeniasis/cysticercosis

Clinical	History of epileptic seizures Passage of *Taenia* proglottids in stool Palpable subcutaneous and muscle cysticerci Neuroimaging modalities (MRI, CT scan, X-ray) Ultrasound-guided aspiration of cyst fluid
Coprology	Faecal microscopy for characteristic *Taenia* eggs and proglottids Coproantigen ELISA Egg detection by DNA hybridisation techniques Copro-polymerase chain reaction (PCR)
Serological/biochemistry	EITB (antibody detection) ELISA (antigen detection) LFA assays CSF-cytochemistry
Molecular diagnostics	PCR targeting *Taenia* DNA LAMP PCR qPCR

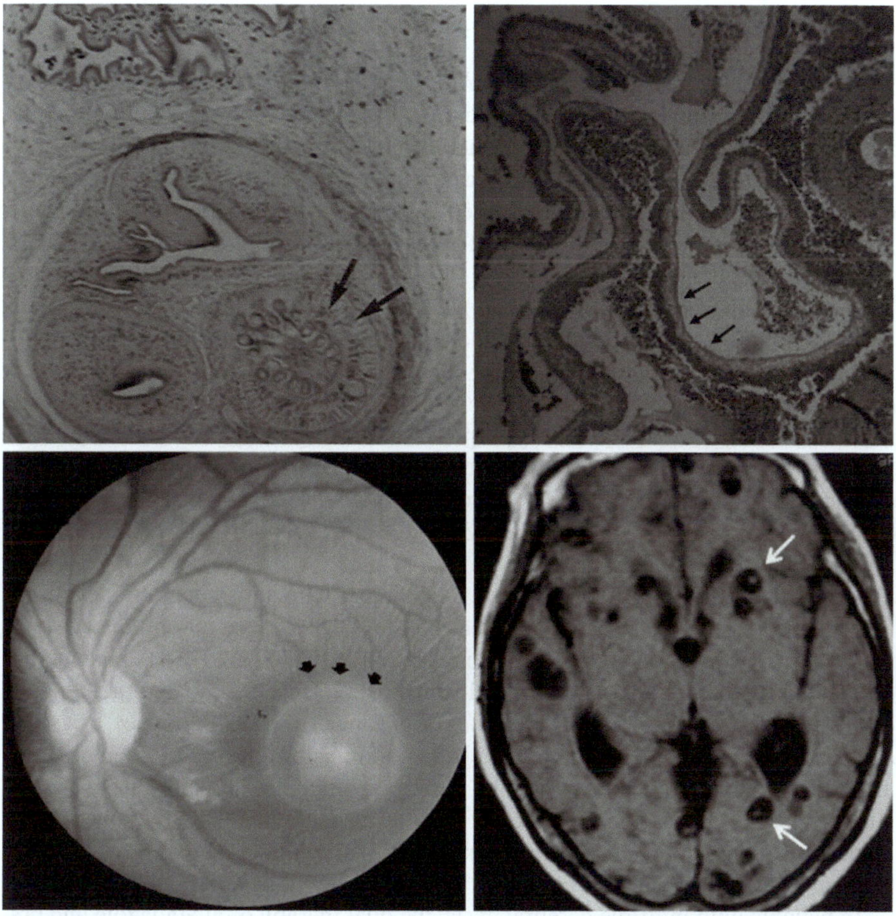

Fig. 35.4 Diagnostic measures for neurocysticercosis. Upper panel displaying biopsy findings with the characteristic rostellum of *Taenia solium* (large arrows) and three-layered membranes observed in racemose cysticerci (small arrows). Lower panel shows the typical appearance of subretinal cysticercus (arrowheads) and the "hole-with-dot" imaging produced by the scolex (white arrows). (From Del Brutto et al., *J. Neurol. Sci.* 372: 202–210, 2017)

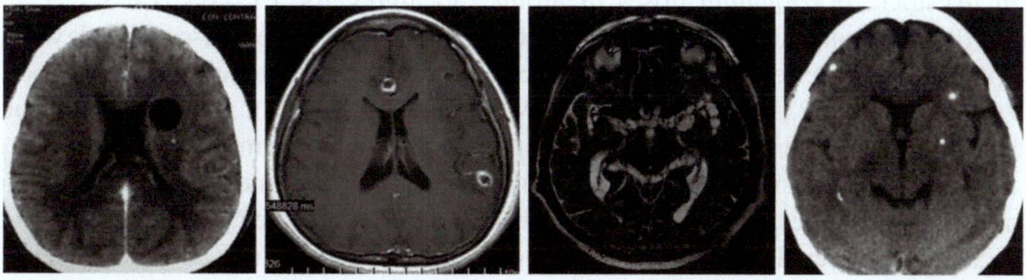

Fig. 35.5 Major neuroimaging criteria for neurocysticercosis, including cystic lesion without discernible scolex (left), ring-enhancing lesions (centre left), multilobulated cysts in basal subarachnoid cisterns (centre right), and typical parenchymal brain calcifications (right). (From Del Brutto et al., J. Neurol. Sci. 372: 202–210, 2017)

Treatment

Praziquantel and niclosamide are the primary anthelmintic drugs employed for the treatment of intestinal taeniasis (Haby et al., 2020). The recommended regimen for niclosamide is a single dose of 2 g for adults and children over the age of 6, administered post-breakfast, followed by a laxative after 2 h. For younger populations, the dosing is scaled—1 g for children aged 2–6 years and 500 mg for those under 2 years. Praziquantel is administered at a dosage of 5–10 mg/kg (Pawlowski et al., 2005; Symeonidou et al., 2018).

However, in the case of human neurocysticercosis, management typically involves a multimodal approach that includes antiseizure medications and corticosteroids alongside the potential use of albendazole or praziquantel, with surgery considered in select circumstances (Mendlovic et al., 2021). The use of corticosteroids aims to mitigate inflammatory responses following the lysis of active cysts due to antiparasitic treatment. Recent studies indicate that the combination of albendazole and praziquantel is markedly more efficacious in reducing viable brain cysts than monotherapy with albendazole, without a corresponding increase in adverse effects (Zhao et al., 2016).

The critical role of corticosteroids in conjunction with anti-metacestodal therapy is evidenced by their necessity in decreasing the risk of severe inflammatory shock that can arise from parasitic death (Raibagkar and Berkowitz, 2018). A recent clinical trial highlighted that a regimen of albendazole (15 mg/kg/day) and praziquantel (50 mg/kg/day) over 10 days significantly enhanced the resolution of cysts and decreased seizure incidence compared to the administration of either agent alone, all while maintaining a favourable safety profile. Dexamethasone was co-administered at 0.1 mg/kg/day, starting 1 day prior to antiparasitic therapy, supplemented by antiepileptics for 7 days prior to starting the antiparasitic regimen (Garcia et al., 2014; Garcia et al., 2016).

In patients with multiple cerebral cysts, the dual therapy of albendazole and praziquantel has demonstrated superior cysticidal efficacy over albendazole monotherapy, without producing any significant adverse effects (Garcia et al., 2016). Hence, this combination therapy can be recommended, particularly in cases of parenchymal NCC as it induces better modulation of immunoinflammatory responses by ensuring complete destruction of the cysts, and thus minimises the chances of residual calcifications (White Jr et al., 2018). While, for management of cysticercosis in livestock, praziquantel doses of around 50 mg/kg for 15 days (Flisser et al., 1990), or albendazole administered at 30 mg/kg/day for three consecutive days are optimal (Gonzalez et al., 1995). For porcine cysticercosis (C. tenuicollis), mebendazole is dosed at 25 mg/kg for 5 days (Vanparijs et al., 1987), while fenbendazole is efficacious against C. cellulosae infections in pigs when administered at 5 mg/kg for 7 days (Mkupasi et al., 2013). Nevertheless, a single treatment with oxfendazole at 30 mg/kg has been shown to effectively eliminate all muscle cysticerci in infected pigs (Gonzales et al., 1996). In companion animals, such as dogs and cats, praziquantel remains the gold standard anthelminthic agent, given its high efficacy against both mature and immature adult stages of taeniid cestodes present in the intestines.

References

Aung, A. K., & Spelman, D. W. (2016). Taenia solium taeniasis and cysticercosis in Southeast Asia. *The American Journal of Tropical Medicine and Hygiene, 94*(5), 947.

Armstrong, H. (1888). A case of Cysticercus cellulosae of brain in a native coolly. *Indian Medical Gazette, 23*, 252.

Alvarez, J. I., Rivera, J., & Teale, J. M. (2008). Differential release and phagocytosis of tegument glycoconjugates in neurocysticercosis: implications for immune evasion strategies. *PLoS Neglected Tropical Diseases, 2*(4), e218. https://doi.org/10.1371/journal.pntd.0000218

Bandyopadhyay, D., & Sen, S. (2009). Disseminated cysticercosis with huge muscle hypertrophy. *Indian Journal of Dermatology, 54*(1), 49–51. https://doi.org/10.4103/0019-5154.48987

Butala, C., Brook, T. M., Majekodunmi, A. O., & Welburn, S. C. (2021). Neurocysticercosis: Current perspectives on diagnosis and management. *Frontiers 8*, 256. https://doi.org/10.3389/fvets.2021.615703.

Brunetti, E., & White, A. C., Jr (2012). Cestode infestations: hydatid disease and cysticercosis. *Infectious disease clinics of North America, 26*(2), 421–435. https://doi.org/10.1016/j.idc.2012.02.001

Bhattacharya, D., Das, A., Sarkhel, S. P., & Satyaprakash, K. (2024). Milk-borne parasitic zoonoses. In *The microbiology, pathogenesis and zoonosis of milk borne diseases* (pp. 295–304). Academic Press.

Bowman, N., Donroe, J., Gilman, R. (2006). Cestodes. In: Ortega, Y.R. (eds) Foodborne Parasites. Food Microbiology And Food Safety Series. Springer, Boston, MA. https://doi.org/10.1007/0-387-31197-1_8

Braae, U. C., Devleesschauwer, B., Gabriël, S., Dorny, P., Speybroeck, N., Magnussen, P., Torgerson, P., & Johansen, M. V. (2016). CystiSim - An Agent-Based Model for Taenia solium Transmission and Control. *PLoS Neglected Tropical Diseases, 10*(12), e0005184. https://doi.org/10.1371/journal.pntd.0005184

Carabin, H., & Traoré, A. A. (2014). Taenia solium taeniasis and cysticercosis control and elimination through community-based interventions. *Current Tropical Medicine Reports, 1*, 181–193.

Carpio, A. (2002). Neurocysticercosis: An update. *The Lancet Infectious Diseases, 2*(12), 751–762.

Coyle, C. M. (2014). Neurocysticercosis: An update. *Current Infectious Disease Reports, 16*, 1–9.

Cox, F. E. G. (2002). History of human parasitology. Clinical Microbiology Reviews, 15(4), 595–612. https://doi.org/10.1128/cmr

Carpio, A., Fleury, A., Romo, M. L., & Abraham, R. (2018). Neurocysticercosis: the good, the bad, and the missing. *Expert Review of Neurotherapeutics, 18*(4), 289–301. https://doi.org/10.1080/14737175.2018.1451328

Cabaret, J., Geerts, S., Madeline, M., Ballandonne, C., & Barbier, D. (2002). The use of urban sewage sludge on pastures: the cysticercosis threat. *Veterinary Research, 33*(5), 575–597. https://doi.org/10.1051/vetres:2002040

Coyle, C. M., & Tanowitz, H. B. (2009). Diagnosis and treatment of neurocysticercosis. *Interdisciplinary Perspectives on Infectious Diseases, 2009*, 180742. https://doi.org/10.1155/2009/180742

CystiTeam Group (2019) The World Health Organization 2030 goals for Taenia solium insights and perspectives from transmission dynamics modelling. *Gates Open Research 3*(1546) https://doi.org/10.12688/gatesopenres.13068.2

Centers for Disease Control and Prevention (CDC). (1993). Recommendations of the International Task Force for Disease Eradication. *MMWR Recommendations and Reports, 42*, 1–38.

Cysvax, (2026). Cysvax product details. Let's vaccinate and stop spread of cysticercosis (pork tapeworm infection). Available: https://cysticercosis.in/cysvax-product-details.

Carpio, A., Fleury, A., Romo, M. L., Abraham, R., Fandiño, J., Durán, J. C., Cárdenas, G., Moncayo, J., Leite Rodrigues, C., San-Juan, D., Serrano-Dueñas, M., Takayanagui, O., & Sander, J. W. (2016). New diagnostic criteria for neurocysticercosis: Reliability and validity. *Annals of Neurology, 80*(3), 434–442. https://doi.org/10.1002/ana.24732

Cox, F. E. G. (1993). Milestones in parasitology. *Parasitology Today, 9*(10), 347–348. https://doi.org/10.1016/0169-4758(93)90076-R
Degiorgio, C. M., Houston, I., Oviedo, S., & Sorvillo, F. (2002). Deaths associated with cysticercosis: Report of three cases and review of the literature. *Neurosurgical Focus FOC, 12*(6), 1–4. https://doi.org/10.3171/foc.2002.12.6.3
Del Brutto, O. H. (2022b). Human Neurocysticercosis: An Overview. *Pathogens, 11*(10), 1212. https://doi.org/10.3390/pathogens11101212
Del Brutto, O. H. (2012). Neurocysticercosis in Western Europe: A re-emerging disease? *Acta Neurologica Belgica, 112*(4), 335–343.
Del Brutto, O. H. (2022a). Neurocysticercosis in the tropics (and beyond). *Current Tropical Medicine Reports, 9*(4), 119–129.
Del Brutto, O. H., & García, H. H. (2014). *Cysticercosis of the human nervous system*. Springer.
Del Brutto, O. H., Nash, T. E., White, A. C., Jr., Rajshekhar, V., Wilkins, P. P., Singh, G., Vasquez, C. M., Salgado, P., Gilman, R. H., & Garcia, H. H. (2017). Revised diagnostic criteria for neurocysticercosis. *Journal of the Neurological Sciences, 372*, 202–210.
Dixon, M. A., Winskill, P., Harrison, W. E., & Basáñez, M. G. (2021). Taenia solium taeniasis/cysticercosis: From parasite biology and immunology to diagnosis and control. *Advances in Parasitology, 112*, 133–217.
Deplazes, P., Eichenberger, R. M., & Grimm, F. (2019). Wildlife-transmitted Taenia and Versteria cysticercosis and coenurosis in humans and other primates. International journal for parasitology. *Parasites and wildlife, 9*, 342–358. https://doi.org/10.1016/j.ijppaw.2019.03.013
Deplazes, P., Rinaldi, L., Rojas, C. A., Torgerson, P. R., Harandi, M. F., Romig, T., Antolova, D., Schurer, J. M., Lahmar, S., Cringoli, G., Magambo, J., Thompson, R. C. A., & Jenkins, E. J. (2017). Global distribution of alveolar and cystic echinococcosis. *Advances in Parasitology, 95*, 315–493. https://doi.org/10.1016/bs.apar.2016.11.001.
Dumendiak, S., Halajian, A., Mekonnen, Y. T., Aschenborn, O., Camacho, G. J., Mackenstedt, U., Romig, T., & Wassermann, M. (2025). Hidden diversity of cestodes in wild African carnivores: II. Taeniidae - new findings and an updated checklist. *International journal for parasitology. Parasites and wildlife, 28*, 101116. https://doi.org/10.1016/j.ijppaw.2025.101116
Deręgowska, A., Solarska-Ściuk, K., Wnuk, M., & Schwarzbacherová, V. (2025). Pig as an Interorgan Communication Model and Its Applications in Biomaterials, Drug Delivery, and Bioengineering Research. *Comprehensive Physiology, 15*(5), e70043. https://doi.org/10.1002/cph4.70043
Del Brutto, O. H., Salgado, P., Lama, J., Del Brutto, V. J., Campos, X., Zambrano, M., & García, H. H. (2015). Calcified neurocysticercosis associates with hippocampal atrophy: a population-based study. *The American Journal of Tropical Medicine and Hygiene, 92*(1), 64–68. https://doi.org/10.4269/ajtmh.14-0453
Dell, A., Haslam, S. M., Morris, H. R., & Khoo, K. H. (1999). Immunogenic glycoconjugates implicated in parasitic nematode diseases. *Biochimica et Biophysica Acta, 1455*(2-3), 353–362. https://doi.org/10.1016/s0925-4439(99)00064-2
Dhiman, R., Devi, S., Duraipandi, K., Chandra, P., Vanathi, M., Tandon, R., & Sen, S. (2017). Cysticercosis of the eye. *International Journal of Ophthalmology, 10*(8), 1319–1324. https://doi.org/10.18240/ijo.2017.08.21
Del Brutto, O. H., Rajshekhar, V., White, A. C., W. Tsang, V. C., Nash, T. E., Takayanagui, O. M., Schantz, P. M., W. Evans, C. A., Flisser, A., Correa, D., Botero, D., Allan, J. C., Sarti, E., Gonzalez, A. E., Gilman, R. H. & García, H. (2001). Proposed diagnostic criteria for neurocysticercosis. *Neurology, 57* (2), 177–183. https://doi.org/110.1212/WNL.57.2.177.
Del Brutto, O. H. (2012). Diagnostic criteria for neurocysticercosis, revisited. *Pathogens and Global Health, 106*(5), 299–304. https://doi.org/10.1179/2047773212Y.0000000025
Rajshekhar, V. (2016). Neurocysticercosis: Diagnostic problems & current therapeutic strategies. *The Indian Journal of Medical Research, 144*(3), 319–326. https://doi.org/10.4103/0971-5916.198686
Dorny, P., Dermauw, V., Van Hul, A., Trevisan, C., & Gabriël, S. (2017). Serological diagnosis of Taenia solium in pigs: No measurable circulating antigens and antibody response follow-

ing exposure to Taenia saginata oncospheres. *Veterinary Parasitology, 245*, 39–41. https://doi.org/10.1016/j.vetpar.2017.08.008

Eom, K. S., Rim, H. J., & Jeon, H. K. (2020). Taenia asiatica: Historical overview of taeniasis and cysticercosis with molecular characterization. *Advances in Parasitology, 108*, 133–173. https://doi.org/10.1016/bs.apar.2019.12.004

Eichenberger, R. M., Thomas, L. F., Gabriël, S., Bobić, B., Devleesschauwer, B., Robertson, L. J., Saratsis, A., Torgerson, P. R., Braae, U. C., Dermauw, V., & Dorny, P. (2020). Epidemiology of taenia saginata taeniosis/cysticercosis: A systematic review of the distribution in East, Southeast and South Asia. *Parasites & Vectors, 13*, 1–11.

Eom, K. S., Jeon, H. K., & Rim, H. J. (2009). Geographical distribution of Taenia asiatica and related species. The Korean journal of parasitology, 47 Suppl(Suppl), S115–S124. https://doi.org/10.3347/kjp.2009.47.S.S115

Flisser, A., Correa, D., Avilla, G., & Marvilla, P. (2005). Biology of Taenia solium, Taenia saginata and Taenia saginata asiatica. In K. D. Murrell (Ed.), WHO/FAO/OIE Guidelines for the surveillance, prevention and control of taeniosis/cysticercosis (pp. 1–9). OIE (World Organisation for Animal Health). OIE (World Organisation for Animal Health), WHO (World Health Organization), FAO (Food and Agriculture Organization), Paris, 156 pp. https://iris.who.int/bitstream/handle/10665/43291/9290446560_eng.pdf?sequence=1

Flisser, A. (2013a). Epidemiology of Neurocysticercosis in Mexico: From a Public Health Problem to Its Control. InTech. https://doi.org/10.5772/53839

Flisser, A. (2013b). Taenia solium, Taenia saginata and Taenia asiatica. In R. G. Labbé & S. *García (Eds.), Guide to Foodborne Pathogens* (pp. 317–328). John Wiley & Sons, Ltd. Chichester. https://doi.org/10.1002/9781118684856.ch20

Flisser, A., Craig, P., & Ito, A. (2011). Cysticercosis and taeniosis: Taenia solium, Taenia saginata and Taenia asiatica. In P. Torgerson & D. Brown (Eds.), *Oxford Textbook of Zoonoses: Biology, Clinical Practice, and Public Health Control* (pp. 627–644). Oxford University Press.

Fallah, A. A., Makhtumi, Y., & Pirali-Kheirabadi, K. (2016). Seasonal study of parasitic contamination in fresh salad vegetables marketed in Shahrekord, Iran. *Food Control, 60*, 538–542.

Fan, X., Zhang, Y., Ouyang, R., Luo, B., Li, L., He, W., Liu, M., Jiang, N., Yang, F., Wang, L., & Zhou, B. (2021). Cysticercus cellulosae Regulates T-Cell Responses and Interacts With the Host Immune System by Excreting and Secreting Antigens. *Frontiers in Cellular and Infection Microbiology, 11*, 728222. https://doi.org/10.3389/fcimb.2021.728222

Flisser, A., Gonzalez, D., Shkurovich, M., Madrazo, I., Correa, D., Rodriguez-Carbajal, J., Cohen, S., Rodriguez-del-Rosal, E., Collado, M., & Fernandez, B. (1990). Praziquantel treatment of porcine brain and muscle Taenia solium cysticercosis. 1. Radiological, physiological and histopathological studies. *Parasitology Research, 76*(3), 263–269. https://doi.org/10.1007/BF00930823

Flisser, A., Correa, D., & Evans, C. A. W. (2002). Taenia solium cysticercosis: New and revisited immunological aspects. In G. Singh & S. Prabhakar (Eds.), Taenia solium Cysticercosis: From Basic to Clinical Science (pp. 15–24). CABI Publishing, Wallingford, UK. https://doi.org/10.1079/9780851996288.0015.

Fan, P. C., Chung, W. C., & Wu, J. C. (1994). Experimental infection of an isolate of Taenia solium from Hainan in domestic animals. *Journal of Helminthology, 68*(3), 265–266. https://doi.org/10.1017/s0022149x00014450

Flisser, A. (2013c). State of the Art of Taenia solium as Compared to Taenia asiatica. *Parasites, Hosts and Diseases, 51*(1), 43–49.

Gomez-Puerta, L., Vargas-Calla, A., Castillo, Y., Lopez-Urbina, M. T., Dorny, P., Garcia, H. H., Gonzalez, A. E., O'Neal, S. E., & Cysticercosis Working Group in Peru (2019). Evaluation of cross-reactivity to Taenia hydatigena and Echinococcus granulosus in the enzyme-linked immunoelectrotransfer blot assay for the diagnosis of porcine cysticercosis. *Parasites & Vectors, 12*(1), 57. https://doi.org/10.1186/s13071-018-3279-5

Garcia, H. H. (2018). Neurocysticercosis. *Neurologic Clinics, 36*(4), 851–864.

García, H. H., Gonzalez, A. E., Evans, C. A., & Gilman, R. H. (2003). Taenia solium cysticercosis. *The Lancet, 362*(9383), 547–556.

Gabriël, S., Dorny, P., Mwape, K. E., Trevisan, C., Braae, U. C., Magnussen, P., Thys, S., Bulaya, C., Phiri, I. K., Sikasunge, C. S., Makungu, C., Afonso, S., Nicolau, Q., & Johansen, M. V. (2017). Control of Taenia solium taeniasis/cysticercosis: The best way forward for sub-Saharan Africa?. *Acta Tropica, 165*, 252–260. https://doi.org/10.1016/j.actatropica.2016.04.010

Garcia, H. H., Nash, T. E., & Del Brutto, O. H. (2014). Clinical symptoms, diagnosis, and treatment of neurocysticercosis. *The Lancet Neurology, 13*(12), 1202–1215.

Garcia, H. H., Gonzalez, A. E., & Gilman, R. H. (2020). Taenia solium cysticercosis and its impact in neurological disease. *Clinical Microbiology Reviews, 33*(3), 10–1128.

Goeze, J. A. E. (1784). *Neueste Entdeckung: Dass die Finnen im Schweinefleische keine Drüsenkrankheit sondern wahre Blasenwürmer sind.* J.G. Heller.

Grove, D. (2014). Tapeworms, lice, and prions: a compendium of unpleasant infections. Oxford university press.

González, L. M., Montero, E., Puente, S., López-Velez, R., Hernández, M., Sciutto, E., Harrison, L. J., Parkhouse, R. M., & Gárate, T. (2002). PCR tools for the differential diagnosis of Taenia saginata and Taenia solium taeniasis/cysticercosis from different geographical locations. *Diagnostic Microbiology and Infectious Disease, 42*(4), 243–249. https://doi.org/10.1016/s0732-8893(01)00356-x

Garg, R. K., Garg, P., Paliwal, V. K., & Pandey, S. (2025). Vision loss in neurocysticercosis: A systematic review of case reports and series. *Journal of Clinical Neurology, 21*(2), 137–145.. https://doi.org/10.3988/jcn.2024.0565

Gripper, L. B., & Welburn, S. C. (2017). Neurocysticercosis infection and disease-A review. *Acta Tropica, 166*, 218–224. https://doi.org/10.1016/j.actatropica.2016.11.015

Garcia, H. H., Lescano, A. G., Gonzales, I., Bustos, J. A., Pretell, E. J., Horton, J., Saavedra, H., Gonzalez, A. E., Gilman, R. H., & Cysticercosis Working Group in Peru (2016). Cysticidal Efficacy of Combined Treatment With Praziquantel and Albendazole for Parenchymal Brain Cysticercosis. *Clinical Infectious Diseases, 62*(11), 1375–1379. https://doi.org/10.1093/cid/ciw134

Garg, R. K. (2002). Single enhancing computerized tomography-detected lesion in immunocompetent patients. *Neurosurgical Focus, 12*(6), e4. https://doi.org/10.3171/foc.2002.12.6.5

Garg, R.K., Singh, M.K., & Misra, S. (2000). Single-enhancing CT lesions in Indian patients with seizures: a review. *Epilepsy Research, 38*(2–3), 91–104. https://doi.org/10.1016/s0920-1211(99)00092-3

Gauci, C., Jayashi, C., & Lightowlers, M. W. (2013). Vaccine development against the Taenia solium parasite: the role of recombinant protein expression in Escherichia coli. *Bioengineered, 4*(5), 343–347. https://doi.org/10.4161/bioe.23003

Garcia, H. H., O'Neal, S. E., Noh, J., & Handali, S. (2018). Laboratory diagnosis of neurocysticercosis (Taenia solium). Journal of Clinical Microbiology, 56(9), e00424-18. https://doi.org/10.1128/jcm.00424-18. https://doi.org/10.1128/jcm.00424-18

Garcia, H. H., Verastegui, M. R., Arroyo, G., Bustos, J. A., & Gilman, R. H. (2022). New animal models of neurocysticercosis can help understand epileptogenesis in neuroinfection. *Frontiers in Molecular Neuroscience, 15*, 1039083.

Grewal, J. S., Kaur, S., Bhatti, G., Ganguly, N. K., Mahajan, R. C., & Malla, N. (2000). Kinetics of humoral & cellular immune responses in experimental cysticercosis in pigs infected with Taenia solium. *The Indian Journal of Medical Research, 111*, 43–49.

Gonzalez, A. E., Garcia, H. H., Gilman, R. H., Lopez, M. T., Gavidia, C., McDonald, J., Pilcher, J. B., & Tsang, V. C. (1995). Treatment of porcine cysticercosis with albendazole. *The American Journal of Tropical Medicine and Hygiene, 53*(5), 571–574. https://doi.org/10.4269/ajtmh.1995.53.571

Galán-Puchades, M. T., & Fuentes, M. V. (2000). The Asian Taenia and the possibility of cysticercosis. *The Korean Journal of Parasitology, 38*(1), 1–7. https://doi.org/10.3347/kjp.2000.38.1.1

Garg R. K. (2004). Diagnostic criteria for neurocysticercosis: some modifications are needed for Indian patients. *Neurology India, 52*(2), 171–177.

Garcia, H. H., Gonzales, I., Lescano, A. G., Bustos, J. A., Zimic, M., Escalante, D., Saavedra, H., Gavidia, M., Rodriguez, L., Najar, E., Umeres, H., Pretell, E. J., & Cysticercosis Working Group in Peru (2014). Efficacy of combined antiparasitic therapy with praziquantel and albendazole for neurocysticercosis: a double-blind, randomised controlled trial. *The Lancet. Infectious diseases, 14*(8), 687–695. https://doi.org/10.1016/S1473-3099(14)70779-0

Gonzales, A. E., Garcia, H. H., Gilman, R. H., Gavidia, C. M., Tsang, V. C., Bernal, T., Falcon, N., Romero, M., & Lopez-Urbina, M. T. (1996). Effective, single-dose treatment or porcine cysticercosis with oxfendazole. *The American Journal of Tropical Medicine and Hygiene, 54*(4), 391–394. https://doi.org/10.4269/ajtmh.1996.54.391

Haby, M. M., Sosa Leon, L. A., Luciañez, A., Nicholls, R. S., Reveiz, L., & Donadeu, M. (2020). Systematic review of the effectiveness of selected drugs for preventive chemotherapy for Taenia solium taeniasis. *PLoS Neglected Tropical Diseases, 14*(1), e0007873. https://doi.org/10.1371/journal.pntd.0007873

Hakeem, S. Y., Rashid, A., Khuroo, S., & Bali, R. S. (2012). Taenia saginata: A Rare Cause of Gall Bladder Perforation. *Case Reports in Surgery, 2012*, 572484. https://doi.org/10.1155/2012/572484

Hurła, M., Pikor, D., Kościelecka, K., Drelichowska, A., Banaszek, N., & Paul, M. (2024). Neurocysticercosis—Diagnostic Mystery: Current Status for Europe. *BioMed, 4*(3), 302–313. https://doi.org/10.3390/biomed4030024

Horak, I. G., Marchiondo, A. A., Colwell, D. D., Crafford, D., & Kok, D. J. (2019). Platyhelminthes. Parasiticide Screening: Volume 2: In Vitro and In Vivo Tests with Relevant Parasite Infection/Infestation Rearing Methods, and Host 1–133. https://doi.org/10.1016/B978-0-12-8165775.00006-5

Huerta, M., de Aluja, A. S., Fragoso, G., Toledo, A., Villalobos, N., Hernández, M., Gevorkian, G., Acero, G., Díaz, A., Alvarez, I., Avila, R., Beltrán, C., Garcia, G., Martinez, J. J., Larralde, C., & Sciutto, E. (2001). Synthetic peptide vaccine against Taenia solium pig cysticercosis: successful vaccination in a controlled field trial in rural Mexico. *Vaccine, 20*(1-2), 262–266. https://doi.org/10.1016/s0264-410x(01)00249-3

Hoeppli, R. (1956). The knowledge of parasites and parasitic infections from ancient times to the 17th century. *Experimental Parasitology, 5*(4), 398–419. https://doi.org/10.1016/0014-4894(56)90024-8

Heyneman, D. (1996). Cestodes. In S. Baron (Ed.), *Medical Microbiology* (4th ed.). University of Texas Medical Branch at Galveston. https://www.ncbi.nlm.nih.gov/books/NBK8399/

Hossain, M. S., Shabir, S., Toye, P., Thomas, L. F., & Falcone, F. H. (2023). Insights into the diagnosis, vaccines, and control of Taenia solium, a zoonotic, neglected *parasite. Parasites & vectors, 16*(1), 380. https://doi.org/10.1186/s13071-023-05989-6

Hoberg E. P. (2002). Taenia tapeworms: their biology, evolution and socioeconomic significance. *Microbes and infection, 4*(8), 859–866. https://doi.org/10.1016/s1286-4579(02)01606-4

Hossain, M. S., Shabir, S., Toye, P., Thomas, L. F., & Falcone, F. H. (2023). Insights into the diagnosis, vaccines, and control of Taenia solium, a zoonotic, neglected parasite. *Parasites & Vectors, 16*(1), 380.

Hou, J., Chen, W., Chen, R., He, C., Ma, Y., & Qu, J. (2023). A rare case of human taeniasis caused by Taenia saginata with species undetermined cysticercosis. *Parasitology, 150*(3), 240–247. https://doi.org/10.1017/S003118202200169X

Ito, A., & Budke, C. M. (2021). Genetic Diversity of Taenia solium and its Relation to Clinical Presentation of Cysticercosis. *The Yale Journal of Biology and Medicine, 94*(2), 343–349.

Ito, A., Nakao, M., & Wandra, T. (2003). Human taeniasis and cysticercosis in Asia. *The Lancet, 362*(9399), 1918–1920.

Jansen, F., Dorny, P., Gabriël, S., Dermauw, V., Johansen, M. V., & Trevisan, C. (2021). The survival and dispersal of Taenia eggs in the environment: What are the implications for transmission? A systematic review. *Parasites & Vectors, 14*, 1–16.

Jabbar, A., Swiderski, Z., Mlocicki, D., Beveridge, I., & Lightowlers, M. W. (2010). The ultrastructure of taeniid cestode oncospheres and localization of host-protective antigens. *Parasitology, 137*(3), 521–535. https://doi.org/10.1017/S0031182009991260

Johnson, K. S., Harrison, G. B., Lightowlers, M. W., O'Hoy, K. L., Cougle, W. G., Dempster, R. P., Lawrence, S. B., Vinton, J. G., Heath, D. D., & Rickard, M. D. (1989). Vaccination against ovine cysticercosis using a defined recombinant antigen. *Nature, 338*(6216), 585–587. https://doi.org/10.1038/338585a0

Jayshree, R. S., & Mahadevan, A. (2022). Non invasive tests for diagnosis of parasitic infections of CNS. *Indian Journal of Pathology and Microbiology, 65*(Supplement), S164–S175. https://doi.org/10.4103/ijpm.ijpm_1123_21.

Kaur, R., Arora, N., Rawat, S. S., Keshri, A. K., Sharma, S. R., Mishra, A., … Prasad, A. (2021). Vaccine for a neglected tropical disease Taenia solium cysticercosis: fight for eradication against all odds. *Expert Review of Vaccines, 20*(11), 1447–1458. https://doi.org/10.1080/14760584.2021.1967750

Kabululu, M. L., Ngowi, H. A., Mlangwa, J. E. D., Mkupasi, E. M., Braae, U. C., Colston, A., Cordel, C., Poole, E. J., Stuke, K., & Johansen, M. V. (2020). TSOL18 vaccine and oxfendazole for control of Taenia solium cysticercosis in pigs: A field trial in endemic areas of Tanzania. *PLoS Neglected Tropical Diseases, 14*(10), e0008785. https://doi.org/10.1371/journal.pntd.0008785

Krishnaswamy, C. S. (1912). Case of Cysticercus cellulosae. *Indian Medical Gazette, 47*, 43–44.

Knapp, J., Nakao, M., Yanagida, T., Okamoto, M., Saarma, U., Lavikainen, A., & Ito, A. (2011). Phylogenetic relationships within Echinococcus and Taenia tapeworms (Cestoda: Taeniidae): an inference from nuclear protein-coding genes. *Molecular Phylogenetics and Evolution, 61*(3), 628–638. https://doi.org/10.1016/j.ympev.2011.07.022

Kungu, J. M., Dione, M. M., Ejobi, F., Ocaido, M., & Grace, D. (2017). Risk factors, perceptions and practices associated with Taenia solium cysticercosis and its control in the smallholder pig production systems in Uganda: a cross-sectional survey. *BMC Infectious Diseases, 17*(1), 1. https://doi.org/10.1186/s12879-016-2122-x

Kraft, R. (2007). Cysticercosis: an emerging parasitic disease. *American Family Physician, 76*(1), 91–96.

Khandpur, S., Kothiwala, S. K., Basnet, B., Nangia, R., Venkatesh, H. A., & Sharma, R. (2014). Extensive disseminated cysticercosis. *Indian Journal of Dermatology, Venereology and Leprology, 80*(2), 137.

Lightowlers, M. W., Rolfe, R., & Gauci, C. G. (1996). Taenia saginata: vaccination against cysticercosis in cattle with recombinant oncosphere antigens. *Experimental Parasitology, 84*(3), 330–338. https://doi.org/10.1006/expr.1996.0121

Lightowlers, M. W., Assana, E., Jayashi, C. M., Gauci, C. G., & Donadeu, M. (2015). Sensitivity of partial carcass dissection for assessment of porcine cysticercosis at necropsy. *International Journal for Parasitology, 45*(13), 815–818. https://doi.org/10.1016/j.ijpara.2015.08.004

Lightowlers M. W. (2010). Fact or hypothesis: concomitant immunity in taeniid cestode infections. *Parasite Immunology, 32*(8), 582–589. https://doi.org/10.1111/j.1365-3024.2010.01227.x

Lopura, L. D., Gelan, E. A., Mengiste, D. G., & Yehualawork, S. F. (2025). Taenia saginata infection: A rare case of jejunal perforations from Ethiopia: A case report. *International Journal of Surgery Case Reports, 127*, 110974. https://doi.org/10.1016/j.ijscr.2025.110974

Lee, C., Noh, J., O'Neal, S. E., Gonzalez, A. E., Garcia, H. H., Cysticercosis Working Group in Peru, & Handali, S. (2019). Feasibility of a point-of-care test based on quantum dots with a mobile phone reader for detection of antibody responses. *PLoS Neglected Tropical Diseases, 13*(10), e0007746. https://doi.org/10.1371/journal.pntd.0007746

Lightowlers, M. W., Garcia, H. H., Gauci, C. G., Donadeu, M., & Abela-Ridder, B. (2016). Monitoring the outcomes of interventions against Taenia solium: options and suggestions. *Parasite Immunology, 38*(3), 158–169. https://doi.org/10.1111/pim.12291

Liu, G., Liang, P., Zhang, S., Guo, A., Wang, L., Zheng, Y., & Luo, X. (2017). TsPKA-r: a potential immunodiagnostic antigen for the detection of porcine cysticercosis. *Acta Tropica, 171*, 80–85. https://doi.org/10.1016/j.actatropica.2017.03.026

Lobo, F. D., Vatsala, K. B., A S, D. A., & Rai, S. (2021). Cysts that Still Persist: A Case Series of Cysticercosis on Histopathological Evaluation. Cysts that Still Persist: A Case Series of Cysticercosis on Histopathological Evaluation. *Turk Patoloji Dergisi, 37*(3), 254–257. https://doi.org/10.5146/tjpath.2020.01518

Lumsden, R. D. (1975). The Tapeworm Tegument: A Model System for Studies on Membrane Structure and Function in Host-Parasite Relationships. *Transactions of the American Microscopical Society, 94*(4), 501–507. https://doi.org/10.2307/3225522

Lavikainen, A., Haukisalmi, V., Lehtinen, M. J., Henttonen, H., Oksanen, A., & Meri, S. (2008). A phylogeny of members of the family Taeniidae based on the mitochondrial cox1 and nad1 gene data. *Parasitology, 135*(12), 1457–1467. https://doi.org/10.1017/S003118200800499X

Lunney, J. K., Van Goor, A., Walker, K. E., Hailstock, T., Franklin, J., & Dai, C. (2021). Importance of the pig as a human biomedical model. *Science Translational Medicine, 13*(621), eabd5758. https://doi.org/10.1126/scitranslmed.abd5758

McManus D. P. (2008). Taenia in the gastrointestinal tract. *The New England Journal of Medicine, 358*(3), 311. https://doi.org/10.1056/NEJMc072882

Minani, S., Gasogo, A., Ntirandekura, J.-B., Trevisan, C., & Gabriël, S. (2025). Stakeholders' practices in the pork value chain: Potential risks for Taenia solium and Toxoplasma gondii transmission in Burundi. *Preventive Veterinary Medicine, 239*, 106492. https://doi.org/10.1016/j.prevetmed

MacArthur, W. P. (1934). Cysticercosis as a cause of epilepsy in man. *Transactions of the Royal Society of Tropical Medicine and Hygiene, 26*, 525–528.

Mendlovic, F., Fleury, A., & Flisser, A. (2021). Zoonotic Taenia infections with focus on cysticercosis due to Taenia solium in swine and humans. *Research in Veterinary Science, 134*, 69–77.

Murrell, K. D. (2013). Zoonotic foodborne parasites and their surveillance. *Revue Scientifique et Technique, 32*(2), 559–569.

Moudgil, A. D., Nehra, A. K., & Moudgil, P. (2025). Taenia asiatica: Mitochondrial signatures based analysis of an emerging public health threat in India. *New Microbes and New Infections, 63*, 101562. https://doi.org/10.1016/j.nmni.2024.101562

Murrell K.D. (2005). Epidemiology of taeniosis and cysticercosis. In FAO/WHO/OIE Guidelines for the surveillance, prevention and control of taeniosis/cysticercosis (K.D. Murrell, ed.). OIE, Paris, 27– 34.

Minciullo, P. L., Cascio, A., Isola, S., & Gangemi, S. (2016). Different clinical allergological features of Taenia solium infestation. *Clinical and Molecular Allergy: CMA, 14*, 18. https://doi.org/10.1186/s12948-016-0056-x

Mehta, S., Verma, P. K., Rana, R., & Bhat, N. K. (2021). Subconjunctival cysticercosis - Is surgery always indicated? Role of medical management-literature review and report of two cases. *Journal of Family Medicine and Primary Care, 10*(9), 3519–3521. https://doi.org/10.4103/jfmpc.jfmpc_1461_20

Menezes, S. A., & Tasca, T. (2024). Extracellular vesicles in parasitic diseases - from pathogenesis to future diagnostic tools. *Microbes and Infection, 26*(4), 105310. https://doi.org/10.1016/j.micinf.2024.105310

Mubanga, C., Van Damme, I., Trevisan, C., Schmidt, V., Phiri, I. K., Zulu, G., Noh, J., Handali, S., Mambo, R., Chembensofu, M., Masuku, M., Reynders, D., Jansen, F., Bottieau, E., Magnussen, P., Winkler, A. S., Dorny, P., Mwape, K. E., & Gabriël, S. (2021). Evaluation of an Antibody Detecting Point of Care Test for Diagnosis of Taenia solium Cysticercosis in a Zambian Rural Community: A Prospective Diagnostic Accuracy Study. *Diagnostics, 11*(11), 2121. https://doi.org/10.3390/diagnostics11112121

Morales, J., Martínez, J. J., Manoutcharian, K., Hernández, M., Fleury, A., Gevorkian, G., Acero, G., Blancas, A., Toledo, A., Cervantes, J., Maza, V., Quet, F., Bonnabau, H., de Aluja, A. S., Fragoso, G., Larralde, C., & Sciutto, E. (2008). Inexpensive anti-cysticercosis vaccine: S3Pvac expressed in heat inactivated M13 filamentous phage proves effective against naturally acquired Taenia solium porcine cysticercosis. *Vaccine, 26*(23), 2899–2905. https://doi.org/10.1016/j.vaccine.2008.03.042

Manjunathachar, H. V., Saravanan, B. C., Joshi, C., Mohmad, A., Aravind, M., Jacob, S. S., & Sankar, M. (2025). Development of point-of-care immunodiagnostic test for Taenia solium cysticercosis in pigs. *Research in Veterinary Science, 182*, 105466.

Marie, C., & Petri, W. A., Jr. (2025). Taenia asiatica (Asian tapeworm) infection. In MSD Manual Professional Version. MSD Manuals. https://www.msdmanuals.com/professional/infectious-diseases/cestodes-tapeworms/taenia-asiatica-asian-tapeworm-infection

Meena, D., Gupta, M., Jain, V. K., & Arya, R. K. (2016). Isolated intramuscular cysticercosis: Clinicopathological features, diagnosis and management - A review. *Journal of Clinical Orthopaedics and trauma, 7*(Suppl 2), 243–249. https://doi.org/10.1016/j.jcot.2016.06.016

Miller, H. M., Jr. (1931). Immunity of the albino rat to superinfestation with Cysticercus fasciolaris. *Journal of Preventive Medicine, 5*, 453–464.

Mwape, K. E., & Gabriël, S. (2014). The parasitological, immunological, and molecular diagnosis of human taeniasis with special emphasis on Taenia solium taeniasis. *Current Tropical Medicine Reports, 1*(4), 173–180.. https://doi.org/10.1007/s40475-014-0028-5

Nash, T. E., Ware, J. M., Coyle, C. M., & Mahanty, S. (2019). Etanercept to Control Inflammation in the Treatment of Complicated Neurocysticercosis. *The American Journal of Tropical Medicine and Hygiene, 100*(3), 609–616. https://doi.org/10.4269/ajtmh.18-0795

Noh, J., Rodriguez, S., Lee, Y. M., Handali, S., Gonzalez, A. E., Gilman, R. H., Tsang, V. C., Garcia, H. H., & Wilkins, P. P. (2014). Recombinant protein- and synthetic peptide-based immunoblot test for diagnosis of neurocysticercosis. *Journal of Clinical Microbiology, 52*(5), 1429–1434. https://doi.org/10.1128/JCM.03260-13

Nkouawa, A., Sako, Y., Nakao, M., Nakaya, K., & Ito, A. (2009). Loop-mediated isothermal amplification method for differentiation and rapid detection of Taenia species. *Journal of Clinical Microbiology, 47*(1), 168–174. https://doi.org/10.1128/JCM.01573-08

Nkouawa, A., Sako, Y., Okamoto, M., & Ito, A. (2016). Simple identification of human Taenia species by multiplex loop-mediated isothermal amplification in combination with dot enzyme-linked immunosorbent assay. *American Journal of Tropical Medicine and Hygiene, 94*, 1318–1323.

Ng-Nguyen, D., Stevenson, M. A., Dorny, P., Gabriël, S., Vo, T. V., Nguyen, V. T., Phan, T. V., Hii, S. F., & Traub, R. J. (2017). Comparison of a new multiplex real-time PCR with the Kato Katz thick smear and copro-antigen ELISA for the detection and differentiation of Taenia spp. in human stools. *PLoS Neglected Tropical Diseases, 11*(7), e0005743. https://doi.org/10.1371/journal.pntd.0005743

Nakao, M., Yanagida, T., Okamoto, M., Knapp, J., Nkouawa, A., Sako, Y., & Ito, A. (2010). State-of-the-art Echinococcus and Taenia: phylogenetic taxonomy of human-pathogenic tapeworms and its application to molecular diagnosis. *Infection, Genetics and Evolution, 10*(4), 444–452. https://doi.org/10.1016/j.meegid.2010.01.011

Nash, T. E., & Garcia, H. H. (2011). Diagnosis and treatment of neurocysticercosis. *Nature Reviews Neurology, 7*(10), 584–594.

Ng-Nguyen, D., Stevenson, M. A., & Traub, R. J. (2017). A systematic review of taeniasis, cysticercosis and trichinellosis in Vietnam. *Parasites & Vectors, 10*, 1–15.

Ngwili, N., Sentamu, D. N., Korir, M., Adriko, M., Beinamaryo, P., Dione, M. M., Kaducu, J. M., Mubangizi, A., Mwinzi, P. N., Thomas, L. F., & Dixon, M. A. (2023). Spatial and temporal distribution of Taenia solium and its risk factors in Uganda. *International Journal of Infectious Diseases, 129*, 274–284. https://doi.org/10.1016/j.ijid.2023.02.001

Nelson, G. S., Pester, F. R., & Rickman, R. (1965). The significance of wild animals in the transmission of cestodes of medical importance in Kenya. T*ransactions of the Royal Society of Tropical Medicine and Hygiene, 59*(5), 507–524. https://doi.org/10.1016/0035-9203(65)90153-7

Nash, T. E., Bustos, J. A., Garcia, H. H., & Cysticercosis Working Group in Perú (2017). Disease Centered Around Calcified Taenia solium Granuloma. *Trends in Parasitology, 33*(1), 65–73. https://doi.org/10.1016/j.pt.2016.09.003

Omeragić, J., Alagić, D., Šerić-Haračić, S., & Kapo, N. (2023). Epidemiology of taeniosis/cysticercosis in humans and animals. In One Health - From Local to Global Health. IntechOpen. https://doi.org/10.5772/intechopen.110727

Oliver, J. H. (1871). *Seventh annual report of the Sanitary Commissioner (1870) of the Government of India* (pp. 82–83). Government Printing House.

Ortiz, J. B., Uys, M., Seguino, A., & Thomas, L. F. (2024). Foodborne helminthiasis. *Current Clinical Microbiology Reports, 11*(3), 153–165.

O'Neal, S., Winthrop, K., & Gonzalez, A. (2011). Cysticercosis control: bringing advances to the field. *Journal of Global Infectious Diseases, 3*(2), 156–160. https://doi.org/10.4103/0974-777X.81693

Okello, A. L., & Thomas, L. F. (2017). Human taeniasis: current insights into prevention and management strategies in endemic countries. *Risk Management and Healthcare Policy, 10*, 107–116. https://doi.org/10.2147/RMHP.S116545

O'Connell, E. M., Harrison, S., Dahlstrom, E., Nash, T., & Nutman, T. B. (2020). A Novel, Highly Sensitive Quantitative Polymerase Chain Reaction Assay for the Diagnosis of Subarachnoid and Ventricular Neurocysticercosis and for Assessing Responses to Treatment. *Clinical infectious diseases: an official publication of the Infectious Diseases Society of America, 70*(9), 1875–1881. https://doi.org/10.1093/cid/ciz541

Pretell, E. J., Martinot, C., Jr, Garcia, H. H., Alvarado, M., Bustos, J. A., Martinot, C., & Cysticercosis Working Group in Peru (2005). Differential diagnosis between cerebral tuberculosis and neurocysticercosis by magnetic resonance spectroscopy. *Journal of Computer Assisted Tomography, 29*(1), 112–114. https://doi.org/10.1097/01.rct.0000149959.63294.8f

Prasad, K. N., Verma, A., Srivastava, S., Gupta, R. K., Pandey, C. M., & Paliwal, V. K. (2011). An epidemiological study of asymptomatic neurocysticercosis in a pig farming community in northern India. *Transactions of the Royal Society of Tropical Medicine and Hygiene, 105*(9), 531–536. https://doi.org/10.1016/j.trstmh.2011.06.001

Pawłowski, Z., Allan, J. C., & Meinardi, H. (2005). Control measures for taeniosis and cysticercosis. In K. D. Murrell (Ed.), *WHO/FAO/OIE guidelines for the surveillance, prevention and control of taeniosis/cysticercosis* (pp. 73–99). OIE (World Organisation for Animal Health).

Pinheiro, A. A. S., Torrecilhas, A. C., Souza, B. S. F., Cruz, F. F., Guedes, H. L. M., Ramos, T. D., Lopes-Pacheco, M., Caruso-Neves, C., & Rocco, P. R. M. (2024). Potential of extracellular vesicles in the pathogenesis, diagnosis and therapy for parasitic diseases. *Journal of Extracellular Vesicles, 13*(8), e12496. https://doi.org/10.1002/jev2.12496

Pray, I. W., Muro, C., Gamboa, R., Vilchez, P., Wakeland, W., Pan, W., Lambert, W. E., Garcia, H. H., O'Neal, S. E., & Cysticercosis Working Group in Peru (2019). Seasonal patterns in risk factors for Taenia solium transmission: a GPS tracking study of pigs and open human defecation in northern Peru. *Parasites & Vectors, 12*(1), 352. https://doi.org/10.1186/s13071-019-3614-5

Pabst R. (2020). The pig as a model for immunology research. *Cell and Tissue Research, 380*(2), 287–304. https://doi.org/10.1007/s00441-020-03206-9

Parija, S. C., & Ponnambath, D. K. (2013). Laboratory diagnosis of Taenia asiatica in humans and animals. *Tropical Parasitology, 3*(2), 120–124. https://doi.org/10.4103/2229-5070.122127

Perroncito, E. (1877). On the tenacity of life of the cysticercus in the flesh of oxen and on the rapid development of the corresponding Taenia mediocanellata in the human body. *The Veterinarian, 50*, 817–818.

pickler, A. R. (2020). Taeniasis, cysticercosis and coenurosis. Retrieved from http://www.cfsph.iastate.edu/ DiseaseInfo/factsheets.php.

Prasad, K. N., Prasad, A., Verma, A., & Singh, A. K. (2008). Human cysticercosis and Indian scenario: A review. *Journal of Biosciences, 33*, 571–582.

Prodjinotho, U. F., Lema, J., Lacorcia, M., Schmidt, V., Vejzagic, N., Sikasunge, C., Ngowi, B., Winkler, A. S., & Prazeres da Costa, C. (2020). Host immune responses during Taenia solium Neurocysticercosis infection and treatment. *PLoS Neglected Tropical Diseases, 14*(4), e0008005. https://doi.org/10.1371/journal.pntd.0008005

Qadeer, A., Wajid, A., Rafey, H. A., Nawaz, S., Khan, S., Rahman, S. U., Alzahrani, K. J., Khan, M. Z., Alsabi, M. N. S., Ullah, H., Safi, S. Z., Xia, Z., & Zahoor, M. (2024). Exploring extracellular vesicles in zoonotic helminth biology: implications for diagnosis, therapeutic and delivery. *Frontiers in Cellular and Infection Microbiology, 14*, 1424838. https://doi.org/10.3389/fcimb.2024.1424838

Rausch, R. L. (1994). Family Taeniidae Ludwig, 1886. In L. F. Khalil, A. Jones, & R. A. Bray (Eds.), *Keys to the cestode parasites of vertebrates* (pp. 665–682). CAB International.

Raether, W., & Hänel, H. (2003). Epidemiology, clinical manifestations and diagnosis of zoonotic cestode infections: an update. *Parasitology Research, 91*(5), 412–438. https://doi.org/10.1007/s00436-003-0903-9

Ríos-Valencia, D. G., Ambrosio, J., Tirado-Mendoza, R., Carrero, J. C., & Laclette, J. P. (2023). What about the Cytoskeletal and Related Proteins of Tapeworms in the Host's Immune Response? An Integrative Overview. *Pathogens (Basel, Switzerland), 12*(6), 840. https://doi.org/10.3390/pathogens12060840

Restrepo, B. I., Alvarez, J. I., Castaño, J. A., Arias, L. F., Restrepo, M., Trujillo, J., Colegial, C. H., & Teale, J. M. (2001). Brain granulomas in neurocysticercosis patients are associated with a Th1 and Th2 profile. *Infection and Immunity, 69*(7), 4554–4560. https://doi.org/10.1128/IAI.69.7.4554-4560.2001

Ratcliffe, C., Adan, G., Marson, A., Solomon, T., Saini, J., Sinha, S., & Keller, S. S. (2023). Neurocysticercosis-related seizures: Imaging biomarkers. *Seizure, 108*, 13–23. https://doi.org/10.1016/j.seizure.2023.04.005

Rodríguez-Leyva, I., Cantú-Flores, K., Domínguez-Frausto, A., Vaudano, A. E., Archer, J., Bernhardt, B., Caciagli, L., Cendes, F., Chinvarun, Y., Federico, P., Gaillard, W. D., Kobayashi, E., Ogbole, G., Rampp, S., Wang, I., Wang, S., & Concha, L. (2023). Neurocysticercosis and epilepsy: Imaging and clinical characteristics. *Epileptic Disorders, 25*(1), 94–103. https://doi.org/10.1002/epd2.20060

Rajput, S., Ramalakshmi, V., & Rajpoot, A. (2025). Cutaneous cysticercosis mimicking an epidermal inclusion cyst: *A rare case. Cureus, 17*(5), e84328. https://doi.org/10.7759/cureus.84328

Raibagkar, P., & Berkowitz, A. L. (2018). The Many Faces of Neurocysticercosis. *Journal of the Neurological Sciences, 390*, 75–76. https://doi.org/10.1016/j.jns.2018.04.018

Sithole, M. I., Bekker, J. L., Tsotetsi-Khambule, A. M., & Mukaratirwa, S. (2019). Ineffectiveness of meat inspection in the detection of Taenia solium cysticerci in pigs slaughtered at two abattoirs in the Eastern Cape Province of South Africa. *Veterinary Parasitology, Regional Studies and Reports, 17*, 100299. https://doi.org/10.1016/j.vprsr.2019.100299

Salazar-Anton, F., Tellez, A., & Lindh, J. (2012). Evaluation of an immunodot blot technique for the detection of antibodies against Taenia solium larval antigens. *Parasitology Research, 110*, 2187–2191. https://doi.org/10.1007/s00436-011-2747-z

Sadr, S., Hajjafari, A., Sazmand, A., Santucciu, C., Masala, G., Soroushianfar, M., Nazemian, S., Rahdar, A., Pandey, S., Guettari, M., & Borji, H. (2025). Nanobiosensors for revolutionizing parasitic infections diagnosis: a critical review to improve global health with an update on future challenges prospect. *European Journal of Medical Research, 30*(1), 484. https://doi.org/10.1186/s40001-025-02685-2

Sako, Y., Nakao, M., Ikejima, T., Piao, X. Z., Nakaya, K., & Ito, A. (2000). Molecular characterization and diagnostic value of Taenia solium low-molecular-weight antigen genes. *Journal of Clinical Microbiology, 38*(12), 4439–4444. https://doi.org/10.1128/JCM.38.12.4439-4444.2000

Sitali, M. C., Schmidt, V., Mwenda, R., Sikasunge, C. S., Mwape, K. E., Simuunza, M. C., da Costa, C. P., Winkler, A. S., & Phiri, I. K. (2022). Experimental animal models and their use in understanding cysticercosis: A systematic review. *PloS One, 17*(7), e0271232. https://doi.org/10.1371/journal.pone.0271232

Singh, B. B., Khatkar, M. S., Gill, J. P., & Dhand, N. K. (2017). Estimation of the health and economic burden of neurocysticercosis in *India. Acta Tropica, 165*, 161–169. https://doi.org/10.1016/j.actatropica.2016.01.017

Singh, A. K., Singh, S. K., Singh, A., Gupta, K. K., Khatoon, J., Prasad, A., Rai, R. P., Gupta, R. K., Tripathi, M., Husain, N., & Prasad, K. N. (2015). Immune response to Taenia solium cysticerci after anti-parasitic therapy. *International Journal for Parasitology, 45*(12), 749–759. https://doi.org/10.1016/j.ijpara.2015.05.007

Singhi P. (2011). Neurocysticercosis. *Therapeutic Advances in Neurological Disorders, 4*(2), 67–81. https://doi.org/10.1177/1756285610395654

Singh, A. K., Prasad, K. N., Prasad, A., Tripathi, M., Gupta, R. K., & Husain, N. (2013). Immune responses to viable and degenerative metacestodes of Taenia solium in naturally infected swine. *International Journal for Parasitology, 43*(14), 1101–1107. https://doi.org/10.1016/j.ijpara.2013.07.009

Shrestha, B., Shrestha, P., Gajurel, B. P., Kandel, G., & Shah, L. (2025). Disseminated Neurocysticercosis With Intraventricular and Cisternal Extension Without Hydrocephalus: A Case Report. *Clinical Case Reports, 13*(4), e70354. https://doi.org/10.1002/ccr3.70354

Saelens, G., Robertson, L., & Gabriël, S. (2022). Diagnostic tools for the detection of taeniid eggs in different environmental matrices: A systematic review. *Food and waterborne parasitology, 26*, e00145. https://doi.org/10.1016/j.fawpar.2022.e00145

Stewart, I. E. (1951). Helminths in History. *The Scientific Monthly, 72*(6), 345–352. http://www.jstor.org/stable/20278

Singhi, P., & Saini, A. G. (2019). Pediatric neurocysticercosis. *The Indian Journal of Pediatrics, 86*, 76–82.

Symeonidou, I., Arsenopoulos, K., Tzilves, D., Soba, B., Gabriël, S., & Papadopoulos, E. (2018). Human taeniasis/cysticercosis: A potentially emerging parasitic disease in Europe. *Annals of Gastroenterology, 31*(4), 406.

Smith J. L. (1994). Taenia solium Neurocysticercosis 1. *Journal of Food Protection, 57*(9), 831–844. https://doi.org/10.4315/0362-028X-57.9.831

Stelzle, D., Makasi, C. E., Schmidt, V., Van Damme, I., Trevisan, C., Ruether, C., Fleury, A., Noh, J., Handali, S., Dorny, P., Magnussen, P., Zulu, G., Mwape, K. E., Bottieau, E., Gabriël, S., Ngowi, B. J., Winkler, A. S., & SOLID collaborators (2024). Evaluation of a point-of-care test for the diagnosis of Taenia solium neurocysticercosis in rural southern Tanzania: a diagnostic accuracy study. The Lancet. *Infectious Diseases, 24*(1), 98–106. https://doi.org/10.1016/S1473-3099(23)00378-X

Singh, G., & Sander, J. W. (2019). Historical perspective: The British contribution to the understanding of neurocysticercosis. Journal of the *History of the Neurosciences, 28*(3), 332–344. https://doi.org/10.1080/0964704X.2018.1564523

Ta, R., & Blond, B. N. (2022). The prevalence of and contributors to neurocysticercosis in endemic regions. *Journal of the Neurological Sciences, 441*, 120393. https://doi.org/10.1016/j.jns.2022.120393

Takayanagui, O. M., Odashima, N. S., Bonato, P. S., Lima, J. E., & Lanchote, V. L. (2011). Medical management of neurocysticercosis. *Expert Opinion on Pharmacotherapy, 12*(18), 2845–2856. https://doi.org/10.1517/14656566.2011.634801

Taylor, M. A., Coop, R. L., & Wall, R. (2015). Veterinary helminthology. In *Veterinary Parasitology* (pp. 1–109). John Wiley & Sons, Ltd. https://doi.org/10.1002/9781119073680.ch1

Tyson, E. (1683). Lumbricus teres, or some anatomical observations on the round worm bred in human bodies. *Philosophical Transactions of the Royal Society of London, 13*(147), 154–161.

Tanowitz, H. B., Weiss, L. M., & Wittner, M. (2001). Tapeworms. *Current Infectious Disease Reports, 3*(1), 77–84. https://doi.org/10.1007/s11908-001-0062-z

Trevisan, C., Damme, I. V., Ngowi, B., Schmidt, V., Stelzle, D., Møller, K. S., Kabululu, M., Makasi, C. E., Magnussen, P., Bottieau, E., Abatih, E., Johansen, M. V., Ngowi, H., Ndawi, B., Mwape, K. E., Zulu, G., Dorny, P., Winkler, A. S., Gabriël, S., & on behalf of the SOLID Consortium. (2021). Trial Design of a Prospective Multicenter Diagnostic Accuracy Study of a Point-of-Care Test for the Detection of Taenia solium Taeniosis and Neurocysticercosis in Hospital-Based Settings in Tanzania. *Diagnostics, 11*(9), 1528. https://doi.org/10.3390/diagnostics11091528

Varcasia, A., Tamponi, C., Ahmed, F., Cappai, M. G., Porcu, F., Mehmood, N., Dessì, G., & Scala, A. (2022). Taenia multiceps coenurosis: A review. *Parasites & Vectors, 15*(1), 84.

Verster, A. (1969). A taxonomic revision of the genus Taenia Linnaeus, 1758 s. str. Onderstepoort *Journal of Veterinary Research, 36*(1), 3–58.

Venkat, B., Aggarwal, N., Makhaik, S., & Sood, R. (2016). A comprehensive review of imaging findings in human cysticercosis. *Japanese Journal of Radiology, 34*, 241–257.

Verastegui, M., Gilman, R. H., Arana, Y., Barber, D., Velásquez, J., Farfán, M., Chile, N., Kosek, J. C., Kosek, M., Garcia, H. H., Gonzalez, A., & Cysticercosis Working Group in Peru (2007). Taenia solium oncosphere adhesion to intestinal epithelial and Chinese hamster ovary cells in vitro. *Infection and Immunity, 75*(11), 5158–5166. https://doi.org/10.1128/IAI.01175-06

Vanparijs, O., Hermans, L., Lauwers, H. (1987). Cysticercus Fasciolaris in Mice: A Laboratory Model for Selecting New Drugs on Cysticercosis. In: Geerts, S., Kumar, V., Brandt, J. (eds)

Helminth Zoonoses. Current Topics in Veterinary Medicine and Animal Science, vol 43. Springer, Dordrecht. https://doi.org/10.1007/978-94-009-3341-5_15

Van Damme, I., Trevisan, C., Kabululu, M., Stelzle, D., Makasi, C. E., Schmidt-Urbaneja, V., Mwape, K. E., Mubanga, C., Zulu, G., Møller, K. S., Jansen, F., Reynders, D., Noh, J., Handali, S., Bottieau, E., Winkler, A. S., Dorny, P., Magnussen, P., Gabriël, S., & Ngowi, B. (2025). Evaluation of a rapid lateral flow assay for the detection of taeniosis and cysticercosis at district hospital level in Tanzania: A prospective multicentre diagnostic accuracy study. *PLoS Neglected Tropical Diseases, 19*(3), e0012310. https://doi.org/10.1371/journal.pntd.0012310

World Organisation for Animal Health (WOAH). (2021). Cysticercosis (including infection with Taenia solium). In Manual of Diagnostic Tests and Vaccines for Terrestrial Animals (Chapter 3, p. 3) https://www.woah.org/fileadmin/Home/fr/Health_standards/tahm/3.10.03_CYSTICERCOSIS.pdf

Wong, L. W., Ong, K. S., Khoo, J. R., Goh, C. B. S., Hor, J. W., & Lee, S. M. (2020). Human intestinal parasitic infection: a narrative review on global prevalence and epidemiological insights on preventive, therapeutic and diagnostic strategies for future perspectives. *Expert Review of Gastroenterology & Hepatology, 14*(11), 1093–1105. https://doi.org/10.1080/17474124.2020.1806711

Wilson, C., Nonga, H. E., Mdegela, R. H., Makingi, G., Stelzle, D., Mkupasi, E. M., Winkler, A. S., & Ngowi, H. A. (2025). Effect of Community Health Education Intervention on Prevalence and Pig Farmers' Knowledge, Attitudes, and Practices Related to Porcine Cysticercosis in Selected Areas of Tanzania. *Veterinary Medicine International, 2025*, 9929531. https://doi.org/10.1155/vmi/9929531

Wilcox, R. S., Bowman, D. B., Barr, S. C., & Euclid, J. M. (2009). Intestinal obstruction caused by Taenia taeniaeformis infection in a cat. *Journal of the American Animal Hospital Association, 45*(2), 93–96. https://doi.org/10.5326/0450093.

World Health Organization. (1997). PAHO/WHO Informal Consultation on the Taeniosis/Cysticercosis Complex, 23-25 August, 1995, Brasília, Brazil. In PAHO/WHO Informal Consultation on the Taeniosis/Cysticercosis Complex, 23-25 August, 1995, Brasília, Brazil (pp. 20-20).

World Health Organization & Thomas, L. F. (2015). *Landscape analysis: Control of Taenia solium*. Geneva: World Health Organization. https://apps.who.int/iris/handle/10665/164359.

Wang, L. Q., Liu, T. L., Liang, P. H., Zhang, S. H., Li, T. S., Li, Y. P., Liu, G. X., Mao, L., & Luo, X. N. (2020). Characterization of exosome-like vesicles derived from Taenia pisiformis cysticercus and their immunoregulatory role on macrophages. *Parasites & Vectors, 13*(1), 318. https://doi.org/10.1186/s13071-020-04186-z

World Health Organization. (2021). WHO guidelines on management of Taenia solium neurocysticercosis [Internet]. Geneva: World Health Organization; 2021. ANNEX 4, EVIDENCE PROFILES. Available from: https://www.ncbi.nlm.nih.gov/books/NBK573857/

White, A. C., Jr, Coyle, C. M., Rajshekhar, V., Singh, G., Hauser, W. A., Mohanty, A., Garcia, H. H., & Nash, T. E. (2018). Diagnosis and Treatment of Neurocysticercosis: 2017 Clinical Practice Guidelines by the Infectious Diseases Society of America (IDSA) and the American Society of Tropical Medicine and Hygiene (ASTMH). *Clinical infectious diseases: an official publication of the Infectious Diseases Society of America, 66*(8), e49–e75. https://doi.org/10.1093/cid/cix1084

World Health Organization. (2022). Taeniasis/cysticercosis. Available online: https://www.who.int/news-room/fact-sheets/detail/taeniasis-cysticercosis.

Wu, Y. D., Dai, G. D., Li, L., Littlewood, D. T. J., Ohiolei, J. A., Zhang, L. S., Guo, A. M., Wu, Y. T., Ni, X. W., Shumuye, N. A., Li, W. H., Zhang, N. Z., Fu, B. Q., Fu, Y., Yan, H. B., & Jia, W. Z. (2022). Expansion of Cyclophyllidea Biodiversity in Rodents of Qinghai-Tibet Plateau and the "Out of Qinghai-Tibet Plateau" Hypothesis of Cyclophyllideans. *Frontiers in Microbiology, 13*, 747484. https://doi.org/10.3389/fmicb.2022.747484

Yadav, R., & Parihar, P. (2025). A rare case of muscular cysticercosis: Incidental discovery in a patient presenting with inguinal swelling. *Radiology Case Reports, 20*(6), 2729–2733. https://doi.org/10.1016/j.radcr.2025.02.058

Zhang, D., Qi, Y., Cui, Y., Song, W., Wang, X., Liu, M., Cai, X., Luo, X., Liu, X., & Sun, S. (2021). Rapid Detection of Cysticercus cellulosae by an Up-Converting Phosphor Technology-Based Lateral-Flow Assay. *Frontiers in Cellular and Infection Microbiology, 11*, 762472. https://doi.org/10.3389/fcimb.2021.762472

Zhou, M., Zhang, N., Zhang, M., & Ma, G. (2020). Culture, eating behavior, and infectious disease control and prevention. *Journal of Ethnic Foods, 7*, 1–7.

Zemmour, I., Durieux, M. F., Herault, E., Rouges, C., Šoba, B., Mercier, A., Ariey, F., Preux, P. M., Yera, H., & On Behalf Of Collaborators Group (2023). Neurocysticercosis Diagnosis in a Non-Endemic Country: France. *Pathogens (Basel, Switzerland), 12*(10), 1205. https://doi.org/10.3390/pathogens12101205

Zhao, B. C., Jiang, H. Y., Ma, W. Y., Jin, D. D., Li, H. M., Lu, H., Nakajima, H., Huang, T. Y., Sun, K. Y., Chen, S. L., & Chen, K. B. (2016). Albendazole and Corticosteroids for the Treatment of Solitary Cysticercus Granuloma: A Network Meta-analysis. *PLoS Neglected Tropical Diseases, 10*(2), e0004418. https://doi.org/10.1371/journal.pntd.0004418

Echinococcosis

36

Maninder Singh

Abstract

Echinococcosis has a long history dating back to ancient times. The disease has been controlled in five nations while still remains a challenge for the world. Echinococcosis is one of the 17 neglected diseases targeted for control or elimination by 2050 by World Health Organization. This disease is an excellent example requiring One Health approach for its control and prevention.

Keywords

Hydatid disease · *Echinococcus* · Parasitic zoonoses

Synonyms

- Infection of *Echinococcus granulosus*: hydatid disease, cystic echinococcosis
- Infection of *Echinococcus multilocularis*: alveolar echinococcosis
- Infection of *Echinococcus vogeli* (polycystic echinococcosis) and *Echinococcus oligarthus* (unicystic echinococcosis): neotropical echinococcosis

Aetiology

Echinococcosis is a parasitic tapeworm disease of humans and animals and is caused by larvae of cestode *Echinococcus* (family Taeniidae). Being the cestode, the parasite has segmented body, the last segment containing the gravid uterus responsible

M. Singh (✉)
Department of Veterinary Public Health and Epidemiology, College of Veterinary Science, Guru Angad Dev Veterinary and Animal Sciences University, Rampura Phul, Bathinda, India

R. Biswas et al. (eds.), *Concept of One Health and Zoonoses*,
https://doi.org/10.1007/978-981-95-6627-3_36

Table 37.1 Definitive and intermediate hosts of various *Echinococcus* species

Echinococcus species	Definitive host	Intermediate host	Zoonosis
E. granulosus sensu stricto	Dog, wolf, jackal	Sheep, goat, cattle, pigs	Yes
E. canadensis	Dog, wolf	Pig, camel, cervids	Yes
E. ortleppi	Dog	Cattle	Yes
E. felidis	Lion	Zebra, buffalo, antelope	Not reported
E. equinus	Dog	Horse, cervids	Not reported
E. multilocularis	Fox, wolf, dog, cat, raccoon dog	Rodents, lagomorphs	Yes
E. oligarthus	Wild felids	Opossum, agouti	Yes
E. vogeli	Dog	Paca	Yes
E. shiquicus	Tibetan fox	Tibetan plateau pika	Not reported

for release of eggs. The life cycle of *Echinococcus* involves two types of hosts: definitive hosts and intermediate hosts. Definitive hosts are the hosts that carry the adult parasites and are carnivores. The intermediate hosts are the hosts in which larval stages of parasite develop and are herbivores. The humans are the accidental hosts and act as intermediate host when infected. There are various species of *Echinococcus*, among which four are known to cause disease in humans, viz. *E. granulosus*, *E. multilocularis*, *E. vogeli* and *E. oligarthus*. Earlier *E. granulosus* was divided into ten distinct genotypes (G1–G10) on the basis of mitochondrial cytochrome oxidase C gene. However, recently the genotypes have been assigned into five species as *E. granulosus* sensu stricto (G1, G2 and G3), *Echinococcus equinus* (G4 strain), *Echinococcus ortleppi* (G5 strain), *Echinococcus canadensis* (G6–G10) and *Echinococcus felidis* (lion strain); among these *E. granulosus* sensu stricto, *E. ortleppi* and *E. canadensis* are zoonotic in nature.

The definitive and intermediate hosts for various *Echinococcus* species are indicated in Table 36.1.

Epidemiology

The epidemiology of human echinococcosis varies with the *Echinococcus* species. Cystic echinococcosis caused by *E. granulosus* is most widespread. It is estimated that around one million people are affected by echinococcosis worldwide annually causing around 19,300 deaths and loss of 871,000 disability-adjusted life years (DALYs). The prevalence of disease varies with the region but is endemic in South America, Russia, Middle East, Western China, the Mediterranean region and eastern Africa. In parts of Argentina, Peru, East Africa, Central Asia and China, prevalence in humans as high as 5–10% is reported. Nevertheless, Iceland, New Zealand, Tasmania, Falklands and Cyprus have controlled the disease by applying public health policies. Cystic echinococcosis is rare in North America. Cystic echinococcosis is endemic in India too. The high number of cases are reported in Andhra Pradesh, Tamil Nadu and Jammu and Kashmir. Alveolar echinococcosis caused by *E. multilocularis* is restricted to the Northern Hemisphere with northern and central parts of Europe, Asia and North America. Neotropical echinococcosis caused by *E. vogeli* and *E. oligarthus* are restricted to Central and South America. The annual incidence of cystic echinococcosis ranges from 1 to 200 per 100,000, while that of alveolar echinococcosis ranges from 0.03 to 1.2 per 100,000. The mortality rate of cystic echinococcosis is much lower (2–4%) than alveolar echinococcosis (90%).

As per 2020 surveillance report of European Centre for Disease Control and Prevention, there were 529 confirmed cases of echinococcosis in European Union; among them 243 were of *E. granulosus*, 114 of *E. multilocularis* and 172 of unknown species. However, in 2022, the number of cases increased to 722 (EFSA & ECDC, 2023). At any point of time, more than one million people are having echinococcosis globally.

In Uganda, the overall prevalence of cystic echinococcosis reported is 21.9% in sheep, 15.2% in zebu cattle, 53.5% in goats and 2.1% in Ankole cattle (Omadang et al., 2023). In Kashmir valley of India, the prevalence in sheep slaughter reported is 4.04% (Beigh et al., 2021); however 21% prevalence in bovine carcasses is observed in Punjab state, India (Singh et al., 2020).

Transmission

The existence of life cycle of *Echinococcus* species depends upon the predator-prey association, and the disease stops once this cycle breaks. The definitive hosts, i.e. carnivores (canids and felids), harbour hundreds to thousands of adult *Echinococcus* tapeworms in their intestines. The adults of *E. granulosus* are of size 3–7 mm, while those of *E. multilocularis* are 1.2–4.5 mm. In the intestine, the gravid uterus of last segment of *E. granulosus* releases eggs which are secreted in the faeces. When the eggs are ingested by the intermediate host, i.e. herbivores (ungulates, rodents and lagomorphs), the eggs hatch in their intestine and release oncospheres. Oncospheres pass through the portal and lymphatic vessels to reach in the liver. In the liver, oncospheres develop to larvae (the metacestode stage of the parasite) leading to formation of cysts. Apart from liver, cysts from the lungs, brain and bones are reported. In the cysts, protoscoleces (the fertile forms) are produced which, when ingested by the definitive host, evaginate their scoleces in the latter's intestine and attach to the intestinal wall to generate mature adult worm. The experiments have shown the maturation of *Echinococcus granulosus* eggs in the intestine of dogs occurs in 36–40 days, and segments are matured by the 50th day (Yang et al., 2023).

In public settings, the life cycle of the disease progresses as the organs carrying the cysts are thrown by the butchers which are easily accessed by the dogs. Further, the water and feed contamination by the eggs through faecal contamination carries over the infection from dogs to the herbivores. Therefore, the disease is higher in the areas where dogs have easy access to the offals of slaughtered animals.

Humans act as intermediate hosts when they accidentally get exposed to the eggs of the parasite through contaminated feed and water. Similar to the herbivores, the metacestode cyst formation in the liver, lungs and other organs occurs. As the disease progresses very slow, the clinical manifestations of disease are reported in the later stages of life.

Pathogenesis

The clinical implications of the echinococcosis are limited to the intermediate host. In the intermediate hosts, the metacestode cysts are formed in various organs especially the liver and lungs. In these organs, the oncospheres reach through the portal and lymphatic vessels. In case of cystic echinococcosis, the cyst formation and expansion occur as concentric expansion. The cyst has a germinal layer, which is surrounded by acellular laminated layer which is further surrounded by the fibrous capsule (adventitial layer) formed due to host immune response. The germinal layer produces cystic fluid, protoscoleces and brood capsule of daughter cysts. The daughter cyst may pass across the laminated layer of parent cyst and pushed outwards. The cysts may contain multiple litres of fluid. On ingestion by the definite host, the protoscoleces, the infective stage, evaginate and attach with the intestinal mucosa by the scolex.

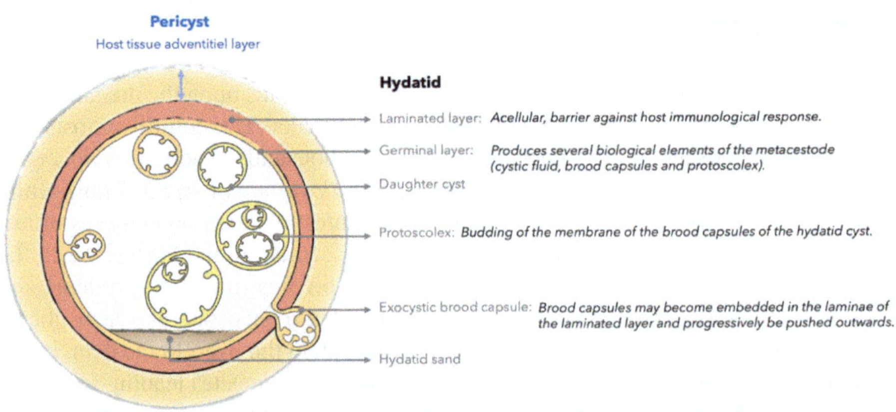

Structure of *Echinococcus granulosus* cyst (Source: Calame et al. Role of the radiologist in the diagnosis and management of the two forms of hepatic echinococcosis. *Insights into Imaging*, 2022, 13:68. https://doi.org/10.1186/s13244-022-01190-y. This article is licensed under a Creative Commons Attribution 4.0 International License, which permits use, sharing, adaptation, distribution and reproduction in any medium or format, as long as you give appropriate credit to the original author(s) and the source http://creativecommons.org/licenses/by/4.0/)

In case of alveolar echinococcosis, the pathogenesis differs from cystic echinococcosis. On reaching the liver, oncospheres multiply by the progressive outgrowth from germinal layer as the germinal layer is not protected by any fibrous layer. The germinal layer is fragile, and there is no clear demarcation between the cyst and the normal tissue. Being the growth invasive in nature with no clear demarcations, multilocular appearance emerges with the thickening of the tissue leading to tumour-like structure. Alveolar echinococcosis progresses slower than cystic echinococcosis. The mortality rates are higher in alveolar echinococcosis.

Clinical Signs

In cystic echinococcosis, hydatid cysts are mainly reported in the liver and lungs with the sporadic reports for the spleen, kidney, bones, heart and brain. The symptoms depend upon the organ involved. In case of hydatid cysts in the liver, symptoms range from asymptomatic to hepatomegaly, pain, nausea, vomiting and obstructive jaundice. The cystic pressure on the surrounding structures may lead to venous obstruction and portal hypertension. In case of involvement of lungs, cough, chest pain, haemoptysis and dyspnoea are seen. Cystic echinococcosis is generally reported in adults of age 30–40 years, while alveolar echinococcosis is the disease of old age (more than 50 years). The signs of alveolar echinococcosis are not specific and includes pain in right upper quadrant, weight loss and obstructive jaundice.

Diagnosis

Animals

Definitive Hosts

Necropsy is generally performed for the studies in the wild animals and where the domestic carnivores are killed humanely. It is performed to demonstrate the adults in the small intestine of dead animals as the differentiation between *Taenia* egg and *Echinococcus* eggs is not possible on the basis of phenotypic characteristics of eggs. The small intestine is divided into four parts and is screened for detection of *Echinococcus* worms. *Echinococcus granulosus* is usually found in the first third of the small intestine of dogs, while *E. multilocularis* is found in the mid/posterior region. The approach has high sensitivity. Sedimentation and counting technique for the worms can be employed.

Coproantigen tests: Adult *Echinococcus* worms in the small intestine release both surface or secretory molecules, which can be detected by coproELISA. Both polyclonal and monoclonal antibodies are available to detect either of these antigens. Few commercial ELISA kits are available. The test usually detects *Echinococcus* genus and not species.

PCR: The parasite releases eggs in the faecal matter, and the DNA of eggs can be detected by PCR. Various molecular targets are used for detection and differentiation of *Echinococcus* species.

Serological tests: Serological ELISA is available for detection of antibodies generated against secretory and somatic antigens. The test shows cross-reactions with *Taenia* species. The coproantigen detection tests (mentioned earlier) have higher sensitivity than serological tests.

Intermediate Host

Necropsy: For prevalence and surveillance purposes, necropsy in the intermediate host is a commonly employed test. It indirectly helps in ascertaining the existing situation of echinococcosis in a region. During necropsy, the organs of slaughtered intermediate hosts are examined for the presence of metacestode stage in the visceral organs especially the liver and lungs. The organs are palpated and incised to reveal metacestode. The cystic contents are further evaluated to determine the fertile nature of protoscoleces. As the presence of metacestode in intermediate host is age dependent, it is important to determine the number of young animals slaughtered in a slaughter house. The DNA from cysts collected is extracted and genotyped by PCR for genus and species detection.

Serological Tests

Similar to the definitive host, ELISA is available for detection of antibodies in intermediate host. It is used for surveillance and epidemiological studies in the endemic areas. However, the test has lower sensitivity and specificity and thus is not preferred over the necropsy examination.

Humans

In humans, diagnosis is based on imaging and serological tests. Among imaging techniques, ultrasonography is most commonly used. On the basis of ultrasonography, the World Health Organization has classified the cystic lesions of cystic echinococcosis into six types. Apart from ultrasonography, CT scan and MRI are used in diagnosis. These techniques are better in localizing the exact position of lesion and therefore are more helpful in clinical management of patient.

Apart from imaging techniques, serology using ELISA is used in the primary diagnosis as well as during follow-up of treatment. The test is sensitive and specific in nature. Among locations of cysts, the cysts in the liver tend to have higher seropositivity than cysts in the lung.

Control and Prevention

Globally, control and prevention of echinococcosis are focused on deworming in definite hosts to kill adult parasites, meat inspection in intermediate host to detect metacestode stage and thus preventing its further ingestion by definitive hosts, targeted surveillance in livestock and human beings and health education among the butchers and communities to promote good slaughtering practices and good hygienic practices. The control of echinococcosis is a classic example of One Health as multiple institutions are involved. Iceland, New Zealand, Tasmania, Falklands and Cyprus are the five countries where cystic echinococcosis has been controlled using well-defined controlled programmes. The programmes have been run for <15 years to >50 years.

The regular monthly deworming of dogs with praziquantel has been known to reduce prevalence of echinococcosis in hyper-endemic areas of China. For control of infection in intermediate hosts, EG95 vaccine has been successfully developed. The vaccine is approved for use in Morocco and is manufactured in Argentina, Morocco and China. Two subcutaneous injections, 1 month apart, are injected which provide sufficient protection against future exposure to *Echinococcus* eggs. Vaccine does not treat the existing infection; therefore, young animals are vaccinated at the time of their weaning. In Rio Negro area of Argentina, use of EG95 vaccine reduced the prevalence of echinococcosis in sheep from 26.2% to 7.8%. For definitive hosts, no such vaccine is available which is known to reduce the fecundity of parasite and reduce or prevent the infection in dogs.

In humans, the disease is treatable and preventable. The treatment is based on surgical interventions with adjunct antiparasitic therapy. Percutaneous therapy called puncture-aspiration-injection-reaspiration (PAIR) is also an option. On an average 2.2% deaths are reported in patients undergoing surgery for cystic echinococcosis, and in 6.5% of patients relapse is reported. Prevention is based on hygienic food consumption as the major risk of transmission is through contaminated food and water.

References

Beigh, A., Darzi, M., Bashir, S., Dar, P., Ganai, N., Malik, S., & Bhat, B. (2021). Epidemiological and molecular characterization of Echinococcus granulosus isolated from small ruminants in Kashmir Valley, India. *Iranian Journal of Parasitology, 16*(3), 357–365. https://doi.org/10.18502/ijpa.v16i3.7088

Calame, P., Weck, M., Busse-Cote, A., Brumpt, E., Richou, C., Turco, C., Doussot, A., Bresson-Hadni, S., & Delabrousse, E. (2022). Role of the radiologist in the diagnosis and management of the two forms of hepatic echinococcosis. *Insights Into Imaging, 13*(1), 68. https://doi.org/10.1186/s13244-022-01190-y

European Food Safety Authority (EFSA), & European Centre for Disease Prevention and Control (ECDC). (2023). The European Union one health 2022 zoonoses report. *EFSA Journal: European Food Safety Authority, 21*(12), e8442. https://doi.org/10.2903/j.efsa.2023.8442

Larrieu, E., Mujica, G., Araya, D., Labanchi, J. L., Arezo, M., Herrero, E., Santillán, G., Vizcaychipi, K., Uchiumi, L., Salvitti, J. C., Grizmado, C., Calabro, A., Talmon, G., Sepulveda, L., Galvan, J. M., Cabrera, M., Seleiman, M., Crowley, P., Cespedes, G., García Cachau, M., et al. (2019). Pilot field trial of the EG95 vaccine against ovine cystic echinococcosis in Rio Negro, Argentina: 8 years of work. *Acta Tropica, 191*, 1–7. https://doi.org/10.1016/j.actatropica.2018.12.025

Omadang, L., Chamai, M., Ejobi, F., Erume, J., Oba, P., & Ocaido, M. 2023. Prevalence of cystic echinococcosis among livestock in pastoral and agro-pastoral areas in Uganda. *Parasitology*, 1–9. Advance online publication. https://doi.org/10.1017/S0031182023001154

Singh, H., Aulakh, R. S., Sharma, R., & Singh, B. B. (2020). Prevalence and molecular characterisation of Echinococcus granulosus in disposed of bovine carcasses in Punjab, India. *Journal of Parasitic Diseases: Official Organ of the Indian Society for Parasitology, 44*(3), 521–527. https://doi.org/10.1007/s12639-020-01223-7

Yang, L., Yang, Y., Yu, W., Wang, Q., Zhong, B., Hua, K., Liu, Y., & Huang, Y. (2023). Prevalence of *Echinococcus* infections in wild carnivores based on copro—DNA tests in Serthar County of Sichuan Province. *Zhongguo xue xi chong bing fang zhi za zhi = Chinese Journal of Schistosomiasis Control, 35*(5), 492–496. https://doi.org/10.16250/j.32.1374.2023063

Toxoplasmosis

37

Vikrant Sudan

Abstract

Toxoplasma gondii, the tissue cyst forming coccidium, is one of the polyxenous parasites known to date, which has a facultative heteroxenous life cycle and can infect probably all warm-blooded animals ranging from birds to mammals including human beings. From the zoonotic point of view, *T. gondii* is one of the most important protozoan parasites causing abortion or a congenital disease in the intermediate hosts.

Keywords

Cat zoonoses · Toxoplasma · Hydrocephalus

History

- The first comprehensive description of *T. gondii* tachyzoites in the spleen, liver, and blood of *Ctenodactylus gundi*, a species of North African rodents, was provided by Nicolle and Manceaux 1908.
- They named the genus *Toxoplasma* in 1909, and *T. gondii* became the type species of the genus.
- Dr. Jatinder Pal Dubey, USDA, a scientist of Indian origin, has worked exclusively on *T. gondii*.

V. Sudan (✉)
Department of Veterinary Parasitology, College of Veterinary Science, Rampura Phul, Guru Angad Dev Veterinary and Animal Sciences University, Punjab, India

R. Biswas et al. (eds.), *Concept of One Health and Zoonoses*,
https://doi.org/10.1007/978-981-95-6627-3_37

Aetiology

The parasite *Toxoplasma*, specifically *Toxoplasma gondii*, is a facultative intracellular tissue cyst-forming coccidian. It uses cats and their relatives as definitive hosts, while virtually all other animals, including humans, serve as intermediate hosts. *Toxoplasma gondii* can infect all warm-blooded animals, targeting every nucleated cell. Unlike many parasites, it does not produce any toxins. The name *Toxoplasma* (derived from "Toxon" meaning arc and "plasma" meaning life or form) refers to the crescent, arc, or comma shape of its tachyzoite stage. The genus is characterized by asexual reproduction, known as merogony or schizogony, occurring in both intermediate and definitive hosts. In definitive hosts, merogony results in the formation of meronts/schizonts and gamonts, followed by isosporan-type oocysts in the intestinal epithelium, with sporulation occurring outside the host in the environment. In intermediate hosts, merogony leads to the formation of tissue stages, including tachyzoites and bradyzoites, within host cells.

Transmission and Developmental Stages

The parasite is transmitted through three main routes:

1. Congenitally
2. Through carnivorism
3. Via the faecal-oral route

Toxoplasma has three infective stages:

1. Tachyzoites within pseudocysts
2. Bradyzoites within tissue cysts
3. Sporozoites within sporulated oocysts

1. *Tachyzoites*: Tachyzoites are the rapidly multiplying sporozoites within host cells. They enter host cells either through active penetration or by phagocytosis. Inside the host cell, tachyzoites multiply through repeated endodyogeny until the cell is filled with crescent-shaped zoites, all enclosed in a parasitophorous vacuole. This formation is termed a clone or pseudocyst, as there is no well-defined parasitic membrane surrounding the group of zoites.
2. *Bradyzoites*: Bradyzoites are the slowly dividing forms within host cells. Thousands of bradyzoites are enclosed in a tissue cyst with a well-defined membrane, representing the resting stage of the parasite within the host cell. These tissue cysts are spherical, with their shape and size varying with the chronicity of the infection. Persistent tissue cysts are typically located in neural and muscular tissues such as the brain, eyes, and skeletal and cardiac muscles, though they can also be found in visceral organs like the lungs, liver, and kidneys. Intact tissue cysts do not harm the host and can survive for the host's entire life.

3. *Sporozoites*: Sporozoites are found within the isosporan oocysts, infecting definitive hosts through contaminated feed or water. The transmission of *T. gondii* involves two developmental cycles.

Life Cycle

1. *Entero-epithelial Cycle*: This cycle occurs in felines, the definitive hosts. When they ingest any infective materials such as pseudocysts, tissue cysts, or sporulated oocysts, it leads to asexual reproduction by endodyogeny and endopolygeny in the intestinal epithelium. This results in the formation of merozoites, gamonts, and oocysts. A series of asexual generations, labelled from type A to E, are produced in the intestines, multiplying to form gametes and oocysts. Cats shed oocysts 3–5 days, 9–11 days, and 21–24 days after ingesting tissue cysts, pseudocysts, and oocysts, respectively. Tissue cysts containing bradyzoites are the most infective to felines.
2. *Extra-intestinal Cycle*: This cycle occurs in non-felines, the intermediate hosts, when they ingest sporulated oocysts, infected meat, or through placental transmission. Upon ingestion of sporulated oocysts with feed or water, the oocysts initiate an enteric infection, releasing sporozoites that multiply intracellularly in the intestinal epithelium. They spread to regional lymph nodes and various organs via circulation, forming pseudocysts containing tachyzoites in various host cells, such as fibroblasts, reticular cells, myocardial cells, hepatocytes, and neurons. Accumulation of tachyzoites eventually destroys these cells, leading to further infection of fresh cells. The presence of tachyzoites indicates an acute visceral infection. If the host overcomes this acute stage, the tachyzoites transform into bradyzoites, forming tissue cysts, particularly in the heart, eyes, and skeletal muscles, indicating the chronic stage of the infection. The cycle completes when tissue cysts are ingested by felines.

During carnivorism, meat containing tissue cysts is digested by proteolytic enzymes, releasing bradyzoites. These bradyzoites enter host cells, form tachyzoites, and eventually encyst in tissues, awaiting predation by felines. Congenital infection can occur when a previously uninfected host becomes infected during pregnancy. The organism multiplies in the placenta and then in foetal tissues. Although infection can occur at any stage of gestation, the foetus is most seriously affected if the dam is infected during the first half of gestation.

Pathogenesis

Toxoplasmosis is a mysterious disease, ranging from asymptomatic infections to acute, fatal conditions. Animals show varying degrees of susceptibility, with asymptomatic toxoplasmosis being very common and clinical signs rarely appearing in affected animals. The cause of death in a host is often due to necrosis of the intestines and mesenteric lymph nodes, caused by the intracellular growth of the organism, long before other organs are affected. Although a host may die from acute toxoplasmosis, more often it recovers, acquiring immunity. By the third week of infection, tachyzoites disappear from the visceral organs and localize as cysts in neural and muscular tissues, leading to chronic infection. This immunity is strong and persists for the host's lifetime. However, if immunity fails, tachyzoite formation can resume, causing fatal effects once again. Common lesions of chronic toxoplasmosis include encephalitis, pneumonia, and eye lesions. In immunocompromised individuals, toxoplasma encephalitis (TE) is common. While serum antibodies to *T. gondii* have been demonstrated in many domestic animals worldwide, isolating the organism from tissues has been rarely successful. In India, antibodies have been detected in sheep, goats, pigs, cattle, camels, dogs, cats, and chickens, and the organism has been isolated from pigs, sheep, rodents, and chickens.

Clinical Sings in Animals

Toxoplasmosis in Cats Clinical manifestations of toxoplasmosis in cats are rare, even though cats are the primary hosts of the organism. It is noted that 90% of cats that have previously shed oocysts during a primary infection will not shed oocysts again upon reinfection, unless there is a disruption in local intestinal immunity. Cats with a primary infection may show mild symptoms such as enteritis, fever, lethargy, pneumonia, retinitis, and encephalitis. Congenital infection is also common in cats.

Toxoplasmosis in Dogs The clinical form of toxoplasmosis in dogs is often fatal and is usually observed in conjunction with canine distemper virus, which causes immunosuppression. There are three major clinical signs: neuromuscular symptoms, including progressive paresis and paralysis; respiratory symptoms, including dyspnoea; and gastrointestinal symptoms, including diarrhoea. Other significant complications include pneumonia, myositis, and encephalitis.

Toxoplasmosis in Sheep and Goats *Toxoplasma gondii* causes abortion, stillbirth, and neonatal mortality in small ruminants, particularly sheep and goats, to which it is most pathogenic. Early embryonic death, foetal mummification, abortion, stillbirth, and the birth of weak lambs and kids are common. White necrotic clusters of tachyzoites are often found on the foetal cotyledons. Severe damage to cotyledons (placental necrosis), leading to anorexia, is a significant cause of foetal death. Abortion can occur at any stage of gestation and in ewes of any age that acquire the infection during pregnancy, though it is less common in subsequent pregnancies. Animals that abort typically return to heat and conceive normally. In goats, abortion can occur within 21 days after infection, accompanied by symptoms such as fever, anorexia, respiratory distress, and death from enteritis. The organisms persist in the edible tissues for a long time, with tissue cysts found in the thigh muscles, heart, liver, kidney, brain, and diaphragm, with a higher concentration in the liver and muscles.

Toxoplasmosis in Pigs The primary source of infection for pigs is the environment, through the ingestion of oocysts. Newborn piglets can acquire the infection congenitally and are highly susceptible. The infection is mostly subclinical, presenting with pulmonary disorders and incoordination of movements. In pigs, symptoms such as pyrexia, lacrimation, nasal discharge, tachypnoea, prostration, and abortion have been observed. Mortality is higher in young pigs.

Toxoplasmosis in Cattle and Horses Cattle and horses are among the most resistant species to *T. gondii.* It remains uncertain whether *T. gondii* is a primary cause of abortion in cattle. Although they are generally more resistant to toxoplasmosis compared to other livestock species, young calves can develop fatal bronchopneumonia around 20 days after infection.

Toxoplasmosis in Birds The organism has been identified in various bird species such as farmyard chickens, pigeons, crows, canaries, turkeys, quails, and others. The disease typically manifests as a chronic condition, leading to symptoms such as emaciation, pallor, blindness, nervous signs, and ultimately death.

Toxoplasmosis in Humans

There are two types of human toxoplasmosis:

1. *Congenitally Acquired Toxoplasmosis*: This occurs when a woman acquires the infection at any time during pregnancy. The greatest damage occurs when infection happens during the first trimester. Infection commonly occurs through ingestion of undercooked or raw meat, especially pork products containing tissue cysts, or through exposure to oocysts defecated by cats around or inside houses. Babies born with congenital toxoplasmosis may be born dead or alive with severe symptoms such as hydrocephalus, retino-choroiditis, cerebral calcification, and encephalitis, collectively known as the "tetrad of signs". Some babies may be born without symptoms but develop issues like learning disabilities, dyslexia, and gradual vision loss as they grow older. Ocular damage involving the posterior chamber of the eye is a common consequence of congenital toxoplasmosis.
2. *Postnatally Acquired Toxoplasmosis*: This occurs when individuals come into contact with infectious material during their lifetime. There are four types of symptoms observed: lymphadenopathy (enlargement of lymph nodes, especially cervical), exanthematous (skin rash), cerebral (involving the brain), and ophthalmic (involving the eyes). Lymphadenopathy is the most commonly observed form, accompanied by symptoms such as fever, fatigue, muscle pain, sore throat, and eye lesions. Individuals who have undergone heart or organ transplants, as well as those with cancer or AIDS, are at higher risk of developing toxoplasmosis due to their compromised immune systems. Toxoplasma encephalitis (TE) is a common complication in AIDS patients, where tachyzoites repli-

cate freely in the brain, causing cell lysis and necrotic foci, leading to neurological complications. Infection is most likely when the CD4+ count is less than 0.1×10^3.

Toxoplasma seroprevalence is highest in France (90% in adults) due to raw meat consumption, followed by the UK (80%) and the USA (60%) due to environmental contamination with oocysts. In India, approximately 20% of children and 75–80% of adults are infected with *Toxoplasma*, with the highest seroprevalence (77%) reported in the Kumaon region. Common neurological deficits include altered consciousness, cranial nerve palsies, movement disorders, neuropsychiatric manifestations, and cerebellar signs. A study from South India reported 10% incidence of TE in HIV patients at the 4th International Congress on AIDS.

Diagnosis

Diagnosis of toxoplasmosis cannot rely solely on clinical signs due to their variability and mild nature. Oocyst production is limited to a 2-week period, after which shedding ceases.

Diagnosis is achieved through biological, histological, and serological methods:

(A) Detection of the organism in tissues collected post-mortem or through biopsy and inoculation of suspected materials into mice for bioassay.

(B) Microscopic examination of impression smears from suspected tissues/organs using Giemsa staining.

(C) Immunohistochemical staining (immunoperoxidase test).

(D) Serology methods include:

   1. *Sabin-Feldman dye test (DT)*, once considered the gold standard, detects antibodies by modifying tachyzoite cell walls so they do not take up methylene blue dye at pH 11. However, live antigen use has limited its popularity.
   2. Indirect hemagglutination test (IHAT), latex agglutination test (LAT), enzyme-linked immunosorbent assay (ELISA), modified agglutination test (MAT), and immunofluorescence assays (IFAT) are commonly used, with modifications allowing detection of IgM antibodies.
   3. MAT is preferred for *Toxoplasma* diagnosis in all species.
   4. Carbon immunoassay (CIA) employs India ink to stain antigen-antibody complexes for microscopic visualization.

(E) Molecular detection: Advances in polymerase chain reaction (PCR) facilitate detection of toxoplasmic DNA, even in autolyzed tissues or aborted foetuses.

Humans Continuous monitoring of pregnant women for IgM and IgG antibodies during pregnancy is crucial to assess the risk of foetal transmission, using serological tests like ELISA (TORCH ELISA). Foetal infection can be detected through ultrasound, amniocentesis, and PCR.

Treatment

(A) Monensin or lasalocid at 0.02% added to cat feed effectively prevents oocyst shedding.
(B) Clindamycin and roxithromycin have shown effectiveness as antibiotics in experimental toxoplasmosis infections.
(C) The combination of sulphadiazine (100 mg/kg) and pyrimethamine (1 mg/kg) is the preferred treatment for human toxoplasmosis.
(D) Spiramycin (3 g/day) is a safe drug for pregnant women, concentrating at the placenta to prevent further foetal infection. Folinic acid at 5 mg/kg is also administered to prevent bone marrow suppression.

Control and Prevention

Avoid feeding raw meat to cats. Dispose of carcasses of rodents, birds, and aborted foetuses properly. Control flies, cockroaches, and rodents in the environment. Hygiene measures include washing fruits and vegetables thoroughly before consumption, using separate utensils for cooking meat, wearing gloves while gardening, and ensuring meat is thoroughly cooked.

There is no vaccine available for human toxoplasmosis. The disease is highly significant in terms of zoonotic transmission, particularly concerning the risk to pregnant women and the transfer to babies.

Immunization "Toxovac" is a live toxoplasma vaccine marketed in the UK to protect breeding ewes and prevent sheep abortion. The vaccine contains live tachyzoites of the S48 strain. It is administered as a 2-mL intramuscular injection in the neck region, starting from the 5th month of age or 4 months prior to mating. It is not used in sheep less than 6 weeks old or in pregnant ewes. Consumption of meat from vaccinated animals (chevon) is not recommended for human consumption for 42 days after vaccination.

Immunity Antibodies (humoral immune response) target and destroy extracellular organisms, while cell-mediated immune responses are responsible for eliminating intracellular protozoans. *Toxoplasma gondii* evades the humoral immune response by residing within macrophages. During cell-mediated immune responses, T cells produce lymphokines that activate macrophages, leading to the intracellular killing of the parasites. Intermediate hosts develop lifelong immunity as a result of these immune responses.

References

Khairullah, A., Kurniawan, S., Widodo, A., Effendi, M., Hasib, A., Silaen, O., Ramandinianto, S., Moses, I., Riwu, K., Yanestria, S., Samodra, M., & Afnani, D. (2024). A comprehensive review of toxoplasmosis: Serious threat to human health. *Open Public Health Journal, 17*, e18749445281387. https://doi.org/10.2174/0118749445281387240202094637

Matta, S. K., Rinkenberger, N., Dunay, I. R., & Sibley, L. D. (2021). Toxoplasma gondii infection and its implications within the central nervous system. *Nature Review Microbiology, 19*, 467–480. https://doi.org/10.1038/s41579-021-00518-7

Ybañez, R. H. D., Ybañez, A. P., & Nishikawa, Y. (2020). Review on the current trends of toxoplasmosis serodiagnosis in humans. *Frontiers in Cellular and Infectious Microbiology, 10*, 204. https://doi.org/10.3389/fcimb.2020.00204

Larva Migrans

38

Souti Prasad Sarkhel, Sunanda Sarkhel, and Tuhina Banerjee

Abstract

Larva migrans is a zoonotic helminthosis in humans caused by the abnormal migration of animal nematode larvae within visceral or cutaneous tissues that typically occurs after the inadvertent ingestion of infective eggs or larvae through contaminated food and water. As larvae undergo somatic migration through various internal tissues and organs, they trigger host inflammatory responses, often resulting in a range of clinical symptoms commonly called as larva migrans syndrome. The primary causative agents for visceral larva migrans include *Toxocara* spp., *Angiostrongylus* spp., *Gnathostoma spinigerum*, and *Baylisascaris procyonis*, while the cutaneous form of larva migrans results from the percutaneous migration of animal hookworm larvae. Children are particularly vulnerable due to their tendency for geophagia and poor understanding of hygiene. While many cases remain asymptomatic, heavy parasitic loads or infections affecting vital organs can result in substantial morbidity. The diagnosis of larval migraine is primarily based on serology-based testing, with elevated immunoglobulin E (IgE) levels serving as adjunct markers that may indicate disease severity and outcomes. Therapeutic management typically includes the administration of anthelmintics, alongside corticosteroids, for symptomatic relief of itching and inflammation. However, effective prevention strategies must emphasise reducing

S. P. Sarkhel (✉)
Department of Veterinary Parasitology, FVAS, Institute of Agricultural Sciences, Banaras Hindu University, South Campus, Mirzapur, Uttar Pradesh, India
e-mail: souti.sarkhel@bhu.ac.in

S. Sarkhel
Department of Optometry and Vision Sciences, Amity University, Gurgaon, Haryana, India

T. Banerjee
Department of Microbiology, Institute of Medical Sciences, Banaras Hindu University, Varanasi, Uttar Pradesh, India

R. Biswas et al. (eds.), *Concept of One Health and Zoonoses*,
https://doi.org/10.1007/978-981-95-6627-3_38

exposure to infection sources, implementing routine deworming protocols for pets, and fostering public health education initiatives.

Keywords

Ocular larva migrans · Visceral larva migrans · Cutaneous larva migrans · Larva migrans · Eosinophilia · Toxocarosis · Zoonoses

Introduction

Larva migrans is a zoonotic infection that occurs when the larvae of parasitic roundworms of animals inadvertently invade a nonnatural host species, known as an aberrant host, such as humans. This infection usually occurs after the accidental ingestion of infective eggs or larvae, often resulting from the consumption of raw or undercooked tissues from infected paratenic hosts or contaminated food and water. When infective larvae migrate within non-specific host species, it typically leads to erratic movement through internal soft tissues and visceral organs, often resulting in a range of inflammatory immune responses that inhibit the further development of invading larvae (Roberts et al., 2009). In these incidental and aberrant hosts, the larval stages of diverse parasitic species exhibit limited development and migrate primarily through organs such as the liver, lungs, kidneys, and brain, resulting in a wide range of clinical outcomes, from asymptomatic cases to severe manifestations. The erratic migration of these larvae can trigger a variety of clinical symptoms and contribute to pathogenesis, leading to persistent and pronounced eosinophilia, with ocular involvement potentially resulting in granulomatous retinitis, a condition frequently misdiagnosed as retinoblastoma (Fan et al. 2015), which may, unfortunately, lead to unintended enucleation of the eye. Moreover, the aberrant migration of larvae into the central nervous system can result in eosinophilic meningitis (Diaz, 2010), characterised by an elevated eosinophil count in the cerebrospinal fluid (CSF), particularly when the eosinophil percentage exceeds 10% of the total white cell count in the CSF (Graeff-Teixeira et al. 2009). Several nematode species are implicated in human larva migrans syndrome. Visceral, ocular, and neural larva migrans, as well as eosinophilic meningitis, are frequently associated with *Angiostrongylus cantonensis*, *Gnathostoma spinigerum*, *Baylisascaris procyonis*, and *Toxocara* spp (Diaz, 2009; Dangoudoubiyam and Kazacos, 2009). Cutaneous larva migrans, characterised by inflammatory skin lesions around the migratory tracks of the larvae in the skin, are typically caused by animal hookworms, including *Ancylostoma braziliense*, *Ancylostoma caninum*, *Uncinaria stenocephala*, *Bunostomum phlebotomum*, and *Strongyloides stercoralis* (animal strains), as well as *Gnathostoma* spp. (Kim et al. 2006; Hoff et al., 2008). Globally, larva migrans infections are increasingly recognised as significant contributors to morbidity and mortality (Clément-Rigolet et al., 2004), particularly in vulnerable populations (Kanecki et al., 2025). Delays in diagnosis or incorrect assessments, often owing to limited awareness, can result in severe health consequences, including fatalities (Nikopoulou et al., 2025), especially in children, immunocompromised, and elderly individuals. Consequently, a comprehensive understanding of the aetiology, symptoms, and signs of these infections is crucial for prompt diagnosis and intervention to safeguard human health and prevent potentially life-threatening complications or accidental vision loss in affected patients (Ma et al., 2018).

History

In 1921, Fulleborn became the first to identify the zoonotic potential of *Toxocara*, a genus of parasitic nematodes, specifically emphasising the implications of *Toxocara canis* larvae in human infections. He described the association between these larvae and the development of granulomatous lesions, which predominantly occur in the liver, lungs, kidneys, and brain of affected individuals (Bowman, 2020). Later, in 1950, Wilder recognised ocular larva migrans (OLMs) as another significant clinical manifestation of toxocarosis (Wilder, 1950). In 1952, Beaver and colleagues acknowledged the role of *T. canis* larvae in cases characterised by persistent circulating eosinophilia, hepatomegaly, pulmonary infiltration, fever, cough, pneumonitis, and hyperglobulinemia. These symptoms arise from the invasion of nematode larvae into the internal organs of young children, leading to the introduction of the term "visceral larva migrans" (VLMs) (Beaver et al., 1952; Despommier, 2003).

Epidemiology

Larva migrans remains a neglected zoonotic disease (Reichert et al., 2018), despite its significant impact on public health in developing countries, particularly in India. Human toxocarosis, which is caused primarily by the ingestion of embryonated *Toxocara* spp. eggs from close contact with pet dogs or cats, is the main cause of visceral larva migrans (VLMs) and ocular larva migrans (OLMs) (Wu and Bowman, 2020). Additionally, other parasitic species, such as *Angiostrongylus cantonensis*, *Gnathostoma spinigerum*, *Toxascaris leonina*, *Capillaria hepatica*, *Ascaris suum*, *Lagochilascaris minor* and *Baylisascaris procyonis*, are also commonly associated with these conditions (Sprent, 1971; Soulsby et al., 1987; Maruyama et al., 1996; Gavin et al., 2005; Herman and Chiodini, 2009; Wang et al., 2012; Kazacos, 2016; Campos et al., 2017). In humans, lesions typical of cutaneous larva migrans may arise from infections caused by the human hookworms *Ancylostoma duodenale* and *Necator americanus* (Beaver, 1956; Corte et al., 2013). Additionally, *Strongyloides stercoralis* can cause a form of cutaneous larva migrans upon percutaneous transmission, commonly referred to as larva currens due to the characteristic rapid migratory movement of the invading larvae. VLM is predominantly observed in tropical and developing countries where improved sanitary conditions are lacking (Yu et al., 2012). In contrast, ocular larva migrans (OLMs) occur when parasitic larvae invade the eye, whereas neural larva migrans (NLMs) involve these larvae penetrating the brain and/or spinal cord (Kazacos, 2001). Actually, cutaneous larva migrans (CLM) is a dermatosis caused by the percutaneous migration of parasitic nematode larvae of animal hosts, primarily from percutaneous transmission of animal hookworm species. Close contact with environmental hookworm larvae populations in areas such as pastures or parks frequented by pet dogs or cats can lead to percutaneous infection through exposed skin, resulting in elevated, erythematous lesions along migratory tracks. These lesions are most commonly observed on the feet and arms of affected individuals (Reichert et al., 2018). In recent years, ocular form of larva migrans infections, known for their potential to cause significant morbidity and poor clinical outcomes like blindness, unilateral vision impairment, etc. (Hare and Franco-Paredes, 2014), have spread beyond their traditional endemic areas. This expansion can be attributed to factors such as increased tourism, conflicts, and shifting migration patterns. With the rise in international travel, the prevalence of these infections has increased in new regions (Shinkar et al., 2005). Furthermore, the prolonged environmental persistence of hookworm larvae in tropical and subtropical regions is driven by various factors, including tropical climates, inadequate hygiene and sanitation practices, the presence of stray dogs, open defecation by dogs, and improper faecal disposal. This has led to cutaneous larva migrans (CLM), reported as the most common skin condition among travellers returning from tropical destinations (Caumes, 2000; Bouchaud et al., 2000). Despite the reported high prevalence of ancylostomiasis and toxocarosis among canine populations across various regions in India, there remains a significant deficiency in epidemiological data regarding human infections caused by larva migrans. This gap in information underscores the urgent need for comprehensive epidemiological studies focused on zoonotic animal nematodes. Such research is vital for improving our understanding and guiding public health strategies aimed at mitigating the risks associated with these parasitic zoonoses.

Risk Factors

Humans can act as paratenic hosts for various animal nematode species, with worms capable of persisting for extended periods, sometimes exceeding 10 years in cases of human toxocarosis (Girdwood, 1986). In human toxocarosis, the ingestion of embryonated *Toxocara* eggs leads to larval hatching and subsequent penetration of the intestinal wall. After penetrating the intestinal wall, the third-stage larvae undergo somatic migration inside the human body through the tissues and visceral organs. They travel from the liver via the portal circulation to the lungs and then enter the systemic circulation. However, if larvae encounter blood vessels smaller than their diameter, they may actively penetrate vessel walls and migrate aimlessly into surrounding tissues. Migration through internal tissues and organs in nonnatural hosts elicits inflammatory reactions and granuloma formation (Soulsby et al., 1987), often resulting in larval death. However, inflammation in vital organs, especially neural tissues, the eyes, and the heart, can lead to severe morbidity and even mortality (Nikopoulou et al., 2025). Toddlers and young children, mainly those aged 3–5 years, are especially vulnerable to VLM because of their soil-eating habits and exposure to environments frequented by definitive hosts such as dogs and cats. Exposure to contaminated soil in parks and peridomestic areas, combined with a lack of understanding of hygiene practices, increases the risk for individuals. While relatively uncommon, various animal species have been documented as potential paratenic hosts for different *Toxocara* species, including cattle, lamb, rabbits, ducks, chickens, quails, and ostriches (Pahari and Sasmal, 1990). Consequently, the consumption of raw or undercooked meat from these animals can lead to visceral larva migrans in humans (Ma et al., 2018). Although many infections remain asymptomatic, the involvement of vital organs can result in severe health consequences. In humans, *Toxocara* spp. larvae do not develop into adult forms, and the symptoms of larva migrans often lack distinctive features. In contrast, the larval stages of *Baylisascaris* spp. can grow and develop to a greater extent in paratenic and human hosts, leading to much more severe consequences for infected patients (Nathwani et al., 1992). In human toxocarosis, conventional diagnostic methods, such as stool examinations, which are effective for identifying adult worms in patent infections, and tissue biopsies, which are used in cases of visceral larva migrans (VLMs), often prove ineffective. This condition poses significant diagnostic challenges, compounded by the lack of specific clinical symptoms and the difficulty of detecting the parasite owing to the migratory nature of the invading larvae, which further complicates sensitive detection. In cases of ocular toxocarosis, which primarily affects one eye in young children, the clinical manifestations can include posterior and peripheral retinochoroiditis, optic papillitis, endophthalmitis, diffuse unilateral subacute neuroretinitis, and, occasionally, the presence of motile subretinal nematodes (Park et al., 2000; Sabrosa and Souza, 2001; Gupta and Tripathy, 2023). While many parasites typically target organs other than the eyes, the unique structure and relatively weak immune environment of the eye make it susceptible to infections (Klotz et al., 2000), and the intraocular envrionment shields the invading parasites from the host's immune machineries. Furthermore, the interactions among the parasite, its excretory-secretory products, and the host immune system affect the outcome of the infection. Research reports indicate that the seroprevalence of anti-*Toxocara* antibodies is significantly greater in children than adults, a finding supported by various studies (Mazur-Melewska et al., 2012; Woodhall and Fiore, 2014; Bokharaei-Salim et al., 2024; Lima et al., 2025). This increased prevalence in children is primarily attributed to their tendency for inadvertent or frequent pica behaviour, geophagia, and their closer interactions with pet dogs and cats. Zoonotic parasites, which primarily infect animals, can also impact human health, particularly in light of the growing popularity of organic food and free-range farming, which bring livestock closer to human habitats. The presence of companion animals in households increases the likelihood of zoonotic transmission of parasites to humans (Robertson et al., 2000). Parasitic infections can originate from various sources, with the oral ingestion of contaminated water and soil being the primary routes of transmission. Additional potential sources include the consumption of raw or undercooked freshwater fish, crabs, and game meats, as well as food contamination by cockroaches and houseflies. Close contact with household pets that are not regularly dewormed further elevates the risk of transmission of various animal nematode species (Stull et al., 2015).

Aetiology

Larva migrans is a parasitic condition characterised by the aberrant migration of nematode larvae through host internal tissues and organs. Classification is typically performed according to the primary anatomical sites involved, which include visceral larva migrans (VLMs), ocular larva migrans (OLMs), neural larva migrans (NLMs), and cutaneous larva migrans (CLMs) (Sharma et al., 2015). The predominant causative agents of VLM and OLM in humans are *Toxocara* spp. (Alter and Turcios, 2009), which are typically transmitted through the ingestion of embryonated eggs acquired through close contact with domestic dogs and cats. In contrast, cutaneous larva migrans occur when animal hookworm larvae penetrate the skin, causing creeping eruptions that are characteristic of this condition. On the other hand, other zoonotic parasites, such as *Angiostrongylus* spp., *Gnathostoma* spp., and *Baylisascaris procyonis*, are also frequently implicated in the pathogenesis of fatal visceral larva migrans (Otranto and Deplazes, 2019). These zoonotic parasitic species, which serve as the primary etiological agents of larva migrans complex disease in humans, are further detailed in Table 38.1, providing a comprehensive overview of their sources, natural hosts, and associated clinical manifestations.

Visceral Larva Migrans (VLMs)

Roundworms from *Toxocara* species, which belong to the Ascarididae family, are identified as the primary cause of human toxocarosis and are the main agents responsible for VLM in humans (Wu and Bowman, 2020). The most prevalent species among these are *T. canis* (dog ascarid), *Toxocara cati* (feline ascarid), *Toxocara vitulorum* (cattle ascarid), and *Toxocara pteropodis* (bat ascarid) (Gorp et al., 1987; Macpherson, 2013). Humans serve as paratenic or accidental hosts, primarily acquiring infection during childhood through geophagia (soil consumption), pica (the craving for nonfood substances), or onychophagia (nail biting) (Kroten et al., 2018). Infection can also occur through the consumption of raw visceral meat from potential paratenic hosts such as quails, chickens, ducks, pigs, cattle, and rabbits (Pahari and Sasmal, 1990; Xu and Han, 2024; Ebrahimipour et al., 2024). Additionally, the potential dispersion of eggs/ infective larvae by house flies (Pegg, 1971), and widespread environmental contamination with infectious eggs contribute to the transmission of the disease (Papajová et al., 2008; Bowman, 2021), resulting in diverse clinical conditions that can vary from asymptomatic cases to severe illnesses.

Baylisascaris procyonis is an ascarid parasite primarily found in raccoons and belongs to the family Ascarididae, superfamily Ascaridoidea, order Ascaridida, class Chromadorea, and phylum Nematoda. These parasites produce highly resilient eggs that can persist in the environment (Kazacos, 2001; Page, 2013). When these eggs are inadvertently ingested through contaminated fruits, food, or water, they can result in severe conditions for accidental intermediate hosts, including humans, particularly affecting children and leading to ocular diseases (OLMs or diffuse

Table 38.1 Key ascaridoid, strongylid, and spirurid helminths associated with visceral larva migrans in humans

Sr. No.	Name of the worm	Distribution	Natural final host	Natural IMH	Reported paratenic host	Predilection site in DH	Predilection site in humans	Infective stage for humans	Disease caused
1.	*Angiostrongylus* (syn. *Parastrongylus*) *cantonensis*	Southeast Asia	Rodents	Gastropods	Crayfishes, crabs, toads, frogs, monitor lizards	Pulmonary arteries	L3 invade CNS	L3 larvae	Eosinophilic meningoencephalitis or neural angiostrongylosis of man cerebral angiostrongyliasis
2.	*Angiostrongylus* (syn. *Parastrongylus*) *costaricensis*	The Americas, The Caribbean, Japan	Rodents	Terrestrial slug *Sarasinula plebeia*, gastropods	?	Mesenteric arteries	Ileocaecocolic branches of anterior mesenteric artery in man	L3 larvae	Abdominal angiostrongylosis mimicking appendicitis, eosinophilic enteritis
3.	*Ascaris suum*	Worldwide	Pigs, wild boars	X	Cattle, chicken	Small intestine	Liver, lungs	Larvated eggs containing L3	Eosinophilic pneumonia
4.	*Toxocara canis*	Worldwide	Dogs, wild canids	X	Rodents, chicken, duck, pigeon, quails, pigs, humans	Small intestine	Viscera	Embryonated eggs containing L3, raw or rare meats	VLM, OLM, cerebral larva migrans/ neurotoxocarosis/ larva migrans visceralis, larva migrans ocularis
5.	*Toxocara mystax* (syn. *T. cati*)	Worldwide	Cat	X	Rodents, chicken, duck, pigeon, quails, pigs, humans	Small intestine	Viscera	Embryonated eggs containing L3, raw or rare meats	

(continued)

Table 38.1 (continued)

Sr. No.	Name of the worm	Distribution	Natural final host	Natural IMH	Reported paratenic host	Predilection site in DH	Predilection site in humans	Infective stage for humans	Disease caused
6.	*Toxascaris leonina*	Worldwide	Dog, cat, fox	X	Rodents	Small intestine	Viscera	Embryonated eggs containing L2	VLM
7.	*Toxocara vitulorum*	Worldwide	Cattle, water buffalo	X	X	Small intestine	Viscera	Embryonated eggs containing L3	VLM, OLM, cerebral larva migrans/ neurotoxocarosis/ larva migrans visceralis, larva migrans ocularis
8.	*Baylisascaris procyonis*	North America	Raccoon, dogs (?)	–	Dog, cat, rodents, lagomorphs, emu, ostrich	Small intestine	Brain, CNS	Larvated eggs containing L2	CNS, brain affections (eosinophilic meningoencephalitis)
9.	*Baylisascaris columnaris*		Skunks	–	?		Viscera, brain, CNS		Larva migrans (NLM, OLM, VLM)
10.	*Baylisascaris melis*	Europe	Badgers	–	?		Brain, CNS		
11.	*Lagochilascaris minor*	Americas	Cat	Rodents	X	Pharynx, oesophagus, stomach	Tonsils, nose, neck, rarely in the brain	Wild rodent meat containing encysted L3 larvae	VLM, subcutaneous abscesses in the cervical region, neck, ear, jaw, tonsils

12.	*Gnathostoma spinigerum*	Southeast Asia, Africa	Cat, dog, mink, polecat, wild carnivores	1st IMH: copepods (aquatic crustaceans), 2nd IMH: small mammals, birds, reptiles, fish, amphibians	Fish, snakes, poultry, frogs	Stomach	Subcutis, viscera, brain, CNS, eyes	L3 larvae	Cutaneous/ocular/ nervous gnathostomosis
13.	*Gnathostoma hispidum*	Japan, Spain							
14.	*Gnathostoma doloresi*	Japan							
15.	*Gnathostoma nipponicum*	Japan							
16.	*Gnathostoma binucleatum*	Latin America							
17.	*Strongyloides stercoralis* animal strain	Worldwide	Dog, cat, fox, humans	–	–	Small intestine	Cutaneous tissue, visceral organs	L1 larvae	Cutaneous larva migrans, larva currens ("racing larva")
18.	*Gongylonema pulchrum (syn. Gongylonema scutatum)*	Worldwide	Herbivores, primates, humans	Coprophagous beetles, cockroaches	–	Oesophagus, rumen	Oral cavity, oesophagus	L3 larvae in dung beetles, cockroaches	Cutaneous, oesophageal and ocular
19.	*Capillaria hepatica*	Southeast Asia, East Asia	Rodents, lagomorphs, squirrel, mole, shrew	X	X	Liver	Liver	Embryonated eggs	Hepatic capillariasis

unilateral subacute neuroretinitis (DUSN)) or clinical neural larva migrans (NLMs) (Yoshida et al., 2016). The larvae of *B. procyonis* grow during their migration, which sets them apart from *Toxocara* larvae (Dangoudoubiyam and Kazacos, 2009). Like *Toxocara canis* larvae, these larvae can invade the central nervous system of paratenic hosts. Even a mere 1–3 larvae in the brain can lead to potentially fatal consequences in various small animals (Kazacos, 2016). Furthermore, the larvae of *B. procyonis* are larger and more neurotropic (Gavin et al., 2005), with their capacity for neural migration resulting in fatal diseases across more than 130 animal species, including groundhogs, rabbits, ground squirrels, various birds, and even humans (Kazacos et al., 2013).

Gnathostoma, known as the stomach worm of both domestic and wild carnivores, are small roundworms belonging to the class Secernentea, order Spirurida, and family Gnathostomatidae. Adult worms are rusty coloured, and usually measure approximately 2–3 cm in length (Rusnak and Lucey, 1993), while the L3 larvae vary in colour from milky white to pinkish cream, and usually measure 3–4 mm in length (Herman and Chiodini, 2009; Liu et al., 2020). These parasites are responsible for foodborne zoonotic diseases. Their life cycle involves copepod crustaceans as the first intermediate host, during which they develop into third-stage larvae. Fish and amphibians serve as the second intermediate hosts, where the parasite progresses to an advanced third larval stage upon the ingestion of infected copepods. Cats, dogs, and other animals acquire a patent infection by consuming infected fish (Thiangtrongjit et al., 2021). Humans are not the definitive hosts; rather, they serve as paratenic hosts when they ingest water contaminated with infected copepods, consume raw fish dishes or undercooked meat from infected paratenic hosts, handle infected fish without gloves, or undergo transplacental migration from an infected mother. The parasite does not develop into adults but matures somewhat as advanced L3 larvae do, primarily in the muscles or visceral organs. This can lead to significant health issues, as they may cause migratory swellings in the skin during aberrant migration in the human face, limbs, or torso. Severe outcomes may occur if they reach vital organs such as the brain, heart, or eyes (Diaz, 2015; Liu et al., 2020; Thiangtrongjit et al., 2021).

Ocular Larva Migrans (OLMs)

Toxocara retinitis, also frequently known as ocular larva migrans (OLMs), is caused primarily by larvae of *T. canis* and *T. cati* (Mirdha & Khokar, 2002; Fomda et al., 2007). In rare cases, as reported mainly from the Americas, OLM can also be caused by *Baylisascaris procyonis*, the raccoon roundworm (Brasil et al., 2006). Visceral larva migrans (VLM) and OLM represent the primary clinical manifestations of toxocarosis. Ocular toxocarosis is commonly observed in older children and young adults. Ocular larvae transfer (OLM) typically affects one eye (unilateral), with symptoms including decreased visual acuity, strabismus, or leukocoria (Yokoi et al., 2003; Stewart et al., 2005; Shields, 1984). Bilateral involvement is exceedingly rare (Benitez del Castillo et al., 1995). The most frequent initial signs are impaired visual acuity and leukocoria (Sharma et al., 2015; Hare and Franco-Paredes, 2014).

Cutaneous Larva Migrans (CLM)

Cutaneous larva migrans, commonly known as "creeping eruptions", typically occur in humans because of the migration of nematode larvae (Brenner and Patel, 2003). This condition is characterised by linear, tortuous, erythematous, and intensely pruritic eruptions on the skin. The primary causative agent is the hookworm larvae of animals, particularly *Ancylostoma braziliense*. However, other nematode larvae from animals, such as *Ancylostoma caninum*, *Uncinaria stenocephala*, *Bunostomum phlebotomum*, *Strongyloides stercoralis* (animal strain), and *Gnathostoma* spp., have also been reported in clinical cases. Additionally, the larvae of *Ancylostoma caninum*, *Ancylostoma ceylanicum*, *Ancylostoma tubaeforme*, and *Necator americanus*, which typically mature in humans, can cause transient creeping eruptions at their penetration sites in sensitised individuals (Brenner and Patel, 2003; Le Joncour et al., 2012; Calvopina et al., 2024).

Clinical Signs and Symptoms

Visceral larva migrans, which are primarily associated with infections from *Toxocara* species, pose a serious zoonotic risk with significant public health implications. Transmission typically occurs through the faecal-oral route, most commonly resulting from the accidental ingestion of contaminated food items, such as fruits and vegetables, or water contaminated with the eggs of the dog roundworm *Toxocara canis* and, to a lesser extent, with eggs of *Toxocara cati*, the roundworm of cats (Yoshida et al., 2016; Xu and Han, 2024). In highly contaminated environments, particularly those with substantial populations of stray dogs and cattle, the risk of food contamination with *Toxocara* eggs is exacerbated by the dispersion of embryonated eggs/ infective larvae by houseflies. Furthermore, mangoes contaminated with *T. pteropodis* eggs origin from excreta of pteropod bat species can lead to acute hepatitis or gastroenteritis symptoms in children (Moorhouse, 1982). Moreover, the consumption of raw liver from potential paratenic hosts may result in human toxocarosis (Choi et al., 2008). However, bad habits in young children (<5 years old), such as geophagia (soil ingestion) and onychophagia (nail biting), can contribute to toxocarosis, especially in areas with significant soil contamination (Kroten et al., 2018). Human toxocarosis is widespread, impacting approximately 20% of the global population, with seroprevalence rates exceeding 80% in certain tropical and subtropical regions of Africa and Asia (Lötsch et al., 2017; Rostami et al., 2019; Kong and Peng, 2020). Human toxocarosis manifests in several clinical forms, each with distinct characteristics, such as systemic (classical VLM, incomplete VLM), compartmentalised (OLM, NLM), covert toxocarosis, and asymptomatic (Auer and Walochnik, 2020; Kanecki et al., 2025). While *Toxocara* infections in humans mostly remain asymptomatic, some individuals may experience moderate to severe systemic, neurological, or ocular manifestations (Ahn et al., 2014; Cunningham and Zierhut, 2021). Interestingly, infections in nonnatural hosts, such as cats that ingest *T. canis* embryonated eggs, can lead to disease manifestations similar to those observed in humans. These infections can cause persistent eosinophilia and the formation of large eosinophilic granulomas in vital organs such as the kidneys and liver, as well as severe medial hypertrophy of the pulmonary vessels in the lungs (Lim, 2008; Bowman, 2020; Nikopoulou et al., 2025; Lauriente et al., 2025). The clinical manifestations of toxocarosis in humans exhibit a broad spectrum, ranging from asymptomatic cases to significant organ damage (Pawlowski, 2001; Rubinsky-Elefant et al., 2010; Yoon et al., 2018). This variability is influenced by many factors such as the age of the host, parasite load, specific sites of larval migration, and magnitude of the host inflammatory response (Rubinsky-Elefant et al., 2010; Ahn et al., 2014).

Visceral Larva Migrans/(Overt Toxocarosis)

The clinical signs and symptoms vary on the basis of the infective larval load, the specific sites affected by wandering larvae, and the frequency of reinfections (Xinou et al., 2003; Baalbaki et al., 2020). Key clinical features include leukocytosis, persistent and severe (hyper-) eosinophilia, and persistently elevated IgE levels. Additional symptoms may include enlargement of the liver and spleen, inflammation, and swelling of regional lymph nodes, as well as respiratory issues such as cough and wheezing, which may present as asthmatic manifestations. This form can also lead to recurrent eosinophilic pneumonia (Loeffler syndrome), convulsions, abdominal pain, digestive disturbances, loss of appetite, nausea, intermittent fever, headaches, body aches, limb pains, neurological disorders, behavioural changes, altered sleep patterns, and epileptic episodes (Magnaval et al., 2001; Gao and Liu, 2019; Sim et al., 2019; Auer and Walochnik, 2020). Severe infections can cause extensive damage to vital organs, including the CNS, lungs, liver, and kidneys, potentially resulting in unfavourable outcomes (Carpio et al., 2016); however, the disease is often self-limiting (de Almeida Carvalho and Rocha, 2014). Although VLM is more prevalent than OLM in older patients, severe cases are more commonly observed in children than in adults (Yoshida et al., 2016).

Ocular Larva Migrans (OLM)/Ocular Toxocarosis

In contrast to visceral larva migrans, OLMs predominantly affect individuals over the age of 12 (Gupta and Tripathy, 2023; Jowsey and McLeod, 2023). In this case, the afflictions are usually linked to a single larva rather than multiple larvae (Taylor, 2006; Ryan and Durand, 2011), which may cause a stronger immune reaction in the host due to a lower infection level (Paul et al., 2009). This longer survival may be attributed to delayed encapsulation. Major signs of OLM include granulomatous endophthalmitis, fibrous traction bands, retinitis, uveitis, retinal detachment (which can be mistaken for retinoblastoma), loss of visual acuity, and loss of vision (Otranto and Eberhard, 2011). Unlike VLM, leukocytosis and peripheral eosinophilia is rare in this form (Jowsey and McLeod, 2023). OLMs can manifest in three classic patterns: posterior granulomas, typically submacular or peripapillary with prominent retinal striae or folds; peripheral granulomas, often accompanied by a vitreoretinal traction band extending toward the posterior pole; and severe, unilateral intermediate or panuveitis, termed nematode or *Toxocara* endophthalmitis. Additionally, atypical presentations of OLM have been documented (Pivetti-Pezzi, 2009; Cunningham and Zierhut, 2021).

Covert Toxocarosis (CT)

Covert toxocarosis is the most prevalent form of the disease and is characterised by non-specific signs and symptoms that do not align neatly with defined categories such as visceral larva migrans (VLMs), ocular larva migrans (OLMs), or neural larva migrans (NLMs) (Chen et al., 2018; Inagaki et al., 2023). While many mild infections remain asymptomatic, they can occasionally present with vague symptoms, such as general malaise, urticaria, arthralgia, and loss of appetite, often accompanied by intermittent fever (Taylor et al., 1988; Despommier, 2003; Magnaval et al., 2022). Patients may also test positive for the infection serologically while experiencing transient symptoms, including headaches, chronic cough, and abdominal cramps. This condition typically originates from the host's immune response to parasite antigens in specific organs, leading to organ-specific immunopathological changes (Gillespie, 1987; Taylor et al., 1987; Magnaval et al., 2001).

Asymptomatic Toxocarosis

Asymptomatic toxocarosis refers to a form of human infection in which there are no clinical manifestations, and it is typically identified only through serological testing. In some cases, this condition may be accompanied by eosinophilia (de Almeida Carvalho and Rocha, 2014; Kim et al., 2014; Chen et al., 2018). In human toxocarosis, the clinical signs often present as non-specific, which can lead to significant underdiagnosis; therefore, the actual prevalence of human toxocarosis remains uncertain. Sero-epidemiological studies suggest that *Toxocara* infection may be more widespread than currently diagnosed, with a slight increase in serological positivity observed with age. This increase may be attributed to the long-term survival of larvae within host tissues, which continiously stimulate the immune system even after the initial infection (Taylor et al., 1988; Maizels, 2013; Fan et al., 2013; Rostami et al., 2019).

Cutaneous Larva Migrans (CLM)

Cutaneous larva migrans (CLMs) typically begin with a small reddish papule at the penetration site that develops into an itchy, serpiginous track with accompanying rashes (van Gasselt and van de Sandt, 2000). The condition usually progresses at a rate of less than 1–2 cm per day (Liu and Weller, 1993), with the parasite actively moving through the stratum spinosum layer of the epidermis (Balfour et al., 2002). The larva currens lesion, resulting from the invasion and migration of *Strongyloides stercoralis* animal strain larvae within the epidermal layer, typically advances at a rate of 5 to 10 cm per hour (Sałamatin et al., 2023). The onset of symptoms can vary depending on the infecting species; for example, *Ancylostoma braziliense* may cause symptoms within an hour, whereas *Uncinaria stenocephala* may take several days before papular lesions become apparent (Maxfield and Crane, 2023). In some cases, CLM can result in more severe reactions, such as vesiculobullous disease or papulopustular inflammation in hair follicles, although these reactions are less common (Heukelbach and Feldmeier, 2008).

Diagnosis

The clinical manifestations of visceral larva migrans (VLMs) are influenced primarily by the location and load of parasitic larvae within the host. Once the larvae hatch from the infective eggs within the small intestine of an aberrant host such as a human, they penetrate the intestinal lining, enter the circulation, and subsequently wander throughout the body. This active migration elicits a delayed-type hypersensitivity (DTH) response in these hosts, akin to that observed in paratenic hosts, resulting in the formation of granulomatous lesions that encapsulate the migrating larvae (Despommier, 2003). Experimental models suggest that these juvenile larvae have a particular affinity for neural tissues, likely owing to mechanisms that allow them to evade or resist host immune responses within the immune-privileged sites such as the eyes and central nervous system (CNS) (Sprent, 1956; Sprent, 1958; Glickman and Summers, 1983; Glickman and Shofer, 1987). This neurological involvement may present as sensory or motor deficits, although clinical improvement can often be achieved following effective anthelmintic therapy (Sircar et al., 2016; Sánchez et al., 2018). Serological diagnostic techniques are essential for identifying covert infections and visceral larva migrans (VLMs), particularly through antibody-enzyme-linked immunosorbent assays (Ab-ELISAs) that utilise excretory-secretory (ES) antigens (Magnaval et al., 2001; Mohamad et al., 2009; Fillaux and Magnaval, 2013; Ahn et al., 2014; Wilkins, 2014). These methods demonstrate superior sensitivity compared with those used for ocular larva migrans (OLMs). In cases where cerebral impairment is suspected, cerebrospinal fluid (CSF) can also be analysed through these diagnostic approaches (Vidal et al., 2003). Recent studies have highlighted the importance of immunoglobulin E (IgE) antibodies in the diagnosis of human toxocarosis, particularly ocular toxocarosis (OT). Patients diagnosed through a combination of clinical evaluation and serological techniques often present with elevated IgE levels, indicating its potential as a valuable supplementary diagnostic marker (Fillaux and Magnaval, 2013; Ahn et al., 2014; Nikopoulou et al., 2025). Importantly, a reduction in IgE levels following therapy may serve as an indicator of successful treatment (Magnaval et al., 1992).

In contrast, while peripheral eosinophilia is often linked to systemic toxocarosis, it is less frequently observed in ocular toxocarosis (OT) (Gillespie et al., 1993). These findings suggest that eosinophil counts may not have the same diagnostic importance as IgE levels or ELISA results (Rubinsky-Elefant et al., 2010; Elefant et al., 2006). Nevertheless, the presence of eosinophilia could indicate potential correlations between the systemic and ocular forms of the disease, highlighting the necessity for comprehensive evaluations and personalised treatment approaches. Seropositivity rates for toxocarosis have been reported among asymptomatic individuals, especially in endemic regions where active screening for *Toxocara*-specific antibodies was conducted (Magnaval et al., 2022). Prevalence estimates vary significantly, ranging from 4% in Italy to 35% among Amazonian natives (Oliart-Guzmán at al., 2014; Fan et al., 2015). While the direct identification of larvae can be quite challenging, immunodiagnostic tests are reliable for diagnosis. Among these tests, ELISA is particularly favoured because of its high sensitivity and specificity (Jacob and Lorber, 2015). In the case of cutaneous larva migrans (CLM), diagnosis relies primarily on clinical and epidemiological investigations, which include assessing the patient's travel history to endemic regions and examining the distinctive characteristics of the rash. This information is essential for differentiating CLM from other migratory infections (Bava et al., 2011; Maxfield and Crane, 2023). Blood tests are typically unnecessary, as eosinophilia occurs in less than 40% of cases and lacks specificity for CLM (Hochedez and Caumes, 2007). Potential complications include cellulitis, which is often associated with *Streptococcus pyogenes*, allergic reactions, and Loeffler's syndrome (Brenner and Patel, 2003; Podder et al., 2016). Consequently, a comprehensive history and physical examination are crucial for an accurate diagnosis, particularly since some patients may exhibit elevated eosinophil counts and increased IgE levels. Furthermore, differential diagnosis is essential to exclude conditions with overlapping presentations, such as scabies, erythema chronica migrans, larva currens, dermatophyte infections, and phytophotodermatitis (Kokollari et al., 2015; Osman et al., 2024; Currie et al., 2025). This process should involve analysing distinguishing features and employing suitable differentiation techniques during patient evaluation. However, the detection and confirmation of larva migrans predominantly rely on serological assays targeting antibodies against native antigens from parasites. These antigens may include crude extract antigens, excretory-secretory products, or partially purified fractions derived from adult worms or larvae. The use of recombinant antigens in diagnostic protocols has gained traction, with several proteins, such as low-molecular-weight (>35 kDa) and high-molecular-weight TES-Ag, as well as recombinant TES antigens (rTES-30, rTES-120) and the 29, 31 kDa proteins of *Angiostrongylus cantonensis* and the 20, 21, and 24 kDa components of *Gnathostoma spinigerum*, which are beneficial for antibody-based assays (Fillaux and Magnaval, 2013; Wilkins, 2014; dos Santos et al., 2019; Melo et al., 2022; Fonseca et al., 2022; Kundi et al., 2023). Furthermore, advanced PCR methods that target mitochondrial genes (mt-DNA), such as cytochrome oxidase I (COI), along with nuclear gene regions, such as internal transcribed spacer regions (ITS-1, ITS-2), 5.8S, 18S, and 28S, are crucial for identifying various nematode species (Nisa et al., 2022). However, obtaining

DNA samples from suspected cases can be challenging; nonetheless, successful acquisition may be attempted by isolating larvae from biopsy samples (Fillaux & Magnaval, 2013).

Treatment

The treatment for visceral larva migrans (VLMs) focuses on eliminating migrating larvae and managing inflammation. Effective anthelmintic medications include albendazole, thiabendazole, mebendazole, fenbendazole, diethylcarbamazine, and ivermectin (Horton, 2000; Strube et al., 2013; Kazacos, 2016; Chen et al., 2018; Chai et al., 2021; Magnaval et al., 2022). The selection of the appropriate drug, along with its dosage and treatment duration, depends on the severity of the infection and the patient's condition (Nikopoulou et al., 2025). Albendazole is the preferred treatment for toxocarosis, with commonly recommended regimens involving either a 5-day course at a dosage of 10 mg/kg daily in two divided doses or 400 mg twice daily for 5 days (Hotez, 1995; Kim et al., 2010). For systemic toxocarosis, a longer course of albendazole (400 mg twice a day for 7–14 days) is considered standard and is often more effective than thiabendazole (Stürchler et al., 1989). However, mebendazole is a second-line treatment because of its poor absorption outside the gastrointestinal tract and may be given at higher doses over 21 days. Diethylcarbamazine, at a rate of 3–4 mg/kg daily for 21 days, starting at 25 mg daily and gradually increasing, has also demonstrated effectiveness in treating systemic toxocarosis (Magnaval, 1995). Whereas for cerebral larva migrans, albendazole should be prescribed at a dose of 800 mg daily (or 15 mg/kg in pediatric patients) for an extended duration, usually for 21–28 days (Chai et al., 2021). Furthermore, symptomatic management, particularly with corticosteroids, can alleviate severe allergic reactions associated with infection (Ahn et al., 2014; Kim et al., 2016). In cases of ocular larva migrans, surgical interventions such as vitrectomy are typically preferred and are often combined with anthelminthic chemotherapy and corticosteroids to curb intraocular inflammation and mitigate complications such as vitreous opacification (Small et al., 1989; Ahn et al., 2014). While the efficacy of anthelmintics against intraocular *Toxocara* larvae remains uncertain owing to a lack of pharmacokinetic studies (Ahn et al., 2014), and individual variability (Jung et al., 1990; Magnaval et al., 2022), existing studies suggest that combination of these medications with corticosteroids may yield positive outcomes (Othman, 2012).

Control

Effective management of larva migrans necessitates a comprehensive approach that combines the targeted eradication of *Toxocara* spp. and other zoonotic nematodes, with comprehensive measures to minimise the reduction of human exposure through public health awareness campaigns. Furthermore, the implementation of a consistent deworming protocol for female dogs, queens, puppies, and kittens is essential for decreasing environmental contamination with *Toxocara* eggs. Given the remarkable resilience of *Toxocara* ova compared with *Ascaris* spp., thorough cleaning and proper disposal of faecal matter are critical in controlling the environmental presence of *Toxocara* ova. Educating pet owners about the parasite's life cycle, transmission pathways, and associated health risks is crucial for raising awareness and ensuring compliance. Furthermore, limiting pet access to community parks and recreational areas can significantly diminish the risk of transmission, particularly among children. Finally, promoting strict personal hygiene practices, including regular grooming and sanitation for both pets and humans, is vital for preventing *Toxocara* transmission and safeguarding public health.

References

Ahn, S. J., Ryoo, N. K., & Woo, S. J. (2014). Ocular toxocarosis: Clinical features, diagnosis, treatment, and prevention. *Asia Pacific Allergy, 4*(3), 134–141.

Alexandre, N., Sabrosa, Cunha, E., & Souza, d. (2001). Nematode infections of the eye: Toxocariasis and diffuse unilateral subacute neuroretinitis. *Current Opinion in Ophthalmology, 12*(6), 450–454. https://doi.org/10.1097/00055735-200112000-00010

Alter, S. J., & Turcios, N. L. (2009) *Pulmonary manifestations of pediatric diseases pulmonary manifestations of parasitic diseases* (pp. 274–294). Elsevier.

Auer, H., & Walochnik, J. (2020). Toxocariasis and the clinical spectrum. *Advances in Parasitology, 109*, 111–130.

Baalbaki, M., El Najjar, M., Atweh, S., & El Ayoubi, N. K. (2020). Toxocara infection in the differential diagnosis of multiple sclerosis in the Middle East multiple. *Sclerosis Journal – Experimental Translational and Clinical, 6*(1). https://doi.org/10.1177/2055217319855757

Balfour, E., Zalka, A., & Lazova, R. (2002). Cutaneous larva migrans with parts of the larva in the epidermis. *Cutis-New York, 69*(5), 368–370.

Bava, J., Gonzalez, L. G., Seley, C. M., López, G. P., & Troncoso, A. (2011). A case report of cutaneous larva migrans in Argentina. *Asian Pacific Journal of Tropical Biomedicine, 1*(1), 81–82. https://doi.org/10.1016/S2221-1691(11)60074-9

Beaver, P. C., Snyde, C. H., & Carrera, G. M. (1952). Chronic eosinophilia due to visceral larva migrans. *Pediatrics, 9*, 7–19.

Beaver, P. C. (1956). Larva migrans. *Experimental Parasitology, 5*(6), 587–621. https://doi.org/10.1016/0014-4894(56)90032-7

Benitez del Castillo, J. M., Herreros, G., Guillen, J. L., Fenoy, S., Bañares, A., & Garcia, J. (1995). Bilateral ocular toxocarosis demonstrated by aqueous humor enzyme-linked immunosorbent assay. *American Journal of Ophthalmology, 119*(4), 514–516. https://doi.org/10.1016/s0002-9394(14)71241-3

Bokharaei-Salim, F., Khanaliha, K., Ghadimi, T., Babaei, R., Sadeghi, M., & Salemi, B. (2024). Seroprevalence and molecular investigation of toxocariasis in working children and control group in Tehran. *Iran Clinical Epidemiology and Global Health, 27*, 101572. https://doi.org/10.1016/j.cegh.2024.101572

Bouchaud, O., Houzé, S., Schiemann, R., Durand, R., Ralaimazava, P., Ruggeri, C., & Coulaud, J.-P. (2000). Cutaneous larva migrans in travelers: A prospective study, with assessment of therapy with ivermectin. *Clinical Infectious Diseases, 31*(2), 493–498. https://doi.org/10.1086/313942

Bowman, D. D. (2020). History of Toxocara and the associated larva migrans. In *Advances in parasitology* (Vol. 109, pp. 17–38). https://doi.org/10.1016/bs.apar.2020.01.037. .

Bowman, D. D. (2021). Ascaris and Toxocara as foodborne and waterborne pathogens. *Research in Veterinary Science*, 1351–1357. https://doi.org/10.1016/j.rvsc.2020.12.017

Brasil, O. F., Lewis, H., & Lowder, C. Y. (2006). Migration of Baylisascaris procyonis into the vitreous. *The British Journal of Ophthalmology, 90*(9), 1203–1204. https://doi.org/10.1136/bjo.2006.095323. PMID: 16929065; PMCID: PMC1857416.

Brenner, M. A., & Patel, M. B. (2003). Cutaneous larva migrans: The creeping eruption. *Cutis, 72*(2), 111–115. PMID: 12953933.

Calvopina, M., Lozano-Alvarez, K., Enriquez-Morillo, S., & Cordova-Calisto, I. (2024). Vesiculobullous cutaneous larva Migrans in the absence of domestic dogs and cats. *Successful Treatment with Oral Ivermectin Tropical Medicine and Infectious Disease, 9*(5), 106. https://doi.org/10.3390/tropicalmed9050106

Campos, D. M. B., Barbosa, A. P., Oliveira, J. A., Tavares, G. G., Cravo, P. V. L., et al. (2017). Human lagochilascariasis—A rare helminthic disease. *PLoS Neglected Tropical Diseases, 11*(6), e0005510. https://doi.org/10.1371/journal.pntd.0005510

Caumes, E. (2000). Treatment of cutaneous larva Migrans. *Clinical Infectious Diseases, 30*(5), 811–814. https://doi.org/10.1086/313787

Carpio, A., Romo, M. L., Parkhouse, R. M. E., Short, B., & Dua, T. (2016). Parasitic diseases of the central nervous system: Lessons for clinicians and policy makers. *Expert Review of Neurotherapeutics, 16*(4), 401–414. https://doi.org/10.1586/14737175.2016.1155454

Chai, J. Y., Jung, B. K., & Hong, S. J. (2021). Albendazole and Mebendazole as anti-parasitic and anti-cancer agents: An update. *The Korean Journal of Parasitology, 59*(3), 189–225.

Chen, J., Liu, Q., Liu, G.-H., Zheng, W.-B., Hong, S.-J., Sugiyama, H., Zhu, X.-Q., & Elsheikha, H. M. (2018). Toxocariasis: A silent threat with a progressive public health impact infectious diseases of. *Poverty, 7*(1). https://doi.org/10.1186/s40249-018-0437-0

Choi, D., Lim, J. H., Choi, D.-C., Paik, S. W., Kim, S.-H., & Huh, S. (2008). Toxocariasis and ingestion of raw cow liver in patients with eosinophilia. *The Korean Journal of Parasitology, 46*(3), 139. https://doi.org/10.3347/kjp.2008.46.3.139

Corte, L. D., da Silva, M. V. S., & Souza, P. R. M. (2013). Simultaneous larva Migrans and larva Currens caused by Strongyloides stercoralis. *A Case Report Case Reports in Dermatological Medicine, 2013*, 1–3. https://doi.org/10.1155/2013/381583

Cunningham, E. T., & Zierhut, M. (2021). Ocular Toxocariasis. *Ocular Immunology and Inflammation, 29*(7–8), 1243–1245. https://doi.org/10.1080/09273948.2021.2018825

Currie, B. J., Hoopes, J., & Cumming, B. (2025). Cutaneous larva Migrans refractory to therapy with Ivermectin. *Case Report and Review of Implicated Zoonotic Pathogens Epidemiology Anthelmintic Drug Resistance and Therapy Tropical Medicine and Infectious Disease, 10*(6), 163. https://doi.org/10.3390/tropicalmed10060163

de Almeida Carvalho, E. A., & Rocha, R. L. (2014). Visceral larva Migrans syndromes associated with Toxocariasis: Epidemiology, clinical and laboratory aspects of human Toxocariasis. *Current Tropical Medicine Reports, 1*, 74–79. https://doi.org/10.1007/s40475-013-0011-6

de Melo, L. C. V., de Souza, F. C. R., de Oliveira Baccin, A., da Mota, D. J. G., Pereira-Chioccola, V. L., & Pinto, P. L. S. (2022). Immunoanalysis of different antigenic preparations of Angiostrongylus cantonensis for neuroangiostrongyliasis diagnosis improvement. *Memórias do Instituto Oswaldo Cruz*, 117. https://doi.org/10.1590/0074-02760220086

Despommier, D. (2003). Toxocarosis: Clinical aspects, epidemiology, medical ecology, and molecular aspects. *Clinical Microbiology Reviews, 16*(2), 265–272.

Diaz, J. H. (2009). Recognizing and reducing the risks of helminthic eosinophilic meningitis in travelers: Differential diagnosis disease management prevention and control. *Journal of Travel Medicine, 16*(4), 267–275. https://doi.org/10.1111/j.1708-8305.2009.00305.x

Diaz, J. H. (2010). The helminthic eosinophilic meningitides: Emerging zoonotic parasitic diseases worldwide. *Tropical Medicine and Health, 38*(4), 115–126.

Diaz, J. H. (2015). Gnathostomiasis: An emerging infection of raw fish consumers in Gnathostoma Nematode-endemic and nonendemic countries. *Journal of Travel Medicine, 22*(5), 318–324. https://doi.org/10.1111/jtm.12212

dos Santos, L. M., Donassolo, R. A., Berne, M. E., Leite, F. P. L., da Costa Avila, L. F., Scaini, C. J., Moreira, Â. N., & Conceição, F. R. (2019). The serodiagnostic potential of recombinant proteins TES–30 and TES–120 in an indirect ELISA in the diagnosis of toxocariasis in cattle horses and sheep PLOS. *ONE, 14*(3), e0213830. https://doi.org/10.1371/journal.pone.0213830

dos Santos Vaccaro Lima, V., Ferreira, I. B., da Silva Rapchan, G. G., Lescano, S. Z., Fonseca, G. R. e., Giuffrida, R., Kmetiuk, L. B., Biondo, A. W., & Santarém, V. A. (2025). Toxocariasis in children: Seroprevalence after 15 years in a major city in Brazil. *Frontiers in Pediatrics, 13*. https://doi.org/10.3389/fped.2025.1663016

Dwight D. Bowman, 4 - Helminths, Dwight D. Bowman, Georgis' parasitology for veterinarians (11th), W.B. Saunders, 2021b, Pages 135–260,. ISBN 9780323543965, https://doi.org/10.1016/B978-0-323-54396-5.00013-1.

Ebrahimipour, M., Parsa, H., Kianifard, N., Nasibi, S., & Rostami, A. (2024). Understanding the research and practical needs required to control toxocariasis in Iran. *Parasite Epidemiology and Control, 25*, e00351. https://doi.org/10.1016/j.parepi.2024.e00351

Edward, T., & Durand, R. M. Tropical infectious diseases. In *Principles pathogens and practice ocular disease* (pp. 991–1016). Elsevier.

Elefant, G. R., Shimizu, S. H., Sanchez, M. C., Jacob, C. M., & Ferreira, A. W. (2006). A serological follow-up of toxocariasis patients after chemotherapy based on the detection of IgG, IgA, and IgE antibodies by enzyme-linked immunosorbent assay. *Journal of Clinical Laboratory Analysis, 20*(4), 164–172. https://doi.org/10.1002/jcla.20126

Fan, C.-K., Liao, C.-W., & Cheng, Y.-C. (2013). Factors affecting disease manifestation of toxocarosis in humans: Genetics and environment. *Veterinary Parasitology, 193*(4), 342–352. https://doi.org/10.1016/j.vetpar.2012.12.030

Fan, C.-H., Holland, C. V., Loxten, K., & Barghouth, U. (2015). Cerebral Toxocariasis: Silent progression to neurodegenerative disorders? *Summary Clinical Microbiology Reviews, 28*(3), 663–686. https://doi.org/10.1128/CMR.00106-14

Fillaux, J., & Magnaval, J.-F. (2013). Laboratory diagnosis of human toxocariasis. *Veterinary Parasitology, 193*(4), 327–336. https://doi.org/10.1016/j.vetpar.2012.12.028

Fomda, B. A., Ahmad, Z., Khan, N. N., Tanveer, S., & Wani, S. A. (2007). Ocular toxocarosis in a child: A case report from Kashmir, North India. *Indian Journal of Medical Microbiology, 25*(4), 411–412. https://doi.org/10.4103/0255-0857.37352

Fonseca, G. R. E., Corral, M. A., Paula, F. M., Meisel, D. M. C. L., Gryschek, R. C. B., & Lescano, S. A. Z. (2022). Toxocara canis 30-35 kDa excretory-secretory antigen is an important marker in mice challenged by inocula containing different parasite load levels. *Revista do Instituto de Medicina Tropical de Sao Paulo, 64*, e9. https://doi.org/10.1590/S1678-9946202264009

Gao, Y.-L., & Liu, Z.-H. (2019). Cutaneous larva Migrans with Löeffler's syndrome. *The American Journal of Tropical Medicine and Hygiene, 100*(3), 487–488. https://doi.org/10.4269/ajtmh.18-0406

Gavin, P. J., Kazacos, K. R., & Shulman, S. T. (2005). Baylisascariasis. *Clinical Microbiology Reviews, 18*(4), 703–718. https://doi.org/10.1128/CMR.18.4.703-718.2005

Girdwood, R. W. A. (1986). Human toxocariasis. *Journal of Small Animal Practice, 27*, 649–654.

Glickman, L. T., & Summers, B. A. (1983). Experimental Toxocara canis infection in cynomolgus macaques (Macaca fascicularis). *American Journal of Veterinary Research, 44*(12), 2347–2354. PMID: 6660626. https://doi.org/10.2460/ajvr.1983.44.12.2347

Glickman, L. T., & Shofer, F. S. (1987). Zoonotic visceral and ocular larva migrans. *The Veterinary Clinics of North America. Small Animal Practice, 17*(1), 39–53. https://doi.org/10.1016/s0195-5616(87)50604-0. PMID: 3551310.

Gillespie, S. H. (1987). Human toxocariasis. *Journal of Applied Bacteriology, 63*(6), 473–479. 10.1111/jam.1987.63.issue-6. https://doi.org/10.1111/j.1365-2672.1987.tb02716.x

Gillespie, S. H., Dinning, W. J., Voller, A., & Crowcroft, N. S. (1993). The spectrum of ocular toxocariasis. *Eye (London, England), 7*(Pt 3), 415–418. https://doi.org/10.1038/eye.1993.82. PMID: 8224297.

Gorp, K. V., Mangelschots, M., & Brandt, J. (1987). Toxocara Vitulorum: A possible agent of larva Migrans in humans ? In S. Geerts, V. Kumar, & J. Brandt (Eds.), *Helminth Zoonoses*.

Current topics in veterinary medicine and animal science (Vol. 43). Springer. https://doi.org/10.1007/978-94-009-3341-5_22

Graeff-Teixeira, C., da Silva, A. C. A., & Yoshimura, K. (2009). Update on eosinophilic meningoencephalitis and its clinical relevance. *Clinical Microbiology Reviews, 22*(2), 322–348. https://doi.org/10.1128/CMR.00044-08

Gupta, A., & Tripathy, K. (2023). Ocular Toxocariasis. In *StatPearls [Internet]*. StatPearls Publishing. Available from: https://www.ncbi.nlm.nih.gov/books/NBK576384/

Hare, A. Q., & Franco-Paredes, C. (2014). Ocular larva Migrans: A severe manifestation of an unseen epidemic current. *Tropical Medicine Reports, 1*(1), 69–73. https://doi.org/10.1007/s40475-013-0004-5

Herman, J. S., & Chiodini, P. L. (2009). Gnathostomiasis another emerging imported disease. *Clinical Microbiology Reviews, 22*(3), 484–492. https://doi.org/10.1128/CMR.00003-09

Heukelbach, J., & Feldmeier, H. (2008). Epidemiological and clinical characteristics of hookworm-related cutaneous larva migrans. *The Lancet Infectious Diseases, 8*(5), 302–309. https://doi.org/10.1016/S1473-3099(08)70098-7

Hochedez, P., & Caumes, E. (2007). Hookworm-Related Cutaneous Larva Migrans. *Journal of Travel Medicine, 14*(5), 326–333. https://doi.org/10.1111/j.1708-8305.2007.00148.x

Hoff, N.-P., Mota, R., Groffik, A., & Hengge, U. R. (2008). Kutane Larva migrans Cutaneous larva migrans. *Der Hautarzt, 59*(8), 622–626. https://doi.org/10.1007/s00105-008-1514-2

Horton, J. (2000). Albendazole: A review of anthelmintic efficacy and safety in humans. *Parasitology, 121*(S1), S113–S132.

Hotez, P. J. (1995). Toxocara canis. In F. D. Burg, E. R. Wald, J. R. Ingelfinger, & P. A. Polin (Eds.), *Gellis and Kaganís current pediatric therapy* (15th ed., pp. 683–684). W. B. Saunders Pubs.

Inagaki, K., Williams, J. M., Hobbs, C. V., & Palmer, A. (2023). Ten facts about Toxocariasis. *Journal of the Mississippi State Medical Association, 64*(1).

Jacob, J., & Lorber, B. (2015). Diseases transmitted by man's best friend: The dog. *Microbiology Spectrum, 3*. https://doi.org/10.1128/microbiolspec.iol5-0002-2015

Jowsey, G. W., & McLeod, G. X. (2023). A delayed diagnosis of ocular toxocariasis presenting as total monocular retinal detachment in an immunocompetent 57-year-old male. *IDCases, 32*, e01764. https://doi.org/10.1016/j.idcr.2023.e01764

Jung, H., Hurtado, M., Sanchez, M., Medina, M. T., & Sotelo, J. (1990). Plasma and CSF levels of albendazole and praziquantel in patients with neurocysticercosis. *Clinical Neuropharmacology, 13*(6), 559–564. https://doi.org/10.1097/00002826-199012000-00008

Kanecki, K., Lewtak, K., Szot, P., Goryński, P., Kosińska, I., Kanecka, A., & Nitsch-Osuch, A. (2025). Visceral larva migrans among hospitalized patients in Poland: Analysis of national hospital discharge data from 2014 to 2023. *Scientific Reports, 15*(1). https://doi.org/10.1038/s41598-025-18345-8

Kazacos, K. R. (2001). Baylisascaris and related species. In W. M. Samuel, M. J. Pybus, & A. A. Kocan (Eds.), *Parasitic diseases of wild mammals* (2nd ed., pp. 301–341). Iowa State University Press.

Kazacos, K. R., Jelicks, L. A., & Tanowitz, H. B. (2013). Baylisascaris larva migrans. *Handbook of Clinical Neurology, 114*, 251–262. https://doi.org/10.1016/B978-0-444-53490-3.00020-0

Kazacos, K. R. (2016). *Baylisascaris larva Migrans: U.S. Geological Survey circular* (Vol. 1412, p. 122 p., 3 appendixes). https://doi.org/10.3133/cir1412

Kim, K., Weiss, L. M., & Tanowitz Murray, H. B. *Nadel's Textbook of Respiratory Medicine Parasitic Infections* (pp. 682–698.e8). Elsevier.

Kim, T.-H., Lee, B.-S., & Sohn, W-M. (2006). Three clinical cases of cutaneous larva migrans. *The Korean Journal of Parasitology, 44*(2), 145–149. https://doi.org/10.3347/kjp.2006.44.2.145

Kim, M.-H., Jung, J.-W., Kwon, J.-W., Kim, T.-W., Kim, S.-H., Cho, S.-H., Min, K.-U., Kim, Y.-Y., & Chang, Y.-S. (2010). A case of recurrent Toxocariasis presenting with Urticaria. *Allergy Asthma and Immunology Research, 2*(4), 267. https://doi.org/10.4168/aair.2010.2.4.267

Kim, H. S., Jin, Y., Choi, M.-H., Kim, J.-H., Lee, Y. H., Yoon, C. H., Hwang, E.-H., Kang, H., Ahn, S.-Y., Kim, G. J., & Hong, S.-T. (2014). Significance of serum antibody test for Toxocariasis in

healthy healthcare examinees with eosinophilia in Seoul and Gyeongsangnam-do Korea. *Journal of Korean Medical Science, 29*(12), 1618. https://doi.org/10.3346/jkms.2014.29.12.1618
Klotz, S. A., Penn, C. C., Negvesky, G. J., & Butrus, S. I. (2000). Fungal and parasitic infections of the eye. *Clinical Microbiology Reviews, 13*(4), 662–685. https://doi.org/10.1128/CMR.13.4.662-685.2000
Kokollari, F., Gërçari, A., Blyta, Y., Daka, Q., & Krasniqi-Daka, A. (2015). Falls in diagnosis of cutaneous larva Migrans-a case report from Kosovo. *Medical archives (Sarajevo, Bosnia and Herzegovina), 69*(4), 271–273. https://doi.org/10.5455/medarh.2015.69.271-273
Kong, L., & Peng, H.-J. (2020). *Toxocara and Toxocariasis current epidemic situation of human toxocariasis in China* (pp. 433–448). Elsevier.
Kroten, A., Toczylowski, K., Oldak, E., et al. (2018). Toxocarosis in children: Poor hygiene habits and contact with dogs is related to longer treatment. *Parasitology Research, 117*, 1513–1519. https://doi.org/10.1007/s00436-018-5833-7
Kundi, P., Alaskarov, E., & Ulusoy, S. (2023). Angiostrongylus cantonensis (the rat lungworm) infection and hearing loss. In A. E. Arısoy, E. S. Arısoy, N. Bayar Muluk, C. Cingi, & A. G. Correa (Eds.), *Hearing loss in congenital, neonatal and childhood infections. Comprehensive ENT*. Springer. https://doi.org/10.1007/978-3-031-38495-0_62
Lauriente, T., Nicolson, H., Merkeley, H., Enrique, A., David, G., & Harris, J. (2025). Toxocariasis in 62-year-old man with eosinophilia. *Canadian Medical Association Journal, 197*(42), E1449–E1454. https://doi.org/10.1503/cmaj.250714
Le Joncour, A., Lacour, S. A., Lecso, G., Regnier, S., Guillot, J., & Caumes, E. (2012). Molecular characterization of Ancylostoma braziliense larvae in a patient with hookworm-related cutaneous. *Larva Migrans The American Society of Tropical Medicine and Hygiene, 86*(5), 843–845. https://doi.org/10.4269/ajtmh.2012.11-0734
Lim, J. H. (2008). Toxocariasis of the liver: Visceral larva migrans. *Abdominal Imaging, 33*, 151–156. https://doi.org/10.1007/s00261-007-9325-y
Liu, L. X., & Weller, P. F. (1993). STRONGYLOIDIASIS and other intestinal NEMATODE infections. *Infectious Disease Clinics of North America, 7*(3), 655–682. https://doi.org/10.1016/S0891-5520(20)30548-1
Liu, G. H., Sun, M. M., Elsheikha, H. M., et al. (2020). Human gnathostomiasis: A neglected food-borne zoonosis. *Parasites and Vectors, 13*, 616. https://doi.org/10.1186/s13071-020-04494-4
Lötsch, F., Vingerling, R., Spijker, R., & Grobusch, M. P. (2017). Toxocariasis in humans in Africa—A systematic review. *Travel Medicine and Infectious Disease, 20*, 15–25. https://doi.org/10.1016/j.tmaid.2017.10.009
Ma, G., Holland, C. V., Wang, T., Hofmann, A., Fan, C. K., Maizels, R. M., Hotez, P. J., & Gasser, R. B. (2018). Human toxocarosis. *The Lancet Infectious Diseases, 18*(1), e14–e24.
Macpherson, C. N. L. (2013). The epidemiology and public health importance of toxocariasis: A zoonosis of global importance. *International Journal for Parasitology, 43*(12–13), 999–1008. https://doi.org/10.1016/j.ijpara.2013.07.004
Magnaval, J. F., Fabre, R., Maurières, P., Charlet, J. P., & de Larrard, B. (1992). Evaluation of an immunoenzymatic assay detecting specific anti-Toxocara immunoglobulin E for diagnosis and posttreatment follow-up of human toxocariasis. *Journal of Clinical Microbiology, 30*. https://doi.org/10.1128/jcm.30.9.2269-2274.1992
Magnaval, J. F. (1995). Comparative efficacy of diethylcarbamazine and mebendazole for the treatment of human toxocarosis. *Parasitology, 110*(5), 529–533.
Magnaval, J.-F., Glickman, L. T., Dorchies, P., & Morassin, B. (2001). Highlights of human toxocariasis. *The Korean Journal of Parasitology, 39*(1), 1. https://doi.org/10.3347/kjp.2001.39.1.1
Magnaval, J.-F., Bouhsira, E., & Fillaux, J. (2022). Therapy and prevention for human toxocarosis. *Microorganisms, 10*(2), 241. https://doi.org/10.3390/microorganisms10020241
Maizels, R. M. (2013). Toxocara canis: Molecular basis of immune recognition and evasion. *Veterinary Parasitology, 193*(4), 365–374.
Marina, C., Rigolet, C., Danis, M., & Caumes, E. (2004). La gnathostomose une maladie exotique de plus en plus souvent importée dans les pays occidentaux. *La Presse Médicale, 33*(21), 1527–1532. https://doi.org/10.1016/S0755-4982(04)98978-6

Maruyama, H., Nawa, Y., Noda, S., Mimori, T., & Choi, W.-Y. (1996). An outbreak of visceral larva migrans due to Ascaris suum in Kyushu Japan. *The Lancet, 347*(9017), 1766–1767. https://doi.org/10.1016/S0140-6736(96)90844-2

Maxfield, L., & Crane, J. S. (2023). Cutaneous Larva Migrans. In *StatPearls [internet]; StatPearls publishing: Treasure Island.*

Mazur-Melewska, K., Mania, A., Figlerowicz, M., Kemnitz, P., Służewski, W., & Michalak, M. (2012). The influence of age on a clinical presentation of Toxocara spp. infection in children. *Annals of Agricultural and Environmental Medicine, 19*(2), 233–236.

Mirdha, B. R., & Khokar, S. K. (2002). Ocular toxocarosis in a north Indian population. *Journal of Tropical Pediatrics, 48*(6), 328–330. https://doi.org/10.1093/tropej/48.6.328

Mohamad, S., Azmi, N. C., & Noordin, R. (2009). Development and evaluation of a sensitive and specific assay for diagnosis of human toxocariasis by use of three recombinant antigens (TES-26, TES-30USM, and TES-120). *Journal of Clinical Microbiology, 47*(6), 1712–1717. https://doi.org/10.1128/JCM.00001-09

Moorhouse, D. E. (1982). Toxocariasis. A possible cause of the Palm Island mystery disease. *The Medical Journal of Australia, 1*(4), 172–173. PMID: 7078487.

Nathwani, D., Laing, R. B. S., & Currie, P. F. (1992). Covert Toxocariasis—A cause of recurrent abdominal pain in childhood Summary. *International Journal of Clinical Practice, 46*(4), 271–271. https://doi.org/10.1111/ijcp.v46.4. https://doi.org/10.1111/j.1742-1241.1992.tb10365.x

Nikopoulou, A., Sideras, L., Peristeri, A. M., Ligda, P., Akritidou, O., & Sotiraki, S. (2025). Human Toxocariasis: Presenting a case report of eosinophilic myocarditis associated with Toxocara canis infection and a literature review. *Microbiology Research, 16*(3), 58. https://doi.org/10.3390/microbiolres16030058

Nisa, R. U., Tantray, A. Y., & Shah, A. A. (2022). Shift from morphological to recent advanced molecular approaches for the identification of nematodes. *Genomics, 114*(2), 110295. https://doi.org/10.1016/j.ygeno.2022.110295

Oliart-Guzmán, H., Delfino, B. M., Martins, A. C., Mantovani, S. A., Braña, A. M., Pereira, T. M., Branco, F. L., Ramalho, A. A., Campos, R. G., Fontoura, P. S., de Araujo, T. S., de Oliveira, C. S., Muniz, P. T., Rubinsky-Elefant, G., Codeço, C. T., & da Silva-Nunes, M. (2014). Epidemiology and control of child toxocariasis in the western Brazilian Amazon – a population-based study. *The American Journal of Tropical Medicine and Hygiene, 90*(4), 670–681. https://doi.org/10.4269/ajtmh.13-0506

Osman, S., Tarnari, N., Ahsan, A., & Ahmed, K. A. H. M. (2024). Cutaneous larva Migrans (CLM) may not be easy to diagnose: A case report and narrative review. *Oxford Medical Case Reports, 2024*(4), omae025. https://doi.org/10.1093/omcr/omae025

Othman, A. A. (2012). Therapeutic battle against larval toxocariasis: Are we still far behind? *Acta Tropica, 124*(3), 171–178. https://doi.org/10.1016/j.actatropica.2012.08.003

Otranto, D., & Eberhard, M. L. (2011). Zoonotic helminths affecting the human eye. *Parasites & Vectors, 4*(1), 41.

Otranto, D., & Deplazes, P. (2019). Zoonotic nematodes of wild carnivores. *International Journal for Parasitology: Parasites and Wildlife*, 9370–9383. https://doi.org/10.1016/j.ijppaw.2018.12.011

Page, L. K. (2013). Parasites and the conservation of small populations: The case of Baylisascaris procyonis. *International Journal for Parasitology: Parasites and Wildlife, 2*, 203–210. https://doi.org/10.1016/j.ijppaw.2013.05.003

Pahari, T. K., & Sasmal, N. K. (1990). Infection of Japanese quail with Toxocara canis larvae and establishment of patent infection in pups. *Veterinary Parasitology, 35*(4), 357–364. https://doi.org/10.1016/0304-4017(90)90140-7

Papajová, I., Juriš, P., Szabová, E., Venglovský, J., Sasáková, N., Šefčíková, H., Martinez, J., & Gáboň, T. (2008). Decontamination by anaerobic stabilisation of the environment contaminated with enteronematode eggs Toxocara canis and Ascaris suum. *Bioresource Technology, 99*(11), 4966–4971. https://doi.org/10.1016/j.biortech.2007.09.044

Park, S.-P., Park, I., Park, H.-Y., Lee, S.-U., Huh, S., & Magnaval, J.-F. (2000). Five cases of ocular toxocariasis confirmed by serology the. *Korean Journal of Parasitology, 38*(4), 267. https://doi.org/10.3347/kjp.2000.38.4.267

Paul, M., Stefaniak, J., Twardosz-Pawlik, H., et al. (2009). The co-occurrence of Toxocaraocular and visceral larva migrans syndrome: A case series. *Cases Journal, 2*, 6881. https://doi.org/10.1186/1757-1626-0002-0000006881

Pawlowski, Z. (2001). Toxocariasis in humans: Clinical expression and treatment dilemma. *Journal of Helminthology, 75*(4), 299–305. https://doi.org/10.1017/S0022149X00701465

Pegg, E. J. (1971). Infection of dogs by Toxocara canis carried by flies. *Parasitology, 62*(3), 409–414. https://doi.org/10.1017/S0031182000077568

Pivetti-Pezzi, P. (2009). Ocular Toxocariasis. *International Journal of Medical Sciences, 6*(3), 129–130. https://doi.org/10.7150/ijms.6.129

Podder, I., Chandra S., & Gharami, R. C. (2016). Loeffler's syndrome following cutaneous larva migrans: An uncommon sequel. *Indian Journal of Dermatology, 61*(2), 190. https://doi.org/10.4103/0019-5154.177753

Reichert, F., Pilger, D., Schuster, A., Lesshafft, H., de Oliveira, S. G., Ignatius, R., & Feldmeier, H. (2018). Epidemiology and morbidity of hookworm-related cutaneous larva migrans (HrCLM): Results of a cohort study over a period of six months in a resource-poor community in Manaus Brazil PLOS. *Neglected Tropical Diseases, 12*(7), e0006662. https://doi.org/10.1371/journal.pntd.0006662

Roberts LS, Schmidt GD, Janovy JJr. (2009) Gerald D. Schmidt and Larry S. Roberts' foundations of parasitology. 8th edition. McGraw-Hill Higher Education, Boston, USA

Robertson, I. D., Irwin, P. J., Lymbery, A. J., & Thompson, R. C. A. (2000). The role of companion animals in the emergence of parasitic zoonoses. *International Journal for Parasitology, 30*(12–13), 1369–1377. https://doi.org/10.1016/S0020-7519(00)00134-X

Rostami, A., Ma, G., Wang, T., Koehler, A. V., Hofmann, A., Chang, B. C. H., Macpherson, C. N., & Gasser, R. B. (2019). Human toxocariasis—A look at a neglected disease through an epidemiological 'prism'. *Infection Genetics and Evolution, 74*, 104002. https://doi.org/10.1016/j.meegid.2019.104002

Rubinsky-Elefant, G., Hirata, C. E., Yamamoto, J. H., & Ferreira, M. U. (2010). Human toxocariasis: Diagnosis, worldwide seroprevalences and clinical expression of the systemic and ocular forms. *Annals of Tropical Medicine & Parasitology, 104*(1), 3–23. https://doi.org/10.1179/136485910X12607012373957

Rusnak, J. M., & Lucey, D. R. (1993). Clinical Gnathostomiasis: Case report and review of the English-language literature. *Clinical Infectious Diseases, 16*(1), 33–50. http://www.jstor.org/stable/4456848

Ryan, E. T., & Durand, M. (2011). *Tropical Infectious diseases: Principles pathogens and practice ocular disease*. Elsevier pp. 991–1016.

Sałamatin, R., Knysz, B., Paszta, W., Lelonek, E., Matos, O., & Wesołowska, M. (2023). Cutaneous larva migrans: A one health perspective on familial infection among tourists returning from Southeast Asia. In *Clinical cosmetic and investigational dermatology volume 16* (pp. 3375–3382). https://doi.org/10.2147/CCID.S425885

Sánchez, S. S., García, H. H., & Nicoletti, A. (2018). Clinical and magnetic resonance imaging findings of Neurotoxocariasis. *Frontiers in Neurology, 9*, 53. https://doi.org/10.3389/fneur.2018.00053

Sharma, R., Singh, B. B., & Gill, J. P. S. (2015). Larva migrans in India: Veterinary and public health perspectives. *Journal of Parasitic Diseases, 39*(4), 604–612. https://doi.org/10.1007/s12639-013-0402-6

Shields, J. A. (1984). Ocular toxocarosis. A review. *Survey of Ophthalmology, 28*(5), 361–381. https://doi.org/10.1016/0039-6257(84)90242-x

Shinkar, R. M., Stocks, R., & Thomas, E. (2005). Cutaneous larva migrans, creeping eruption, sand worm. *Archives of Disease in Childhood, 90*(10), 998. https://doi.org/10.1136/adc.2005.073197. PMID: 16177155; PMCID: PMC1720113.

Sim, D. W., Son, D. J., Cho, E., Choi, S. K., Shin, S. S., & Jun, C. H. (2019). What are the clinical features and etiology of eosinophilic liver infiltration? *Gut and Liver, 13*(2), 183–190. https://doi.org/10.5009/gnl18266

Sircar, A. D., Abanyie, F., Blumberg, D., Chin-Hong, P., Coulter, K. S., Cunningham, D., Huskins, W. C., Langelier, C., Reid, M., Scott, B. J., Shirley, D. A., Babik, J. M., Belova, A., Sapp, S. G., McAuliffe, I., Rivera, H. N., Yabsley, M. J., & Montgomery, S. P. (2016). Raccoon roundworm infection associated with central nervous system disease and ocular disease – six states, 2013–2015. *MMWR. Morbidity and Mortality Weekly Report, 65*(35), 930–933. https://doi.org/10.15585/mmwr.mm6535a2. PMID: 27608169.

Small, K. W., McCuen, B. W., De Juan, E., & Machemer, R. (1989). Surgical management of retinal retraction caused by toxocarosis. *American Journal of Ophthalmology, 108*, 10–14.

Soulsby, E. J. L., Geerts, S., Kumar, V., Brandt, J., & Zoonoses, H. (1987). *Larva Migrans in perspective* (pp. 137–148). Springer.

Sprent, J. F. A. (1956). The life history and development of Toxocara cati (Schrank 1788) in the domestic cat. *Parasitology, 46*(1–2), 54–78. https://doi.org/10.1017/S0031182000026342

Sprent, J. F. A. (1958). Observations on the development of Toxocara canis (Werner, 1782) in the dog. *Parasitology, 48*(1–2), 184–209. https://doi.org/10.1017/S0031182000021168

Sprent, J. F. (1971). Speciation and development in the genus Lagochilascaris. *Parasitology, 62*(1), 71–112.

Sriveny, D., & Kazacos, K. R. (2009). Differentiation of larva Migrans caused by Baylisascaris procyonis and Toxocara species by Western blotting ABSTRACT. *Clinical and Vaccine Immunology, 16*(11), 1563–1568. https://doi.org/10.1128/CVI.00251-09

Stewart, J. M., Cubillan, L. D., & Cunningham, E. T., Jr. (2005). Prevalence, clinical features, and causes of vision loss among patients with ocular toxocarosis. *Retina, 25*(8), 1005–1013. https://doi.org/10.1097/00006982-200512000-00009

Strube, C., Heuer, L., & Janecek, E. (2013). Toxocara spp. infections in paratenic hosts. *Veterinary Parasitology, 193*(4), 375–389. https://doi.org/10.1016/j.vetpar.2012.12.033

Stull, J. W., Brophy, J., & Weese, J. S. (2015). Reducing the risk of pet-associated zoonotic infections. *Canadian Medical Association Journal, 187*(10), 736–743. https://doi.org/10.1503/cmaj.141020

Stürchler, D., Schubarth, P., & Gualzata, M. (1989). Thiabendazole v. albendazole in treatment of toxocarosis: A clinical trial. *Annals of Tropical Medicine and Parasitology, 83*, 473–478.

Taylor, M. R. H., Keane, C. T., O'connor, P., Anthony Girdwood, R. W., & Smith, H. (1987). Clinical features of covert Toxocariasis. *Scandinavian Journal of Infectious Diseases, 19*(6), 693–696. https://doi.org/10.3109/00365548709117206

Taylor, M. R. H., O'Connor, P., Keane, C. T., Mulvihill, E., & Celia, H. (1988). The expanded spectrum of Toxocaral disease. *The Lancet, 331*(8587), 692–695. https://doi.org/10.1016/S0140-6736(88)91486-9

Taylor, M. R. (2006). Ocular toxocariasis. In *Toxocara: The enigmatic parasite* (pp. 127–144). CABI. https://doi.org/10.1079/9781845930264.0127

Thiangtrongjit, T., Nogrado, K., Ketboonlue, T., Malaitong, P., Adisakwattana, P., & Reamtong, O. (2021). Proteomics of Gnathostomiasis: A way forward for diagnosis and treatment development. *Pathogens, 10*(9), 1080. https://doi.org/10.3390/pathogens10091080

van Gasselt, M. W., & van de Sandt, M. (2000). Serpiginous plaques on the leg. *Archives of Dermatology, 136*(12), 1559–1564. https://doi.org/10.1001/archderm.136.12.1559-c

Vidal, J. E., Sztajnbok, J., & Seguro, A. C. (2003). Eosinophilic meningoencephalitis due to Toxocara Canis: Case report and review of the literature. *The American Journal of Tropical Medicine and Hygiene, 69*(3), 341–343. https://doi.org/10.4269/ajtmh.2003.69.341

Wang, Q. P., Wu, Z. D., Wei, J., Owen, R. L., & Lun, Z. R. (2012). Human Angiostrongylus cantonensis: An update. *European Journal of Clinical Microbiology & Infectious Diseases, 31*(4), 389–395.

Wilder, H. C. (1950). *Nematode endophthalmitis* (pp. 99–109). Transactions of the American Academy of Ophthalmology and Oto-laryngology.

Wilkins, P. P. (2014). Immunodiagnosis of human Toxocariasis and prospects for improved diagnostics. *Current Tropical Medicine Reports, 1*, 44–51. https://doi.org/10.1007/s40475-013-0001-8

Woodhall, D. M., & Fiore, A. E. (2014). Toxocariasis: A review for pediatricians. *Journal of the Pediatric Infectious Diseases Society, 3*(2), 154–159. https://doi.org/10.1093/jpids/pit066

Wu, T., & Bowman, D. D. (2020). Visceral larval migrans of Toxocara canis and Toxocara cati in non-canid and non-felid hosts. *Advances in Parasitology, 109*, 63–88.

Xinou, E., Lefkopoulos, A., Gelagoti, M., Drevelegas, A., Diakou, A., Milonas, I., & Dimitriadis, A. S. (2003). CT and MR imaging findings in cerebral Toxocaral disease. *American Journal of Neuroradiology, 24*(4), 714–718.

Xu, J., & Han, Q. (2024). Prevalence infection and risk to human beings of Toxocara canis in domestic food-producing animals. *Veterinary Sciences, 11*(2), 83. https://doi.org/10.3390/vetsci11020083

Yokoi, K., Goto, H., Sakai, J., & Usui, M. (2003). Clinical features of ocular toxocarosis in Japan. *Ocular Immunology and Inflammation, 11*(4), 269–275. https://doi.org/10.1076/ocii.11.4.269.18266

Yoon, S.-Y., Baek, S., Park, S. Y., Shin, B., Kwon, H.-S., Cho, Y. S., Moon, H.-B., & Kim, T.-B. (2018). Clinical course and treatment outcomes of toxocariasis-related eosinophilic disorder. *Medicine, 97*(37), e12361. https://doi.org/10.1097/MD.0000000000012361

Yoshida, A., Hombu, A., Wang, Z., & Maruyama, H. (2016). Larva migrans syndrome caused by Toxocara and Ascaris roundworm infections in Japanese patients. *European journal of clinical microbiology & infectious diseases : official publication of the European Society of Clinical Microbiology, 35*(9), 1521–1529. https://doi.org/10.1007/s10096-016-2693-x

Yu, T., Zhao, L. N., Fan, M. J., Wu, H., & Chen, Q. K. (2012). Visceral larva migrans associated with earthworm and gecko ingestion: A case report. *Journal of Medical Case Reports, 6*, 210. https://doi.org/10.1186/1752-1947-6-210. PMID: 22809241; PMCID: PMC3408347.

Diphyllobothriasis

39

Tuhina Banerjee, Aradhana Singh,
and Souti Prasad Sarkhel

Abstract

Diphyllobothriasis is a parasitic infection caused by cestodes of the Diphyllobothriidae family, commonly called as broad or fish tapeworms. This infection poses a significant public health challenge, with an estimated 20 million people worldwide affected by various species of *Diphyllobothrium* (syn. *Dibothriocephalus*) that can infect humans. The primary mode of transmission is through the consumption of raw or undercooked fish. In humans, while *Dibothriocephalus latus* (formerly *Diphyllobothrium latum*) is the most frequently identified infection species, other species, such as *Dibothriocephalus dendriticus* (formerly *Diphyllobothrium dendriticum*), *Dibothriocephalus nihonkaiensis* (formerly *Diphyllobothrium nihonkaiense*), *Adenocephalus pacificus* (formerly *Diphyllobothrium pacificum*), *Diphyllobothrium balaenopterae*, and *Diphyllobothrium stemmacephalum*, can also cause zoonotic disease. Unlike taeniid tapeworms, these parasites exhibit a complex life cycle that involves aquatic intermediate hosts and various terrestrial definitive hosts, including canines, felines, ursids, avians, and even humans. Diphyllobothriasis, while frequently asymptomatic, may present gastrointestinal manifestations, including vomiting, diarrhoea, abdominal pain, and anaemia, and in rare instances, chronic

T. Banerjee (✉)
Department of Microbiology, Institute of Medical Sciences, Banaras Hindu University, Varanasi, Uttar Pradesh, India
e-mail: drtuhina105@bhu.ac.in

A. Singh
Ashworth Laboratories, Institute of Immunology and Infection Research, School of Biological Sciences, University of Edinburgh, Edinburgh, UK

S. P. Sarkhel
Department of Veterinary Parasitology, FVAS, Institute of Agricultural Sciences, Banaras Hindu University, South Campus, Mirzapur, Uttar Pradesh, India

R. Biswas et al. (eds.), *Concept of One Health and Zoonoses*,
https://doi.org/10.1007/978-981-95-6627-3_39

infections can result in intestinal obstruction. The recent trend of consuming raw fish dishes has especially contributed to an increase in human infections and outbreaks, even in regions that are not typically endemic to this parasite.

Keywords

Dibothriocephalus latus · Diversity · Zoonosis · Broad fish tapeworms · Plerocercoid · Sparganosis

Introduction

Diphyllobothriasis, a parasitic infection caused by cestodes commonly known as broad or fish tapeworms, has garnered increasing attention because of a surge in reported cases, particularly in nonendemic regions (Waeschenbach et al., 2017). Recent research integrating both morphological and molecular techniques has resulted in the reclassification of the majority of diphyllobothriid tapeworms infecting terrestrial and aquatic tetrapod vertebrate hosts (Waeschenbach et al., 2017; Scholz et al., 2019). Members of the family Diphyllobothriidae, belonging to the order Diphyllobothriidea, are notable for their size, representing some of the largest tapeworms (Scholz et al., 2009; Scholz et al., 2019). The life cycle of these cestodes, intricately linked to aquatic environments, is distinctly different from that of cyclophyllidean cestodes. These tapeworms exhibit a complex, multihost life cycle, with adults typically parasitizing a broad spectrum of terrestrial mammals and marine animals (Scholz et al., 2019).

Among all the diphyllobothriid tapeworms, *Dibothriocephalus latus* (formerly known as *Diphyllobothrium latum*) stands out as the most prevalent species infecting humans (Scholz & Kuchta, 2016; Durrani et al., 2023); however, human infections are also attributed to six other species from three genera: *Dibothriocephalus*, *Diphyllobothrium*, and *Adenocephalus*. The confirmed species that are known to infect humans, which have been thoroughly characterized both morphologically and through molecular techniques, include *Dibothriocephalus dendriticus* (previously classified as *Diphyllobothrium dendriticum*), *Dibothriocephalus nihonkaiensis* (formerly known as *Diphyllobothrium nihonkaiensis*), *Adenocephalus pacificus* (formerly *Diphyllobothrium pacificum*), *Diphyllobothrium balaenopterae*, and *Diphyllobothrium stemmacephalum* (Mathison & Pritt, 2019).

In humans, diphyllobothriasis usually occurs incidentally, primarily due to the consumption of raw or undercooked fish or fish products containing viable third-stage (plerocercoids) larvae (Cong & Elsheikha, 2021). This often results in asymptomatic cases or mild symptoms, such as abdominal discomfort, vitamin B12 deficiency, weight loss, vomiting, and diarrhoea (Dick, 2007; McConnaughey, 2014; Kitaoka et al., 2020; Cong & Elsheikha, 2021). Typically, infection in humans is recognized only when the mature tapeworm is expelled in the faeces (Ito & Budke, 2014).

Aetiology and Host

Diphyllobothriid tapeworms primarily infect piscivorous animals, including dogs, cats, bears, fish-eating birds, gulls, fur seals, and sea lions, as well as humans (Kuchta et al., 2015; Scholz et al., 2019; Singh & Banerjee, 2022). This underscores their zoonotic potential, as diphyllobothriasis in humans and other definitive hosts typically results from the consumption of raw or inadequately cooked infected fish (Cong & Elsheikha, 2021). The larvae of diphyllobothriid tapeworms usually pass through two different intermediate hosts before reaching their definitive host. The first intermediate host is typically freshwater crustaceans, such as cyclops, mesocyclops, or diaptomus (including *Acanthodiaptomus*, *Diaptomus*, *Arctodiaptomus*, *Eudiaptomus*, *Boeckella*, and *Eurytemora*) (Torres et al., 2004; Jones, 2015; Singh & Banerjee, 2022). The second intermediate host generally consists of freshwater fish, including pike, trout, salmon, perch, or marine fish. Perch and pike are common for *Dibo. latus* and salmon for *Dibo. nihonkaiense* (Dick & Choudhury, 1996; Dick et al., 2001). While these tapeworms do not exhibit strict specificity for their definitive hosts, they are found primarily in a variety of carnivorous mammals, including felids, canids, bears, and mustelids, in addition to humans (Scholz et al., 2019). Notably, only *Dibo. dendriticus* and Dibo. ditremus have been documented to infect birds (Kuchta et al., 2013). For *A. pacificus*, pinnipeds, dogs, and jackals serve as definitive hosts. Among the species associated with cetaceans, *Diph. stemmacephalum* is present in dolphins and porpoises, whereas *Diph. balaenopterae* has been identified in whales (Scholz et al., 2019). In nature, transmission to mammalian or avian hosts occurs when these animals consume fish containing viable infectious larvae of diphyllobothriidean tapeworms, thereby facilitating the maturation of the tapeworms within the intestines of their definitive hosts.

Taxonomy

Historically, cestodes that parasitize humans have been categorized into two primary groups: Cyclophyllidea, with notable genera including *Taenia*, *Dipylidium*, and *Hymenolepis*, and Pseudophyllidea, encompassing genera such as *Dibothriocephalus*, *Adenocephalus*, and *Diphyllobothrium* (Mathison and Pritt, 2018). However, as noted by Kuchta et al. (2008), the classification of Pseudophyllidea has been revised to include two principal subgroups: *Bothriocephalidea* and *Diphyllobothriidea*. Importantly, the Diphyllobothriidea group comprises parasites that specifically affect humans (Scholz et al., 2019). Among the diphyllobothriid tapeworms, *Dibothriocephalus latus* (previously known as *Diphyllobothrium latum*) is the most common species infecting humans (Scholz & Kuchta, 2016). Nonetheless, infections in humans can also be caused by six other species from three different genera: *Dibothriocephalus*, *Diphyllobothrium*, and *Adenocephalus*. The other validated species that have been well characterized both morphologically and through molecular methods include *Dibo. dendriticus* (once known as *Diphyllobothrium dendriticum*), *Dibo.*

nihonkaiensis (formerly *Diphyllobothrium nihonkaiensis*), *Adenocephalus pacificus* (previously *Diphyllobothrium pacificum*), *Diphyllobothrium balaenopterae*, and *Diph. stemmacephalum*, (Scholz et al., 2019).

Epidemiology

Dibothriocephalus latus worm is recognized primarily as a parasite of humans (Von Bonsdorff, 1977; Radačovská et al., 2022), with various fish-eating mammals and birds acting as its competent reservoir hosts (Dick et al., 2001; Scholz et al., 2009). Current estimates indicate that the number of human diphyllobothriasis cases exceeds 20 million worldwide (Scholz et al., 2009). While the disease can affect individuals of any age and sex, the majority of identified cases are concentrated in middle-aged populations (Arizono et al., 2009). Among the zoonotic diphyllobothriid worms, various species of *Dibothriocephalus* demonstrate a cosmopolitan distribution (Scholz & Kuchta, 2016), with endemic occurrences reported in North America, Eurasia, and, more recently, South America (Chai et al., 2005). Recent studies have shown a decline in the prevalence of diphyllobothriasis in much of Europe and North America (Scholz et al., 2009; Kuchta et al., 2023), however the parasite may persist in fish eating reservoir hosts. Historically, regions with high endemicity, such as Finland and Alaska, have also experienced a reduction in incidence over the past few decades. Conversely, an increase in reported cases has been observed in South America (Semenas et al., 2024). New incidences of diphyllobothriasis have emerged in various global areas, including Japan, Siberia, parts of Europe, and Korea (Dick, 2007; Ando et al., 2021). In subtropical and tropical regions of Asia, *Dibothriocephalus latus* has been identified in countries such as China, Taiwan, Malaysia, India, and Pakistan (Scholz et al., 2009; Betson et al., 2020; Fan et al., 2023). Notably, *Dibo. nihonkaiense* is endemic to the northern Pacific Ocean, with cases frequently reported in Eastern Asian regions where the consumption of raw fish is culturally significant (Ikuno et al., 2018). Meanwhile, *Dibo. dendriticus* is primarily found in arctic regions, and a considerable number of cases of *A. pacificus* have been reported along the Pacific coast of South America (Roth et al., 2018; Autier et al., 2019). In developing nations, significant pollution of freshwater sources, such as lakes and pools, with human night soil through untreated sewage systems has been recognized as a major contributor to the increased infectivity in fish and other paratenic hosts, as well as the widespread prevalence of the disease in these regions (Gustinelli et al., 2016).

Morphology

Within Diphyllobothriidea, species that parasitize humans possess specialized groove-like structures at their anterior ends called "bothria" Generally, cyclophyllidean tapeworms, with which the adult worms attach to the intestinal wall mucosa of their definitive host. Like other cestode parasites, diphyllobothriids absorb nutrients through their body surface, lined with a living outer body layer known as the tegument, which serves both absorptive and secretory roles (Bogitsh et al., 2005). Generally, cyclophyllidean tapeworms are distinguished by their scolex, which features four muscular suckers and a retractable rostellum that is often armed with spines. The eggs of cyclophyllideans typically lack an operculum and are shed in an embryonated state, containing a hexacanth embryo, which is a hooked oncosphere larva that is immediately infectious to an intermediate host. Within the intermediate host, the infectious oncosphere larva transforms into a highly variable larval form, referred to as a cysticercus, hydatid cyst, tetrathyridium, etc., on the basis of its anatomical structure and morphology (Chervy, 2002; Adl & Mathison, 2019; Mathison & Sapp, 2021). In contrast, pseudophyllidean parasites possess a scolex with two longitudinal muscular grooves referred to as bothria. The ribbon-like tapeworms are the longest human parasites and have a body consisting of three main regions: the scolex or the holdfast organ, which attaches to the intestinal mucosa of the definitive host; the neck, which is the base of the strobila; and the strobila itself, which is the primary body made up of segments called proglottids (Scholz & Kuchta, 2024). Pseudophyllidean tapeworms release yellow, ovoid, operculate eggs through a midline uterine pore on the ventral surface of the proglottids (Wittner et al., 2011; Saari et al., 2019) (Fig. 39.1). These eggs measure approximately 66 × 44 µm (Horák et al., 2019), and usually mature at temperatures between 16°C and 20°C for 8–12 days (Dupouy-Camet & Peduzzi, 2004), leading to the emergence of free swimming coracidium larvae, characterized by a ciliated embryophore that surrounds the oncosphere larva.

Life Cycle

Tapeworms are parasitic worms that primarily inhabit the small intestine of their definitive host. These organisms have complex life cycles that involve multiple hosts, with the definitive host becoming infected by consuming the larval stage encysted within the tissues of an intermediate host. The life cycle is typically indirect, and characterized by both heterogony and heteroxeny. In their suitable definitive hosts, pseudophyllidean tapeworms mature and release eggs through a midline uterine pore located ventrally; these eggs are typically ovoid, operculate, and yellow, are approximately 70 µm in size, and contain a living zygote (Kuchta et al., 2013). After release, the eggs usually mature/ embryonate at temperatures between 16°C and 20°C within 8–12 days. They hatch in aquatic environments upon contact with water, releasing ciliated coracidia, each containing an oncosphere larva ensheathed in a ciliated embryonic envelope. Small copepod crustaceans (zooplankton) mistakenly ingest coracidia as plankton, leading to infection, during which the larvae develop into procercoids within the body cavity of the copepods. When infected copepods are consumed by freshwater fish, the procercoid larvae develop into plerocercoids, approximately 1–2 cm in size, which represent the infective stage for fish-eating mammals. Predatory fish, which act as paratenic hosts, can accumulate a large number of plerocercoids within their musculature or

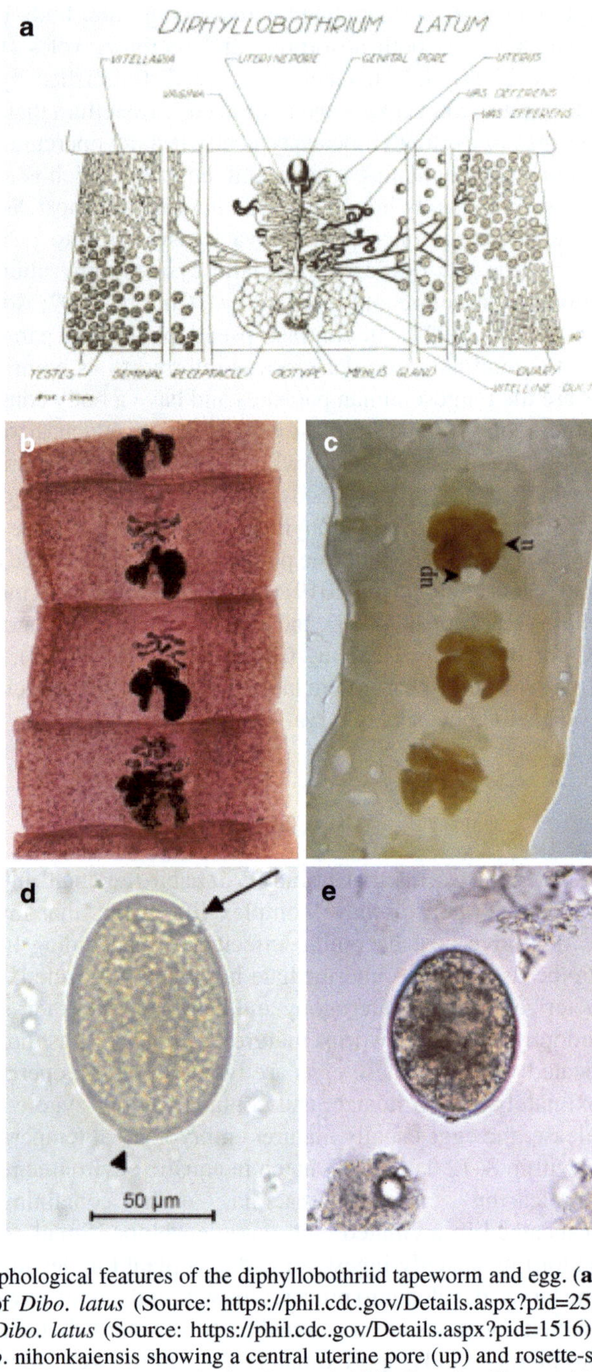

Fig. 39.1 Morphological features of the diphyllobothriid tapeworm and egg. (**a**) Major anatomical structures of *Dibo. latus* (Source: https://phil.cdc.gov/Details.aspx?pid=25315). (**b**) Gravid proglottids of *Dibo. latus* (Source: https://phil.cdc.gov/Details.aspx?pid=1516). (**c**) Gravid proglottids of *Dibo*. nihonkaiensis showing a central uterine pore (up) and rosette-shaped uterus (u), as indicated by arrowheads (source: Autier, B., et al. 2019. Parasites Vectors 12, 267. https://doi.org/10.1186/s13071-019-3516-6). (**d**) Egg of *Dibo. nihonkaiense* (Source: Greigert, V., et al. 2020. Parasitol Res 119, 513–518. https://doi.org/10.1007/s00436-019-06566-6). (**e**) Egg of *Dibo. latus* (Source: https://phil.cdc.gov/Details.aspx?pid=1541)

viscera due to the ingestion of numerous smaller fish harbouring these larvae (Jeon & Eom, 2024).

In the case of diphyllobothriid tapeworms, the infective stage for their suitable definitive host species is the plerocercoid larva. Transmission to humans or other definitive hosts typically occurs through the consumption of raw or inadequately cooked infected fish, allowing the plerocercoid larvae to enter the small intestine, where they mature within 5–6 weeks (Scholz et al., 2009). Adult tapeworms reside in the ileum and jejunum of humans and usually consist of 3000–4000 proglottids (Wittner et al., 2011). These worms are highly fecund, releasing millions of eggs daily (Schmutzhard & Helbok, 2014). Unembryonated eggs are excreted in human faeces, leading to the contamination of aquatic environments such as rivers, lakes, and ponds.

The life cycle of a diphyllobothriid tapeworm is illustrated in Fig. 39.2. Upon release, the eggs generally begin to embryonate upon contact with water within a timeframe of 8–12 days. In this phase, a spherical, ciliated embryo known as a coracidium develops inside the eggshell, where the ciliated embryophore (inner membrane) layer encases the first larval stage, commonly known as the oncosphere larva. The mature coracidia then escape into the water through the opened operculum. For the life cycle to continue successfully, the coracidia must be ingested by an intermediate host, such as freshwater crustaceans, within 12 h. Once inside the host, the coracidium sheds its cilia in the intestine, penetrates the intestinal wall, and moves

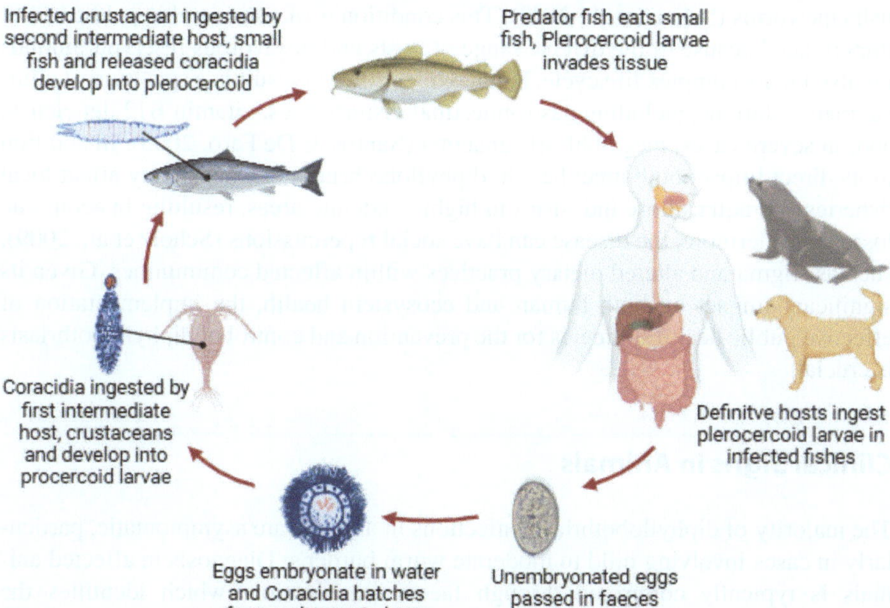

Fig. 39.2 Illustration of the complex life cycle of a diphyllobothriid tapeworm that is characterized by different developmental stages and a range of diverse intermediate and definitive hosts

to the body cavity (hemocoel) of the crustacean. Over the next 2–3 weeks, it matures into a procercoid, the second larval stage. This procercoid does not have a fully developed anterior end for attachment but possesses posterior appendages with embryonic hooks. Typically, each crustacean host harbours a maximum of two procercoids.

Upon ingestion of the infected copepod by a second intermediate fish host, the procercoid is liberated into the intestinal lumen. The procercoid subsequently penetrates the intestinal wall and either encysts within the body cavity or infiltrates the muscular and visceral organs of the fish. Within a period of 1–3 weeks, the procercoid differentiates into the third larval stage, referred to as the plerocercoid. This stage is identifiable by its flattened, white morphology and distinct, irregular, unsegmented wrinkles. Additionally, the anterior end exhibits a depression indicative of the future head of the organism. The plerocercoid represents the infective stage of the larva; when a large piscivorous fish preys upon the second intermediate host, it subsequently reinvades the musculature of the predatory fish and re-encyst. The plerocercoid larvae are transmitted to definitive hosts upon the consumption of raw or partially cooked infected fish.

Zoonotic Importance

Diphyllobothriasis is a zoonotic infection that arises from the consumption of raw or inadequately cooked fish infected with the third larval stage (plerocercoids) of fish tapeworms (Scholz et al., 2009). This condition is of considerable public health importance because of the diverse range of hosts and piscivorous reservoir animals involved in its complex life cycle. Human infection can result in a spectrum of clinical manifestations, including gastrointestinal disturbances, vitamin B12 deficiency, and, in severe cases, megaloblastic anaemia (Santos & De Faro, 2005). In addition to its direct impact on human health, diphyllobothriasis can negatively affect local fisheries and aquaculture industries in highly endemic areas, resulting in economic losses. Furthermore, the disease can have social repercussions (Scholz et al., 2009), such as stigma and altered dietary practices within affected communities. Given its significant impact on both human and ecosystem health, the implementation of effective public health strategies for the prevention and control of diphyllobothriasis is crucial.

Clinical Signs in Animals

The majority of diphyllobothriasis infections in animals are asymptomatic, particularly in cases involving mild to moderate worm burdens. Diagnosis in affected animals is typically confirmed through faecal examination, which identifies the characteristic eggs. Unlike cyclophyllideans, proglottids do not detach (apolysis), and therefore are not shed in the faeces. Clinical symptoms in animals can include gastrointestinal disturbances such as diarrhoea, vomiting, weight loss, and abdominal discomfort (Conboy, 2009; Taylor et al., 2015). Furthermore, segments of proglottids may be excreted by the animals, particularly in instances of heavy infections. Importantly, if left untreated, the adult worm may persist within the host for several years.

Clinical Importance in Humans

Diphyllobothriasis is frequently asymptomatic in humans in the majority of cases. When symptomatic, clinical manifestations are primarily gastrointestinal, classically including vomiting, diarrhoea, abdominal pain, and anaemia. Other reported symptoms include nausea, dizziness, fatigue, myalgia, dyspepsia, mild abdominal cramping, and epigastric pain. Bowel obstruction and the expulsion of worm segments during emesis have also been reported. Rarely, chronic infections may lead to intestinal obstruction. *Dibothriocephalus* species can affect multiple organ systems, resulting in diverse manifestations beyond the gastrointestinal tract. These include central nervous system symptoms (e.g. paraesthesia, demyelinating symptoms, headaches, and encephalopathy), respiratory issues (e.g. dyspnoea), further gastrointestinal complications (e.g. constipation, subacute appendicitis, cholecystitis, and cholangitis), haematological disturbances (e.g. pancytopenia, eosinophilia, and pernicious anaemia), ocular conditions (e.g. optic neuritis), and dermatological symptoms (e.g. glossitis, allergic reactions, and pallor). While human infections are often subclinical, both human and animal infections can result in vitamin B12 deficiency, potentially leading to macrocytic, megaloblastic anaemia of the pernicious type (Ramana et al., 2011; Jimenez et al., 2012; Lee et al., 2015; Montgomery & Richards Jr, 2018; Sharma et al., 2018; Sun et al., 2025). A well-established association exists between *D. latus* infection and vitamin B12 deficiency, likely due to the parasite's high rate of vitamin B12 absorption compared with that of the human gut (Nyberg et al., 1961; Scholz et al., 2009). Although *D. nihonkaiense* infections in humans are typically mild, the protracted period required for worm segment expulsion can place a significant emotional and economic burden on patients and their families (Scholz et al., 2009). The diagnosis of diphyllobothriasis is achieved primarily through the identification of eggs in faecal samples. These eggs resemble those of *Fasciola* spp. but are approximately half the size of *F. hepatica* eggs (Cong & Elsheikha, 2021). Consequently, sedimentation is the recommended method for faecal examination.

Prevention and Control in Animals and Humans

To prevent diphyllobothriasis, it is crucial to avoid eating raw or undercooked fish. The fish if properly cooked to an internal temperature of at least 63°C (145°F) effectively eliminate *Dibothriocephalus* larvae (CDC, 2024; Ziarati et al., 2022). Additionally, freezing fish at −20°C (−4°F) for 7 days or at −35 °C (−31°F) for 15 h effectively kills the larvae (Scholz et al., 2009; Williams et al., 2020). Preventing contamination of drinking water sources with the faecal matter also helps control sparganosis infection. Other preventive and control measures include implementing inspection protocols in fish markets and fisheries to identify and remove infected fish from the food supply; promoting safe handling practices in both commercial and home kitchens to avoid cross-contamination; educating the public, especially in highly endemic areas, about infection sources and the importance of proper fish preparation; and establishing surveillance systems to identify outbreaks and their sources. Additionally, in endemic regions, removing the predation and carrion feeding of cats and dogs is essential. For treating animal diphyllobothriasis, the animals can be treated with praziquantel at a dose of 35 mg/kg orally or niclosamide at 250 mg/kg. Infected fish can be made safe either by cooking or deep freezing at −10°C for 72 h or −18°C for 24 h (Scholz et al., 2009).

Treatment

The majority of diphyllobothriasis cases are effectively managed on an outpatient basis. For human diphyllobothriasis, praziquantel is recommended as two doses of 7.5 mg/kg administered orally or subcutaneously on consecutive days. However, the Centres for Disease Control and Prevention (CDC), USA, advises a single oral dose of 5–10 mg/kg praziquantel as the preferred treatment for both children and adults (Ohnishi & Kato, 2003; Scholz et al., 2009). The drug functions by interfering with the worm's calcium metabolism, and tegumental vacuolization resulting in spastic paralysis of the worm (Chai, 2013). Although the safety profile of praziquantel in pregnant women and children has not been firmly established, instances of accidental administration during mass distribution programmes indicate no adverse effects on foetal outcomes in pregnant women (Adam et al., 2004; Olveda et al., 2016). Common side effects of praziquantel include headache, weakness, dizziness, fever, and abdominal pain (Rim, 1986; Chai, 2013). However, in many countries other than the United States, niclosamide is used as an alternative therapy, typically given as a single 2-g dose for adults and 1 g for children (Scholz et al., 2009). The mode of action of niclosamide involves the uncoupling of oxidative phosphorylation. Side effects related to niclosamide are quite rare, as it is not absorbed through the gastrointestinal tract (Kadri et al., 2018). In contrast, for treating sparganosis, praziquantel may be given orally at doses of 75 mg/kg/day for 7 days, although surgical intervention may be necessary in certain cases (Kim et al., 1996; Chai, 2013).

References

Adam, I., Elwasila El, T., & Homeida, M. (2004). Is praziquantel therapy safe during pregnancy? *Transactions of the Royal Society of Tropical Medicine and Hygiene, 98*(9), 540–543. https://doi.org/10.1016/j.trstmh.2004.01.001

Adl, S. M., & Mathison, B. A. (2019). Chapter 135: Taxonomy and classification of human eukaryotic parasites. In K. C. Carroll, M. A. Pfaller, M. L. Landry, et al. (Eds.), *Manual of clinical microbiology* (pp. 2620–2641). ASM Press.

Ando, Y., Ono, Y., & Ono, S. (2021). Diphyllobothriasis from eating sushi. *The American Journal of Tropical Medicine and Hygiene, 104*(6), 1953.

Arizono, N., Yamada, M., Nakamura-Uchiyama, F., & Ohnishi, K. (2009). Diphyllobothriasis associated with eating raw Pacific Salmon. *Emerging Infectious Diseases, 15*(6), 866–870. https://doi.org/10.3201/eid1506.090132

Autier, B., Belaz, S., Degeilh, B., et al. (2019). Dibothriocephalus nihonkaiensis: An emerging foodborne parasite in Brittany (France)? *Parasites and Vectors, 12*, 267. https://doi.org/10.1186/s13071-019-3516-6

Betson, M., Alonte, A. J. I., Ancog, R. C., Aquino, A. M. O., Belizario, V. Y., Jr., Bordado, A. M. D., et al. (2020). Zoonotic transmission of intestinal helminths in Southeast Asia: Implications for control and elimination. *Advances in Parasitology, 108*, 47–131.

Bogitsh, B., Carter, C., & Oeltmann, T. (2005). General characteristics of the Cestoda. In *Human Parasitology* (4th ed., pp. 217–235). Academic Press: ; ISBN 9780124159846. https://doi.org/10.1016/B978-0-12-415915-0.00012-1

Centers for Disease Control and Prevention. Available from: CDC—DPDx—Diphyllobothriasis. Accessed on 29 May 2024. https://www.cdc.gov/dpdx/diphyllobothriasis/

Chai, J. Y. (2013). Praziquantel treatment in trematode and cestode infections: An update. *Infection & chemotherapy, 45*(1), 32–43. https://doi.org/10.3947/ic.2013.45.1.32

Chai, J.-Y., Darwin, K., Alan, M., & Lymbery, J. (2005). Fish-borne parasitic zoonoses: Status and issues. *International Journal for Parasitology, 35*(11–12), 1233–1254. https://doi.org/10.1016/j.ijpara.2005.07.013

Conboy, G. (2009). Cestodes of dogs and cats in North America. *Veterinary Clinics: Small Animal Practice, 39*(6), 1075–1090. https://doi.org/10.1016/j.cvsm.2009.06.005

Cong, W., & Elsheikha, H. M. (2021). Biology, epidemiology, clinical features, diagnosis, and treatment of selected Fishborne parasitic Zoonoses. *The Yale Journal of Biology and Medicine, 94*(2), 297–309.

Dick, T. A., & Choudhury, A. (1996). Parasites, diseases and disorders. In J. F. Craig (Ed.), *Pike* (Fish and fisheries series) (Vol. 19). Springer. https://doi.org/10.1007/978-94-015-8775-4_7

Dick, T. A., Nelson, P. A., & Choudhury, A. (2001). Diphyllobothriasis: Update on human cases, foci, patterns and sources of human infections and future considerations. *Southeast Asian Journal of Tropical Medicine and Public Health, 32*, 59–76.

Dick, T. A. (2007). Diphyllobothriasis: The Diphyllobothrium latumHuman infection conundrum and reconciliation with a worldwide zoonosis. In K. D. Murrell & B. Fried (Eds.), *Food-borne parasitic Zoonoses* (World class parasites) (Vol. 11). Springer. https://doi.org/10.1007/978-0-387-71358-8_4

Dupouy-Camet, J. E., & Peduzzi, R. (2004). Current situation of human diphyllobothriasis in Europe. *Eurosurveillance, 9*(5), 5–6.

Durrani, M. I., Basit, H., & Blazar, E. (2025). Diphyllobothriasis (fish tapeworm infection) [updated 2023 Nov 25]. In *StatPearls [internet]*. StatPearls Publishing. Available from: https://www.ncbi.nlm.nih.gov/books/NBK540971/

Fan, C. K., Barčák, D., Scholz, T., Sonko, P., Orosová, M., Su, K. E., Chang, C. C., Lee, Y. J., Kuchta, R., & Oros, M. (2023). Human diphyllobothriosis in Taiwan: A review of cases and molecular evidence of Dibothriocephalus nihonkaiensis. *Food and Waterborne Parasitology, 33*, e00213.

Greigert, V., Brunet, J., Pfaff, A. W., et al. (2020). Locally acquired infection with *Dibothriocephalus nihonkaiense* (=*Diphyllobothrium nihonkaiense*) in France: The importance of molecular diagnosis. *Parasitology Research, 119*, 513–518. https://doi.org/10.1007/s00436-019-06566-6

Gustinelli, A., Menconi, V., Prearo, M., Caffara, M., Righetti, M., Scanzio, T., Raglio, A., & Fioravanti, M. L. (2016). Prevalence of Diphyllobothrium latum (Cestoda: Diphyllobothriidae) plerocercoids in fish species from four Italian lakes and risk for the consumers. *International Journal ofFood Microbiology, 235*, 109–112. https://doi.org/10.1016/j.ijfoodmicro.2016.06.033

Horák, I., Marchiondo, A., & Colwell, D. (2019). Platyhelminthes. In A. Marchiondo, L. Cruthers, & J. Fourie (Eds.), *Parasiticide screening* (Vol. 2, pp. 1–133). Academic Press. https://doi.org/10.1016/B978-0-12-816577-5.00006-5

Ikuno, H., Akao, S., & Yamasaki, H. (2018). Epidemiology of Diphyllobothrium nihonkaiense diphyllobothriasis, Japan, 2001–2016. *Emerging Infectious Diseases, 24*(8), 1428.

Ito, A., & Budke, C. M. (2014). Culinary delights and travel? *A review of zoonotic cestodiases and metacestodiases Travel Medicine and Infectious Disease, 12*(6), 582–591. https://doi.org/10.1016/j.tmaid.2014.06.009

Jeon, H. K., & Eom, K. S. (2024). Cestodes and cestodiasis. In *Molecular Medical Microbiology* (pp. 2941–2963). Academic Press. https://doi.org/10.1016/B978-0-12-818619-0.00044-7

Jimenez, J. A., Rodriguez, S., Gamboa, R., Rodriguez, L., Garcia, H. H., & Cysticercosis Working Group in Peru. (2012). Diphyllobothrium pacificum infection is seldom associated with megaloblastic anemia. *The American Journal of Tropical Medicine and Hygiene, 87*(5), 897–901. https://doi.org/10.4269/ajtmh.2012.12-0067

Jones, S. R. M. (2015). *Foodborne parasites in the food supply web transmission dynamics of foodborne parasites in fish and shellfish* (pp. 293–315). Elsevier.

Kadri, H., Lambourne, O. A., & Mehellou, Y. (2018). Niclosamide, a drug with many (re)purposes. *ChemMedChem, 13*(11), 1088–1091. https://doi.org/10.1002/cmdc.201800100

Kim, D. G., Paek, S. H., Chang, K. H., Wang, K. C., Jung, H. W., Kim, H. J., Chi, J. G., Choi, K. S., & Han, D. H. (1996). Cerebral sparganosis: Clinical manifestations, treatment, and outcome. *Journal of Neurosurgery, 85*(6), 1066–1071. https://doi.org/10.3171/jns.1996.85.6.1066

Kitaoka, H., Takamizawa, K., & Shimizu, N. (2020). Raw fish and diphyllobothriasis infection. *QJM: An International Journal of Medicine, 113*(9), 695–696.

Kuchta, R., Brabec, J., Kubáčková, P., & Scholz, T. (2013). Tapeworm Diphyllobothrium dendriticum (Cestoda)—Neglected or emerging human parasite? *PLoS Neglected Tropical Diseases, 7*(12), e2535. https://doi.org/10.1371/journal.pntd.0002535

Kuchta, R., Radačovská, A., Bazsalovicsová, E. Č., & Králová-Hromadová, I. (2023). *Ups and downs of infections with the broad fish tapeworm Dibothriocephalus latus in Europe (part II) and Asia from 1900 to 2020* (pp. 1–69). Elsevier.

Kuchta, R., Scholz, T., Brabec, J., & Bray, R. A. (2008). Suppression of the tapeworm order Pseudophyllidea (Platyhelminthes: Eucestoda) and the proposal of two new orders Bothriocephalidea and Diphyllobothriidea. *International Journal for Parasitology, 38*(1), 49–55. https://doi.org/10.1016/j.ijpara.2007.08.005

Kuchta, R., Scholz, T., Brabec, J., & Narduzzi-Wicht, B. (2015). Diphyllobothrium, Diplogonoporus and Spirometra. In *Biology of foodborne parasites*. Section III Important Foodborne Parasites.

Lee, S. H., Park, H., & Yu, S. T. (2015). Diphyllobothrium latum infection in a child with recurrent abdominal pain. *Korean Journal of Pediatrics, 58*(11), 451–453. https://doi.org/10.3345/kjp.2015.58.11.451

Lenta, C. (2002). The terminology of larval cestodes or metacestodes. *Systematic Parasitology, 52*(1), 1–33. https://doi.org/10.1023/A:1015086301717

Mathison, B. A., & Pritt, B. S. (2018). A systematic overview of zoonotic helminth infections in North America. *Laboratory Medicine, 49*(4), e61–e93. https://doi.org/10.1093/labmed/lmy029

Mathison, B. A., & Pritt, B. S. (2019). Medical parasitology taxonomy update, 2016–2017. *Journal of Clinical Microbiology, 57*. https://doi.org/10.1128/jcm.01067-18

Mathison, B. A., & Sapp, S. G. H. (2021). An annotated checklist of the eukaryotic parasites of humans exclusive of fungi and algae. *ZooKeys*, 10691-313. https://doi.org/10.3897/zookeys.1069.67403

McConnaughey, M. (2014). Life cycle of parasites. In *Biomedical science* (pp. 1–15). Elsevier. https://doi.org/10.1016/B978-0-12-801238-3.05115-1

Montgomery, S. P., & Richards, F. O., Jr. (2018). Diphyllobothrium, Dipylidium, and Hymenolepis species. In S. S. Long, C. G. Prober, & M. Fisher (Eds.), *Principles and practice of pediatric infectious diseases* (pp. 1394–1397). Elsevier. https://doi.org/10.1016/B978-0-323-40181-4.00279-6

Nyberg, W, Ralph, G., Matti, S., & Von Bonsdorff Bertel, W. (1961). Serum vitamin B12Levels and incidence of tapeworm anemia in a population heavily infected with Diphyllobothrium Latum the. *American Journal of Clinical Nutrition, 9*(5), 606–612. https://doi.org/10.1093/ajcn/9.5.606

Ohnishi, K., & Kato, Y. (2003). Single low-dose treatment with praziquantel for Diphyllobothrium nihonkaiense infections. *Internal medicine (Tokyo, Japan), 42*(1), 41–43. https://doi.org/10.2169/internalmedicine.42.41

Olveda, R. M., Acosta, L. P., Tallo, V., Baltazar, P. I., Lesiguez, J. L., Estanislao, G. G., Ayaso, E. B., Monterde, D. B., Ida, A., Watson, N., McDonald, E. A., Wu, H. W., Kurtis, J. D., & Friedman, J. F. (2016). Efficacy and safety of praziquantel for the treatment of human schistosomiasis during pregnancy: A phase 2, randomised, double-blind, placebo-controlled trial. *The Lancet. Infectious diseases, 16*(2), 199–208. https://doi.org/10.1016/S1473-3099(15)00345-X

Radačovská, A., Čisovská Bazsalovicsová, E., Šoltys, K., Štefka, J., Minárik, G., Gustinelli, A., et al. (2022). Unique genetic structure of the human tapeworm Dibothriocephalus latus from the alpine lakes region—A successful adaptation? *Parasitology, 149*(8), 1106–1118. https://doi.org/10.1017/S0031182022000634

Ramana, K., Rao, S., Vinaykumar, M., Krishnappa, M., Reddy, R., Sarfaraz, M., Kondle, V., Ratnamani, M., & Rao, R. (2011). Diphyllobothriasis in a nine-year-old child in India: A case report. *Journal of Medical Case Reports, 5*, 332. https://doi.org/10.1186/1752-1947-5-332

Rim, H. J. (1986). The current pathobiology and chemotherapy of clonorchiasis. *The Korean Journal of Parasitology, 24*(Suppl), 1–141. https://doi.org/10.3347/kjp.1986.24.Suppl.1

Roth, D., Arbetman, M., Flores, V., Semenas, L., & Viozzi, G. (2018). Diphyllobothriidea in the north area of the Andean Patagonia: Epidemiology in urban dogs, morphometrical and molecular identification, with comments on wild carnivores. *Veterinary Parasitology: Regional Studies and Reports, 14*, 161–169.

Saari, S., Näreaho, A., & Nikander, S. (2019). Cestoda (tapeworms). In *Canine parasites and parasitic diseases* (pp. 55–81). Academic Press. https://doi.org/10.1016/B978-0-12-814112-0.00004-0

Santos, F. L. N., & De Faro, L. B. (2005). The first confirmed case of Diphyllobothrium latum in Brazil. *Memórias do Instituto Oswaldo Cruz, 100*, 585–586. https://doi.org/10.1590/S0074-02762005000600013

Schmutzhard, E., & Helbok, R. (2014). Rickettsiae, protozoa, and opisthokonta/metazoa. *Handbook of Clinical Neurology, 121*, 1403–1443. https://doi.org/10.1016/B978-0-7020-4088-7.00096-1

Scholz, T., Garcia, H. H., Kuchta, R., & Wicht, B. (2009). Update on the human broad tapeworm (genus Diphyllobothrium). *Including Clinical Relevance. Clin Microbiol Rev, 22*. https://doi.org/10.1128/cmr.00033-08

Scholz, T., & Kuchta, R. (2016). Fish-borne, zoonotic cestodes (Diphyllobothrium and relatives) in cold climates: A never-ending story of neglected and (re)-emergent parasites. *Food and Waterborne Parasitology, 4*, 23–38. https://doi.org/10.1016/j.fawpar.2016.07.002

Scholz, T., Kuchta, R., & Brabec, J. (2019). Broad tapeworms (Diphyllobothriidae), parasites of wildlife and humans: Recent progress and future challenges. International journal for parasitology. *Parasites and wildlife, 9*, 359–369. https://doi.org/10.1016/j.ijppaw.2019.02.001

Scholz, T., & Kuchta, R. L. (2024). Diphyllobothriidea Kuchta et al., 2008 (order): The broad tapeworms. In S. L. Gardner & S. A. Gardner (Eds.), *Concepts in animal parasitology*. Zea Books. https://doi.org/10.32873/unl.dc.ciap022

Semenas, L., Arbetman, M., Viozzi, G., et al. (2024). Human diphyllobothriasis in Argentina: Assessing the epidemiological significance from historical records and reports of new cases. *Parasitology Research, 123*, 205. https://doi.org/10.1007/s00436-024-08220-2

Sharma, K., Wijarnpreecha, K., & Merrell, N. (2018). Diphyllobothrium latum mimicking subacute appendicitis. *Gastroenterology Research, 11*(3), 235–237. https://doi.org/10.14740/gr989w

Singh, A., & Banerjee, T. (2022). Diphyllobothriasis. In S. C. Parija & A. Chaudhury (Eds.), *Textbook of parasitic Zoonoses. Microbial Zoonoses*. Springer. https://doi.org/10.1007/978-981-16-7204-0_31

Singh, A., & Banerjee, T. (2022). Diphyllobothriasis. In *Textbook of parasitic zoonoses* (pp. 327–336). Springer Nature Singapore.

Sun, J., Guo, J., Zhou, Y., Chen, S., & Lu, Y. (2025). First human case of Diphyllobothriosis due to Dibothriocephalus Dendriticus infection - China, November 2023. *China CDC weekly, 7*(16), 544–548. https://doi.org/10.46234/ccdcw2025.089

Taylor, M. A., Coop, R. L., & Wall, R. L. (2015). Parasites of dogs and cats. In *Veterinary parasitology* (4th ed., pp. 599–677). John Wiley & Sons, Ltd. https://doi.org/10.1002/9781119073680.ch12

Torres, P., Villalobos, L., Woelfl, S., & Puga, S. (2004). Identification of the copepod intermediate host of the introduced broad fish tapeworm Diphyllobothrium latum in southern Chile. *Journal of Parasitology, 90*(5), 1190–1193. https://doi.org/10.1645/GE-3332RN

Von Bonsdorff, B. (1977). 19782901376, English, Thesis, 0–12–724050-0, *Diphyllobothriasis* in man., (xii + 189 pp.), Academic Press Inc. (London) Ltd., 24/28 Oval Road, London NW1 7DX., Diphyllobothriasis in man.

Waeschenbach, A., Brabec, J., Scholz, T., Littlewood, D. T. J., & Kuchta, R. (2017). The catholic taste of broad tapeworms–multiple routes to human infection. *International Journal for Parasitology, 47*(13), 831–843. https://doi.org/10.1016/j.ijpara.2017.06.004

Williams, M., Hernandez-Jover, M., & Shamsi, S. (2020). A critical appraisal of global testing protocols for zoonotic parasites in imported seafood applied to seafood safety in Australia. *Foods (Basel, Switzerland), 9*(4), 448. https://doi.org/10.3390/foods9040448

Wittner, M., White, A. C., & Tanowitz, H. B. (2011). Taenia and other tapeworm infections. In *Tropical infectious diseases* (pp. 839–847). Elsevier. https://doi.org/10.1016/B978-0-7020-3935-5.00121-X

Ziarati, M., Zorriehzahra, M. J., Hassantabar, F., Mehrabi, Z., Dhawan, M., Sharun, K., Emran, T. B., Dhama, K., Chaicumpa, W., & Shamsi, S. (2022). Zoonotic diseases of fish and their prevention and control. *The Veterinary Quarterly, 42*(1), 95–118. https://doi.org/10.1080/01652176.2022.2080298

Part V

Mycotic Zoonoses

Superficial Mycoses: Dermatophytosis

40

Ajay Pratap Singh

Abstract

Superficial fungal infections, commonly known as dermatomycoses, predominantly involve the body's keratin-rich structures such as the outermost skin layer (stratum corneum), hair shafts, and nail plates. Dermatophytosis, a widespread and highly contagious superficial skin infection, is a major concern globally, affecting both humans and animals alike. Dermatophytosis is a widespread and often stubborn superficial fungal infection caused by dermatophytes—a group of fungi specifically evolved to inhabit keratinized tissues. Their ability to colonize the skin, hair, and nails contributes to the recurring nature of this condition. Dermatophytosis is an important occupational health risk for individuals in close contact with animals. Dairymen, animal handlers, livestock farmers, veterinarians, and pet owners are particularly vulnerable, as their frequent interaction with infected animals increases the risk of contracting this disease.

Keywords

Dermatophytosis · Skin infection · Superficial fungal infections

History

The earliest documented case of a dermatophyte infection dates back to 30 A.D., when Roman physician Aulus Cornelius Celsus described a purulent scalp condition in his medical treatise *De Re Medicina*. Celsus identified a severe inflammatory

A. P. Singh (✉)
Department of Veterinary Microbiology, College of Veterinary Science & Animal Husbandry, U.P. Pandit Deen Dayal Upadhyaya Pashu Chikitsa Vigyan Vishwavidyalaya Evam Go-Anusandhan Sansthan, Mathura, Uttar Pradesh, India

R. Biswas et al. (eds.), *Concept of One Health and Zoonoses*,
https://doi.org/10.1007/978-981-95-6627-3_40

form of tinea, which we now recognize as "**kerion**", a type of tinea capitis caused by dermatophytes. William Dampier (1651–1715), during his voyages to the Pacific, observed and documented a chronic dermatophytosis with concentric scaly lesions locally referred as "**Tokelau**," now known as *tinea imbricata*, affecting Polynesian and Micronesian populations. The term "ringworm" was coined by Saxon as early as the sixteenth century to describe these infections. The term "dermatophyte" itself was coined in 1837 by the German scientist Felix Von Hogrewe. The fungal elements of favus were first observed in 1837 by Robert Remak, a physician and researcher at Berlin University. Schönlein's seminal 1839 work provided the first definitive evidence that "**favus**" resulted from fungal infection. During his medical practice in Paris (1841), Hungarian doctor David Gruby successfully isolated the fungal pathogen responsible for favus while characterizing various dermatophytoses such as tinea favosa, along with both ectothrix and endothrix trichophytosis and microsporiosis. Two years later (1843), Gruby established the *Microsporum* genus and identified *Microsporum audouinii* through microscopic examination of patient samples. In 1845, Per Hendrik Malmsten described *Trichophyton*, the second genus of dermatophytes. The taxonomic classification of dermatophytes was completed in 1910 when Raymond Sabouraud introduced *Epidermophyton* as the third and final genus within this group. In the 1920s, the development of selective media, such as Sabouraud's dextrose agar, greatly enhanced the ability to culture and identify dermatophytes in clinical settings. The taxonomic foundation for dermatophytes was solidified by Emmons' 1934 classification into *Trichophyton*, *Microsporum*, and *Epidermophyton*. Nearly a quarter-century later (1958), Gentles revolutionized treatment approaches by introducing griseofulvin as the first effective systemic antifungal agent against superficial mycoses.

Etiological Agent

The cutaneous fungal infection dermatophytosis is caused by members of the Arthrodermataceae family (Onygenales order), which includes 7 established genera containing approximately 40 species. Current classification divides these dermatophytes into anthropophilic (human pathogens), zoophilic (animal pathogens), and geophilic (environmental) varieties based on their ecological preferences.

Anthropophilic dermatophytes	Zoophilic dermatophytes	Geophilic dermatophytes
Trichophyton rubrum *Trichophyton tonsurans* *Epidermophyton floccosum* *Trichophyton digitale* *Trichophyton schoenleinii*	*Microsporum canis* *Nannizzia persicolor* *Nannizzia nana* *Trichophyton equinum* *Trichophyton mentagrophytes* *Trichophyton verrucosum*	*Nannizzia gypsea*

Human dermatophytoses are primarily caused by anthropophilic and zoophilic species, with rare cases involving geophilic dermatophytes. Anthropophilic species have evolved to specialize in human hosts, often demonstrating site-specific tropism. These infections are clinically classified by their anatomical location:

(i) Tinea pedis ("athlete's foot"): foot infections
(ii) Tinea capitis: scalp infections
(iii) Tinea unguium: nail infections (onychomycosis)
(iv) Tinea barbae: beard area infections
(v) Tinea faciei: facial infections
(vi) Tinea corporis: body surface infections
(vii) Tinea manuum: hand infections
(viii) Tinea cruris: groin infections

Major Pathogenic Species

In Human Infections The *Trichophyton rubrum* complex accounts for most cases. Other prevalent species include *T. mentagrophytes*, *T. tonsurans*, and *Trichophyton violaceum.*

In Companion Animals Dogs and cats are most commonly affected by *Microsporum canis*. Other significant pathogens include *Nannizzia gypsea* (formerly *Microsporum gypseum*), *T. mentagrophytes*, and occasionally *M. persicolor*.

In Livestock

Cattle, sheep, and goats: *Trichophyton verrucosum* predominates, with occasional infections by *M. canis*, *N. gypsea*, *T. mentagrophytes*, and *T. equinum*.

Horses: *Trichophyton equinum* is primary, though *M. canis*, *M. equinum*, *N. gypsea*, *T. mentagrophytes*, and *T. verrucosum* may occur.

In Other Species

Rodents, rabbits, and hedgehogs: primarily *T. mentagrophytes,* voles; *M. persicolor* has been documented.

Poultry: *Microsporum gallinae* is the main agent, while *Trichophyton simii* shows particular prevalence in Indian poultry populations.

Epidemiology

Globally, approximately 20–25% of the population is affected by superficial mycoses, with dermatophytosis being the most prevalent type. It predominantly affects children, particularly those under the age of six, with younger individuals are generally more prone to dermatophytosis, with children commonly developing lesions on their scalps, while adults are more likely to experience infections on glabrous (hairless) skin. Seasonal factors also play a role, as the incidence of dermatophytosis is higher during late autumn and winter, when environmental conditions favor fungal growth and transmission. In adults, tinea unguium (nail infection) is more common in older individuals, particularly those with comorbid conditions such as diabetes or vascular disorders, as these impair immune function and circulation. Toenail infections are more frequent in men, whereas women are more prone to infections in their fingernails. Men have a higher overall incidence of dermatophyte infections. Low socioeconomic status significantly contributes to the prevalence of dermatophytosis. Overcrowding, large family sizes, and inadequate hygiene in such environments increase exposure to dermatophytes. Immune deficiencies and immunosuppression, often due to conditions like AIDS, organ transplants, or prolonged use of immunosuppressive medications, further facilitate the spread of dermatophyte infections. Other risk factors include type 2 diabetes, anemia, vascular diseases, and physical inactivity. Globally, *T. rubrum* accounts for 50–90% of dermatophyte infections, making it the most common species. However, in India, the *T. mentagrophytes* complex is the leading cause. Among childhood fungal infections, tinea capitis shows the highest incidence, where *T. tonsurans* infections constitute the overwhelming majority (95%) of reported cases. *T. mentagrophytes* had the widest host range of all the dermatophytes.

Host Range

Anthropophilic Dermatophytes Anthropophilic dermatophytes demonstrate specific evolutionary adaptations for human infection and represent the principal causative agents of dermatophytosis in human populations.

Zoophilic Dermatophytes Zoophilic dermatophytes primarily infect animals, which serve as their natural hosts. While these fungi are predominantly associated with species such as dogs, cats, and humans, they occasionally affect livestock, including cattle, buffaloes, sheep, goats, pigs, and horses. Sporadic isolated infections have also been reported in laboratory animals such as rabbits, guinea pigs, and mice, as well as in lagomorphs. Additionally, dermatophyte infections have been documented in porcupines, hedgehogs, red kangaroos, voles, and bats, highlighting the broad host range of these pathogens.

Geophilic Dermatophytes These dermatophytes are typically found in the soil. Infections caused by geophilic dermatophytes are rare and are not transmitted between humans or from animals to humans. Instead, humans acquire these infections through direct contact with contaminated soil.

Transmission

Dermatophytosis spreads principally through:

(i) Direct interpersonal contact
(ii) Zoonotic transfer from animal reservoirs

(iii) Indirect spread via:
- Contaminated grooming equipment
- Shared bedding/furniture
- Environmental spores in living spaces

Key Animal Reservoirs Domestic species maintain significant dermatophyte reservoirs, particularly in settings with close human-animal contact. Major carrier species include:

- Companion animals: cats, dogs, rabbits
- Livestock: sheep, goats, pigs, horses, donkeys
- Poultry: ducks, chickens

Among animal hosts, domestic cats pose particular transmission risks due to high carriage rates of pathogenic species, close contact behaviors with humans, environmental contamination through shed hairs and scales, and shared use of grooming tools and resting areas.

Pathogenesis

Both human and animal infections begin following cutaneous exposure to arthroconidia, the specialized fungal elements that mediate host invasion and subsequent disease development. The infectious process begins when arthroconidia establish attachment to host tissues via surface-expressed adhesins that recognize specific carbohydrate moieties, along with fibrillar projections that mediate physical anchorage. Mannan glycoproteins in the fungal cell wall mediate this adhesion process. The germination of arthroconidia and subsequent hyphal growth enable the organism to invade keratin-rich structures, including stratified layers of the epidermis, hair follicle units, and specialized keratin projections (spines/quills). To overcome host barriers and extract nutrients, these fungi secrete multiple enzyme classes:

- Proteinases for keratin breakdown
- Lipolytic enzymes for membrane penetration
- ECM-degrading enzymes (elastase/collagenase)
- Phosphatases for phosphate acquisition

The host's first line of defense against dermatophytes involves two key keratinocyte-derived components: sphingoid bases (sphingosines) with direct antifungal properties and sebaceous fatty acids creating an acidic, antimicrobial barrier. However, dermatophytes counter these barriers by secreting proteases such as metallo-endoproteases (fungalysins) and serine-subtilisins, which digest the keratin network, facilitating tissue penetration. Keratin degradation begins with sulfitolysis of disulfide bridges, followed by proteolysis by endoproteases and exoproteases, releasing peptides and amino acids. These proteases are often host specific, optimizing the pathogen's ability to infect. Mannans may further contribute to infection by decreasing keratinocyte proliferation and suppressing the lymphoproliferative response of mononuclear leukocytes. Fungal metabolites exacerbate the infection by causing erythema, vesicle, or pustule formation, accompanied by pruritus, leading to clinical signs of dermatophytosis. Dermatophytes effectively adapt to different pH conditions during infection to degrade keratin and obtain nutrients. Under acidic conditions, specific proteases become active, initiating the breakdown of keratin. The process generates metabolites that alter the surrounding pH to alkaline, subsequently activating proteases that function best in alkaline environments. Various dermatophytes secrete immunosuppressive substances, including xanthomegnin, hemolysins, and aflatoxin-like compounds, which contribute to evading the host immune response and facilitating infection.

Clinical Finding

Human The medical community employs a topographical classification system where: The prefix "tinea" indicates a dermatophytosis, and the suffix specifies the involved anatomical site (e.g., "corporis" for body, "pedis" for feet). The most common areas of infection include the head, neck, trunk, and limbs. *T. verrucosum* tends to affect the head and upper body, whereas *T. mentagrophytes* was observed in the scalp and neck. Dermatophytosis typically manifests as localized skin lesions. These lesions may appear as ringlike structures or scaling patches, often associated with itching and irritation. The infection can be confined to one area or affect multiple body regions. In many cases, dermatophyte infections lead to visible skin changes, including significant scaling and thickening of the epidermis. However, in some instances, the infection may be asymptomatic, presenting with minimal symptoms or manifesting only as pruritus (itchiness).

Kerion (Nodular Dermatophytosis) Kerion represents a severe, inflammatory form of tinea capitis resulting from an exaggerated delayed-type hypersensitivity (Type IV) reaction to dermatophyte antigens. This immune-mediated response predominantly involves Th1 and Th17 cell activation, with elevated levels of pro-inflammatory cytokines including IL-17, IL-22, and IFN-γ. These mediators drive the characteristic intense inflammation that can lead to scarring alopecia if untreated. Clinically, kerion presents as a tender, circumscribed nodule with overlying erythema, scaling, and patchy hair loss. The inflammatory process begins as granulomatous dermatitis and progresses to suppurative pyogranuloma formation, evidenced histologically by dense dermal infiltrates of neutrophils, lymphocytes, and multinucleated giant cells, frequently accompanied by follicular destruction. While zoophilic species like *T. verrucosum* and *T. mentagrophytes* have traditionally been implicated, recent epidemiological studies reveal an increasing prevalence of anthropophilic strains, particularly *T. tonsurans*, in urban populations.

Dermatophytid (or "id") Reactions Dermatophytid reactions represent a distinct immunological response, occurring in approximately 4–5% of patients with active dermatophyte infections. The prevalence varies significantly among specific populations, affecting 7–8% of pediatric cases, 10–12% of immunocompetent individuals with extensive tinea corporis, and 15–18% of infections caused by zoophilic strains. These secondary eruptions typically manifest as acute, pruritic, eczematous lesions appearing symmetrically on the hands, feet, or trunk—sites distant from the primary fungal infection. The pathophysiology involves a localized delayed-type hypersensitivity (DTH) reaction to circulating fungal antigens. Key mechanisms include:

- Antigen dissemination, whereby keratinolytic enzymes from the primary infection site release fungal proteins into systemic circulation
- T-cell mediated inflammation, characterized by Th1/Th17 polarization and elevated levels of IL-17, IL-22, and IFN-γ
- Molecular mimicry, particularly cross-reactivity between fungal antigens (e.g., Tri r 2 from *T. rubrum*) and human keratinocyte proteins

Emerging research highlights the role of microbial ecology in modulating these reactions. Cutaneous dysbiosis, marked by reduced *Staphylococcus epidermidis* colonization, appears to exacerbate the inflammatory response, while alterations in the gut-skin axis may influence reaction severity. These findings underscore the complex interplay between fungal pathogens, host immunity, and commensal microbiota in dermatophytid pathogenesis.

Animals

Farm Animals In cattle, dermatophytosis typically presents as circular, grayish-white crusty lesions commonly located on the head, neck, and shoulders. These lesions are associated with thickened skin and areas of alopecia (hair loss). They may expand, coalesce, and become dry and scaly, with young animals showing more prominent lesions and faster spread of the infection. In sheep and goats, the condition manifests as circular or irregular patches of alopecia, especially on the head, ears, and legs, accompanied by scaling and crust formation. Lesions may become extensive in immunocompromised or stressed animals. Horses often develop circular, scaly patches of hair loss on the girth, saddle area, or face, sometimes accompanied by mild to severe pruritus. Chronic cases in horses may result in thickened, crusted skin. In pigs, red, inflamed lesions covered with grayish or yellow crusts are common, typically affecting the head, back, and hindquarters, with lesions occasionally coalescing to form larger affected areas.

Pet Animals Dermatophytosis in both dogs and cats often presents with similar cutaneous signs due to the fungi's affinity for keratinized structures. Although species-specific nuances exist—such as cephalic predominance and ear margin scaling in cats, or kerion lesions being more frequent in dogs—the core dermatological patterns frequently overlap, warranting similar diagnostic approaches. Feline dermatophytosis typically presents with a range of cutaneous manifestations that often follow predictable patterns of distribution. The disease most commonly begins on the head region (68% of cases), particularly affecting the muzzle, periocular areas, and ear pinnae, with ear margin scaling being pathognomonic for *Microsporum canis* infections. Initial lesions often appear as irregular papulopustular eruptions accompanied by blackheads (comedones) and progress to form expanding annular areas of alopecia with characteristic epidermal collarettes. Secondary changes including scaling, crusting, erythema, and hyperpigmentation frequently develop as the infection progresses. Approximately 40% of cases demonstrate pedal involvement, presenting with digital erythema and hyperpigmentation that may progress to onychogryphosis (thickened, deformed claws) in chronic infections. In severe cases, patients may develop nodular dermatitis with draining tracts. While these lesions can technically occur anywhere on the body, they predominantly originate in areas of frequent grooming and environmental contact—particularly the face, ears, and paws—before spreading to other regions including the limbs, trunk, and tail. The characteristic cephalic and pedal distribution patterns reflect several factors: the cat's grooming behaviors which spread the infection from primary sites; the higher density of sebaceous glands in facial regions; and the frequent environmental contact of these anatomical areas. As the infection progresses without treatment, localized lesions often generalize, with the potential to involve nearly the entire cutaneous surface in severe, untreated cases.

Laboratory Animals Commonly exhibit clinical signs such as alopecia, scaling, crust formation, and pruritus. In guinea pigs, the condition typically presents as Scaly alopecic lesions distributed over the orbital margins, facial skin, and dorsum of the nose, which can become generalized in severe cases. Rabbits often show erythema, scaling, and early alopecia in mild infections, progressing to crust formation and extensive alopecia in advanced cases. Hedgehogs may initially develop scaling around the base of their spines, while chronic infections lead to alopecia, scab and crust formation, thickened pinnae, and loss of quills. In murine models, favus typically presents as grayish to yellow, crust-like scutula or patchy lesions on the skin.

Diagnosis

1. **Direct Microscopical Examination of Skin Scrapings, Scales, Hairs:** The technique entails the microscopic evaluation of hair specimens to detect fungal structures, including fungal spore and hyphae. Hair samples are typically collected from affected areas, treated with a potassium hydroxide (KOH) solution to dissolve keratin, and then examined microscopically to enhance the visibility of fungal elements such as arthroconidia and hyphae.

2. **Fungal Culture for Definitive Diagnosis of Dermatophytosis:** While direct microscopy provides rapid detection, fungal culture remains the gold standard for the definitive diagnosis of dermatophytosis, allowing species identification and antifungal susceptibility testing. It involves growing the suspected dermatophyte from clinical samples (such as skin scrapings, hair, or nails) on specialized media. Sabouraud's dextrose agar (SDA) is a widely used medium for culturing dermatophytes. Dermatophyte test media (DTM) is a selective medium designed to inhibit the growth of saprophytic fungi and bacteria through the inclusion of cycloheximide, gentamicin, and chlortetracycline. DTM is commonly used for diagnosing ringworm in both humans and animals. Mycobiotic agar, another specialized medium, contains cycloheximide and chloramphenicol to suppress non-fungal contaminants, allowing dermatophytes to grow.
3. **Molecular Techniques**: While fungal culture remains the gold standard for dermatophyte identification, molecular techniques such as PCR and real-time PCR provide rapid, species-specific detection by targeting fungal DNA in clinical specimens. These methods demonstrate superior diagnostic performance, with sensitivity ranging from 85% to 100% and specificity between 90% and 100%, outperforming conventional microscopy and culture. Furthermore, recent development of pan-dermatophyte PCR assays targeting conserved genomic regions (e.g., ITS or 28S rRNA) offer additional advantages by detecting mixed infections and nonviable fungal elements that might be missed by traditional methods.
4. **Dermatophyte Antigen Detection**: Enzyme-linked immunosorbent assay (ELISA) and lateral flow immunoassays have emerged as noninvasive alternatives, detecting dermatophyte antigens (e.g., trichophytin) in clinical samples within minutes. Recent studies report ELISA sensitivities of 80–95% for *T. rubrum* and *Microsporum* spp., offering point-of-care potential. Though less species-specific than PCR, these immunoassays provide scalable, culture-independent options for resource-limited settings.
5. **Wood's lamp examination** is a simple, noninvasive diagnostic tool used to detect dermatophyte infections, particularly *Microsporum canis*. Under ultraviolet (UV) light emitted by the Wood's lamp, certain dermatophytes fluoresce, allowing for visual identification of infected areas. This fluorescence is due to the production of specific metabolites, such as pteridine, by the fungus.

Prevention and Control

Preventive and control measures for dermatophytosis in both animals and humans are crucial to minimize the spread of the infection. Key strategies include:

Hygiene and Sanitation Regular cleaning and disinfection of animal housing, equipment, and grooming tools to reduce fungal spores in the environment. Frequent cleaning of shared spaces, such as barns, stables, and animal pens, to limit the spread of dermatophytes. Ensuring that animals are not kept in overcrowded or unsanitary conditions, as these factors can facilitate fungal growth.

Monitoring and Early Detection Regularly inspect animals for signs of dermatophytosis, including hair loss, scaling, or crusting, and promptly isolate affected animals to prevent transmission. Routine veterinary checks, especially in high-risk animals like cattle, sheep, goats, and pets, to identify early infections.

Personal Protective Equipment (PPE) Farm workers and individuals handling infected animals should wear gloves, long sleeves, and boots to reduce the risk of direct contact with fungal spores. After handling infected animals, workers should wash hands thoroughly with soap and water and disinfect clothing and equipment.

Environmental Control Fungal spores can persist in the environment, so regularly disinfecting the living areas of infected animals and replacing bedding material is important. Use of antifungal solutions in areas with high traffic or high-risk populations (e.g., animal shelters or farms).

Quarantine and Isolation Newly introduced or sick animals should be quarantined and closely monitored for signs of dermatophytosis to prevent outbreaks. Isolate any infected animals from healthy ones until they are no longer contagious.

Educating Animal Handlers and Farmers Providing education about the signs, symptoms, and prevention of dermatophytosis can help animal handlers and farmers take early action to control outbreaks. Teaching proper handling, sanitation, and the use of protective gear helps reduce the spread of infection.

References

Aly, R. (1994). Ecology and epidemiology of dermatophyte infections. *Journal of the American Academy of Dermatology, 31*(3), S21–S25.

Degreef, H. (2008). Clinical forms of dermatophytosis (ringworm infection). *Mycopathologia, 166*, 257–265.

Hainer, B. L. (2003). Dermatophyte infections. *American Family Physician, 67*(1), 101–109.

Negroni, R. (2010). Historical aspects of dermatomycoses. *Clinics in Dermatology, 28*(2), 125–132.

Nenoff, P., Krüger, C., Ginter-Hanselmayer, G., & Tietz, H. J. (2014). Mycology–an update. Part 1: Dermatomycoses: Causative agents, epidemiology and pathogenesis. *JDDG: Journal der DeutschenDermatologischen Gesellschaft, 12*(3), 188–210.

Seebacher, C., Bouchara, J. P., & Mignon, B. (2008). Updates on the epidemiology of dermatophyte infections. *Mycopathologia, 166*, 335–352.

Segal, E., & Elad, D. (2021). Human and zoonotic dermatophytoses: Epidemiological aspects. *Frontiers in Microbiology, 12*, 713532.

Weitzman, I., & Summerbell, R. C. (1995). The dermatophytes. *Clinical Microbiology Reviews, 8*(2), 240–259.

Systemic Mycoses

41

Ripan Biswas

Abstract

Systemic mycoses are fungal infections that can impact internal organs and spread throughout the body. Many of these infections are zoonotic, meaning they can be transmitted between animals and humans. Common examples include histoplasmosis, blastomycosis, cryptococcosis, and coccidioidomycosis. These infections are typically contracted by inhaling fungal spores from the environment, especially from soil contaminated with animal droppings. While some individuals may remain asymptomatic or experience mild symptoms, others can develop severe illness affecting the lungs, skin, bones, and central nervous system (CNS). Early diagnosis and prompt treatment with antifungal medications are crucial for effectively managing systemic mycotic zoonoses.

Keywords

Invasive fungal disease · Opportunistic mycoses · Histoplasmosis · Blastomycosis · Cryptococcosis · Coccidioidomycosis

Histoplasmosis

The discovery of *Histoplasma capsulatum* is credited to Dr. Samuel Taylor Darling, so histoplasmosis, a well-known systemic mycotic zoonosis, is also known as Darling disease. In 1905, he identified the organism in tissue samples from patients in Panama who had died from a disease resembling tuberculosis (Kauffman, 2007). *Histoplasma capsulatum*, a dimorphic fungus, exists in two distinct forms. In the

R. Biswas (✉)
Department of Veterinary Public Health and Epidemiology, West Bengal University of Animal and Fishery Sciences, Kolkata, West Bengal, India

R. Biswas et al. (eds.), *Concept of One Health and Zoonoses*,
https://doi.org/10.1007/978-981-95-6627-3_41

environment, it thrives as a mold with filamentous hyphae that produce spores. Upon entering the human body, it undergoes a transformation, assuming a yeast form characterized by single, budding cells. This remarkable ability to switch between morphologies in response to environmental conditions is a hallmark of dimorphic fungi. *Histoplasma capsulatum* belongs under ascomycetes family, and it exists in the environment with two mating types, like "+" and "-" but, human mostly get sick from the "-" type. *Histoplasma capsulatum* produces two distinct types of conidia: the macroconidia and the microconidia. The macroconidia, also known as tuberculate conidia, range from 8 to 15 μm in diameter and feature a thick cell wall with distinctive surface projections. In contrast, the microconidia are much smaller, smooth structures measuring 2–4 μm in diameter. It is these smaller microconidia that represent the infectious form of the fungus. Upon entering the alveoli, the inhaled microconidia transform in response to the body's temperature, developing into unicellular yeast forms typically 2–5 μm in diameter. *Ryp1*, *Ryp2*, and *Ryp* genes are responsible for the transition and growth of yeast cells. In the environment, *H. capsulatum* exists as a mold that forms white to tan colonies on Sabouraud dextrose agar when cultured at temperatures between 25°C and 30°C.

Epidemiology

Histoplasmosis is primarily transmitted through the inhalation of spores from contaminated soil, often in areas where there are large bird or bat populations. *Histoplasma capsulatum* has traditionally been classified into three distinct varieties based on their geographic distribution, clinical presentation, and morphological characteristics. The most common variety worldwide is *H. capsulatum* var. *capsulatum*, which causes classic histoplasmosis. *Histoplasma capsulatum* var. *duboisii*, another human pathogen, is restricted to Central and West Africa. Additionally, *H. capsulatum* var. *farciminosum*, an animal pathogen, is responsible for equine histoplasmosis. Histoplasmosis is prevalent across North and Central America and widely distributed throughout South America, including in Brazil, Ecuador, Venezuela, Paraguay, Uruguay, and Argentina. Asia has also seen a high prevalence of histoplasmosis, with China, particularly along the Yangtze River region, reporting the most cases. Additionally, significant numbers of cases have been documented in Thailand, South Korea, and India. While epidemiological data on histoplasmosis in Africa is limited, there have been reports of the disease occurring in countries such as Uganda, South Africa, and Tanzania. Histoplasmosis is infrequently reported in Europe, with most cases linked to travel to endemic regions. This infection is considered an AIDS-defining illness because of its frequent association with HIV-infected individuals. The major fungal infections associated with COVID-19 include candidiasis, aspergillosis, mucormycosis, pneumocystosis, cryptococcosis, and histoplasmosis. Nine reported cases of histoplasmosis have been associated with COVID-19 to date, with three in Brazil, two in the United States, two in Argentina, and one in India.

Pathogenesis

Following inhalation of microconidia or mycelial fragments, the fungus enters the alveoli of the host's lungs and undergoes dimorphic conversion to the yeast form triggered by the higher temperature and oxygen tension within the host. Once the fungus assumes the yeast form, it can be engulfed by macrophages and then escape destruction within these cells, enabling the infection to spread to other organs characterizing histoplasmosis in its invasive form. The primary immune response to *Histoplasma* involves cell-mediated immunity driven by T cells, incorporating both innate and adaptive mechanisms.

Clinical Manifestation

Most individuals (<1%) infected with histoplasmosis experience minimal symptoms. Typically, *H. capsulatum* infections are self-limited and mild, often going unnoticed by healthcare providers.

Acute pulmonary histoplasmosis—Children are most commonly affected by acute pulmonary histoplasmosis upon their first exposure. Patients with acute pulmonary histoplasmosis experience an abrupt onset of illness, presenting with symptoms like fever, chills, malaise, difficulty breathing, cough, and chest pain. The disease is often mistaken for other common respiratory infections. Approximately 5% of patients with acute self-limited pulmonary histoplasmosis also experience rheumatologic and/or dermatologic manifestations which include erythema nodosum and erythema multiforme. Most of the cases recover without treatment, but in some pneumonic cases the residual nodules often become calcified, resulting in a "buckshot" appearance throughout the lung field.

Chronic pulmonary histoplasmosis—Older patients with preexisting pulmonary conditions are most susceptible to this manifestation of histoplasmosis. Chronic cavitary pulmonary histoplasmosis occurs primarily in individuals with underlying lung disease, such as emphysema, chronic obstructive pulmonary disease, or old tuberculosis scars. Chronic pulmonary histoplasmosis is believed to develop when inflammation occurs adjacent to preexisting bullae. This leads to thickening of the bullae walls, followed by central necrosis, ultimately resulting in the formation of large cavitary lesions. These cavities exhibit a progressive enlargement, creating a "marching" appearance as they expand over time. Disseminated histoplasmosis, the most severe manifestation of the disease, occurs when the fungus spreads systemically to various organ.

Granulomatous mediastinitis—Granulomatous mediastinal lymphadenitis is an uncommon complication of pulmonary histoplasmosis, characterized by the enlargement of multiple mediastinal lymph nodes to form a mass that often undergoes caseating necrosis. This presentation may pose diagnostic challenges, as the enlargement and necrosis of the mediastinal lymph nodes can resemble malignancy or tuberculosis.

Fibrosing mediastinitis is a rare but serious complication of pulmonary histoplasmosis, distinct from granulomatous mediastinal lymphadenopathy. In this condition, the mediastinal structures become encased within a thick, fibrotic reaction, leading to significant compression and obstruction of the surrounding vasculature and airways.

Disseminated histoplasmosis—This predominantly affects young children and immunosuppressed individuals, particularly those with HIV infection and CD4 cell counts below 150/μL. Another manifestation of disseminate histoplasmosis is adrenal gland involvement, which can result in adrenal insufficiency due to the destruction of the adrenal glands by the fungus. Patients with disseminated histoplasmosis typically present with fever, fatigue, poor appetite, and unintentional weight loss.

Performing a bronchoscopic alveolar lavage (BAL) culture may be considered, although it is unlikely to yield a positive result, as it can help exclude the presence of any other concurrent pathologies. Serological tests, such as complement fixation, immunodiffusion, and enzyme immunoassay, can also provide supportive diagnostic information by detecting antibodies or antigens specific to *Histoplasma*. An ELISA utilizing a yeast cell extract exhibited a sensitivity of 86% and a specificity of 91% for detecting acute pulmonary histoplasmosis when employing antihuman IgG. In contrast, the sensitivity was 66% and the specificity was 100% when using antihuman IgM. PCR-based diagnostic approaches, while promising, are currently limited by the need for further research and development to optimize their performance and establish their clinical utility for reliably diagnosing histoplasmosis infections.

According to the Infectious Disease Society guidelines, if symptoms persist for more than 4 weeks, a 3-month course of itraconazole is typically recommended. For patients with chronic pulmonary histoplasmosis and noncavitary disease, a 6-month therapy is generally sufficient. In contrast, those with cavitary chronic pulmonary histoplasmosis may require treatment for up to a year. Patients with progressive disseminated histoplasmosis typically need induction therapy with amphotericin-B for 2–4 weeks, followed by a year of itraconazole. Disseminated histoplasmosis in the setting of advanced HIV disease poses an even greater challenge, as the severely compromised immune system renders these patients highly susceptible to the most severe and life-threatening manifestations of the infection.

Blastomycosis

Blastomycosis (Gilchrist disease/Chicago disease), a fungal infection caused by the dimorphic fungus *Blastomyces dermatitidis*, was first described in 1894 by Dr. Thomas Caspar Gilchrist, who initially misidentified the causative agent as a protozoan. Later, he and Dr. Arthur Stein correctly classified it as a fungus. Early cases frequently presented with prominent skin lesions, leading to the initial perception of blastomycosis as a localized skin disease. The term "North American blastomycosis" was used to distinguish it from South American blastomycosis, but this distinction is no longer used as the disease has been found in regions beyond North

America. Recent genetic analysis has identified a distinct species, *Blastomyces gilchristii*, that can also cause blastomycosis. Although genetically distinct, *B. dermatitidis* and *B. gilchristii* are physically similar and cause clinically comparable disease. They are often referred to as the *B. dermatitidis/gilchristii* species complex. In recent years, blastomycosis has emerged as an increasingly concerning public health issue, particularly in areas where the causative fungus is endemic, with reports indicating a rise in both the incidence and mortality rates associated with this infection.

Epidemiology

Blastomyces dermatitidis exists in two forms: a mycelial form in the environment and a yeast form when infecting a host. The fungus grows best in moist, organic-rich soil, such as along rivers, around lakes, and in wooded areas. Historically, blastomycosis was believed to be confined to North America, with the disease being most prevalent in the states bordering the Mississippi and Ohio River valleys. The geographical distribution of blastomycosis has expanded to new regions in recent decades, with cases reported outside the traditional endemic areas, such as in Canada, Europe, and Africa. This shift in the epidemiology of blastomycosis is likely multifactorial, with potential contributing factors including changes in climate, soil conditions, and human activities that may have influenced the ecology and habitat of the fungus. The primary reservoir for *Blastomyces dermatitidis/gilchristii* is the environment, specifically soil with decaying organic matter, especially near waterways. *Blastomyces* can infect various mammals, such as dogs, cats, and other wildlife, but these animals are not the primary sources of the fungus. They are considered incidental hosts rather than reservoir hosts. Humans and other animals become infected by inhaling fungal spores from the environment, not by contact with infected animals. Blastomycosis can affect both healthy, immunocompetent individuals and those with compromised immune systems, such as people living with HIV/AIDS, organ transplant recipients, or those undergoing treatment with immunosuppressive medications. Outbreaks are frequently associated with activities that disturb soil, such as construction or excavation projects. Additionally, living in close proximity to waterways and moist, organic-rich soil with acidic pH levels can increase the risk of infection. Blastomycosis is not contagious and does not spread directly from person to person. Instead, infection occurs through the inhalation of fungal spores present in the environment.

Pathogenesis

Blastomycosis pathogenesis commences with the inhalation of *Blastomyces dermatitidis/gilchristii* conidia from the environment. The inhaled conidia then undergo a crucial morphological transition, shifting from the mycelial form to the pathogenic yeast form. This morphological transition is induced by the elevated temperature and increased carbon dioxide concentration within the host's body. The yeast cells then multiply and disseminate within the lungs, leading to the development of pneumonia. A robust cell-mediated immune response is crucial in determining the course of the infection, as it can effectively contain the infection and lead to asymptomatic or mild pulmonary disease. In individuals with compromised immunity, the yeast cells can evade phagocytosis, proliferate, and spread to other bodily sites beyond the lungs, leading to severe disseminated disease that can be life-threatening. Dissemination can occur in various sites, including the skin, bones, genitourinary tract, and central nervous system. The yeast cells can cause granulomatous inflammation and tissue damage in these organs, leading to a range of clinical manifestations. The fungus's ability to transition between mold and yeast forms, a process known as dimorphism, is crucial for its pathogenicity. Specific components in the yeast cell wall, such as BAD-1, help the fungus avoid the immune system and establish infection.

Clinical Manifestation in Humans

The clinical manifestations of blastomycosis in humans vary greatly, ranging from asymptomatic infection to severe, disseminated disease. The lungs are typically the most affected organ, although the infection can also spread to other parts of the body.

Pulmonary Blastomycosis

- Acute Pulmonary Blastomycosis: Symptoms often include flu-like illness with fever, cough, chills, chest pain, and fatigue. Some cases may improve without treatment.
- Chronic Pulmonary Blastomycosis: This can resemble tuberculosis or other chronic lung diseases, with persistent cough, weight loss, night sweats, and shortness of breath.

Extrapulmonary Blastomycosis

Blastomycosis can spread from the lungs to other parts of the body, leading to various symptoms based on the site of infection.

- Skin: Skin lesions like ulcers, lumps, warts, or abscesses are common.
- Bones: Bone involvement can cause pain, swelling, and tenderness.
- Genitourinary Tract: Infection of the prostate, epididymis, or testes can lead to pain and swelling in the affected area.
- Central Nervous System: CNS involvement is serious and can cause meningitis, brain inflammation, or brain abscesses. Symptoms may include headache, fever, stiff neck, seizures, and mental status changes.

Additionally, blastomycosis may less frequently involve other organs, such as the liver, spleen, and adrenal glands.

Pets like dogs and cats can get blastomycosis. The symptoms in animals vary depending on which organs are affected, but respiratory issues are most common. Here are some common symptoms in animals with this infection.

Dogs with blastomycosis may have:

- Breathing problems like coughing, trouble breathing, runny nose, fever, and tiredness.
- Eye issues such as inflammation, blindness, and the eye bulging out.
- Skin lesions that look like bumps, sores, or draining areas. The infection can spread to the bones, causing lameness.
- Other signs like weight loss, poor appetite, and swollen lymph nodes.

Cats with blastomycosis may have:

- Breathing problems like coughing, trouble breathing, and fever
- Skin issues such as lesions and draining areas
- Eye problems
- General symptoms including weight loss, poor appetite, and tiredness

Diagnosis

- The diagnosis of blastomycosis can be challenging given its wide-ranging clinical manifestations and the difficulties associated with isolating the causative fungus.
- Microscopic examination of respiratory secretions, pus, or tissue samples can reveal the characteristic broad-based budding yeast cells of *Blastomyces*. This is a rapid diagnostic test, but sensitivity can vary.
- Fungal culture is the gold standard for diagnosis. Samples are inoculated onto specific media and incubated at 25–30°C, but growth can take several weeks, delaying diagnosis.
- Histopathology can identify the yeast forms in infected tissues by examining stained tissue biopsies.
- Serologic tests like enzyme immunoassays and immunodiffusion can detect antibodies to *Blastomyces* antigens in the patient's serum. However, these tests may be less reliable in individuals with weakened immune systems, as they may have a diminished antibody response to the fungal infection.
- Molecular tests, such as PCR, can rapidly and accurately detect *Blastomyces* DNA.

Treatment

Itraconazole is the preferred treatment for mild to moderate blastomycosis cases, typically administered orally for 6–12 months. In cases of severe, disseminated, or central nervous system involvement, amphotericin B is often utilized, followed by prolonged itraconazole therapy.

Prevention and Control

There are no preventive measures or vaccines for blastomycosis. The fungus lives in the environment, making it hard to avoid exposure, especially in areas where it is common. Be aware of regions where blastomycosis is common. If you live in or visit these areas, know there is a risk. Wear a mask when working in areas where blastomycosis is common, like when gardening, digging, or doing construction, to avoid breathing in dust and soil. If your dog develops respiratory or skin issues and you live in an area with blastomycosis, seek prompt veterinary care.

Cryptococcosis

Cryptococcosis, a fungal infection of immense zoonotic importance, has long been the subject of extensive research and clinical scrutiny, largely due to its remarkable ability to cause life-threatening infections, which disproportionately affect immunocompromised individuals. The genus *Cryptococcus* contains 37 recognized species, although the majority are not considered pathogenic. However, *Cryptococcus neoformans* and *Cryptococcus gattii* are significant pathogens that can infect both humans and animals. In 1894, *Cryptococcus* spp. were first identified by Sanfelice from fermenting peach juice and in the same year by Busse and Buschke from a patient's tibial lesion. Early cases often linked with cancer patients. The first case of cryptococcal meningitis was reported in 1905. Over the last few decades, the incidence of invasive fungal infections, such as cryptococcosis, has increased substantially, which can be largely attributed to the rise of the HIV/AIDS epidemic. *Cryptococcus neoformans* is often associated with soil enriched with pigeon droppings, and *C. gattii* has been linked to eucalyptus trees, particularly decaying hollows in living trees. While these are considered the primary reservoirs for *C. neoformans* and *C. gattii*, both species have been found in a range of other environments. For instance, *C. neoformans* has been isolated from various types of bird droppings and even some decaying wood, while *C. gattii* has been discovered in tree species beyond just eucalyptus. While *Cryptococcus laurentii* and *Cryptococcus albidus* are less common causes of disease, they can still rarely infect immunocompromised patients.

While the widespread adoption of antiretroviral therapy has undoubtedly reduced the global burden of cryptococcosis, this fungal disease continues to be one of the leading causes of mortality among individuals living with HIV/AIDS, particularly in resource-limited regions. The primary route of *Cryptococcus* infection is through the inhalation of desiccated yeast cells or spores, which can then disseminate to the central nervous system, leading to life-threatening meningitis or meningoencephalitis.

Epidemiology

Cryptococcus neoformans is a global cause of cryptococcal meningitis in humans and is primarily associated with desiccated bird droppings in sheltered environments. *Cryptococcus neoformans* and *C. gattii* develop systemic infections in both immunocompromised and immunocompetent individuals over one million people globally each year. Cryptococcosis is a major cause of mortality among individuals living with HIV/AIDS, claiming an estimated 200,000 lives each year. In Western countries, the incidence of cryptococcosis has risen to around 5%, while in sub-Saharan Africa, the rate can reach as high as 30% among individuals with malignancies or those receiving corticosteroids or immunosuppressive therapy. *Cryptococcus neoformans* and *C. gattii* have been isolated from a variety of animal hosts, including cats, dogs, magpies, bandicoot rats, horses, koalas, ferrets, sheep, goats, kiwis, cows, African grey parrots, cockatoos, other parrot species, spinner dolphins, and echidnas. The pathogenesis of *Cryptococcus* is influenced by three key factors: the status of the host's immune defenses, the virulence of the infecting strain, and the size of the infectious inoculum. Evidence suggests that humans and animals typically acquire the infection from the environment where the fungus proliferates abundantly. Notably, *C. neoformans* has been re-isolated after 14 years from the same site, recovered from the dry and aged pigeon excreta. Pigeon droppings can harbor an estimated 50 million viable *Cryptococcus neoformans* cells per gram of dry material. These natural habitats can serve as a source of infection for both humans and animals. Visiting such contaminated sites may pose a significant health risk. Individuals such as zoo attendants, pet bird keepers, bird enthusiasts, pigeon breeders, and those tasked with cleaning historical buildings and old monuments are more likely to be exposed to cryptococcal infection.

Cryptococcosis can naturally occur in both animals and humans, but there is no evidence of transmission between mammals. The zoonotic potential is primarily through environmental exposure rather than direct animal-to-human transmission, but two possible instances of person-to-person transmission have been documented.

Clinical Manifestation in Man

The clinical presentation of *Cryptococcus* infection in humans varies depending on the species, site of infection, and the individual's immune status. Here is an overview.

Cryptococcus neoformans

Pulmonary Cryptococcosis: Immunocompetent individuals are often asymptomatic or present with mild symptoms such as cough, shortness of breath, and chest pain. However, in those with weakened immune systems, the infection can progress to severe pneumonia.

Cryptococcal Meningitis: Cryptococcal meningitis is the most common and serious manifestation, particularly among individuals with weakened immunity. Symptoms often include headache, fever, stiff neck, nausea, vomiting, confusion, and seizures.

Disseminated Cryptococcosis: The infection can spread to other organs, including the skin, bones, and eyes. The symptoms will depend on the specific organ affected.

Cryptococcus gattii

Pulmonary Cryptococcosis: Similar to *C. neoformans* but can cause more severe lung disease even in immunocompetent individuals.

Cryptococcal Meningitis: Can occur in both immunocompromised and immunocompetent individuals.

Brain Abscess: A localized brain infection that can cause focal neurological deficits.

Early diagnosis and prompt treatment are crucial to prevent serious complications of cryptococcosis, especially meningitis.

Clinical Presentation in Animals

Cats: Respiratory (sneezing, nasal discharge, coughing, difficulty breathing), central nervous system (seizures, ataxia, head tilt, blindness), and skin/subcutaneous tissues (nodules, ulcers, draining tracts).

Dogs: Respiratory and central nervous system symptoms similar to cats. Disseminated infection can spread to multiple organs, causing various symptoms.

Cattle: Mastitis (inflammation of the udder) leading to decreased milk production, anorexia, and enlarged lymph nodes.

Other Animals: A range of symptoms depending on the species and site of infection.Some animals may be asymptomatic carriers of *Cryptococcus*. If *Cryptococcus* infection is suspected, consult a veterinarian for diagnosis and treatment.

Diagnosis

Diagnosing cryptococcosis involves a combination of approaches. The common diagnostic tools include:

- Direct Microscopy: India ink staining of cerebrospinal fluid can reveal encapsulated yeast cells, suggesting *Cryptococcus* infection. However, this method has limited sensitivity.
- Culture: Growing the fungus from clinical samples on specialized media is the gold standard for diagnosis, although it can take several days to obtain results.
- Cryptococcal Antigen Test: Detecting CrAg in serum or CSF is a rapid and sensitive method for diagnosing cryptococcal meningitis. Lateral flow assays offer point-of-care testing capabilities.
- Histopathology: Examining tissue biopsies for the presence of *Cryptococcus* can be helpful, particularly in disseminated disease.
- Molecular Methods: PCR and other molecular techniques can identify *Cryptococcus* DNA in clinical samples, providing increased sensitivity and specificity.
- Imaging: Chest X-rays, CT scans, and MRI can reveal abnormalities suggestive of cryptococcal infection in the lungs, brain, or other organs.
- (1→3)-β-D-Glucan Test: Detects BG, a fungal cell wall component, in serum. While not specific to *Cryptococcus*, a positive result can indicate a fungal infection.

- The choice of diagnostic test depends on the suspected site of infection, the patient's immune status, and available resources.

Prevention and Control

Preventing and controlling cryptococcosis require a comprehensive approach targeting individual and environmental factors:

Individual-level prevention:

- Effective HIV/AIDS management, including antiretroviral therapy and pre-exposure prophylaxis, to reduce risk of cryptococcal meningitis
- Managing other immunocompromising conditions to prevent opportunistic infections
- Antifungal prophylaxis for high-risk individuals

Environmental-level prevention:

- Avoiding areas with high concentrations of bird droppings to reduce exposure to *C. neoformans*
- Proper waste management practices
- Avoiding contact with eucalyptus trees and decaying wood in *C. gattii*-endemic areas
- Public health surveillance to identify outbreaks and implement targeted control measures

General hygiene and safety measures:

- Regular handwashing
- Proper food handling and storage practices

Although complete eradication of *Cryptococcus* may be challenging, taking appropriate precautions and managing underlying health conditions can substantially lower the risk.

Developing an effective vaccine for cryptococcosis is challenging. The fungal cell wall is complex, and the pathogen varies. Additionally, a vaccine must protect people with weakened immune systems. Researchers are working to improve vaccine formulations, find powerful adjuvants, and rigorously test promising vaccine candidates.

Treatment

Cryptococcosis treatment depends on the infection site, illness severity, and patient's immune status. For cryptococcal meningitis, induction therapy typically involves amphotericin B and flucytosine, followed by consolidation and maintenance therapy with fluconazole. For pulmonary cryptococcosis, fluconazole is often the first-line treatment, with amphotericin B used for severe cases. Treatment for other forms of cryptococcosis varies based on the affected organ and severity. Monitoring, supportive care, management of elevated intracranial pressure, and careful handling of immune reconstitution inflammatory syndrome are important considerations.

Coccidioidomycosis

Introduction

Coccidioidomycosis, commonly referred to as "valley fever," is a systemic fungal infection caused by the two dimorphic fungi, *Coccidioides immitis* and *Coccidioides posadasii*. Coccidioidomycosis is endemic to dry, warm regions, especially in the southwestern United States, parts of Mexico, and Central and South America. Coccidioidomycosis has a significant public health impact, yet it often goes unrecognized and unreported, posing a silent yet formidable threat.

Epidemiology

Coccidioidomycosis is found in hot, dry regions where the fungi thrive in soil. Activities that disturb the soil, like construction, farming, or windstorms, can release fungal spores into the air, leading to inhalation by people and animals. In recent years, the number of coccidioidomycosis cases has risen significantly, with over

20,000 reported annually in the United States. Factors contributing to this increase include climate change, population growth in endemic areas, and increased soil disruption. Certain groups, such as immunocompromised individuals, pregnant women, and those of African, Filipino, or Native American descent, are at higher risk of developing severe disease.

Pathogenesis

The lifecycle of *Coccidioides* involves two distinct phases: the saprophytic phase in the soil and the parasitic phase in the host. In the environment, the fungus exists as mycelia that produce arthroconidia, which are easily dispersed into the air. When inhaled, these spores transform into spherules in the lungs, which then release endospores, spreading the infection locally or systemically. The host immune response, particularly cell-mediated immunity, plays a critical role in controlling the infection. However, in immunocompromised individuals, the fungus can disseminate to other organs, such as the skin, bones, and central nervous system, leading to severe complications.

Clinical Manifestations

Around 60% of infected people have no symptoms, while the remaining 40% develop mild to severe symptoms. The primary form, acute pulmonary coccidioidomycosis, commonly presents with flu-like symptoms like fever, cough, chest pain, and fatigue. Some patients may develop a rash or erythema nodosum, indicating a strong immune response. In rare cases, the infection can progress to chronic lung disease or spread to other organs, causing meningitis, bone infections, or soft tissue abscesses. Disseminated coccidioidomycosis is linked to high rates of complications and death, especially in those with weakened immune systems.

Diagnosis

Accurate diagnosis of coccidioidomycosis relies on a combination of clinical, radiological, and laboratory findings. Serological tests, such as enzyme immunoassays (EIA) and immunodiffusion, are commonly used to detect antibodies against *Coccidioides*. Culture and histopathological examination of clinical specimens can confirm the diagnosis but require specialized laboratory facilities due to the biohazard risk. Molecular techniques, such as polymerase chain reaction (PCR), are emerging as valuable tools for rapid and specific identification. Imaging studies, including chest X-rays and CT scans, are useful for assessing pulmonary involvement and complications.

Treatment

The management of coccidioidomycosis depends on the severity of the infection. Mild cases often resolve without antifungal therapy. Patients with moderate to severe disease require treatment with antifungal agents like fluconazole or itraconazole. For disseminated or central nervous system infections, amphotericin B is the preferred drug, followed by long-term azole therapy. Treatment duration varies from months to years, depending on the patient's immune status and response. Despite advancements, relapses and treatment failures remain significant challenges.

Prevention and Public Health Implications

Preventing coccidioidomycosis is challenging due to the ubiquitous nature of the fungus in endemic regions. Public health efforts focus on raising awareness among healthcare providers and at-risk populations, particularly those in endemic areas. Individuals are advised to avoid activities that generate dust, such as digging or construction work, and to use protective measures like masks when exposure is unavoidable. Research into vaccines and novel antifungal therapies is ongoing, with promising candidates in preclinical and early clinical trials. Additionally, improved surveillance and reporting systems are essential for better understanding the disease burden and guiding public health interventions.

Conclusion

Coccidioidomycosis is a significant yet often overlooked fungal infection, with the potential to cause severe disease, particularly in vulnerable populations. Its rising incidence, along with the challenges of diagnosis and treatment, underscores the need for increased awareness, research, and public health initiatives. By advancing our understanding of the epidemiology, pathogenesis, and management of coccidioidomycosis, we can mitigate its impact and improve outcomes for affected individuals. As climate change and human activities continue to alter the landscape of infectious diseases, coccidioidomycosis serves as a reminder of the intricate interplay between the environment, pathogens, and human health. Addressing this silent threat requires a multidisciplinary approach, combining scientific innovation, public health strategies, and community engagement.

Bibliography

Barros, N., Wheat, J., & Hage, C. A. (2023). Pulmonary histoplasmosis: A clinical update [Review of Pulmonary histoplasmosis: A clinical update]. *Journal of Fungi, 9*(2), 236. Multidisciplinary Digital Publishing Institute. https://doi.org/10.3390/jof9020236

Coccidioides immitis presenting as a hyphal form in cerebrospinal fluid. (n.d.). Retrieved January 2025, from https://pubmed.ncbi.nlm.nih.gov/9685779

Coccidioidomycosis: A Reemerging Infectious Disease. (n.d.). https://doi.org/10.3201/eid0203.960305.

Dermatologic Manifestations of Coccidioidomycosis: Overview, Patient History, Physical Examination. (2024). https://emedicine.medscape.com/article/1092017-overview

Galgiani, J. N., Ampel, N. M., Catanzaro, A., Johnson, R. H., Stevens, D. A., & Williams, P. (2000). Practice guidelines for the treatment of coccidioidomycosis. *Clinical Infectious Diseases, 30*(4), 658). Oxford University Press. https://doi.org/10.1086/313747

Hospitalizations Associated with Disseminated Coccidioidomycosis, Arizona and California, USA. (n.d.). https://doi.org/10.3201/eid1809.120151

Kauffman, C. A. (2007). Histoplasmosis: A clinical and laboratory update [Review of Histoplasmosis: A clinical and laboratory update]. *Clinical Microbiology Reviews, 20*(1), 115. American Society for Microbiology. https://doi.org/10.1128/cmr.00027-06

Lagrou, K., Van Eldere, J., Keuleers, S., Hagen, F., Merckx, R., Verhaegen, J., Peetermans, W. E., & Boekhout, T. (2005). Zoonotic transmission of Cryptococcus neoformans from a magpie to an immunocompetent patient. *Journal of Internal Medicine, 257*, 385.

Linder, K. A., Kauffman, C. A., & Miceli, M. H. (2023). Blastomycosis: A review of mycological and clinical aspects [Review of Blastomycosis: A review of mycological and clinical aspects]. *Journal of Fungi, 9*(1), 117. Multidisciplinary Digital Publishing Institute. https://doi.org/10.3390/jof9010117

McBride, J. A., Gauthier, G. M., & Klein, B. S. (2017). Clinical manifestations and treatment of blastomycosis [Review of Clinical manifestations and treatment of blastomycosis]. *Clinics in Chest Medicine, 38*(3), 435. Elsevier BV. https://doi.org/10.1016/j.ccm.2017.04.006

Nielsen, K., Cox, G. M., Litvintseva, A. P., Mylonakis, E., Malliaris, S. D., Benjamin, D. K., Giles, S. S., Mitchell, T. G., Casadevall, A., Perfect, J. R., & Heitman, J. (2005). Cryptococcus neoformans α strains preferentially disseminate to the central nervous system during coinfection. *Infection and Immunity, 73*(8), 4922–4923.

Pakasa, N. M., Biber, A., Nsiangana, S., Imposo, D., Sumaili, E. K., Mavoko, H. M., Buitrago, M. J., Barshack, I., & Schwartz, E. A. (2018). African histoplasmosis in HIV-negative patients, Kimpese, Democratic Republic of the Congo. *Emerging Infectious Diseases, 24*(11), 2068). Centers for Disease Control and Prevention. https://doi.org/10.3201/eid2411.180236

Saccente, M., & Woods, G. L. (2010). Clinical and laboratory update on blastomycosis [Review of Clinical and laboratory update on blastomycosis]. *Clinical Microbiology Reviews, 23*(2), 367. American Society for Microbiology. https://doi.org/10.1128/cmr.00056-09

Saubolle, M. A., McKellar, P., & Sussland, D. (2006). Epidemiologic, clinical, and diagnostic aspects of coccidioidomycosis [Review of Epidemiologic, clinical, and diagnostic aspects of coccidioidomycosis]. *Journal of Clinical Microbiology, 45*(1), 26. American Society for Microbiology. https://doi.org/10.1128/jcm.02230-06

Singh, R., & Kumari, A. (2021). Nasal microbiota imbalance as a contributory link in the emergence of COVID-19 associated mucormycosis. *ACS Infectious Diseases, 7*(8), 2211). American Chemical Society. https://doi.org/10.1021/acsinfecdis.1c00371

Singh, K., İlkit, M., Shokohi, T., Tolooe, A., Malík, R., & Seyedmousavi, S. (2018). Cryptococcosis: Emergence of cryptococcus gattii in animals and zoonotic potential. In *Emerging and epizootic fungal infections in animals* (p. 249). Springer eBooks, Springer Nature. https://doi.org/10.1007/978-3-319-72093-7_12

Valdez, A. F., Miranda, D. Z., Guimarães, A. J., Nimrichter, L., & Nosanchuk, J. D. (2022). Pathogenicity & virulence of Histoplasma capsulatum—A multifaceted organism adapted to intracellular environments. *Virulence, 13*(1), 1900–1919.

Yıldız, K., Dokuzeylul, B., Ulgen, S., & Or, M. E. (2016). Blastomycosis: A systematical review. *Research & Reviews: Journal of Veterinary Sciences., 2*, 84–86.

Part VI

Rickettsial Zoonoses

Rickettsial Zoonoses 42

Maninder Singh and Jay Prakash Yadav

Abstract

Bacteria from the order Rickettsiales cause a diverse group of illnesses known as rickettsial infections. These pathogens are transmitted to humans through bites of infected arthropods, such as ticks, fleas, and lice. The symptoms, geographical distribution, and associated arthropod vectors of these diseases vary significantly. Notable examples of rickettsial zoonoses include Rocky Mountain spotted fever, scrub typhus, epidemic typhus, and murine typhus. Typical symptoms of rickettsial infections include fever, headache, rash, and muscle pain. To prevent severe complications, it is crucial to administer appropriate antibiotics promptly. Strategies to reduce the risk of rickettsial infections focus on limiting exposure to vectors. These methods include wearing protective clothing, using insect repellents, and implementing measures to control vector populations in infected areas.

Keywords

Rickettsial diseases · Spotted fever · Scrub typhus · Typhus group · Louse-borne typhus · Sylvatic typhus · Classic typhus · Historical typhus · Jail fever · Red louse disease · Typhus exanthematicus · Tabardillo · Fleckfieber · Murine typhus · Mexican typhus · Flea borne typhus · Shop typhus · Tick-borne typhus fever · Tick fever · Mexican spotted fever · Black fever · Black measles · New World spotted fever · Fiebre maculosa Brasileira/Brazilian spotted fever · Fiebre de Tobia · Bush typhus · Tsutsugamushi disease · River fever · Flood fever

M. Singh · J. P. Yadav (✉)
Department of Veterinary Public Health and Epidemiology, College of Veterinary Science, Guru Angad Dev Veterinary and Animal Sciences University, Rampura Phul, Bathinda, India

R. Biswas et al. (eds.), *Concept of One Health and Zoonoses*,
https://doi.org/10.1007/978-981-95-6627-3_42

Introduction

The Order Rickettsiales comprises gram-negative bacteria that reside exclusively within eukaryotic cells. These microorganisms, initially described by Ricketts in 1909, can colonize a broad spectrum of hosts, including humans, animals, insects, and protists. During its evolution into an intracellular lifestyle, *Rickettsia* has undergone significant genome reduction and the loss of metabolic capabilities. This reductive evolutionary process resulted in the loss of approximately 2135 genes as Rickettsiaceae transitioned from free-living to intracellular organisms. This gene loss makes them dependent on host cells for survival and replication. Additionally, *Rickettsia* species do not reside in host-derived vacuoles; instead, they live directly in the host cell cytoplasm.

The known Rickettsiales families have each developed distinct ecological niches to engage with their respective host-cell environments.

- Family Rickettsiaceae (includes *Rickettsia* and *Orientia*)
- Family Anaplasmataceae (includes *Anaplasma*, *Ehrlichia*, *Neorickettsia*, and *Wolbachia*)
- Family *Candidatus* Midichloriaceae (includes *Candidatus* Midichloria *mitochondrii* and *Candidatus* Lariskella *arthropodarum*)
- Family Deianiraeaceae (includes *Deianiraea*, unlike most Rickettsiales, which are obligate intracellular parasites, the Deianiraeaceae members display a unique extracellular lifestyle)

Family Rickettsiaceae

The genus *Rickettsia* is conventionally divided into two primary categories: spotted fever group (SFG) and typhus group (TG). The majority of known species fall within the SFG, including *Rickettsia rickettsii*, *Rickettsia conorii*, *Rickettsia akari*, *Rickettsia felis*, *Rickettsia africae*, *Rickettsia honei*, and *Rickettsia parkeri*. Historically, classification into these groups is based on physiological traits, such as intracellular location, optimal growth temperature, and cross-reactivity of infected patient serum with somatic antigens of three *Proteus* strains. TG consists of two species, *Rickettsia typhi* and *Rickettsia prowazekii*. *R. typhi* is responsible for murine typhus, a flea-transmitted illness prevalent in warm climates. In contrast, *R. prowazekii* causes epidemic typhus, which occurs during cold months, when heavy clothing and poor hygiene promote lice proliferation.

The genus *Orientia* has the *Orientia tsutsugamushi* species, which is a reclassified species, earlier designated as *Rickettsia tsutsugamushi*, as it is genetically distinct and remains closely related to *Rickettsia* within the same α-1 subgroup. Recently, *Orientia chuto* has been included in this genus (Table 42.1).

Table 42.1 Differences between typhus group and spotted fever group of *Rickettsia*

Characteristics	Typhus group	Spotted fever group
Vectors	Human body lice (*R. prowazekii*), fleas (*R. typhi*)	Mainly ticks but may also include fleas (*Rickettsia felis*) and mites (*R. akari*)
Optimal growth temperature	35°C	32°C
Outer membrane protein	Type B	Types A and B
G-C content (%)	29	32–33
Polymerization of actin	Cannot polymerize actin and found in the cytoplasm of host cells	Can polymerize actin and move into the nuclei of host cells
Transmission	No transovarian transmission in vectors. Mammals are reservoirs (humans for *R. prowazekii*, rodents for *R. typhi*, and cats for *R. felis*)	Transovarian and transstadial transmission in vectors, thus acting as reservoirs also

Table 42.2 Epidemiological features of typhus group Rickettsiae

Disease	Species	Vector	Reservoir(s)	Geographical distribution
Epidemic typhus/ sylvatic typhus/ Brill-Zinsser disease	*R. prowazekii*	Human body louse (*Pediculus humanus corporis*), *Amblyomma* ticks, fleas	Humans, flying squirrels	Worldwide
Endemic typhus/ murine typhus	*R. typhi*	Flea	Rodents, opossums	Worldwide

Typhus Group

"Typhus," a term originating from the Greek word "tuphos" (smoke), has a long history in medical literature. The disease's earliest identification may date back to 1083 in Spain. In 1536, Girolamo Cardano provided the initial description of typhus. A decade later, Fracastoro offered a comprehensive account, clearly differentiating typhus from plague. The transmission of typhus from military to civilian populations was documented in 1676 by Zavorziz in his work titled 'The infection of military camps' (Table 42.2).

Epidemic Typhus

Etiology

Epidemic typhus is caused by *Rickettsia prowazekii*, which is a coccobacillary bacterium that resides within cells and exhibits variable morphology. This microorganism reproduces via binary fission. It is categorized as a class B bioterrorism threat as it can survive in desiccated louse feces and has the potential for aerosolized dissemination.

History

Historically, epidemic typhus has been associated with significant loss of life following conflicts, food shortages, and population movements. The disease's clinical and epidemiological characteristics were initially documented in the 1500s. For two centuries, epidemic typhus was often mistaken for typhoid fever. In 1739, Huxham made the first distinction between the two illnesses. Nearly a century later, in 1836, Gerhard further differentiated the diseases based on the presence of a rash and ulceration of Peyer's patches observed in autopsies. In 1896, Nathan Brill identified sporadic cases of a typhoid-like illness with negative blood cultures, which was subsequently confirmed as epidemic typhus and termed Brill-Zinsser disease, representing the recurrent form of epidemic typhus.

Epidemiology

The reemergence of classic epidemic typhus coincided with World War I. Eastern Europe and Russia witnessed a staggering 25 million suspected cases and 3 million deaths between 1917 and 1925. As World War II unfolded, the illness spread to northern Africa, southern Italy, and central and Eastern Europe. In the aftermath of World War II, the incidence of epidemic typhus began to wane, leading to its categorization as a historical disease. However, *R. prowazekii* infections can resurface in infected individuals years or decades after initial exposure, especially when the immune system is compromised. Epidemic typhus remains a significant global concern among refugee communities. The disease continues to emerge in economically disadvantaged nations, with outbreaks particularly prevalent in regions grappling with severe poverty issues.

Transmission

Body lice, specifically *Pediculus humanus corporis*, are responsible for transmitting this disease between individuals. These parasites become carriers after feeding on an infected person, subsequently releasing large quantities of organisms in their excrement within a week of infection. The transmission process necessitates close physical proximity, as the irritation caused by louse defecation prompts scratching, which can introduce contaminated feces into bite wounds. Alternative infection routes include direct contact of contaminated feces with mucous membranes or inhalation of airborne dried louse feces in crowded settings. Infected lice perish within 1–3 weeks due to digestive tract obstruction and do not pass the organism to their progeny. However, there is possibility of transstadial transmission between lice to lice through ingestion of infected feces. In the United States, the flying squirrel (*Glaucomys volans*) serves as an additional potential reservoir for *R. prowazekii*, with transmission among these animals likely occurring via squirrel lice, fleas, or a combination of both. Prolonged human infection with *R. prowazekii* can result in a subsequent reactivation known as Brill-Zinsser disease.

Clinical Manifestations

Epidemic typhus typically has an incubation period of 10–14 days. The disease's initial symptoms include nausea/vomiting, elevated body temperature, headache, and muscle pain, followed by a rash that appears on the body and extremities. This rash emerges between 2 and 4 days of fever, starting in the axillary folds and upper trunk before spreading outward to the limbs without forming an eschar. Most individuals with epidemic typhus exhibit one or more central nervous system abnormalities such as meningeal irritation and localized or widespread cortical dysfunction. In severe cases, patients may experience symptoms such as myocarditis, pneumonia, thrombocytopenia, jaundice, and abnormal liver function tests. Those who survive epidemic typhus remain lifelong carriers of the pathogen and may experience a milder recurrence known as Brill-Zinsser disease when under stress or immunosuppression. Although the symptoms of this recurrence are generally less severe, these patients can potentially trigger a new epidemic if they become infested with body lice.

Treatment

Early administration of appropriate antibiotics is highly effective in treating the disease. The preferred medication for rickettsial infections is doxycycline, administered orally at a dose of 200 mg/day. For epidemic typhus, the recommended treatment protocol involves either a single 200 mg oral dose of doxycycline, which may be repeated once if necessary, or a brief course of 100–200 mg daily for 3 days.

Prevention and Control

Preventing louse infestations is widely regarded as the most efficient control strategy. Maintaining proper personal hygiene plays a crucial role in averting lice infestation. When dealing with an existing infestation, destroying lice through patient bathing, laundering, or boiling infested garments is recommended. Considering that body lice solely depend on blood for survival and die within 5 days without access to a blood meal, the most straightforward delousing approach involves removing and refraining from wearing infested clothing for a week. An additional effective method for eradicating lice involves the application of a potent insecticide spray such as 10% DDT (dichloro-diphenyl-trichloroethane), 1% malathion, or 1% permethrin.

Endemic Typhus

Etiology

Rickettsia typhi (*Rickettsia mooseri*) is an etiological agent of murine typhus. It is an obligate intracellular bacterium that infects endothelial cells in the mammalian host and the midgut epithelial cells of flea hosts.

History

In 1928, *R. typhi* was identified in Mexico by a team of researchers: Dr. Hermann Mooser, Dr. Maximiliano Ruiz Castañeda, and Dr. Hans Zinsser. They initially referred to it as "Mexican typhus" and named it as *R. mooseri* to honor Hermann Mooser. The first recorded instance of flea-borne typhus (FBT) occurred in 1913 at Grady Hospital in Atlanta. Kenneth F. Maxcy of the US Public Health Service proposed a rodent to be reservoir for FBT and provided the first comprehensive clinical description of "endemic typhus."

Epidemiology

The disease has a global distribution, with particular prevalence in moderate and subtropical regions, including the United States, Australia, Southeast Asia, Mexico, North America, Spain, Africa, and Mediterranean areas, where the disease reservoir (*Rattus* spp.) and vector (*Xenopsylla cheopis* flea) are present. Various wild mammals, such as rodents (*Rattus rattus* and *Rattus norvegicus*) and opossums (*Dydelphis* spp.), along with domestic dogs, serve as reservoirs for *R. typhi*. Reports indicate that travelers and refugees may also transport the infection from endemic regions to disease-free areas. The occurrence of the illness is higher in warmer weather than in colder climates. Individuals engaged in animal husbandry, including veterinarians, animal handlers, and owners, face elevated risks in areas with high flea infestation. Humans may also contract infections from rodents and fleas residing in grain storehouses, waterlogged areas, and food shops.

Transmission

The infectious agent infiltrates the body via two primary routes: through flea fecal matter deposited on abrasions or injuries or through inhalation of desiccated infectious particles that come into contact with the conjunctiva. In vector species, the infection predominantly affects the midgut epithelial cells, perpetuating through horizontal transmission in the flea-rat-flea cycle. In the absence of preferred hosts (typically rats, with occasional instances involving opossums and felines), fleas may turn to humans as alternative blood sources. Once infected, fleas remain carriers for their entire lifespans. Additionally, a minimal level of vertical (transovarian) transmission has been documented.

Clinical Manifestation

The incubation period ranges from 7 to 14 days. The primary symptoms include pyrexia, muscle and joint pain, cephalalgia, and skin eruptions. Maculopapular rash, typically appearing on 5th day and lasting approximately 4 days, is observed in 60–70% of cases. Rashes are observed on the body's trunk and limbs but not on the hands and feet. With appropriate treatment, the symptoms subside within 10–14 days. Approximately 10% of individuals develop organ-specific complications such as liver inflammation, lung infection, meningoencephalitis, or kidney dysfunction. Severe symptoms affecting the respiratory, circulatory, urinary, and nervous systems occur in 2–4% of patients. The mortality rate of treated cases is low, ranging from 0% to 1%. Factors influencing prognosis include patient age, blood disorders, renal insufficiency, electrolyte imbalances (low potassium and sodium levels), decreased albumin in the blood, and delayed commencement of treatment.

Treatment

Doxycycline is the drug of choice for treatment. Prompt treatment is crucial to ensure quick recovery.

Prevention and Control

Reducing the occurrence of endemic typhus is achievable by limiting exposure to rats and their fleas and by maintaining proper hygiene and a clean environment. To keep pets, such as cats and dogs, free from fleas, it is essential to utilize flea control products, including collars, oral medications, or topical treatments. Currently, no effective vaccine exists for endemic typhus. Timely diagnosis and appropriate medication are vital to mitigate severe disease outcomes. In areas where endemic typhus is prevalent, implementing rodent control measures such as rat trapping and application of rodenticides is recommended.

Spotted Fever

Rocky Mountain Spotted Fever

Etiology

The causative agent of this illness is *Rickettsia rickettsii*, which can infect both humans and animals. Hard ticks serve as vectors and reservoirs transmitting pathogen through transovarian and transstadial mechanisms. It is considered as the most virulent disease among all human infectious diseases, mostly in young people residing in North and South America.

History

Initially reported as black measles in Idaho and the Rocky Mountain region during the 1890s, the illness later became known as Rocky Mountain spotted fever (RMSF). In 1906, Howard Ricketts identified RMSF as a bacterial infection transmitted to humans through hard ticks. The disease, originally limited to the Rocky Mountain area, eventually spread to various regions in the Americas, including Argentina, Brazil, Colombia, Costa Rica, Mexico, and Panama.

Epidemiology

RMSF has been identified in various regions across the Americas, including North, Central, and South America, and countries including Argentina, Brazil, Canada, Colombia, Costa Rica, Mexico, and Panama. In the United States, RMSF is the most prevalent tick-borne rickettsial illness, with reports indicating cyclical variations in its occurrence throughout the country. The highest rates of infection have been observed among children under 10 years old, adults between 40 and 64 years of age, and males. Living in areas with forests or tall grasses, as well as interacting with dogs, elevates the likelihood of contracting *R. rickettsii* infection.

Transmission

Ticks belonging to genera such as *Dermacentor*, *Rhipicephalus*, and *Amblyomma* function as both disease vectors and reservoirs. These arthropods maintain and transmit pathogens through transstadial and transovarial routes, perpetuating infections in nature. Ticks become infected by feeding on contaminated animals or humans, subsequently spreading the pathogens to their progeny. Human infection typically occurs when an infected tick attaches and feeds for an extended period, usually 1–2 weeks. During this time, pathogens are transferred from the tick's salivary glands into the human host. Alternative modes of human infection include exposure to infected tick hemolymph while removing ticks (particularly if crushed), blood transfusions, and accidental pathogen introduction through mucous membranes. Furthermore, laboratory workers face potential infection risks from aerosol exposure or parenteral inoculation of the pathogen.

Clinical Manifestation

Following a tick bite, the average incubation period is 7 days, with a range of 2–14 days. Initial symptoms resemble those of other tick-transmitted rickettsial illnesses (anorexia, abdominal pain, fever, headache, malaise, myalgia, nausea, vomiting, and photophobia). It may lead to wide range of body dysfunction that includes cutaneous (rashes), cardiac, gastrointestinal, neurological, ocular, pulmonary, renal, and skeletal muscle. In the first week, only 3% of patients exhibit the characteristic clinical triad of high fever (typically exceeding 38.9 °C), headache, and skin rash. This triad becomes apparent in 60–70% of cases within 2 weeks of infection. Skin eruptions typically manifest 2–5 days after fever onset, beginning as small (1–5 mm in diameter), pale red spots that blanch upon pressure, first appearing on the wrists and ankles before spreading outward to the palms and soles. Subsequently, the rash progresses inward from the wrists and ankles to the arms, legs, and trunk.

Treatment

Doxycycline is the primary drug for treating RMSF infections. Adult patients typically receive 100 mg antibiotic twice daily, administered either orally or through IV. For pediatric cases, a 4-mg dose is recommended, divided into two parts and given every 12 h via oral or IV routes. The duration of treatment spans 3–10 days based upon the intensity of infection. In instances where patients exhibit allergies to doxycycline, chloramphenicol is used in treatment.

Prevention and Control

The main preventive measures against tick bites involve minimizing exposure to tick-populated areas, particularly those with dense vegetation and grass. When avoiding these habitats is not feasible, individuals entering tick-infested environments should opt for light-colored, full-coverage clothing and appropriate footwear. Applying permethrin to garments as a tick repellent may also be beneficial. Any ticks found attached to humans or animals should be carefully extracted. To reduce illness and fatalities associated with the disease, accurate and swift diagnosis is crucial, and susceptible individuals should refrain from visiting endemic regions during peak tick activity (such as April to September in the United States). Currently, no effective vaccine exists for RMSF (Table 42.3).

Scrub Typhus

Etiology

Scrub typhus is a neglected and reemerging vector-borne rickettsial illness caused by obligate intracellular, gram-negative bacteria belonging to the genus *Orientia* (previously classified as *Rickettsia*). Two pathogenic species are known within the *Orientia* genus: *O. tsutsugamushi* and *O. chuto*. *Orientia tsutsugamushi* is the primary pathogenic species, transmitted to humans via bites from trombiculid mite larvae (chiggers). The term "tsutsugamushi" indicates that the disease is primarily endemic to a specific region known as the "tsutsugamushi triangle," which encompasses northern Australia and islands in the Indian and Pacific Oceans. Additionally, the word "scrub" signifies that the disease is endemic to particular vegetation types that sustain the chigger-rodent relationship.

Table 42.3 Other important spotted fever group organisms

Disease	Species	Vector	Reservoir(s)	Geographic distribution
Rickettsiosis	*R. aeschlimannii*	Tick	Unknown	South Africa, Morocco, Mediterranean littoral
African tick-bite fever	*R. africae*	Tick	Ruminants	Sub-Saharan Africa, West Indies
Rickettsial pox	*R. akari*	Mite	House mice, wild rodents	Countries of the former Soviet Union, South Africa, Korea, Turkey, North and South America
Queensland tick typhus	*R. australis*	Tick	Rodents	Australia, Tasmania
Mediterranean spotted fever/ boutonneuse fever	*R. conorii*	Tick	Dogs, rodents	Southern Europe, Southern and Western Asia, Africa, India
Cat flea rickettsiosis	*R. felis*	Flea	Domestic cats, rodents, opossums	Europe, North and South America, Africa, Asia
Mediterranean spotted fever-like disease	*R. massiliae*	Tick	Unknown	France, Greece, Spain, Portugal, Switzerland, Central Africa, Mali
Mediterranean spotted fever-like illness	*R. monacensis*	Tick	Lizards, possibly birds	Europe, North Africa
Maculatum infection	*R. parkeri*	Tick	Rodents	North and South America
Tick-borne lymphadenopathy/*Dermacentor*-borne necrosis	*R. raoultii*	Tick	Unknown	Europe, Asia
Flinders Island spotted fever/Thai tick typhus	*R. honei*	Tick	Rodents, reptiles	Australia, Thailand
Japanese spotted fever	*R. japonica*	Tick	Rodents	Japan
North Asian tick typhus/Siberian tick typhus	*R. sibirica*	Tick	Rodents	Russia, China, Mongolia
Tick-borne lymphadenopathy (TIBOLA)/*Dermacentor*-borne necrosis and lymphadenopathy (DEBONEL)	*R. slovaca*	Tick	Lagomorphs, rodents	Southern and Eastern Europe, Asia

History

Scrub typhus was previously thought to be endemic in the rural and hilly areas of "tsutsugamushi triangle" (northern Australia, Asia, and islands in the Indian and Pacific Oceans) in an area of ~13,000,000 km. But later, it spread to other areas such as Japan, Nepal, Northern New South Wales, Northern Pakistan, Papua New Guinea, South China, South Korea, and Queensland. The disease is mainly prevalent in scrubby areas, i.e., beaches, deserts, forests, and mountains, that support the growth and multiplication of trombiculid mites.

The disease was initially reported in the Japanese folklore, and it was thought to be associated with the jungle mite or chigger which was later named "dangerous bug" (tsutsugamushi). Initially, the disease was described by Hashimoto in 1810. Later, in 1916, Weil and Felix described heterophile antibody agglutination of *Proteus mirabilis* OX-K variant in sera of patients suffering from scrub typhus. In 1931, Ogata isolated the pathogen and named it *Rickettsia tsutsugamushi*. The organism has been reclassified as *Orientia tsutsugamushi*. Scrub typhus poses a threat to approximately 1 billion individuals, with an annual incidence of 1 million cases. If left untreated, the disease has a mortality rate of ~50%.

Transmission

Larval mites, commonly referred to as chiggers, function as both a reservoir and vector for the pathogen. This pathogen is sustained within the mite population through transovarian and transstadial transmission mechanisms. Although these mites predominantly feed on rodents, they occasionally target humans. In endemic areas, particularly wet fields, rodents are instrumental in sustaining chigger population density via the chigger-rodent cycle. Human infection occurs when infected larval trombiculid mites from the genera *Leptotrombidium* and *Schoengastiella* bite individuals. Additionally, the pathogen can be transmitted to fetuses either through vertical transplacental transmission during gestation or perinatal blood-borne infection during childbirth.

Clinical Manifestations

The onset of the disease is characterized by a range of symptoms including elevated body temperature, rigors, respiratory irritation, cranial discomfort, gastrointestinal distress, vomiting, diarrhea, eschar formation, cutaneous eruptions, muscular pain, and generalized lymph node enlargement. Potential severe complications encompass acute respiratory distress syndrome (ARDS), pulmonary inflammation, myocardial inflammation, cardiac insufficiency, hepatic and/or splenic enlargement, acute renal impairment, gastrointestinal hemorrhage, septic shock, meningoencephalitis, mental confusion, and auditory impairment. A hallmark sign of scrub typhus is the development of a necrotic eschar at the site of mite inoculation, which resembles to a cigarette burn. This lesion typically appears in regions where the skin is thin, moist, or creased or where clothing fits snugly. Notably, eschar formation is uncommon among Southeast Asian populations. In approximately 50% of patients, a maculopapular rash emerges around 5th to 6th day of infection, initially manifesting on the trunk before extending to the upper and lower extremities. The facial region, palmar surfaces, and plantar aspects of the feet are generally spared from rash development.

Treatment

Prompt identification and commencement of antibiotics are essential for survival of patients with scrub typhus. Doxycycline is the preferred antibiotic. For those weighing less than 45 kg, the suggested dosage is 45 mg daily and divided into two administrations. Patients over 45 kg should receive 200 mg/day, divided into two doses, for a week-long course. Chloramphenicol can be administered to patients who are allergic to doxycycline, at a dosage of 50–100 mg/kg/day, administered every 6 h, not exceeding 3 g daily. For expectant mothers, a single 500-mg dose of azithromycin is advised for five consecutive days.

Prevention and Control

Currently, no commercial vaccine exists for scrub typhus prevention. As a result, the primary strategy to avoid the disease, particularly for high-risk occupational groups and travelers, is to minimize contact with infected chiggers. Additional preventive measures include the application of insect repellents such as DEET (N,N-diethyl-meta-toluamide), benzyl benzoate, diethyltoluamide, or dimethyl phthalate. Dusting sulfur in residential areas, parks, and other vegetated zones can also help control the breeding of infected chiggers and rodent populations, thereby disrupting the chigger-rodent transmission cycle. For personal protection when entering high-risk areas, individuals can apply permethrin and diethyltoluamide to their clothing and wear light-colored attire, including long-sleeved shirts, long pants, socks, and boots, to prevent chigger bites.

Family Anaplasmataceae

The Anaplasmataceae family, belonging to the Rickettsiales order, comprises four genera: *Anaplasma*, *Ehrlichia*, *Neorickettsia*, and *Wolbachia*. These organisms are small, gram-negative pleomorphic cocci that function as obligate intracellular α-proteobacteria. These pathogens are transmitted by invertebrate vectors. Their transmission cycles persist through vector-wildlife or vector-domestic animal interactions. Studies have shown that four *Anaplasma* species and five *Ehrlichia* species can cause human illnesses, with *Anaplasma phagocytophilum*, *Ehrlichia chaffeensis*, *Ehrlichia ewingii*, and *Ehrlichia muris* being the most prevalent disease-causing species in the United States. These pathogens attach to the membrane receptors of various target cells such as macrophages, monocytes, granulocytes, endothelial cells, and erythrocytes. After phagocytosis, they replicate within membrane-bound vesicles in the cytoplasm of infected cells, forming microcolonies called morulae. The organisms spread from infected to healthy cells through concentration in filopodia or host cells membrane rupture in case of *Ehrlichia* and without the loss of host cell cytoplasm in case of *Anaplasma* (Tables 42.4, 42.5 and 42.6).

Diagnosis of Rickettsioses

To diagnose rickettsial diseases accurately, it is essential to gather a comprehensive patient history, including potential exposure to infection sources or visits to endemic areas. Laboratory testing should be performed in reference diagnostic facilities. When possible, vectors from the bite site or the surrounding area may be submitted for precise identification. The symptoms of rickettsial infections vary from mild and self-resolving to severe and potentially fatal. Common laboratory findings include hyponatremia, transaminitis, and thrombocytopenia. Several diagnostic tests are typically used to identify rickettsial pathogens.

Table 42.4 An overview of family Anaplasmataceae group of organisms

Group	Disease	Species	Vector	Reservoir(s)	Geographic distribution
Anaplasma	Human granulocytic anaplasmosis	*A. phagocytophilum*	Tick	Small mammals, rodents, deer	Worldwide (primarily the United States)
Ehrlichia	Human monocytic ehrlichiosis	*E. chaffeensis*	Tick	Deer, dogs, ruminants, rodents	Worldwide (primarily the United States)
	Ehrlichiosis	*E. muris*	Tick	Deer, rodents	North America, Europe, Asia
	Ehrlichiosis	*E. ewingii*	Tick	Deer, dogs, rodents	North America, Cameroon, Korea
Neoehrlichia	Human neoehrlichiosis	*N. mikurensis*	Tick	Rodents	Europe, Asia
Neorickettsia	Sennetsu fever	*N. sennetsu*	Trematode	Fish	Japan, Malaysia, Asia

Table 42.5 Important species of genus *Anaplasma* and their important characteristics

Species	Disease	Vector	Animal	Distribution
Anaplasma bovis	Bovine monocytic anaplasmosis	Ticks (*Amblyomma*,*Hyalomma*, *Rhipicephalus*)	Cattle	South America, Africa, Iran, India
Anaplasma marginale	Bovine anaplasmosis/ gallsickness	Ticks, biting insects	Cattle	Worldwide
Anaplasma ondiri	Bovine petechial fever/ ondiri disease	Unknown	Cattle	Africa
Anaplasma ovis	Ovine/caprine anaplasmosis	Ticks	Sheep, goats	Worldwide
Anaplasma phagocytophilum	Equine ehrlichiosis Canine granulocytic ehrlichiosis/ anaplasmosis Feline granulocytic ehrlichiosis/ anaplasmosis Pasture fever/ tick-borne fever	*Ixodes* spp.	Horses Dogs Cats Ruminants	America, Europe

Table 42.6 Important species of genus *Ehrlichia*/*Neoehrlichia* and their important characteristics

Species	Disease	Vector	Animal	Distribution
Ehrlichia canis	Canine monocytic ehrlichiosis	*Rhipicephalus sanguineus*	Dogs	Worldwide
Ehrlichia chaffeensis	Canine monocytic ehrlichiosis	*Amblyomma americanum*	Dogs	North America
Ehrlichia ewingii	Canine granulocytic ehrlichiosis	*Amblyomma americanum*	Dogs	North America, Africa
Ehrlichia ruminantium	Heartwater	*Amblyomma* spp.	Cattle, sheep, goats	Africa, Caribbean
Neoehrlichia risticii	Potomac horse fever/ equine monocytic neorickettsiosis	Digenic trematodes	Horses	North America

Staining

The detection of rickettsial organisms in clinical specimen smears can be accomplished using various staining methods. The various stains used include Macchiavello, Castaneda stain, Stamp, and acridine orange.

Culture

Accurate identification and characterization of rickettsial organisms require isolation of the pathogen. However, this process is challenging and time-consuming because of the obligate intracellular nature of organisms. Furthermore, the isolation procedure necessitates BSL3 laboratory facilities for safe handling. Various clinical materials can be used for pathogen isolation, including whole blood, buffy coat, eschar material (crust, swabs, skin biopsies), and entire vectors. The isolation process utilizes different cell lines, including Vero (African Green monkey kidney epithelial cells), HeLa (derived from patient 'Henrietta Lacks') cells, Hep-2 (human epithelial cell lines), MRC-5 (human fetal lung fibroblast cells), human embryonic lung (HEL) cells, L929 (mouse fibroblast cells), and XTC-2 (*Xenopus laevis* cells). Additionally, embryonated chicken eggs or laboratory animals such as guinea pigs can be used for organism isolation.

Serology

In reference diagnostic laboratories, serological testing is often the method of choice due to its efficiency in processing and minimal sample preparation needs. Typically, this test produces positive results within 7–10 days following infection. However, in case of *R. africae*, the positive outcome may take 3–4 weeks.

- *Weil-Felix test*: In the past, this test was used as a preliminary diagnostic method for identifying rickettsial organisms. However, due to its poor sensitivity and specificity, its practical utility is now limited. The test operates on the principle of heterophile agglutination, exploiting the cross-reactivity between certain *Proteus* species and *Rickettsia* species. It utilizes nonmotile strains of *Proteus* (OX19, OX2, and OXK) that share an alkali-stable carbohydrate antigen with specific *Rickettsia* species. Despite its historical use, the test's inadequate accuracy has diminished its relevance in current diagnostic practices. The important serotypes of *Proteus* bacteria and their cross-reactivity with *Rickettsia* organisms are given below (Table 42.7).
- *Enzyme-linked immunosorbent assay (ELISA)*: This test is highly sensitive, specific, and rapid and has high reproducibility. The test was initially developed for the detection of IgG and/or IgM antibodies against typhus group organisms (*R. typhi* and *R. prowazekii*) and later adapted for the detection of antibodies against scrub typhus and spotted fever group of pathogens.
- *Immunofluorescence assay (IFA)*: Regarded as the "gold standard" for rickettsial organism serodiagnosis, this method serves as a reference test in numerous research facilities. It identifies group-specific IgG and/or IgM antibodies associated with the targeted pathogens.
- *Microimmunofluorescence (MIF)*: This method resembles IFA but differs in that the wells are coated with various rickettsial antigens, allowing for concurrent detection of various pathogens.

Table 42.7 Cross reactivity of *Proteus* strains with rickettsial diseases

Disease	*Proteus* strains		
	OX19	OX2	OXK
Epidemic typhus	++	+/–	–
Endemic typhus	++	–	–
Scrub typhus	–	–	++
Rocky mountain spotted fever	+	+	–
Q fever	–	–	–
Rickettsial pox	–	–	–
Indian tick typhus	+	++	–
Other tick borne rickettsial infection	+	++	–

- *Immunoperoxidase assay*: This technique is a variant of IFA that substitutes peroxidase for the fluorochrome. Studies have shown that the test's sensitivity and specificity are comparable to IFA when used for serodiagnosis of certain rickettsial diseases, including scrub typhus, epidemic typhus, and Mediterranean spotted fever.
- *Western blotting*: This technique allows more specific detection of the causative agent. However, it has certain limitations as it requires robust rickettsial antigen collection before test is performed and since it is mainly performed in reference laboratories only.
- *Other serological tests*: Recently other serological tests such as immunodot assays, dipsticks, and immunochromatographic tests are developed for early detection of antibodies against selected pathogens.
- *Molecular techniques*: Various nucleic acid-based techniques, including conventional PCR, real-time PCR, nested PCR, and loop mediated isothermal amplification (LAMP) assays, have been created for quick, accurate, and specific identification of rickettsial pathogens. These methods can be applied to different human specimens such as whole blood, buffy coat, and eschar material (crust, swabs, skin biopsies). Several genes, including citrate synthase (*gltA*), 16S rRNA, outer membrane protein A (*ompA*) and B (*ompB*), periplasmic serine protease (*htrA*), heat shock protein 60 (*groEL*), and gene D (*sca4*), have been employed to detect targeted rickettsial organisms. Vectors like ticks, lice, fleas, and mites can also serve as epidemiological tools for identifying pathogen activity in a given area. Recent advancements in diagnostic techniques for detecting targeted pathogens utilize restriction fragment length polymorphism (RFLP), random amplified polymorphic DNA (RAPD), multilocus sequence typing (MLST), next-generation sequencing, and whole-genome sequencing methods.

References

Abdad, M. Y., Abou Abdallah, R., Fournier, P. E., Stenos, J., & Vasoo, S. (2018). A concise review of the epidemiology and diagnostics of rickettsioses: Rickettsia and Orientia spp. *Journal of Clinical Microbiology, 56*(8), 10–1128.

Banegura, F. K., Munube, G. M., Senyakazana, A. G., & Nzaro, E. (1983). An epidemic of typhus in Ankole, Uganda, 1975–1976. *East African Medical Journal, 60*, 55–59.

Bechah, Y., Capo, C., Mege, J. L., & Raoult, D. (2008). Epidemic typhus. *The Lancet Infectious Diseases, 8*(7), 417–426.

Bedi, J. S., Vijay, D., & Dhaka, P. (2022). *Textbook of Zoonoses*. Wiley.

Bermúdez, C. S. E. (2018). A review of the genus Rickettsia in Central America. *Research and Reports in Tropical Medicine, 9*, 103–112.

Bhandari, M., Singh, R. K., Laishevtcev, A., Mohapatra, T. M., Nigam, M., Mori, E., de Lacerda, B. C. G. V., Coutinho, H. D. M., & Mishra, A. P. (2022). Revisiting scrub typhus: a neglected tropical disease. *Comparative Immunology, Microbiology and Infectious Diseases, 90*, 101888.

Biggs, H. M., et al. (2016). Diagnosis and management of tickborne rickettsial diseases: rocky mountain spotted fever and other spotted fever group rickettsioses, ehrlichioses and anaplasmosis—United States. *MMWR. Recommendations and Reports, 65*(2), 1–44.

Luce-Fedrow, M., Lehman, D., Kelly, K., Mullins, A., Maina, R. S., et al. (2018). A review of scrub typhus (Orientia tsutsugamushi and related organisms): then, now, and tomorrow. *Tropical Medicine and Infectious Disease, 3*, 8.

Part VII

Prion Diseases

Prion Disease: Transmissible Spongiform Encephalopathies (TSEs) 43

Ajay Pratap Singh

Abstract

Prion diseases, also known as transmissible spongiform encephalopathies (TSEs), are a group of invariably fatal neurodegenerative disorders that affect both animals and humans. In animals, these diseases include scrapie in sheep, chronic wasting disease (CWD) in cervids, bovine spongiform encephalopathy (BSE) in cattle, and transmissible mink encephalopathy (TME) in mink. In humans, prion diseases occur as sporadic, genetic, or acquired conditions, with the sporadic Creutzfeldt–Jakob disease (sCJD) being the most prevalent, accounting for approximately 85% of cases. Among the prion diseases of animals, BSE is particularly significant due to its zoonotic potential. This was demonstrated by its role in the emergence of the variant Creutzfeldt–Jakob disease (vCJD) in humans, highlighting the cross-species transmissibility of prions. This zoonotic risk underscores the need for continued surveillance, rigorous research, and proactive public health measures to mitigate the broader implications of prion diseases for both animal and human populations.

Keywords

Transmissible spongiform encephalopathies (TSEs) · Prion protein · BSE · CJD

A. P. Singh (✉)
Department of Veterinary Microbiology, College of Veterinary Science and Animal Husbandry, U.P. Pandit Deen Dayal Upadhyaya Pashu Chikitsa Vigyan Vishwavidyalaya Evam Go-Anusandhan Sansthan, Mathura, Uttar Pradesh, India

R. Biswas et al. (eds.), *Concept of One Health and Zoonoses*,
https://doi.org/10.1007/978-981-95-6627-3_43

History

The earliest recorded prion disease dates back to 1732, when scrapie was first described in merino sheep in Spain, marking the initial documentation of a transmissible neurodegenerative disorder. The origins of human prion disease research trace back to the early twentieth century, with Alfons Jakob's significant contributions between 1921 and 1923. During this period, Jakob described several cases of a rapidly progressive neurodegenerative disorder, which he noted bore similarities to a case reported by Hans Creutzfeldt in 1920 involving a young woman. This convergence of observations by Creutzfeldt and Jakob laid the foundation for what would later be recognized as the Creutzfeldt–Jakob Disease (CJD). This was followed by the seminal work of Cuille and Chelle (1936–1938), who provided the first experimental evidence that scrapie was caused by an infectious agent, initially thought to be a virus, underscoring the unique infectious nature of the disease. Prion diseases were further expanded in scope with the identification of transmissible mink encephalopathy (TME) in 1947, a scrapie-like disease affecting farmed mink. Later, chronic wasting disease (CWD), a similar condition in cervids (e.g., deer and elk), was first recognized in the 1960s, although its prion etiology was confirmed only in 1980, linking these diseases across species. The emergence of bovine spongiform encephalopathy (BSE), commonly known as "mad cow disease," in 1986 marked a pivotal point in prion research. BSE's zoonotic potential was tragically demonstrated with the identification of the first cases of the variant Creutzfeldt–Jakob Disease (vCJD) in humans in the United Kingdom in 1996, signaling a significant public health crisis and highlighting the interspecies transmissibility of prions. In 1997, Stanley Prusiner was awarded the Nobel Prize in Physiology or Medicine for his groundbreaking work in identifying and characterizing the agent responsible for scrapie, a prion disease affecting sheep. Prusiner's research provided definitive evidence that the infectious agent was not a conventional pathogen like a virus or bacterium but rather a misfolded protein. He coined the term "prion" to describe this "proteinaceous infectious particle" and demonstrated its unique ability to propagate by inducing the misfolding of normal cellular prion proteins. Enhanced surveillance measures following the BSE epidemic led to the discovery of "atypical" BSE cases after 2001, characterized by distinct subtypes: L-BSE (BASE) and H-BSE. These variants, primarily identified in older, asymptomatic cattle, exhibited pathological and epidemiological features distinct from the classic C-BSE, suggesting diverse origins and raising critical questions about prion strain diversity.

Etiology

The etiological agent of prion diseases is an abnormal, misfolded form of the prion protein, known as scrapie-associated prion protein (PrP^{Sc}). Unlike conventional pathogens such as bacteria or viruses, prions lack nucleic acids and are composed solely of protein. They originate from the normal cellular prion protein, PrP^{C}, which is present in various tissues, particularly in the central nervous system. Under pathological conditions, PrP^{C} undergoes a conformational change to the misfolded PrP^{Sc} form, which is resistant to proteolysis and capable of inducing similar misfolding in normal PrP^{C} molecules. This self-propagating cycle underlies the progressive accumulation of PrP^{Sc} in neural tissues, leading to the characteristic neurodegeneration seen in prion diseases. PrP^{Sc} aggregates into amyloid fibrils and plaques, causing spongiform changes and neuronal loss. Bovine spongiform encephalopathy (BSE) exists in several subtypes, including the classic form (c-BSE) and two atypical variants: L-type (also known as bovine amyloidotic spongiform encephalopathy or BASE) and H-type. These subtypes are differentiated by analyzing the molecular mass and glycosylation patterns of PrPres using Western blot. The L-type BSE is characterized by a slightly lower molecular mass in the unglycosylated PrPres band compared to c-BSE, while the H-type shows a higher molecular mass in both glycosylated and unglycosylated bands compared to c-BSE.

Epidemiology

Creutzfeldt–Jakob Disease (CJD) CJD affects both men and women equally, with no gender predisposition. The average age of onset for prion diseases is approximately 60 years. While rare, the disease predominantly manifests in middle-aged to older adults and is uncommon in individuals under 40 or over 80 years of age. Sporadic Creutzfeldt–Jakob disease (sCJD) progresses rapidly, with a median survival time of only 5 months following the onset of symptoms. Additionally, 90% of individuals diagnosed with sCJD succumb to the disease within 1 year. Sporadic CJD (sCJD) is the most prevalent form of the Creutzfeldt–Jakob disease, accounting for approximately 85–90% of all cases worldwide. Familial or genetic CJD (fCJD/gCJD), resulting from inherited mutations in the PRNP gene, contributes to about 10% of cases. Acquired prion diseases, including variant CJD (vCJD) and iatrogenic CJD (iCJD), represent 2–5% of all CJD cases globally, highlighting the rarity of these transmission-linked forms.

Scrapie Strain susceptibility is influenced by several factors, primarily the genetic makeup of the sheep and the specific strain of prion responsible for the infection. Studies have shown that different sheep breeds and genotypes exhibit varying levels of susceptibility to different scrapie strains. The most notable factor is the polymorphism at codon 171 of the PRNP gene, where certain genotypes confer greater resistance to scrapie, while others are more susceptible. Purebred flocks had a higher risk compared to commercial flocks. Scrapie outbreaks are more common in certain geographic regions, suggesting that regional factors, such as environmental conditions or farming practices, may influence occurrence of the disease.

Bovine Spongiform Encephalopathy Dairy cows are at higher risk for prion diseases than beef cattle due to feeding practices, longer lifespans, and culling age. Larger herd sizes and geographical differences in regulatory practices and environmental conditions also contribute significantly to disease risk. Young cattle appear to be more susceptible to infection with the classical form than adults, and there is an increased risk of c-BSE in calves born to c-BSE-affected cows.

Host Range

Bovine spongiform encephalopathy (BSE) primarily affects cattle as its natural host. However, experimental studies have demonstrated successful transmission of BSE to sheep and goats via oral and intracerebral routes. In addition, BSE has been transmitted to various other species, including cats, exotic zoo animals such as antelopes and big cats, and laboratory rodents. Scrapie, another transmissible spongiform encephalopathy (TSE), naturally occurs in sheep and goats. Experimentally, scrapie has been transmitted to laboratory rodents, particularly hamsters and mice, providing valuable models for studying the disease. Chronic wasting disease (CWD) occurs naturally in several cervid species, including white-tailed deer, mule deer, Rocky Mountain elk, moose, caribou, and reindeer. Experimental studies have demonstrated that CWD can infect various non-cervid species via routes such as intracranial inoculation. These include nonhuman primates, cattle, swine, raccoons, domestic cats, ferrets, and multiple rodent species.

Transmission

Prion diseases, or transmissible spongiform encephalopathies (TSEs), involve the accumulation of the abnormal prion protein (PrP^{Sc}) in specific tissues, referred to as specified risk materials (SRM). Extensive studies on classical scrapie in sheep, chronic wasting disease (CWD) in deer, and camelid prion disease (CPD) in dromedary camels have demonstrated a wide distribution of infectivity across various SRM tissues in animals naturally infected with the disease. High infectivity is consistently observed in the brain and spinal cord, which are central nervous system tissues. Moderate infectivity levels are found in lymphoreticular tissues, including lymph nodes and spleen. In contrast, peripheral tissues such as muscles and blood carry significantly lower infectivity. Secretions and excretions such as saliva, feces, urine, colostrum, and semen are generally considered noninfective for transmission. Horizontal transmission is considered the primary mode of spread for these diseases, with the disease typically transmitted through direct contact with infected tissues or contaminated environments, particularly through oral intake of infected materials. Vertical transmission, which refers to the transmission of the disease from parent to offspring, is considered doubtful, and there is currently no evidence to support in utero transmission of scrapie in sheep. While prions can be present in maternal tissues, including the placenta, no definitive studies have shown that scrapie can be transmitted directly from mother to fetus.

Even minute quantities of PrP^{Sc}-containing material from infected tissues are capable of transmitting the disease to susceptible hosts. Transmission pathways differ among TSEs and host species. In sheep and deer, horizontal transmission is a major route, facilitated by environmental contamination or direct contact with secretions and excretions such as saliva, feces, urine, and placental tissues. Vertical transmission via intrauterine exposure may also occur. In contrast, BSE and the variant Creutzfeldt–Jakob disease (vCJD) are not transmitted through saliva, emphasizing distinct interspecies transmission dynamics. Experimentally, intracerebral inoculation with high-titer PrP^{Sc} is the most efficient method for studying cross-species transmission. However, natural infections typically result from peripheral exposure to low doses of infectious material, most commonly via the oral route. Ingestion of contaminated feed, forage, or bedding introduces PrP^{Sc} into the gastrointestinal tract, where it propagates. Nasal exposure is also plausible, as inhalation of PrP^{Sc} from contaminated environmental sources has been suggested. BSE exhibits an unusually broad host range compared to other TSE strains, infecting species such as goats, exotic ruminants, domestic cats, and humans. This highlights strain-specific differences in transmissibility and host susceptibility. While PrP^{Sc} has been detected in semen from scrapie-infected rams, there is no conclusive evidence of transmission through natural breeding or artificial insemination. The diverse routes of transmission and tissue tropism of PrP^{Sc} emphasize the complexity of prion pathogenesis and the challenges in controlling these diseases across species.

Pathogenesis

Prion diseases are caused by the misfolding of the normal cellular prion protein (PrP^C) into a pathological isoform, PrP^{Sc}, which initiates a self-propagating cycle of protein misfolding. This conformational change is fundamental to prion infectivity and occurs primarily at the cell surface, facilitated by PrP^C glycosylphosphatidylinositol (GPI) anchor. Experimental evidence suggests that the GPI anchor plays a critical role in PrP^{Sc} formation and infectivity, although its absence in transgenic models leads to PrP amyloid deposition without neurological symptoms, highlighting complex pathways of disease progression. Following oral exposure, PrP^{Sc} crosses the gastrointestinal epithelium via transcytosis by M cells or by dendritic cells that capture and transport the agent to lymphoid tissues such as Peyer's patches, mesenteric lymph nodes, and gut-associated lymphoid tissue (GALT). Within these tissues, follicular dendritic cells in germinal centers play a pivotal role, as they express PrP^C and facilitate the propagation of PrP^{Sc}. From the lymphoid tissues, the prion agent spreads to the enteric nervous system, eventually reaching the central nervous system through the peripheral nervous system. PrP^{Sc} aggregates are detected at the plasma membrane, in vesicles along the endo-lysosomal pathway, and within recycling endosomes. The presence of elevated unesterified cholesterol in prion-infected cells suggests altered membrane dynamics and vesicle trafficking, which may influence the replication and spread of PrP^{Sc}. Despite extensive PrP^{Sc} deposition, the specific mechanisms driving neurotoxicity and brain damage remain unclear. Synaptic loss is an early pathological event, but the exact pathways leading to neuronal vacuolation and spongiform changes involve disrupted cellular checkpoint molecules and prion-induced subversion of normal processes. Interestingly, prion infectivity has been detected in the bone marrow of patients with sCJD and vCJD, raising concerns about prionemia and the potential for disease transmission through this tissue. Cofactor molecules such as RNA and phosphatidylethanolamine have been shown to enhance prion infectivity by several orders of magnitude, suggesting that these molecules play a critical role in the formation of infectious prions. Additionally, different prion strains can coexist within the same brain, potentially masked by dominant pathological phenotypes, further complicating disease mechanisms. While the process of prion propagation is well understood, the molecular mechanisms underlying neurotoxicity and disease progression, particularly in the CNS, require further investigation. Understanding the interplay of prion strain diversity, cofactors, and cellular dynamics is key to unraveling the complexities of prion pathogenesis, which initiates a self-propagating cycle of protein misfolding. This conformational change is fundamental to prion infectivity and occurs primarily at the cell surface, facilitated by PrP^{C}'s glycosylphosphatidylinositol (GPI) anchor. Experimental evidence suggests that the GPI anchor plays a critical role in PrP^{Sc} formation and infectivity, although its absence in transgenic models leads to PrP amyloid deposition without neurological symptoms, highlighting complex pathways of disease progression.

Clinical Symptom

Prion diseases manifest with a wide array of clinical symptoms, varying by species and specific disease type.

Bovine Spongiform Encephalopathy

BSE, a prion disease of bovine, is classified into distinct types based on the molecular characteristics of the abnormal prion protein (PrP^{Sc}) observed after protease digestion and Western blot analysis. These types include:

Classical BSE (cBSE) This is the form initially identified during the BSE epidemic and is primarily linked to the consumption of contaminated feed containing prion-infected material. cBSE is characterized by a specific molecular mass pattern of PrP^{Sc}, which remains consistent across cases. In bovine spongiform encephalopathy (BSE), clinical signs include heightened sensitivity to external stimuli such as light and sound, behavioral changes like depression and nervousness, and motor dysfunction characterized by positional imbalance, difficulty standing or moving, and hind limb paralysis.

Atypical BSE Unlike cBSE, atypical BSE occurs sporadically, typically in older cattle, and does not appear to be associated with contaminated feed. Cattle with atypical BSE will likely have some of the same clinical signs as cattle with classical

BSE. It is further divided into two subtypes based on the molecular mass and glycosylation pattern of PrP^{Sc}: *H-BSE*: This form has a higher molecular mass compared to cBSE. *L-BSE*: This form has a lower molecular mass than cBSE and is also referred to as "bovine amyloidotic spongiform encephalopathy" (BASE) due to its unique histopathological features.

Scrapie

Advanced cases progress to total body paralysis before death. In sheep and goats affected by classical scrapie, behavioral changes such as separation from the herd, weight loss, and an exaggerated response to external stimuli are common. Additional symptoms include bruxism, lip movements, head tremors, ataxia, and intense pruritus, often leading to wool loss. Conversely, in atypical scrapie, symptoms include progressive ataxia, tremors, weight loss, visual impairment, and circular movements but without pruritus or alopecia.

Creutzfeldt–Jakob Disease (CJD)

The Creutzfeldt–Jakob disease (CJD) presents in four main forms:

Sporadic CJD (sCJD) This is non-transmissible and the most common form of CJD, accounting for approximately 85% of cases. Symptoms include rapidly progressive dementia, motor dysfunction (ataxia, gait disturbances), and myoclonus (sudden involuntary muscle jerks). Additional features may include visual disturbances, behavioral changes, and akinetic mutism in advanced stages.

Familial CJD This is a hereditary form caused by mutations in the PRNP gene. Clinical symptoms closely resemble those of sCJD, including progressive dementia, motor dysfunction, and myoclonus. Symptoms may present earlier in life compared to sCJD due to the genetic predisposition.

Iatrogenic CJD This is acquired through exposure to prion-contaminated materials during medical procedures, such as dura mater grafts, corneal transplants, or contaminated surgical instruments. Symptoms depend on the route of exposure but typically include progressive dementia, ataxia, and myoclonus, mirroring the features of sCJD.

Variant CJD (vCJD) A transmissible form linked to BSE, vCJD presents with distinct symptoms, including mood instability, sensory abnormalities, and progressive cognitive decline. Patients may experience paralysis affecting visual, linguistic, and other sensory functions, ultimately leading to immobility and severe cognitive impairment.

Chronic Wasting Disease (CWD)

Chronic wasting disease (CWD) exhibits clinical symptoms similar to other animal prion disorders. Affected deer and elk often display behavioral changes, such as reduced fear of humans or increased aggression, along with neurological abnormalities, hair loss, and severe weight loss (cachexia). Advanced cases commonly involve increased salivation, teeth grinding (bruxism), excessive thirst (polydipsia), and increased urination (polyuria).

Diagnosis

Prion diseases, or transmissible spongiform encephalopathies (TSEs), pose significant diagnostic challenges due to their unique pathogenic mechanisms and the need for highly sensitive and specific detection methods. A range of diagnostic techniques has been developed, leveraging immunological, biochemical, and molecular approaches.

Samples of Choice for Prion Disease Diagnosis

For postmortem diagnosis, CNS tissues are essential as they contain the highest concentration of prion proteins, including the abnormal isoform $PrP^{Sc.}$. The brain is the primary tissue for prion testing, with specific areas like the cortex and cerebellum often analyzed for $PrP^{Sc.}$ deposition. Immunohistochemistry and immunoblotting techniques are commonly used to detect prions in these tissues. Similarly, to the brain, the spinal cord is another key tissue for postmortem diagnosis, though it is less commonly used than brain tissue. In some cases, other tissues like the dura mater (meningeal tissue) may be considered if the standard brain/spinal cord samples are unavailable. For antemortem diagnosis, the use of peripheral samples is an emerging and less invasive approach. Techniques like RT-QuIC and PMCA have expanded the diagnostic potential of these samples, allowing for earlier detection and surveillance. Blood is an increasingly utilized sample for prion disease diagnosis, with real-time quaking-induced conversion (RT-QuIC) and protein misfolding cyclic amplification (PMCA) being able to detect low concentrations of prions in plasma or serum. However, blood-based diagnostics are more commonly used for research purposes, as prions are present in low quantities compared to CNS tissues. Like blood, urine can be used for prion detection, especially using amplification techniques like PMCA. It is a noninvasive sample but typically yields less reliable results than blood or CNS tissues. Recent research has shown that nasal brushings can serve as a reliable noninvasive sample for prion detection, particularly in cases of the sporadic Creutzfeldt–Jakob disease (sCJD). Skin samples, particularly from the back of the neck or forehead, have been identified as an emerging peripheral tissue that may be used in RT-QuIC assays to detect prions. Less commonly, samples like lymphoid tissues (e.g., tonsils) or conjunctival scrapings have been investigated for their potential use in prion diagnostics.

1. Postmortem Diagnostics

The definitive diagnosis of prion diseases is traditionally made postmortem using central nervous system tissues, such as the brain and spinal cord.

- **Immunological Techniques**
  - **Immunoblotting**: Detects the protease-resistant prion protein ($PrP^{Sc.}$) after proteinase K digestion, distinguishing it from the normal prion protein ($PrP^{c.}$).
  - **Enzyme-Linked Immunosorbent Assay (ELISA)**: Used to detect $PrP^{Sc.}$, though its sensitivity may be limited in certain sample types.
  - **Immunohistochemistry**: Visualizes $PrP^{Sc.}$ deposits within tissues, providing both diagnostic and histopathological confirmation.
- **Histopathological Analysis**
  Examines brain tissues for hallmark features of prion diseases, such as spongiform changes, neuronal loss, and gliosis

2. Protein Misfolding Detection Techniques

Prion diseases are characterized by the misfolding of $PrP^{c.}$ into $PrP^{Sc.}$. Advanced methods exploit this unique property to enhance diagnostic accuracy.

Protein Misfolding Cyclic Amplification (PMCA) PMCA mimics prion replication in vitro by amplifying misfolded proteins, enabling the detection of minute amounts of $PrP^{Sc.}$ in biological samples like blood and urine.

Real-Time Quaking-Induced Conversion (RT-QuIC) RT-QuIC is a groundbreaking antemortem diagnostic method that detects disease-associated prion proteins with high sensitivity. This assay uses recombinant prion proteins as substrates to detect misfolded $PrP^{Sc.}$, which triggers the formation of amyloid fibrils monitored in real time. RT-QuIC has been applied to cerebrospinal fluid (CSF), nasal brushings, skin, and eye tissues, achieving nearly 100% accuracy in diagnosing the sporadic Creutzfeldt–Jakob disease (sCJD).

3. Molecular and Advanced Techniques

- **Real-Time Immuno-PCR:** Combines the specificity of antibodies with the sensitivity of polymerase chain reaction (PCR) to detect prions in low concentrations.
- **Biomarker Analysis:** Advances in biomarker research have improved the detection of prion diseases. CSF biomarkers, such as 14-3-3 protein and tau protein, can provide supportive evidence for diagnosis but lack specificity.

Zoonotic Potential

Bovine spongiform encephalopathy (BSE) is the only animal prion disease known to be zoonotic, with a clear link to the variant Creutzfeldt–Jakob disease (vCJD) in humans. In contrast, other animal prion diseases, such as scrapie in sheep, chronic wasting disease (CWD) in deer, elk, and moose, and camel prion disease, have no confirmed direct link to human prion diseases. While bovine spongiform encephalopathy (BSE) has not been naturally reported in sheep, experimental infections have demonstrated that sheep can develop the disease. CWD has been experimentally transmitted to various species under controlled conditions. These include laboratory rodents, sheep (via intracerebral and/or oral routes), cattle, pigs, cats, ferrets, and squirrel monkeys. These studies provide valuable insights into transmissible spongiform encephalopathies (TSEs), which pose unique zoonotic threats due to their ability to cross species barriers under certain conditions. This highlights the critical need for continued research into their pathogenic mechanisms and transmission dynamics.

Kuru

Kuru is a rare, fatal neurodegenerative disorder that was once prevalent among the Fore people of Papua New Guinea. It is classified as a prion disease, caused by the accumulation of abnormal, misfolded prion proteins in the brain, leading to progressive neurological damage. Kuru was transmitted through the practice of ritualistic endocannibalism, where individuals consumed the brains of deceased relatives as part of funeral rites. Symptoms include tremors, loss of coordination, difficulty walking, and eventually severe motor and cognitive decline, culminating in death. The disease gained significant scientific attention as it provided insights into prion biology and transmissible spongiform encephalopathies, such as the Creutzfeldt–Jakob disease. With the cessation of cannibalistic practices, the incidence of Kuru has dramatically declined, making it a rare example of a disease controlled through cultural change.

Control

Effective strategies for preventing and controlling prion diseases focus on limiting transmission, reducing risk factors, and ensuring public health safety. Key measures include the following:

1. **Surveillance and Early Detection:** Monitoring programs for prion diseases in humans and animals are essential. Testing tissues from animals at slaughterhouses and farms identifies infected individuals, while advanced diagnostic tools like cerebrospinal fluid biomarkers and RT-QuIC assays aid early detection in humans.

2. **Biosecurity and Feed Controls:** Banning animal-derived protein in livestock feed and maintaining strict biosecurity practices, including quarantine and culling, reduce the risk of prion spread. Proper husbandry further enhances prevention.
3. **Safe Disposal of High-Risk Materials:** Proper handling and destruction of specified risk materials (SRM), such as brain and spinal tissues, through incineration or alkaline hydrolysis, are critical to limit environmental contamination.
4. **Medical and Surgical Precautions:** Single-use instruments and enhanced decontamination protocols for reusable tools minimize iatrogenic transmission. Screening of blood and tissue donors also reduces the risk of human-to-human spread.
5. **Public Education:** Raising awareness among the public and healthcare workers about transmission pathways and safety protocols helps prevent exposure and promotes adherence to preventive measures.
6. **Research and Development:** Continued research into prion mechanisms and strain diversity supports the development of targeted interventions and improves understanding of transmission risks.
7. **Global Collaboration:** International efforts by organizations like the WHO and OIE ensure consistent surveillance, regulation, and implementation of control measures, mitigating global risks of prion diseases.

References

Aguzzi, A., & Heikenwalder, M. (2006). Pathogenesis of prion diseases: Current status and future outlook. *Nature Reviews Microbiology, 4*(10), 765–775.

Collinge, J. (2001). Prion diseases of humans and animals: Their causes and molecular basis. *Annual Review of Neuroscience, 24*(1), 519–550.

Imran, M., & Mahmood, S. (2011). An overview of animal prion diseases. *Virology Journal, 8*, 1–8.

Johnson, R. T. (2005). Prion diseases. *The Lancet Neurology, 4*(10), 635–642.

Narayan, K. G., Sinha, D. K., & Singh, D. K. (2023). Bovine spongiform encephalitis (BSE)/mad cow disease. In *Veterinary public health & epidemiology*. Springer. https://doi.org/10.1007/978-981-19-7800-5_26

Part VIII

Chlamydial Zoonoses

Chlamydia

44

Maninder Singh

Abstract

Chlamydia are the obligatory intracellular bacteria having unique biphasic life cycle. *Chlamydia* are known to infect through mucous membranes. The bacterium is Gram negative, coccoid belonging to family *Chlamydiaceae* and order *Chlamydiales*. *Chlamydiaceae* has the single genus *Chlamydia*; earlier on the basis of 16S and 23S rRNA sequences, two genera *Chlamydia* and *Chlamydophila* were proposed which was not adopted by *Chlamydia* research committee. Consequently in 2010, the International Committee on the Systematics of Prokaryotes adopted the single genus *Chlamydia*. *Chlamydia* has 12 recognized species which infect different host species. Among these species, *Chlamydia trachomatis* and *Chlamydia pneumoniae* are the human pathogens, while the rest of species infect various domestic and wild animals. Among these species, *Chlamydia psittaci* and *Chlamydia abortus* are the recognized zoonotic pathogens and have been discussed ahead. Other lesser known zoonotic chlamydial species infections have been dealt with at the end of the chapter.

Keywords

Psittacosis · *Chlamydia* · Parrot fever · Parrot disease · Pigeon-fancier's disease · Ornithosis

M. Singh (✉)
Department of Veterinary Public Health and Epidemiology, College of Veterinary Science, Guru Angad Dev Veterinary and Animal Sciences University, Rampura Phul, Bathinda, India

R. Biswas et al. (eds.), *Concept of One Health and Zoonoses*,
https://doi.org/10.1007/978-981-95-6627-3_44

Chlamydia psittaci

History of *C. psittaci*

- "Psittacosis" was first described as human disease during an outbreak in Germany in 1879 among the people who have contact with the recent shipment of exotic birds. About 33% mortality was reported. The etiological agent was not identified from the outbreak.
- The etiological agent of psittacosis was identified during the 1930s. There were pneumonia outbreaks among the bird handlers in the 1930s in the United Kingdom, the United States, Argentina, and Germany; the birds were imported from South American countries and Australia. The parrots were in demand due to colorful plumage for the garments and hats and also as household pets. The contact with infected parrots, led to more than 800 cases globally , and around 100 died. The bacteriological isolation did not lead to any discovery. However, three independent groups of Levinthal, Coles, and Lillie identified small, filterable infectious bodies and named them "Levinthal-Coles-Lillie" bodies.
- In 1966, electron microscopy classified the pathogen as bacterium as, like Gram-negative bacterium, it has its own DNA, RNA, ribosomes, and cell wall.

Etiology

Chlamydia psittaci is a pathogen belonging to family *Chlamydiaceae* and order *Chlamydiales. Chlamydia psittaci* has a single circular chromosome genome with a size of 1.17 MB, GC content of 39%, 1026 genes, and 973 proteins. As a Gram-negative bacterium, the cell wall contains lipopolysaccharides (LPS). Apart from LPS, major outer membrane protein (MOMP) and cysteine-rich proteins are integral components of cell wall. MOMP is encoded by the *ompA* gene. Based on MOMP, six avian (A–F) and two mammalian serotypes (WC and M56) of *C. psittaci* have been described which are host specific. Serotype A is primarily isolated from psittacine birds, serotype B from pigeons, serotype C from ducks and geese, serotype D from turkeys, serotype E from pigeons, and serotype F from parakeets and turkeys. Also, based on molecular genotyping of the *ompA* gene, the genotypes with the same nomenclature as those of serotypes and an additional genotype E/B have been identified. The E/B genotype is present in ducks, turkeys, and pigeons. Non-avian genotypes included WC in cattle and M56 in rodents. A, B, and E/B genotypes are reported in humans also.

All *Chlamydia* species are obligate intracellular in nature and infect eukaryotic host cells. The life cycle is typical with two distinct development stages: elementary body (EB) and reticulate body (RB). EB is the extracellular stage which provides resistance to environmental conditions while infectious in nature and thus plays the role in cell infection. EBs are smaller and spherical having a diameter of 0.2–0.4 μ and dense outer membrane. Chlamydial infection occurs with the attachment of EBs to the cytoplasmic membrane of a cell. After internalization, the bacterium lives in a vacuole and avoids phagocytosis. Within the vacuole, EBs differentiate to RBs (0.6–1.0 μ) which are the active dividing stage of the bacterium in the host. The multiplication occurs through binary fission leading to formation of inclusion bodies within the vacuole. As RBs do not have their own ATPs, they use host ATP for energy requirements. Within 48–72 h, RBs condense to form EBs; the latter are released from host cells to infect other cells.

Epidemiology

Chlamydia psittaci is distributed worldwide. Birds are the reservoir species, and more than 500 bird species are known to be infected by *C. psittaci*. The infection rates are higher in psittacine birds (parakeets, parrots, lories, cockatoos, and budgerigars) belonging to order *Psittaciformes*. Among psittacine birds, prevalence rates are higher in tropical and subtropical regions. Vertical transmission has been reported in birds. The infections are also reported in other bird orders including Anseriformes (ducks), Columbiformes (pigeons), Gruiformes (cranes, coots), and Galliformes (chicken). Among chicken, broiler chicken is known to be infected.

Upon direct or indirect contact with secretions and excretion of infected animals, transmission to other mammalian species occurs. Thirty-two mammalian species are identified to get infected by *C. psittaci*. Infections have been reported in cattle, sheep, goats, horses, pigs, foxes, cats, and rodents. Humans are the accidental hosts and were earlier thought to be dead-end hosts; however, human-to-human transmission of *C. psittaci* has been reported. Although transmission to humans from animals other than birds has not been reported, the transmission of diseases from birds to equines followed by transmission from equines to humans has been reported.

There have been numerous outbreaks of Q-fever in chicken processing workers. There has been an outbreak of community-acquired pneumonia in duck meat processing workers and their close contacts (Zang et al., 2022). Furthermore, there have been numerous small outbreaks or isolated cases globally in the last two decades. The outbreak confirmed the secondary and tertiary human-to-human transmission. The pathogen is classified as category B bioterrorism agent in CDC list.

Transmission

Among birds, transmission occurs through respiratory and intestinal routes. Infected birds, whether symptomatic or asymptomatic, intermittently shed the bacteria, thereby contaminating the environment and feed contents. Further, transmission between different bird species is also reported. For example, in France the detection of duck-associated *C. psittaci* strains in chickens is reported on a mixed poultry farm. Among birds, both horizontal and vertical transmissions are reported. Disease transmission between poultry and wild birds in open-range rearing systems is reported. The ability of *C. psittaci* to disseminate efficiently in host organisms and its rapid replication cycle likely contribute to its successful transmission between birds (.

Humans typically contract *C. psittaci* by inhaling aerosolized particles from respiratory secretions, dried droppings, or feather dust of infected birds. Direct contact with infected birds or their droppings may also lead to infection. However, recent studies have also documented human-to-human transmission. The outbreaks in 2012 in Scotland, in 2013 in Sweden, and in 2020 in China have reported secondary and tertiary human-to-human transmission. Furthermore, in some of these outbreaks, transmission to healthcare workers in hospital settings is reported.

Pathogenesis

The bacterium enters the body infecting epithelial cells and macrophages of host and disseminates rapidly in the host organism, causing systemic disease. After attachment, the elementary bodies (EBs) are internalized into host cells within a membrane-bound vacuole (inclusion). This process avoids fusion with lysosomes, allowing the bacteria to survive. Inside the inclusion, EBs differentiate into reticulate bodies (RBs), which replicate through binary fission. This replication cycle can take 36–72 h. *Chlamydia psittaci* employs various strategies to manipulate host cell functions. It inhibits apoptosis by blocking pro-apoptotic pathways, activating pro-survival pathways, and expressing specific effectors of type III secretion system. Following initial infection in the respiratory tract, *C. psittaci* can disseminate through the bloodstream to other organs, including the liver and spleen, leading to systemic infection (bacteremia). This systemic spread may result in complications such as pneumonia or systemic inflammatory responses.

Clinical Signs

Birds *Chlamydia psittaci* infection in birds can be asymptomatic to symptomatic with clinical signs ranging from asymptomatic to acute and chronic forms. Common clinical signs include weight loss and anorexia, mucopurulent or serous oculonasal discharge, respiratory distress, diarrhea (often with yellowish-green excrement), and lethargy. In severe cases, birds may exhibit systemic lesions including hepatomegaly, splenomegaly, petechial hemorrhages in spleen, and pulmonary edema. Histological examination may reveal airsacculitis, hepatitis, nephritis, and pneumonia. However, asymptomatic infections of *C. psittaci* in various bird species are common.

Humans In humans, symptoms typically develop 5–14 days after exposure. The most common manifestations include fever, chills, dry cough, dyspnea, and myalgia. In severe cases, the infection can progress to pneumonia, respiratory failure, hepatitis, and sepsis. The mortality rates are less than 1%. The bacterium is known to cause fetal and maternal mortality.

Diagnosis

In Birds

Nucleic acid amplification tests (NAATs): Currently, NAATs are the recommended diagnostic tool for avian chlamydiosis, superseding the previous practice of organism isolation. This shift in methodology is due to several factors: the time-consuming nature of isolation, the necessity for high-quality sample collection, the inability to culture certain strains in vitro, and the potential health risks posed to laboratory staff. NAATs offer a quicker, more sensitive, and highly specific alternative for diagnosis. Conventional PCR targeting 16S–23S rRNA or the *ompA* gene is widely used. Real-time PCR targeting *ompA* helps in quantification of pathogen and thus the extent of infection.

Serological tests: Serology alone has limited diagnostic value for identifying ongoing chlamydial infections in birds. This limitation is due to two factors: the high prevalence of these infections in avian populations and the prolonged detectability of antichlamydial antibodies, which can last for several months. Most bird species naturally have high levels of antichlamydial antibodies. Consequently, to accurately diagnose an infection in an individual bird, serological testing must be complemented by either antigen or gene detection methods. Alternatively, analyzing paired serum samples can provide a more accurate diagnosis. The primary serological methods used to detect chlamydial antibodies are (1) various elementary body agglutination (EBA) techniques, (2) Complement Fixation Test (CFT), and (3) ELISA. EBA primarily detects IgM antibodies, allowing for the identification of early infections. However, a negative result does not guarantee the absence of infection due to the test's relatively low sensitivity. Direct CFT detects avian IgG but not IgM, potentially missing recent infections. ELISAs based on recombinant proteins or peptide antigens are available and possess high sensitivity and specificity.

Isolation of pathogen: The isolation of *Chlamydia* can be accomplished through two primary methods: using chicken embryos or cell cultures. Several cell lines are frequently employed for this purpose, including McCoy, buffalo green monkey, HeLa, and Vero. When utilizing chicken embryos, those at 6–7 days of development are selected. The conventional inoculation technique involves introducing 0.5 mL of sample into the yolk sac of embryo. Chlamydial replication leads to embryonic death within a 3–10-day period, accompanied by a distinctive vascular congestion in the yolk sac membranes.

Direct visualization: Chlamydiae can be detected by cytological staining in cloacal or conjunctival swabs, as well as impression smears of organs like the lung, spleen, and air sacs. The stains commonly used are Giemsa, Giménez, modified Giménez, and Macchiavello's stains.

In Humans In humans, the diagnosis is based on serology and nucleic acid amplification tests.

Prevention and Control

In birds: The prevention of *C. psittaci* in birds involves biosecurity and disinfection of premises. Under biosecurity measures, quarantine new birds for at least 30 days before introducing them to existing flocks. Keep domestic birds away from wild birds to reduce risk of infection. Avoid overcrowding in the cages. *Chlamydia psittaci* is susceptible to various disinfectants including bleach and quaternary ammonium compounds. The control of infection in birds relies on application of antibiotics. Clean surfaces thoroughly. For birds, antibiotics remain the exclusive option to control *C. psittaci* infections. *Chlamydia psittaci* is susceptible to several tetracycline-class drugs, such as chlortetracycline and doxycycline. Effective treatment necessitates an extended administration period, with a common recommendation of 45 days for pet birds. There is no vaccine available against *C. psittaci* infection in birds.

In humans: The prevention of *C. psittaci* infection is based on personal hygiene. Bird handlers should wash their hands thoroughly with soap and water after handling birds or their droppings. Use gloves, masks, and protective clothing when handling birds or cleaning their environments to minimize exposure to infectious particles. Bird caretakers should monitor for symptoms of respiratory illness and seek medical attention if they develop symptoms after exposure to birds. As indicated earlier, tetracyclines are the drug of choice for the treatment of *C. psittaci*. Treatments are of 10–14-day duration and are generally successful.

Chlamydia abortus

Chlamydia abortus is a notable zoonotic pathogen capable of causing serious infections in humans, especially pregnant women. It is mainly recognized as the agent responsible for enzootic abortion of ewes/ovine enzootic abortion in small ruminants. The abortion classically occurs during the final 2–3 weeks of gestation, resulting in stillborn lambs and inflamed placentas. Ewes that have experienced an abortion generally do not abort again. During abortion or parturition, infected ewes release large quantities of infectious *C. abortus*, particularly in placental and uterine discharges, creating a source of infection for both animals and humans. In humans, *C. abortus* can invade the placenta, posing a significant threat to pregnant women and potentially leading to septic abortion and stillbirth. Additionally, a case of atypical pneumonia in a 47-year-old male veterinarian has been documented.

The diagnosis of enzootic abortion involves detecting the causal organism's antigen or genetic material in aborted tissues or vaginal discharges from recently aborted females. Genetic material detection targets the 16S–23S rRNA region or pmp genes. This technique can be combined with restriction fragment length polymorphism analysis to distinguish between *C. abortus*, *C. psittaci*, and *Chlamydia pecorum*.

Prevention of *C. abortus* infection involves adopting hygienic measures during parturition to prevent transmission to animals and humans. Veterinarians should use personal protection materials such as face masks, gloves, etc., to prevent occurrence. The infection of *C. abortus* is treatable with tetracyclines. The vaccine is not available for animals and humans.

References

De Boeck, C., Dumont, A., Kalmar, I., & Vanrompay, D. (2015). Longitudinal monitoring for respiratory pathogens in broiler chickens reveals co-infection of Chlamydia psittaci and Ornithobacterium rhinotracheale. *Journal of Medical Microbiology, 64*(Pt 5), 565–574. https://doi.org/10.1099/jmm.0.000047

Mcguigan, C. C., Mcintyre, P. G., & Templeton, K. (2012). Psittacosis outbreak in Tayside, Scotland, December 2011 to February 2012. *Eurosurveillance, 17*(22). https://doi.org/10.2807/ese.17.22.20186-en

Ortega, N., Murcia-Belmonte, A., Buendía, A. J., Salinas, J., Caro, M. R., Álvarez, D., Del Río, L., Cuello, F., & Gallego, M. C. (2015). Isolation of Chlamydia abortus from a laboratory worker diagnosed with atypical pneumonia. *Irish Veterinary Journal, 69*(1). https://doi.org/10.1186/s13620-016-0067-4

Riccio, M. B., Diab, S. S., Chiapparrone, M. L., Uzal, F. A., Cadario, M. E., García, J. P., Cantón, J., Cacciato, C., & Origlia, J. A. (2024). Outbreak of Chlamydia psittaci infection in a commercial psittacine breeding aviary in Argentina. *Animals: An Open Access Journal from MDPI, 14*(13), 1959. https://doi.org/10.3390/ani14131959

Wallensten, A., Runehagen, A., & Fredlund, H. (2014). Multiple human-to-human transmission from a severe case of psittacosis, Sweden, January-February 2013. *Euro Surveillance: Bulletin Europeen Sur Les Maladies Transmissibles = European Communicable Disease Bulletin, 19*(42). https://doi.org/10.2807/1560-7917.es2014.19.42.20937

Zhang, Z., Zhou, H., Cao, H., Ji, J., Zhang, R., Li, W., Guo, H., Chen, L., Ma, C., Cui, M., Wang, J., Chen, H., Ding, G., Yan, C., Dong, L., Holmes, E. C., Meng, L., Hou, P., & Shi, W. (2022). Human-to-human transmission of Chlamydia psittaci in China, 2020: An epidemiological and aetiological investigation. *The Lancet Microbe, 3*(7), e512–e520. https://doi.org/10.1016/s2666-5247(22)00064-7

MIX
Papier aus verantwortungsvollen Quellen
Paper from responsible sources
FSC® C105338

If you have any concerns about our products,
you can contact us on
ProductSafety@springernature.com

In case Publisher is established outside the EU,
the EU authorized representative is:
Springer Nature Customer Service Center GmbH
Europaplatz 3, 69115 Heidelberg, Germany

Printed by Libri Plureos GmbH
in Hamburg, Germany